OT 19:
Operator Theory: Advances and Applications
Vol. 19

Birkhäuser Verlag
Basel · Boston · Stuttgart

Operator Theory and Systems

Proceedings Workshop Amsterdam, June 4–7, 1985

Edited by

H. Bart
I. Gohberg
M. A. Kaashoek

1986

Birkhäuser Verlag
Basel · Boston · Stuttgart

Volume Editorial Office

Department of Mathematics
and Computer Science
Vrije Universiteit
Amsterdam
The Netherlands

CIP-Kurztitelaufnahme der Deutschen Bibliothek

Operator theory and systems : proceedings, workshop,
Amsterdam, June 4–7, 1985 / ed. by H. Bart . . .
– Basel ; Boston ; Stuttgart ; Birkhäuser, 1986.
(Operator theory ; Vol. 19)
ISBN 3-7643-1783-3
NE: Bart, Harm [Hrsg.]; GT

Printed in Germany
ISBN 3-7643-1783-3
ISBN 0-8176-1783-3

CONTENTS

Editorial Introduction

This volume contains the proceedings of the workshop on Operator Theory and its Applications, which was held at the Vrije Universiteit in Amsterdam in the first week of June, 1985, just before the international symposium on the Mathematical Theory of Networks and Systems (MTNS) in Stockholm. The workshop was the third of this kind; the first two were held at Santa Monica (California, U.S.A.) and Rehovot (Israel) in 1981 and 1983, respectively.

The lectures presented in the workshop reviewed recent advances in operator theory and its applications. They gave a reasonable cross-section of the theoretical developments in abstract operator theory as well as in the area of concrete classes of operators. Applications concerned different fields, among them Mathematical Systems Theory, Network Theory and Transport Theory. Special attention was paid to the interplay between complex analysis, linear algebra and operator theory. The majority of the speakers prepared its papers (often in a more extended form) for publication in this volume, and as a result this volume presents an accurate picture of the lectures and discussions in the workshop.

It is a pleasure to thank the Vrije Universiteit and its department of Mathematics and Computer Science for their hospitality, assistance and financial support. Thanks are also due to the Economische Faculteit of the Erasmus Universiteit at Rotterdam for hosting one day of the conference. We gratefully acknowledge the financial support of the Koninklijke Nederlandse Akademie van Wetenschappen, the Vertrouwenscommissie of the Dutch Mathematical Society and the corporate sponsors Control Data B.V., I.B.M. Nederland N.V. and Philips Nederland B.V..

February 1, 1986 H. Bart, I. Gohberg, M.A. Kaashoek

Operator Theory:
Advances and Applications, Vol. 19

MULTIPLICATIVE PROPERTIES IN THE STRUCTURE OF CONTRACTIONS

Gr. Arsene, Zoia Ceaușescu, T. Constantinescu

ABSTRACT. The canonical form of the product of two-by-two matrix contractions is given (Section 4). As preparation, two ways of composing elementary rotations are studied (Sections 1 and 2). Interpretations using multiports are sketched.

1. For two (complex) Hilbert spaces H and H', let $L(H, H')$ denote the set of all (linear, bounded) operators from H into H'. For a contraction $T \in L(H, H')$ (i.e. $\|T\| \leq 1$; we will write $T \in L_1(H, H')$), consider its defect operator $D_T = (I - T^* T)^{\frac{1}{2}}$ and its defect space $\mathcal{D}_T = D_T(H)^-$ (the upper bar denotes the norm closure).

One important point in dilation theory is the study of contractions using special classes of operators (unitaries, isometries or coisometries) on larger spaces (see [18]). In this respect, the use of the following unitary operator is always fundamental. For a contraction $T \in L(H, H')$, denote by $J(T)$ the operator:

$$(1.1)\qquad \begin{cases} J(T) : H \oplus \mathcal{D}_{T^*} \to H' \oplus \mathcal{D}_T \\ J(T) = \begin{bmatrix} T & D_{T^*} \\ D_T & -T^* \end{bmatrix} \end{cases}$$

We will call the unitary operator $J(T)$ the *elementary rotation* associated to the contraction T. It is the core of the minimal unitary dilation of T (see [18, Section I.5]), and several other terminologies from operator theory or systems theory are used for it in the literature.

A systematic study of dilations of contractions asks for the structure of two-by-two matrix contractions. We refer to [8] for the structure presented below, and for a history of the subject. It is worth to add the reference [16] where the use of so-called Redheffer product in the study of matrix contractions was initiated; this was explicitly done in [14]. The mentioned structure is described by the one-to-one correspondence between the set of contractions $\overset{\circ}{A} \in L(H_1 \oplus H_2, K_1 \oplus K_2)$ and the set of 4-tuples $\{A, \Gamma_1, \Gamma_2, \Gamma\}$ with $A \in L_1(H_1, H_2)$, $\Gamma_1 \in L_1(H_2, \mathcal{D}_{A^*})$, $\Gamma_2 \in L_1(\mathcal{D}_A, K_2)$, $\Gamma \in L_1(\mathcal{D}_{\Gamma_1}, \mathcal{D}_{\Gamma_2^*})$, given by the formula

(1.2) $$\mathring{A} = \begin{bmatrix} A & D_{A^*}\Gamma_1 \\ \Gamma_2 D_A & -\Gamma_2 A^* \Gamma_1 + D_{\Gamma_2^*}\Gamma D_{\Gamma_1} \end{bmatrix}.$$

We write in this case:

(1.3) $$\mathring{A} = \mathbf{C}(A, \Gamma_1, \Gamma_2, \Gamma);$$

it is clear that $\mathring{A}^* = \mathbf{C}(A^*, \Gamma_2^*, \Gamma_1^*, \Gamma^*)$. Note that $J(A) = \mathbf{C}(A, I, I, 0)$, with $H_2 = D_{A^*}$ and $K_2 = D_A$. We will call (1.2) the canonical form of the two-by-two matrix contraction $\mathring{A}$.

Moreover, one has a description of the defect space of $\mathring{A}$, namely, the following operator is a unitary one:

(1.4) $$\begin{cases} \alpha(\mathring{A}) = \alpha : D_{\mathring{A}} \to D_{\Gamma_2} \oplus D_\Gamma \\ \alpha(\mathring{A}) = \begin{bmatrix} D_{\Gamma_2} D_A & -(D_{\Gamma_2} A^* \Gamma_1 + \Gamma_2^* \Gamma D_{\Gamma_1}) \\ 0 & D_\Gamma D_{\Gamma_1} \end{bmatrix} \in L(H_1 \oplus H_2, D_{\Gamma_2} \oplus D_\Gamma). \end{cases}$$

The defect space $D_{\mathring{A}^*}$ can be identified with $D_{\Gamma_1^*} \oplus D_{\Gamma^*}$ using $\alpha(\mathring{A}^*) = \alpha_*$.

Among the applications of these facts we mention here only the *triangularization of the defect operators* induced by (1.4); this was used in [5] and [6] for computing some determinants connected with angles in Gaussian processes (and with Szegö-type theorems for them).

Particular cases of this description (i.e. for row – or column – operators, or for $A = 0$; see [9], [3], [4], [17], and so on) are also useful; we list here the case of a row operator, which (together with column case) will be used later. This corresponds to the case where $K_2 = \{0\}$, and thus $\Gamma_2 = 0 \in L(D_A, \{0\})$, $\Gamma = 0 \in L(D_{\Gamma_1}, \{0\})$. Then $A_r \in L_1(H_1 \oplus H_2, K_1)$ iff

$(1.2)_r$ $$A_r = (A, D_{A^*}\Gamma_1),$$

where $A \in L_1(H_1, K_1)$, $\Gamma_1 \in L_1(H_2, D_{A^*})$; moreover, one has the unitary operators:

$(1.4)_r$ $$\begin{cases} \alpha(A_r) : D_{A_r} \to D_A \oplus D_{\Gamma_1} \\ \alpha(A_r) D_{A_r} = \begin{bmatrix} D_A & -A^*\Gamma_1 \\ 0 & D_{\Gamma_1} \end{bmatrix} \in L(H_1 \oplus H_2, D_A \oplus D_{\Gamma_1}), \end{cases}$$

$$(1.4)^*_r \qquad \begin{cases} \alpha(A_r^*) : \mathcal{D}_{A_r^*} \to \mathcal{D}_{\Gamma_1^*} \\ \alpha(A_r^*)D_{A_r^*} = D_{\Gamma_1^*}D_{A^*} \in L(K_1, \mathcal{D}_{\Gamma_1^*}). \end{cases}$$

After this review, let us tackle the question of describing $J(\mathring{A})$ for $\mathring{A} \in L_1(H_1 \oplus H_2, K_1 \oplus K_2)$ in terms of elementary rotations of A, Γ_1, Γ_2, and Γ, where $\mathring{A} = \mathbf{C}(A, \Gamma_1, \Gamma_2, \Gamma)$. We have the following:

PROPOSITION 1.1. *If* $\mathring{A} = \mathbf{C}(A, \Gamma_1, \Gamma_2, \Gamma)$, *then:*

$$(1.5) \qquad [I_{K_1 \oplus K_2} \oplus \alpha]J(\mathring{A})[I_{H_1 \oplus H_2} \oplus \alpha_*^*] =$$

$$= (I_{K_1} \oplus J(\Gamma_2) \oplus I_{\mathcal{D}_\Gamma})(J(A) \oplus J(\Gamma))(I_{H_1} \oplus J(\Gamma_1) \oplus I_{\mathcal{D}_{\Gamma^*}})$$

where $\alpha = \alpha(\mathring{A})$ *and* $\alpha_* = \alpha(\mathring{A}^*)$ *are defined in* (1.4).

PROOF. Both sides of (1.5) act between $H_1 \oplus H_2 \oplus \mathcal{D}_{\Gamma_1^*} \oplus \mathcal{D}_{\Gamma^*}$ and $K_1 \oplus K_2 \oplus \mathcal{D}_{\Gamma_2} \oplus \mathcal{D}_\Gamma$. The right hand side of (1.5) is, after making the products, the matrix:

$$(1.6) \qquad \begin{bmatrix} A & D_{A^*}\Gamma_1 & D_{A^*}D_{\Gamma_1^*} & 0 \\ \Gamma_2 D_A & -\Gamma_2 A^*\Gamma_1 + D_{\Gamma_2^*}\Gamma D_{\Gamma_1} & -(\Gamma_2 A^* D_{\Gamma_1^*} + D_{\Gamma_2^*}\Gamma\Gamma_1^*) & D_{\Gamma_2^*}D_{\Gamma^*} \\ D_{\Gamma_2}D_A & -(D_{\Gamma_2}A^*\Gamma_1 + \Gamma_2^*\Gamma D_{\Gamma_1}) & \Gamma_2^*\Gamma\Gamma_1^* - D_{\Gamma_2}A^*D_{\Gamma_1^*} & -\Gamma_2^* D_{\Gamma^*} \\ 0 & D_\Gamma D_{\Gamma_1} & -D_\Gamma \Gamma_1^* & -\Gamma^* \end{bmatrix}$$

Taking into consideration (1.4), we have that (1.5) follows from (1.6) and the equality:

$$(1.7) \qquad \alpha\mathring{A}^*\alpha_*^* = \begin{bmatrix} -\Gamma_2^*\Gamma\Gamma_1^* + D_{\Gamma_2}A^*D_{\Gamma_1^*} & \Gamma_2^* D_{\Gamma^*} \\ D_\Gamma \Gamma_1^* & \Gamma^* \end{bmatrix} \in L(\mathcal{D}_{\Gamma_1^*} \oplus \mathcal{D}_{\Gamma^*}, \mathcal{D}_{\Gamma_2} \oplus \mathcal{D}_\Gamma).$$

Formula (1.7) is a routine computation. Indeed, it is enough to prove (1.7) for elements $\gamma \in \mathcal{D}_{\Gamma_1^*} \oplus \mathcal{D}_{\Gamma^*}$ of the form

$$(1.8)\qquad \gamma = \alpha_* D_{\mathring{A}^*}(k_1 \oplus k_2) = \begin{bmatrix} D_{\Gamma_1^*} D_{A^*} k_1 - D_{\Gamma_1^*} A \Gamma_2^* k_2 - \Gamma_1 \Gamma^* D_{\Gamma_2^*} k_2 \\ D_{\Gamma^*} D_{\Gamma_2^*} k_2 \end{bmatrix}$$

where $k_1 \in K_1$ and $k_2 \in K_2$. Now, the action of the left hand side of (1.7) to γ is

$$\alpha \mathring{A}^* \alpha_*^* \gamma = \alpha \mathring{A}^* \alpha_*^* \alpha_* D_{\mathring{A}^*}(k_1 \oplus k_2) = \alpha D_{\mathring{A}} \mathring{A}^*(k_1 \oplus k_2),$$

which can be computed using (1.4) and (1.2). The result proves to be equal with the action of the matrix from the right hand side of (1.7) to γ from (1.8). The computations repeatedly use that for a contraction T one has $TD_T = D_{T^*}T$ and $D_T^2 = I - T^*T$. □

REMARKS 1.2. a) The matrix (1.6) appeared in Proposition 1.1 from [15].

b) In the case of a row operator A_r as in $(1.2)_r$ the formula (1.5) becomes

$$(1.5)_r\qquad [I_{K_1} \oplus \alpha(A_r)]J(A_r)[I_{H_1 \oplus H_2} \oplus \alpha(A_r^*)^*] = (J(A) \oplus I_{D_{\Gamma_1}})(I_{H_1} \oplus J(\Gamma_1))$$

which is Lemma 2.2 from [6]. This might give an alternate way of proving Proposition 1.1: write $\mathring{A}$ from (1.2) as a row operator (the two components being some column operators) and apply $(1.5)_r$ and a similar one for columns.

c) The right hand side of (1.5) suggests a sort of simmetry between A and Γ in (1.2). This is also contained in (1.6); indeed, when A, Γ_1, Γ_2 and Γ are pure contractions (T is a pure contraction if $\ker D_T = \{0\}$), then the right hand side of (1.7) is (after intertwining the rows and the columns) exactly $\mathbf{C}(\Gamma^*, \Gamma_1^*, \Gamma_2^*, A^*) \in L(K_2 \oplus K_1, H_2 \oplus H_1)$. It is maybe worth mentioning that if $T \in L_1(H, H')$ is decomposed into the direct sum $T_u \oplus T_p$, where $T_u \in L(\ker D_T, \ker D_{T^*})$ is unitary and $T_p \in L(D_T, D_{T^*})$ is a pure contraction, then

$$(1.9)\qquad J(T) = T_u \oplus J(T_p),$$

with respect to the orthogonal decompositions $H \oplus D_{T^*} = (\ker D_T) \oplus (D_T \oplus D_{T^*})$ and $H' \oplus D_T = (\ker D_{T^*}) \oplus (D_{T^*} \oplus D_T)$.

d) Formula (1.10)' from [8] shows that if $\mathring{A}$ is as in (1.2), then

$$(1.10)\qquad \mathring{A} = (I \oplus \Gamma_2)J(A)(I \oplus \Gamma_1) + (0 \oplus D_{\Gamma_2^*} \Gamma D_{\Gamma_1});$$

here the "imposed" and the "free" parts of $\mathring{A}$ (when A, Γ_1, and Γ_2 are given) are separated. Formula (1.5) gives the reason why (1.10) is true.

e) The elementary rotation of a contraction T can be interpreted as the elementary lattice section

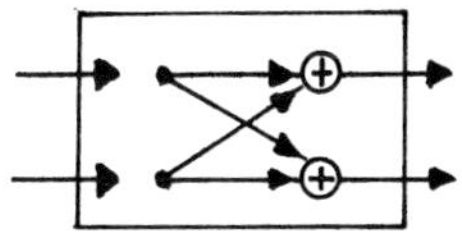

J(T).

Thus, formula (1.5) shows that the elementary rotation of $\mathring{A}$ from (1.2) is

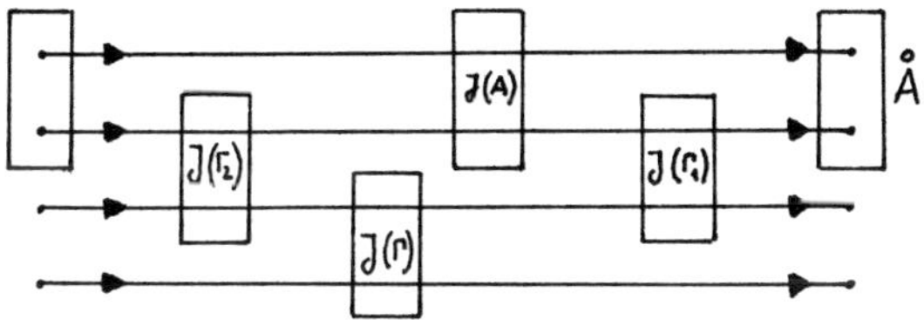

$J(\mathring{A})$.

f) The results of [8] were generalized in [7] to the case of operators (in Pontrjagin spaces) whose "defect operator" has a finite number of negative squares. The facts of this section can be accordingly generalized to that case.

2. Let $T \in L_1(H, H')$, $S \in L_1(H', H'')$, and $R = ST \in L_1(H, H'')$. The aim of this section is to compute J(R) in terms of J(T) and J(S); this will show a different way (from (1.5)) of composing elementary rotations. A close connection with the notion of regular factorization (see [18, Chapter VII]) will appear.

It is natural to try to embed D_R into the direct sum of D_S and D_T. This can be done using the isometry

$$\hat{\gamma}(ST) = \hat{\gamma} : D_R \to D_S \oplus D_T$$

(2.1)

$$\hat{\gamma} D_R = (D_S T,\ D_T)^t ,$$

("t" stands for the matrix transpose). Simple examples (e.g. S = T = 0) show that this operator can be nonunitary. Define

(2.2) $$F(ST) = (\mathbf{Im}\ \gamma)^- = \{D_S Th \oplus D_T h;\ h \in H\}^- \subset D_S \oplus D_T$$

and

(2.3) $$R(ST) = (D_S \oplus D_T) \ominus F(ST) .$$

We have then the unitary operator

(2.4) $$\gamma(ST) = \gamma : D_R \to F(ST), \quad \gamma d = \hat{\gamma} d, \quad d \in D_R .$$

Analogously, D_{R^*} can be identified with $F(T^*S^*) \subset D_{T^*} \oplus D_{S^*}$ through the unitary operator $\gamma_* = \gamma(T^*S^*)$. We have now the following

PROPOSITION 2.1. *For the factorization* R = ST

$$(2.5)\qquad (I_{H''} \oplus \gamma) J(R)(I_H \oplus \gamma_*^*) = (J(S) \oplus I_{D_T})(I_{H'} \oplus J(0))(J(T) \oplus I_{D_{S^*}}) \,|\, H \oplus F(T^*S^*)$$

where the zero operator in right hand side is considered in $L(D_T, D_{S^*})$.

PROOF. We proceed as in the proof of Proposition 1.1. The product from the right hand side is the matrix

$$(2.6)\qquad M = \begin{bmatrix} ST & SD_{T^*} & D_{S^*} \\ D_S T & D_S D_T & -S^* \\ D_T & -T^* & 0 \end{bmatrix} \in L(H \oplus D_{T^*} \oplus D_{S^*}, H'' \oplus D_S \oplus D_T).$$

Let now $x = h \oplus D_{T^*}S^*h'' \oplus D_{S^*}h'' \in H \oplus F(T^*S^*)$, where $h \in H$ and $h'' \in H''$. Then

$$(2.7)\qquad Mx = (Rh + D_{R^*}^2 h'', D_S T(h - R^*h''), D_T(h - R^*h'')).$$

From (2.7) it follows that $M(H \oplus F(T^*S^*)) = H'' \oplus F(ST)$ and that

$$(2.8)\qquad -\gamma R^* \gamma_*^* = \begin{bmatrix} D_S D_{T^*} & -S^* \\ -T^* & 0 \end{bmatrix} \,|\, F(T^*S^*).$$

Formulas (2.1) and (2.2) finish the proof. □

REMARKS 2.2. a) The operator J(0) is the operator of changing the sumands in a direct sum.

b) Replacing J(0) by J(Z) with $Z \in L_1(D_T, D_{S^*})$, the product from the right hand side of (2.5) (replacing $I_{D_{S^*}}$ by $I_{D_{Z^*}}$ and I_{D_T} with I_{D_Z}) is connected with the computation of the elementary rotation of the *contraction* $ST + D_{S^*}ZD_T$. Operators like this last one appeared in the study of positivity for arbitrary block-matrices (see [13]). If Z is factorized through a Hilbert space X as $Z = Z_2Z_1$, one obtains the elementary rotation of the product between the row $(S, D_{S^*}Z_1)$ and the column $(T, Z_2D_S)^t$; see also Section 4.

c) The factorization ST of R is called *regular* if $R(ST) = \{0\}$. This notion was intensively used in the study of invariant subspaces, namely any factorization of the characteristic function of a completely non-unitary contraction provides an invariant subspace for the direct sum of the considered contraction by a unitary operator, which is zero iff the factorization is regular ([18, p.321]). Regular factorizations appeared later in the uniqueness criterium for contractive intertwining dilations (see [1]), and then in the Schur-type analysis of the set of all contractive dilations ([2], [10], [11], [3], [4]). Now, in the setting of Proposition 2.1, it is clear that the contraction

$$Z(S,T) = Z = \begin{bmatrix} -D_S D_{T^*} & S^* \\ T^* & 0 \end{bmatrix} \in L(D_{T^*} \oplus D_{S^*}, D_S \oplus D_T) \tag{2.9}$$

has the property that its pure contractive part acts between $F(T^*S^*)$ and $F(ST)$ (and is canonically unitary equivalent with the pure contractive part of R^*), while its unitary part acts between $R(T^*S^*)$ and $R(ST)$. This is a way of proving the symmetry of regularity with respect to the adjoint operation (see [18, Proposition 3.2 of Chapter VII], [3], [9] for more about this). What Proposition 2.1 says is related with the beginning of this remark: the matrix M "is" the elementary rotation of the product plus a unitary operator which is zero when the factorization is regular.

d) The multiport interpretation of the regular case of Proposition 2.1 is

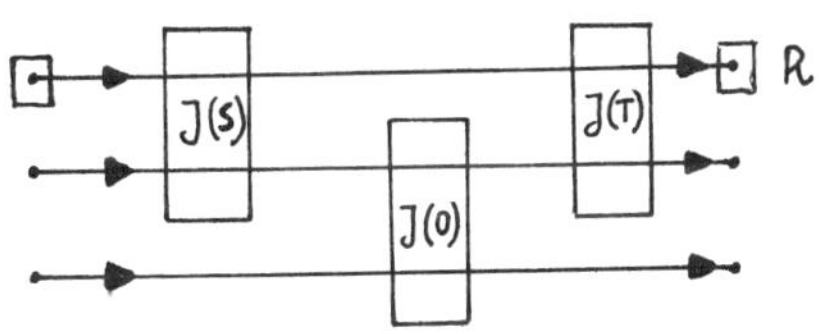

J(R).

3. In this section we indicate the canonical form (as a two-by-two matrix contraction) of the product of a column operator with a row operator. This will help doing this for the case to be considered in Section 4: the canonical form of the product of two-by-two matrix contractions.

Consider the operators:

$$T_r = (T, \; D_{T^*} \Delta_1) \in L_1(H_1 \oplus H_2, \; K) \tag{3.1}$$

where $\Delta_1 \in L(H_2, D_{T^*})$, and

$$S_c = (S, \; \Delta_2 D_S)^t \in L_1(K, \; G_1 \oplus G_2), \tag{3.2}$$

where $\Delta_2 \in L(D_S, G_2)$.

Denote

$$R = ST \in L(H_1, G_1), \text{ and} \tag{3.3}$$

(3.4) $$\overset{\circ}{R} = S_c T_r \in L(H_1 \oplus H_2, G_1 \oplus G_2).$$

Then

(3.5) $$\overset{\circ}{R} = \begin{bmatrix} R & SD_{T^*}\Delta_1 \\ \Delta_2 D_S T & \Delta_2 D_S D_{T^*}\Delta_1 \end{bmatrix}.$$

For writing $\overset{\circ}{R}$ in the canonical form it is clear that we have to use the analysis of defect spaces of R from Section 2. To this end, consider

(3.6) $$P(S,T) = P = P^{\mathcal{D}_S \oplus \mathcal{D}_T}_{F(ST)} \quad \text{and}$$

$(3.6)_*$ $$P_* = P(T^*, S^*).$$

Then, using the operators γ and γ_* from (2.4), define

(3.7) $$\Gamma_1 = \gamma_*^* P_*(\Delta_1, 0)^t : H_2 \to \mathcal{D}_{R^*}$$

(3.8) $$\Gamma_2 = (\Delta_2, 0)\gamma : \mathcal{D}_R \to G_2.$$

From (3.7) it follows

(3.9) $$D_{R^*}\Gamma_1 = SD_{T^*}\Delta_1,$$

which is the canonical form of the (1, 2) entry of $\overset{\circ}{R}$ from (3.5). Analogously,

(3.10) $$\Gamma_2 D_R = \Delta_2 D_S T$$

is the canonical form of the (2, 1) entry of $\overset{\circ}{R}$.

For the (2,2) entry, note first that

(3.11) $$-\Gamma_2 R^* \Gamma_1 = -(\Delta_2, 0)\gamma R^* \gamma_*^* P_*(\Delta_1, 0)^t = (\Delta_2, 0) Z P_*(\Delta_1, 0)^t$$

where we used (2.8) and (2.9). Because

(3.12) $$(\Delta_2, 0)\begin{bmatrix} D_S D_{T^*} & -S^* \\ -T^* & 0 \end{bmatrix}\begin{bmatrix} \Delta_1 \\ 0 \end{bmatrix} = \Delta_2 D_S D_{T^*}\Delta_1,$$

the operator $\Gamma \in L(\mathcal{D}_{\Gamma_1}, \mathcal{D}_{\Gamma_2^*})$ from the canonical form of the (2,2) entry of $\overset{\circ}{R}$ must be obtained from

(3.13) $$(\Delta_2, 0) Z (1 - P_*)(\Delta_1, 0)^t = D_{\Gamma_2^*}\Gamma D_{\Gamma_1}.$$

We will use now the usual method to "identify" the defect operators D_{Γ_1} and

$D_{\Gamma_2^*}$. For any $h_2 \varepsilon H_2$,

(3.14) $$\|D_{\Gamma_1} h_2\|^2 = \|h_2\|^2 - \|\Gamma_1 h_2\|^2 = \|h_2\|^2 - \|\gamma_*^* P_*(\Delta_1 h_2 \oplus 0)\|^2 =$$

$$= \|h_2\|^2 - \|P_*(\Delta_1 h_2 \oplus 0)\|^2 = \|h_2\|^2 - \|\Delta_1 h_2\|^2 + \|(1 - P_*)(\Delta_1 h \oplus 0)\|^2 =$$

$$= \|D_{\Delta_1} h_2\|^2 + \|(1 - P_*)(\Delta_1,\ 0)^t h_2\|^2 .$$

The relation (3.14) shows that the operator :

(3.15) $$\begin{cases} \Phi : H_2 \to D_{\Delta_1} \oplus R(T^* S^*) \\ \Phi = (D_{\Delta_1},\ (1 - P_*)(\Delta_1,\ 0)^t)^t \end{cases}$$

is an isometry; denote $\Phi(H_2) = M$. Then, the operator

(3.16) $$\begin{cases} \phi : D_{\Gamma_1} \to M \\ \phi D_{\Gamma_1} = \Phi \end{cases}$$

is a unitary one.

Similar considerations for $D_{\Gamma_2^*}$ imply that the operator

(3.17) $$\begin{cases} \Phi_* : G_2 \to D_{\Delta_2^*} \oplus R(ST) \\ \Phi_* = (D_{\Delta_2^*},\ (1 - P)(\Delta_2^*,\ 0)^t)^t \end{cases}$$

is an isometry; denoting $\Phi_*(G_2) = M_*$, the operator

(3.18) $$\begin{cases} \phi_* : D_{\Gamma_2^*} \to M_* \\ \phi_* D_{\Gamma_2^*} = \Phi_* \end{cases}$$

is a unitary one.

Now it is easy to verify that the solution for Γ from (3.13) is:

(3.19) $$\Gamma = \phi_*^* P_{M_*}^{D_{\Delta_2^*} \oplus R(ST)} (0 \oplus Z)\phi \ \varepsilon\ L(D_{\Gamma_1}, D_{\Gamma_2^*}) .$$

Indeed, since Φ_* is an isometry, $P_{M_*} = \Phi_* \Phi_*^*$, and then

(3.20) $D_{\Gamma_2^*}\Gamma D_{\Gamma_1} = D_{\Gamma_2^*}\phi_*^*\Phi_*\Phi_*^*(0\oplus Z)\phi D_{\Gamma_1} = \Phi_*^*(0\oplus Z)\Phi = (\Delta_2,\ 0)Z(1-P_*)(\Delta_1,\ 0)^t$.

Thus, we proved:

PROPOSITION 3.1. *With the notation from* (1.3) *we have that* $\overset{\circ}{R} = \mathbf{C}(R,\Gamma_1,\Gamma_2,\Gamma)$, *where* R, Γ_1, Γ_2 *and* Γ *are defined by* (3.3), (3.7), (3.8), *and* (3.19), *respectively.*

The connection with regular factorizations appears also in the following observation:

COROLLARY 3.2. a) *If the factorization* R = ST *from* (3.3) *is regular, then for any choice of* H_2, Δ_1, G_2, Δ_2 *from* (3.1) *and* (3.2) *the operator* Γ *from* (3.19) *is zero.*

b) *If there exist* $H_2 \neq \{0\}$ *and* $G_2 \neq \{0\}$ *such that for any choice of* Δ_1 *and* Δ_2, *the operator* Γ *from* (3.19) *is zero, then the factorization* R = ST *is regular.*

PROOF. a) follows immediately from (3.19), because $R(ST) = \{0\}$ implies $Z = 0$.
b) follows from the fact that $\overline{(I-P)(D_S\oplus\{0\})} = R(S,T)$ (see Lemma 1.1 in [3]). □

REMARK 3.3. The methods from Proposition 3.1 gives the possibility of carrying out the analysis done in Section 1 for the operator $\overset{\circ}{R}$ in (3.4). So, one can obtain descriptions of $D_{\overset{\circ}{R}}$ and $D_{\overset{\circ}{R}^*}$, as well as some equivalences between multiports interpretations of $\overset{\circ}{R}$ resulting from (3.4) and (3.21). We omit the details.

4. We shortly indicate now how to apply the results of Section 3 to the general case of two-by-two matrices.

Let $\overset{\circ}{A} \in L_1(H_1\oplus H_2, K_1\oplus K_2)$, $\overset{\circ}{A} = \mathbf{C}(A,\Gamma_1,\Gamma_2,\Gamma)$ and $\overset{\circ}{B} \in L_1(K_1\oplus K_2, G_1\oplus G_2)$, $\overset{\circ}{B} = \mathbf{C}(B,\Delta_1,\Delta_2,\Delta)$; see (1.3) for notation. Denote

$$\overset{\circ}{C} = \overset{\circ}{B}\overset{\circ}{A} \in L_1(H_1\oplus H_2, G_1\oplus G_2). \tag{4.1}$$

Write $\overset{\circ}{C}$ in the canonical form:

$$\overset{\circ}{C} = \mathbf{C}(C,\Theta_1,\Theta_2,\Theta); \tag{4.2}$$

we indicate below the operators C, Θ_1, Θ_2, and Θ in terms of the stucture operators of $\overset{\circ}{A}$ and $\overset{\circ}{B}$. Having in mind the analysis done in Section 3 we will write $\overset{\circ}{A}$ as a row operator and $\overset{\circ}{B}$ as a column operator.

To this aim, put $\overset{\circ}{A}$ in the form:

(4.3) $$\overset{\circ}{A} = (A_c\,,\ D_{A_c^*}\tilde{\Gamma})\,,$$

where

(4.4) $$A_c = (A\,,\ \Gamma_2 D_A)^t\,,$$

and

(4.5) $$\begin{cases} \tilde{\Gamma} \in L_1(H_2, D_{A_c^*}) \\ \tilde{\Gamma} = \alpha^*(A_c^*)(\Gamma_1,\ \Gamma D_{\Gamma_1})^t\ . \end{cases}$$

This is indeed the case, because $(\Gamma_1 \quad \Gamma D_{\Gamma_1})^t$ is a column contraction acting between H_2 and $D_{A^*} \oplus D_{\Gamma^*}$, while $\alpha^*(A_c^*)$ is a unitary operator acting between $D_{A^*} \oplus D_{\Gamma_2^*}$ and $D_{A_c^*}$; moreover - as in $(1.4)_2$ - $D_{A_c^*}\alpha^*(A_c^*) \in L(D_{A^*} \oplus D_{\Gamma_2^*}, K_1 \oplus K_2)$ has the matrix

(4.6) $$D_{A_c^*}\alpha^*(A_c^*) = \begin{bmatrix} D_{A^*} & 0 \\ -\Gamma_2 A^* & D_{\Gamma_2^*} \end{bmatrix},$$

which implies (4.3).

Similarly

(4.7) $$\overset{\circ}{B} = (B_r\,,\ \tilde{\Delta} D_{B_r})^t\,,$$

where

(4.8) $$B_r = (B\,,\ D_{B^*}\Delta_1)\,,$$

and

(4.9) $$\begin{cases} \tilde{\Delta} \in L_1(D_{B_r}, G_2) \\ \tilde{\Delta} = (\Delta_2,\ D_{\Delta_2^*}\Delta)\alpha(B_r)\,, \end{cases}$$

where

(4.10) $$\alpha(B_r)D_{B_r} = \begin{bmatrix} D_B & -B^*\Delta_1 \\ 0 & D_{\Delta_1} \end{bmatrix}.$$

Now, we can apply Proposition 3.1 to $\overset{\circ}{C} = (B_r\,,\ \tilde{\Delta} D_{B_r})^t(A_c\,,\ D_{A_c^*}\tilde{\Gamma})$. From (3.3) it follows that

(4.11) $$C = B_r A_c = BA + D_{A^*}\Gamma_1\Delta_2 D_B\,.$$

For describing Θ_1, note first that D_{C^*} can be identified with $F(A_c^* B_r^*) \subset D_{A_c^*} \oplus D_{B_r^*}$ via the unitary operator $\gamma(A_c^* B_r^*)$ from (2.4). Using the analysis from $(1.4)_r$ and $(1.4)_r^*$ one has that $D_{A_c^*} \oplus D_{B_r^*}$ can be identified with $(D_{A^*} \oplus D_{\Gamma_2^*}) \oplus D_{\Delta_1^*}$ using the unitary operator $\alpha(A_c^*) \oplus \alpha(B_r^*)$. Composing these identifications we have that if we denote

$$(4.12)\quad \tilde{F}(A_c^* B_r^*) = \{(D_{A^*} B^* g_1 - A \Gamma_2^* \Delta_1^* D_{B^*} g_1) \oplus D_{\Gamma_2^*} \Delta_1^* D_{B^*} g_1 \oplus D_{\Delta_1^*} D_{B^*} g_1 ;\ g_1 \in G_1\} \subset$$
$$\subset D_{A^*} \oplus D_{\Gamma_2^*} \oplus D_{\Delta_1^*},$$

then the operator

$$(4.13)\quad \begin{cases} \tilde{\gamma}_* : D_{C^*} \to \tilde{F}(A_c^* B_r^*) \\ \tilde{\gamma}_* = [\alpha(A_c^*) \oplus \alpha(B_r^*)]\gamma(A_c^* B_r^*) \end{cases}$$

is a unitary one. Denote by $\tilde{P}_*$ the projection of $D_{A^*} \oplus D_{\Gamma_2^*} \oplus D_{\Delta_1^*}$ onto $\tilde{F}(A_c^* B_r^*)$. Then the relations (3.7), (4.3), (4.5), (4.13) and the definition of $\tilde{P}_*$ imply that

$$(4.14)\quad \Theta_1 = \tilde{\gamma}_*^* \tilde{P}_* (\Gamma_1,\ \Gamma D_{\Gamma_1},\ 0)^t \in L_1(H_2, D_{C^*}).$$

The description of Θ_2 is similar. One has to define

$$(4.15)\quad \tilde{F}(B_r A_c) = \tilde{\gamma} F(B_r A_c) \subset D_B \oplus D_{\Delta_1} \oplus D_{\Gamma_2},$$

where

$$(4.16)\quad \tilde{\gamma} = [\alpha(B_r) \oplus \alpha(A_c)]\gamma(B_r A_c),$$

and $\tilde{P}$ the projection of $D_B \oplus D_{\Delta_1} \oplus D_{\Gamma_2}$ onto $\tilde{F}(B_r A_c)$. Then, using (3.8), we obtain

$$(4.17)\quad \Theta_2 = (\Delta_2,\ D_{\Delta_2^*} \Delta,\ 0)\tilde{\gamma} \in L_1(D_C, G_2).$$

For indicating Θ, we have to identify D_{Θ_1} and $D_{\Theta_2^*}$. Using (3.15), (3.16), and the previous analysis of Θ_1, we infer that the operator

$$(4.18)\quad \begin{cases} \tilde{\phi}: D_{\Theta_1} \to \tilde{M} \subset D_\Gamma \oplus D_{A^*} \oplus D_{\Gamma_2^*} \oplus D_{\Delta_1^*} \\ \tilde{\phi} D_{\Theta_1} = (D_\Gamma D_{\Gamma_1},\ (I - \tilde{P}_*)(\Gamma_1,\ \Gamma D_{\Gamma_1},\ 0)^t)^t \end{cases}$$

is a unitary one ($\tilde{M} = \tilde{\phi} D_{\Theta_1}$). Analogous, from (3.17) and (3.18) - and the structure of Θ_2 - it results that the operator

$$(4.19)\qquad \begin{cases} \tilde{\Phi}_* : \mathcal{D}_{\Theta_2^*} \to \tilde{M}_* \subset \mathcal{D}_{\Delta^*} \oplus \mathcal{D}_B \oplus \mathcal{D}_{\Delta_1} \oplus \mathcal{D}_{\Gamma_2} \\ \tilde{\Phi}_* D_{\Theta_2^*} = (D_{\Delta^*} D_{\Delta_2^*},\ (I - \tilde{P})(\Delta_2^*,\ \Delta^* D_{\Delta_2^*},\ 0)^t)^t \end{cases}$$

is a unitary one ($\tilde{M}_* = \tilde{\Phi}^* \mathcal{D}_{\Theta_2^*}$). Consider also the operator:

$$(4.20)\qquad \begin{cases} \tilde{Z} : \mathcal{D}_{A^*} \oplus \mathcal{D}_{\Gamma_2^*} \oplus \mathcal{D}_{\Delta_1^*} \to \mathcal{D}_B \oplus \mathcal{D}_{\Delta_1} \oplus \mathcal{D}_{\Gamma_2} \\ \tilde{Z} = [\alpha(B_r) + \alpha(A_c)] Z(B_r, A_c) \alpha^*(A_c^*) + \alpha^*(B_r^*) \end{cases}$$

(see (2.9)). The explicit matrix of $\tilde{Z}$ is

$$(4.21)\qquad \tilde{Z} = \begin{bmatrix} -(D_B D_{A^*} + B^* \Delta_1 \Gamma_2 A^*) & B^* \Delta_1 D_{\Gamma_2^*} & B^* D_{\Delta_1^*} \\ D_{\Delta_1} \Gamma_2 B^* & -D_{\Delta_1} D_{\Gamma_2^*} & \Delta_1^* \\ D_{\Gamma_2} A^* & \Gamma_2^* & 0 \end{bmatrix}.$$

Using (3.19) we finally have

$$(4.22)\qquad \Theta = \tilde{\Phi}_*^* P_{\tilde{M}_*} (0 \oplus \tilde{Z}) \tilde{\Phi} \in L_1(\mathcal{D}_{\Theta_1}, \mathcal{D}_{\Theta_2^*}).$$

Summing up, we obtain

THEOREM 4.1. *If* $\mathring{A} = \mathbf{C}(A, \Gamma_1, \Gamma_2, \Gamma)$ *and* $\mathring{B} = \mathbf{C}(B, \Delta_1, \Delta_2, \Delta)$, *then* $\mathring{C} = \mathring{B}\mathring{A}$ *has the canonical form* $\mathring{C} = \mathbf{C}(C, \Theta_1, \Theta_2, \Theta)$, *where* C, Θ_1, Θ_2, *and* Θ *are defined by* (4.11), (4.14), (4.17), *and* (4.22), *respectively.*

REMARKS 4.2. a) Corollary 3.2 and Remark 3.3 can be transcribed for the case of this section.

b) The computations connected with angles in Gaussian processes and with Szegö-type theorems ask for the value of the determinants of the defect operators of some products of contractions. As remarked in Section 1, the knowledge of the canonical form of these products implies a triangularization of defect operators, and thus the possibility of iterative computations of their determinants. This was done directly in [5, Theorem 4.5] and in [6, Theorem 3.4] for cases which can now be included in Theorem 4.1.

c) The so-called Schur analysis of contractivity (see [10], [11], [3], [4], [9], [12]) gives algorithms and formulas for computing the general solution of some operatorial extrapolations problems in terms of Schur-type parameters; for this, various manipulations with rotation operators are used. Some of these computations are included in the results of the present paper.

d) A natural question in the Schur analysis of contractivity is the structure of parameters for a product of two solutions; Theorem 4.1 is a step in this respect.

REFERENCES

1. **Ando, T.; Ceaușescu, Zoia; Foiaș, C.:** On intertwining dilations. II, *Acta. Sci. Math. (Szeged)*, **39**(1977), 3-14.
2. **Arsene, Gr.; Ceaușescu, Zoia:** On intertwining dilations. IV, *Tohoku Math. J.*, **30**(1978), 423-438.
3. **Arsene, Gr.; Ceaușescu, Zoia; Foiaș, C.:** On intertwining dilations. VII, *Proc. Coll. Complex Analysis, Joensuu*, Lecture Notes in Math., **747**(1979), 24-45.
4. **Arsene, Gr.; Ceaușescu, Zoia; Foiaș, C.:** On intertwining dilations. VIII, *J. Operator Theory*, **4**(1980), 55-91.
5. **Arsene, Gr.; Constantinescu, T.:** On the structure of Naimark dilation and Gaussian stationary processes, *Integral Equations Operator Theory*, **8**(1985), 181-204.
6. **Arsene, Gr.; Constantinescu, T.:** Structure of positive block-matrices and nonstationary prediction, *J. Funct. Anal.*, to appear.
7. **Arsene, Gr.; Constantinescu, T.; Gheondea, A.:** Lifting of operators and prescribed numbers of negative squares, INCREST Preprint Series, **64**(1985).
8. **Arsene, Gr.; Gheondea, A.:** Completing matrix contractions, *J. Operator Theory*, **7**(1982), 179-189.
9. **Ceaușescu, Zoia:** *Operatorial extrapolations* (Romanian), Thesis, Bucharest, 1980.
10. **Ceaușescu, Zoia; Foiaș, C.:** On intertwining dilations. V, *Acta Sci. Math. (Szeged)*, **40**(1978), 9-32.
11. **Ceaușescu, Zoia; Foiaș, C.:** On intertwining dilations. VI, *Rev. Roumaine Math. Pures Appl.*, **23**(1978), 1471-1482.
12. **Constantinescu, T.:** An algorithm for the operatorial Carathéodory-Féjer problem, in *Spectral theory for linear operators and related topics*, Birkhäuser Verlag, 1984, pp.81-107.
13. **Constantinescu, T.:** Schur analysis of positive block-matrices, preprint INCREST no.**54**(1984).
14. **Foiaș, C.; Frazho, A.E.:** Redheffer products and the lifting of contractions on Hilbert space, *J. Operator Theory*, **11**(1984), 193-196.
15. **Langer, H.; Textorius, B.:** Generalized resolvents of dual pairs of contractions, in *Invariant subspaces and other topics*, Birkhäuser Verlag, 1982, pp.103-118.
16. **Redheffer, R.M. :** Inequalities for a matrix Riccati equation, *J. Math. Mech.*, **8**(1959), 349-377.
17. **Sz.-Nagy, B.; Foiaș, C.:** Forme triangulaire d'un contraction et factorization de la fonction caractéristique, *Acta Sci. Math. (Szeged)*, **28**(1967), 201-212.
18. **Sz.-Nagy, B.; Foiaș, C.:** *Harmonic analysis of operators on Hilbert space*, Amsterdam-Budapest, 1970.

Department of Mathematics, INCREST
Bdul Păcii 220, 79622 Bucharest
Romania.

Operator Theory:
Advances and Applications:

CLASSIFICATION OF SHIFT INVARIANT SUBSPACES OF MATRICES WITH HERMITIAN FORM AND COMPLETION OF MATRICES*

Joseph A. Ball and Israel Gohberg

In this paper we study subspaces of matrices which are invariant under multiplication on the right by the algebra of upper triangular matrices in the presence of an indefinite inner product. For such subspaces a canonical representation is obtained. As an application we solve the problem of completing a specified lower triangular part to a unitary matrix. All middle factors M in the LMU factorization of a completion are described in terms of invariants of the problem. A linear fractional map parametrization of all unitary completions is obtained.

* This work was partially supported by the U.S. National Science Foundation.

TABLE OF CONTENTS

Introduction

The main results of this paper center around unitary completions of matrices wih prescribed entries in the lower triangular positions. Specifically, suppose numbers k_{ij} are specified for $1 \leq j < i \leq n$; the problem is to characterize all nxn matrices $F = [f_{ij}]_{1 \leq i,j \leq n}$ such that i) $f_{ij} = k_{ij}$ for $1 \leq i < j \leq n$ and ii) F is unitary. This problem, with the additional requirement that F admits a factorization F=LU where L is lower triangular and U is upper triangular, was solved for the more general block triangular case in the paper [DG1] (see also [GG2]). In the scalar case, their result was that a solution with this additional requirement exists if and only if all matrices $K_m = [k_{ij}]_{m+1 \leq i \leq n, 1 \leq j \leq m}$ have $\|K_m\| < 1$ for all m $(1 \leq m < n)$. In this situation where $\|K_m\| < 1$ for all m, there was obtained in [BG1] a general description of all the unitary completions via a linear fractional map $G \to (\theta_{11} G + \theta_{12})(\theta_{21} G + \theta_{22})^{-1}$ where θ_{ij} are appropriate size matrices constructed from the given data. The same result was obtained in [BG 2] as an application of a more general lifting theorem. One of the main points of this paper is to obtain an analogous parametrization of the set of all unitary completions for the degenerate case where $\|K_m\| \leq 1$ and equality appears for some m. We are also interested in analyzing the possible middle factors M in the noncanonical LMU factorization of a unitary completion F.

We show in this paper that the Grassmannian approach set up in [BG 1] for the case $\|K_m\| < 1$ for all m can be extended to the more general case considered here, and produces similar results.

Now we shall state our main results for the special scalar case. For this purpose we first extract some additional information from the given data. For (p,q) a pair of integers satisfying $1 \leq q < p \leq n$ and

$\{k_{ij}\}_{1\le j<i\le n}$ a specified set of data as above, let $\Phi_{p,q}(K)$ denote the $(n-p)xq$ matrix

$\Phi_{p,q}(K) = [k_{ij}]_{\substack{p\le i\le n\\ 1\le j\le q}}$. Define integers $\varphi_{p,q}$ by

$$(0.1)\quad \varphi_{p,q} = \dim \operatorname{Ker} (I-\Phi_{p,q}(K)^*\Phi_{p,q}(K))$$

for $1\le q<p\le n$. We may then define integers $r_{p,q}$ (again for $1\le q<p\le n$) by

$$(0.2)\quad r_{p,q} = \varphi_{p,q} - \varphi_{p,q-1} - \varphi_{p+1,q} + \varphi_{p+1,q-1}.$$

(Here $\varphi_{r,s}$ is understood to be 0 if $r=n+1$ or $s=0$.) We then define n-tuples of nonnegative integers $\underline{\nu}^0 = \{\nu_1^0, \ldots, \nu_n^0\}$, $\underline{\nu}^+ = \{\nu_1^+,\ldots, \nu_n^+\}$, $\underline{\nu}^- = \{\nu_1^-, \ldots, \nu_n^-\}$ and $\underline{\nu}^x = \{\nu_1^x,\ldots, \nu_n^x\}$ by

$$(0.3)\quad \nu_p^0 = \sum_{i=p+1}^{n} r_{ip}$$

$$(0.4)\quad \nu_p^+ = 1 - \sum_{i=1}^{p-1} r_{pi}$$

$$(0.5)\quad \nu_p^- = 1 - \sum_{i=p+1}^{n} r_{ip}$$

and

$$(0.6)\quad \nu_p^x = \sum_{i=1}^{p-1} r_{pi}$$

for $1\le p\le n$. For the scalar case which we are discussing here, it turns out that each of the integers in (0.3)–(0.6) is either 0 or 1. For $\underline{\nu} = \{\nu_1,\ldots, \nu_n\}$ any n-tuple of integers, we let $N(\underline{\nu})$ denote the sum $\nu_1+\cdots+\nu_n$. In addition to the data (0.3)–(0.6) extracted as above, we introduce a $N(\underline{\nu}^0)xN(\underline{\nu}^x)$ matrix π with nxn block form

(0.7) $\pi = [\pi_{ij}]_{1\leq i,j\leq n}$

where

$$\pi_{ij} \text{ has size } \nu_i^0 \times \nu_j^x$$

and

$$(0.8)\qquad \pi_{ij} = \begin{cases} 0 & \text{if } i \geq j \\ r_{ji} & \text{if } i < j \end{cases}.$$

(It turns out that π is a permutation matrix)

Given any n-tuple $\underline{\nu} = \{\nu_1, ..., \nu_n\}$, we denote by $\Omega_u(\underline{\nu})$ the class of $N(\underline{\nu})xn$ matrices F having the block upper triangular form

$$F = [F_{ij}]_{1\leq i,j\leq n}$$

where the size of the (i,j)-th block F_{ij} is $\nu_i x1$ and where $F_{ij} = 0$ if $i>j$. Finally, for convenience, when given data $k_{ij}=0$ for $1\leq j<i\leq n$, we set $k_{ij} = 0$ for $1\leq i\leq j\leq n$ and let K denote the strictly lower triangular matrix $K = [k_{ij}]_{1\leq i,j\leq n}$.

The first part of the paper is concerned with finding an appropriate representation for the set of all 2nxn matrices of the form $\begin{bmatrix} I & K \\ 0 & I \end{bmatrix} \begin{bmatrix} X \\ Y \end{bmatrix}$ where X, Y $\epsilon \Omega_u$ and Ω_u is the set of all nxn upper triangular matrices (so Ω_u is shorthand for $\Omega_u(\{1,...,1\})$ in the more general notation introduced above). The following is the required representation theorem.

THEOREM 0.1. Let there be specified the lower triangular entries $\{k_{ij}\}_{1\leq j<i\leq n}$ of the strictly lower triangular nxn matrix K. Extract from this data n-tuples $\underline{\nu}^0$, $\underline{\nu}^+$, $\underline{\nu}^-$, $\underline{\nu}^x$ and the $N(\underline{\nu}^0)xN(\underline{\nu}^x)$ permutation matrix π by (0.1)-(0.8). Then there exists a 2nx2n matrix Θ

of the block form

$$\Theta = \begin{bmatrix} \Theta_{11} & \Theta_{12} & \Theta_{13} & \Theta_{14} \\ \Theta_{21} & \Theta_{22} & \Theta_{23} & \Theta_{24} \end{bmatrix},$$

where each Θ_{ij} has n rows and Θ_{i1}, Θ_{i2}, Θ_{i3}, Θ_{i4} has $N(\underline{\nu}^0)$, $N(\underline{\nu}^+)$, $N(\underline{\nu}^-)$, $N(\underline{\nu}^x)$ columns respectively, such that

$$(0.9) \quad \Theta^* \begin{bmatrix} I_n & 0 \\ 0 & -I_n \end{bmatrix} \Theta = \begin{bmatrix} 0 & 0 & 0 & \pi \\ 0 & I_{N(\underline{\nu}^+)} & 0 & 0 \\ 0 & 0 & -I_{N(\underline{\nu}^-)} & 0 \\ \pi^* & 0 & 0 & 0 \end{bmatrix}$$

and

$$(0.10) \quad \Theta \begin{bmatrix} \Omega_u(\underline{\nu}^0) \\ \Omega_u(\underline{\nu}^+) \\ \Omega_u(\underline{\nu}^-) \\ \Omega_u(\underline{\nu}^x) \end{bmatrix} = \begin{bmatrix} I_n & K \\ 0 & I_n \end{bmatrix} \begin{bmatrix} \Omega_u \\ \Omega_u \end{bmatrix}.$$

Theorem 0.1 enables us to construct a linear fractional map which parametrizes the set of all solutions of the unitary completion problem. To describe the parametrization we need one more piece of notation. For two n-tuples $\underline{\mu} = \{\mu_1, \ldots, \mu_n\}$ and $\underline{\nu} = \{\nu_1, \ldots, \nu_n\}$ of integers, we denote by $\Omega_u(\underline{\mu} x \underline{\nu})$ the class of all $N(\underline{\mu})$ x $N(\underline{\nu})$ matrices F with block upper triangular form $F = [F_{ij}]_{1 \le i,j \le n}$ where the (i,j)-th block F_{ij} has size μ_i x ν_j and $F_{ij} = 0$ if $i > j$. The description of all unitary completions of a given data set $\{k_{ij}\}_{1 \le j < i \le n}$ is given by the following.

THEOREM 0.2. Let $\{k_{ij}\}_{1 \le j < i \le n}$ be a given lower triangular array of complex numbers. Then there exist unitary completions F of $\{k_{ij}\}$ (i.e.

unitary matrices $F = [F_{ij}]_{1\leq i,j\leq n}$ such that $f_{ij} = k_{ij}$ for $j<i$) if and only if each matrix $K_m = [k_{ij}]_{m+1\leq i\leq n, 1\leq j\leq m}$ has $\|K_m\|\leq 1$ for $1\leq m<n$. Moreover, in this case, any matrix F of the form

$$(0.11) \quad F = [\Theta_{11}, \Theta_{12}G+\Theta_{22}]([\Theta_{21}, \Theta_{22}G+\Theta_{23}])^{-1}$$

is a unitary completion of $\{k_{ij}\}$ if G is a unitary matrix in $\Omega_u(\underline{\nu}^+ x \underline{\nu}^-)$;

here $\Theta = \begin{bmatrix} \Theta_{11} & \Theta_{12} & \Theta_{13} & \Theta_{14} \\ \Theta_{21} & \Theta_{22} & \Theta_{23} & \Theta_{24} \end{bmatrix}$ is any matrix as in Theorem 0.1. Conversely, any unitary completion F has the form (0.11) for some unitary matrix G in $\Omega_u(\underline{\nu}^+ x \underline{\nu}^-)$.

It is known (see [GG1]) that any invertible nxn matrix F admits a factorization F=LMU where L is lower triangular, M is a permutation matrix uniquely determined by F, and U is upper triangular. Our final result analyzes the middle factor M in the factorization F = LMU of a unitary completion F.

THEOREM 0.3. For any unitary completion F of a given set of data $\{k_{ij}\}_{1\leq j<i\leq n}$ with factorization F=LMU, the lower triangular entries of the middle factor M are determined from the data $\{k_{ij}\}$ as follows:

$$M_{ij} = r_{ij} \text{ for } i>j$$

where r_{ij} is given by (0.2) and (0.1).

If the unitary completion F is obtained from the unitary matrix $G\in\Omega_u(\underline{\nu}^+ x \underline{\nu}^-)$ as in (0.11), then the upper triangular entries of M are determined by G in the following way: If G has factorization $G = \tilde{L}N\tilde{U}$ (where $\tilde{L}$ is lower triangular, $\tilde{U}$ upper triangular and N a permutation matrix), then N is the matrix remaining from M when all rows and columns which have in their intersection a 1 from the lower triangular part of M are deleted.

As an application we now consider the case where $\|K_m\|<1$ for $1\leq m<n$ in some detail. In this case one easily sees from (0.2)–(0.6) that

$$r_{pq} = 0, \; 1\leq q<p\leq n$$

and

$$\nu_p^+ = 1$$

$$\nu_p^- = 1$$

$$\nu_p^x = 0$$

for $1\leq p\leq n$. From the first part of Theorem 0.3 we see that the lower triangular part of the middle factor M of any unitary completion F = LMU consists entirely of 0's. Since M moreover is a permutation matrix, we see that necessarily M is the identity matrix I. Alternatively, we may conclude that M=I by using the second part of Theorem 0.3 as follows. Since $\nu_p^+ = \nu_p^- = 1$ for $1\leq p\leq n$, we see that $\Omega_u(\underline{\nu}^+ x \underline{\nu}^-) = \Omega_u$ consists of the nxn upper-triangular matrices in the usual sense. But the only unitary upper triangular matrices G are diagonal, and hence factor in the form $\widetilde{L}\widetilde{N}\widetilde{U}$ with N=I. Now by using the recipe in Theorem 0.3 we see that M=I for any unitary completion F. We have obtained the following result which is new even for the scalar case.

COROLLARY 0.4. Suppose $\{k_{ij}\}_{1\leq j<i\leq n}$ is a given set of data such that $\|K_m\|<1$ for $1\leq m<n$. Then all unitary completions F have canonical LU factorization.

All our results concern more general problems than what is described above. Specifically, we consider the case of block matrices, and consider the unitary completion problem for the case where the data is prescribed below some diagonal not necessarily the main diagonal. Also much of the analysis goes through for isometric completions.

Section 1 reviews material from [BG 3] on representation of subspaces of matrices invariant under the action of upper triangular matrices by right multiplication (Ω_u-invariant subspaces). This material from §1 is the starting point for a modification of a representation theorem from [BG 3]; this result is discussed in §2 and is in close analogy with the invariant subspace theorem for the function theory case established in [BH 2]. Section 3 uses this representation theorem to obtain the linear fractional parametrization of all isometric (or contractive) completions. Section 4 goes on to analyze types of middle factors which can occur in the noncanonical LMU factorizations of unitary completions. The theorems of this section are in close analogy with the results of [DG2] and [B] for the function theory case. The proofs are parallel to the developments in [B]. Finally §5 shows how the machinery of this paper can be used to complete the results on lifting theory of [BG 2]. In this way, all the results of this paper (apart from those on LU factorization) extend to the more abstract context of contractive unital representations of the algebra of upper triangular matrices.

1. Invariant subspaces with a Hermitian form

We shall need various classes of block upper and lower triangular matrices set forth in our earlier papers [BG 1-3]. In all the following let n be a fixed natural number. Whenever $\underline{\nu}'=\{\nu_1',...,\nu_1'\}$ and $\underline{\nu}=\{\nu_1,...,\nu_n\}$ are n-tuples of nonnegative integers, we let $\Omega(\underline{\nu}'x\underline{\nu})$ denote the class of nxn block matrices $F=[F_{ij}]_{1\leq i\leq j\leq n}$ where the size of the (i,j)-th block is $\nu_i'x\nu_j$. Thus the size of F as a matrix over the complex numbers $\mathbb{C}$ is $N(\underline{\nu}')xN(\underline{\nu})$ where $N(\underline{\nu})=\sum_{j=1}^{n}\nu_j$. We let $\Omega_u(\underline{\nu}'x\underline{\nu})$ denote the block upper triangular matrices in $\Omega(\underline{\nu}'x\underline{\nu})$, that is, a block matrix $F=[F_{ij}]_{1\le i,j\le n}$ in $\Omega(\underline{\nu}'x\underline{\nu})$ is in $\Omega_u(\underline{\nu}'x\underline{\nu})$ if $F_{ij}=0$ whenever $i>j$. The block upper triangular matrices with block diagonal entries 0 (so $F=[F_{ij}]$ with $F_{ij}=0$ whenever $i\ge j$) we denote by $\Omega_u^0(\underline{\nu}'x\underline{\nu})$. When the roles of block rows and block columns are interchanged in the above, we obtain the classes $\Omega_\ell(\underline{\nu}'x\underline{\nu})$ of block lower triangular and $\Omega_\ell^0(\underline{\nu}'x\underline{\nu})$ of block strictly lower triangular matrices. Let $\Omega_d(\underline{\nu}'x\underline{\nu})$ be the class $\Omega_u(\underline{\nu}'x\underline{\nu})\cap\Omega_\ell(\underline{\nu}'x\underline{\nu})$ of block diagonal matrices. When the right n-tuple $\underline{\nu}$ is equal to $\{1,...,1\}$, we simplify the notation $\Omega_u(\underline{\nu}'x\{1,...,1\})$ to $\Omega_u(\underline{\nu}')$; define $\Omega_u^0(\underline{\nu}')$, $\Omega_\ell(\underline{\nu}')$, $\Omega_\ell^0(\underline{\nu}')$ and $\Omega_d(\underline{\nu}')$ analogously. When both $\underline{\nu}'$ and $\underline{\nu}$ equal $\{1,...,1\}$, we simplify the notation $\Omega_u(\{1,...,1\})$ even further to Ω_u; thus Ω_u is the algebra of (nonblock or scalar) upper triangular nxn matrices. Define Ω_u^0, Ω_ℓ, Ω_ℓ^0 and Ω_d in a similar way.

Denote by E_j the nx1 matrix with all 0's except for a 1 in the j-th row. Let $\mathcal{P}_j$ denote the set of all nx1 matrices (i.e. vectors in $\mathbb{C}^n$) $x=[x_k]_{1\le k\le n}$ with $x_k=0$ if $k>j$; $\mathcal{Q}_j$ we take to be the set of vectors $[x_k]_{1\le k\le n}$ in $\mathbb{C}^n$ with $x_k=0$ if $k<j$. Then

$$\mathcal{P}_j=\Omega_u E_j$$

and

$$\mathcal{Q}_{j+1}=\Omega_{\ell}^{0}E_j.$$

More generally, for an n-tuple $\underline{\nu}=\{\nu_1,\ldots,\nu_n\}$ of nonnegative integers, let $\mathcal{P}_j^{\underline{\nu}}$ denote the set of vectors $x=[x_k]_{1\leq k\leq N(\underline{\nu})}$ in $\mathbb{C}^{N(\underline{\nu})}$ such that $x_k=0$ if $k > \sum_{i=1}^{j} \nu_i$, and $\mathcal{Q}_j^{\underline{\nu}}$ denote the set of such vectors with $x_k=0$ if $k \leq \sum_{i=1}^{j-1} \nu_i$. Then we have the identities

(1.1) $\mathcal{P}_j^{\underline{\nu}}=\Omega_u(\underline{\nu})E_j$

and

(1.2) $\mathcal{Q}_{j+1}^{\underline{\nu}}=\Omega_{\ell}^{0}(\underline{\nu})E_j.$

If $\mathcal{M}$ is a subspace of kxn matrices for some k such that $FU\in\mathcal{M}$ whenever $F\in\mathcal{M}$ and $U\in\Omega_u$, we say that $\mathcal{M}$ is Ω_u-invariant. One of the results from [BG 2] is that any such subspace can be represented as $\mathcal{M}=\Theta\Omega_u(\underline{\nu})$ for a $k\text{x}N(\underline{\nu})$ injective matrix Θ (unique up to an invertible right factor in $\Omega_u(\underline{\nu}\text{x}\underline{\nu})$) for a uniquely determined n-tuple $\underline{\nu}$; in fact one can even arrange that Θ is isometric ($\Theta^*\Theta=I_{N(\underline{\nu})}$).

Now we suppose that we are given an Ω_u-invariant subspace $\mathcal{M}$ of kxn matrices and a kxk Hermitian matrix W. The matrix W induces an inner product $[,]_W$ (in general indefinite and/or degenerate) on kxn matrices X and Y by

$$[X,Y]_W=\mathrm{tr}(Y^*WX).$$

It is natural to say that two such pairs $(\mathcal{M},W)$ and $(\mathcal{M}',W')$ are unitarily equivalent if and only if there is a linear transformation T which maps $\mathcal{M}$

onto $\mathcal{M}'$ in a one-to-one fashion, which commutes with multiplication on the right by an upper triangular matrix, and which transforms the W'-inner product on $\mathcal{M}'$ to the W-inner product on $\mathcal{M}$. It is not difficult to show that two such pairs $(\mathcal{M},W)$ and $(\mathcal{M}',W')$ are unitarily equivalent in this sense if and only if there is a $k\times k'$ matrix Θ such that $\Theta|\operatorname{span}\{\operatorname{Im}X:X\in\mathcal{M}'\}$ is injective with $\mathcal{M}=\Theta\mathcal{M}'$ and $\Theta^*W\Theta=W'$. The problem of unitary equivalence is to find the "simplest pair" $\{\mathcal{M}',W'\}$ which is unitarily equivalent to the given pair $\{\mathcal{M},W\}$. We now recall the solution to this problem presented in [BG 3]. In the next section we use this result to obtain an alternative format for a canonical representative of a given equivalence class which is more suitable for our application to the unitary completion problem in §2.

Let $\mathcal{M}$ be an Ω_u-invariant subspace and W a Hermitian matrix as above. We let $\mathcal{R}_{ij}$ be the quotient space

$$\mathcal{R}_{ij}=\mathcal{P}_{ij}/\mathcal{Q}_{ij}$$

for $1\le i,j\le n$ where

$$\mathcal{P}_{ij}=(\mathcal{M}E_{i-1}^{<\perp>_W}\cap\mathcal{M}E_j)+\operatorname{Ker}W$$

and

$$\mathcal{Q}_{ij}=(\mathcal{M}E_{i}^{<\perp>_W}\cap\mathcal{M}E_j)+(\mathcal{M}E_{i-1}^{<\perp>_W}\cap\mathcal{M}E_{j-1})+\operatorname{Ker}W$$

and in addition

$$(1.3)\qquad \mathcal{R}_{n+1,j}=\frac{\mathcal{M}E_j\cap\operatorname{Ker}W}{\mathcal{M}E_{j-1}\cap\operatorname{Ker}W}.$$

We define the partial rank indices $r_{ij}=r_{ij}(\mathcal{M},W)$ for the pair $(\mathcal{M},W)$ by

$$(1.4)\qquad r_{ij}=\dim\mathcal{R}_{ij}$$

for $1\le i\le n+1$ and $1\le j\le n$, and the partial signature indices

$s_i^+=s_i^+(\mathcal{M},W)$ and $s_i^-=s_i^-(\mathcal{M},W)$

by

(1.5) $s_i^+=$number of positive squares of $\mathcal{R}_{ii}$ in $[,]_W$

and

(1.6) $s_i^-=$number of negative squares of $\mathcal{R}_{ii}$ in $[,]_W$ for $1\leq i\leq n$.

Finally let $\nu_j=\dim \mathcal{M}E_j/\mathcal{M}E_{j-1}$. Then we have the relations

(1.7) $r_{ij}=r_{ji}$ for $1\leq i,j\leq n$

(1.8) $\sum_{i=1}^{n+1} r_{ij}=\nu_j$ for $1\leq j<n$

and

(1.9) $s_i^++s_i^-=r_{ii}$ for $1\leq i\leq n$.

It is not difficult to see that the parital rank indices $r_{ij}(\mathcal{M},W)$ and partial signature indices $s_i^+(\mathcal{M},W),s_i^-(\mathcal{M},W)$ are <u>unitary invariants</u> for the pair $(\mathcal{M},W)$, i.e., if the pair $(\mathcal{M},W)$ is unitarily equivalent to $(\mathcal{M}',W')$ in the sense explained above, then

(1.10) $r_{ij}(\mathcal{M},W)=r_{ij}(\mathcal{M}',W')$,

(1.11) $s_i^+(\mathcal{M},W)=s_i^+(\mathcal{M}',W')$,

(1.12) $s_i^-(\mathcal{M},W)=s_i^-(\mathcal{M}',W')$.

One of the main results from [BG 3] is that this set constitutes a <u>complete</u> set of unitary invariants, i.e. if the pairs $(\mathcal{M},W)$ and $(\mathcal{M}',W')$ are such that (1.10), (1.11) and (1.12) hold, then $(\mathcal{M},W)$ is unitarily equivalent to $(\mathcal{M}',W')$.

To pick out the simplest example of a pair $(\mathcal{M},W)$ with some prescribed partial rank indices $\{r_{ij}:1\leq i\leq n+1,1\leq j\leq n\}$ and partial signature indices $\{s_i^+, s_i^-: 1\le i\le n\}$ (subject to (1.7)-(1.9)), we need to introduce the notion of a $\underline{\nu}$-block Z-subpermutation matrix. A Hermitian matrix $M=[M_{ij}]_{1\le i,j\le n}$ of block class $\Omega(\underline{\nu}\times\underline{\nu})$ we say is a $\underline{\nu}$-block <u>Z-subpermution</u> <u>matrix</u> if the diagonal blocks M_{ii} are themselves diagonal with all nonzero entries equal to either 1 or −1, all nonzero entries in the off-diagonal blocks $M_{ij}(i\ne j)$ are equal to 1, and each (scalar) row and column contains at most one nonzero entry. We introduce partial rank indices $r_{ij}(M)$ and partial signature indices $s_i^+(M)$ and $s_i^-(M)(1\le i,j\le n)$ for a $\underline{\nu}$-block Z-subpermutation matrix $M=[M_{ij}]_{1\le i,j\le n}$ by

(1.13) $r_{ij}(M)=\text{rank } M_{ij}$

(1.14) $s_i^+(M)$ = number of 1's on the diagonal of M_{ii}

(1.15) $s_i^-(M)$ = number of −1's on the diagonal of M_{ii}.

Then the consistency conditions

(1.16) $r_{ij}(M)=r_{ji}(M)$

(1.17) $\sum_{i'=1}^{n} r_{i'j}(M)\le\nu_j$

and

(1.18) $s_i^+(M)+s_i^-(M)=r_{ii}(M)$

for $1\le i,j\le n$ must hold. If we introduce $r_{n+1,j}(M)$ by $r_{n+1,j}=\nu_j-\sum_{i'=1}^{n} r_{i'j}(M)$, then (1.16)-(1.17) are of the same form as (1.7)-(1.9). It is easy to see that given any collection of indices $\{r_{ij}$:

$1\le i\le n+1,\ 1\le j\le n\}$ and $\{s_i^+, s_i^-: 1\le i\le n\}$ satisfying (1.7)-(1.9) then there is a $\underline{\nu}$-block Z-subpermutation matrix M such that $r_{ij}=r_{ij}(M)$, $s_i^+=s_i^+(M), s_i^-=s_i^-(M)$. If we insist that M be in a canonical form by using the "top-most left-most" and "1's before -1's" convention, we can even arrange that such an M is unique (see [BG 3]). We can now state onc of the main results from [BG 3].

THEOREM 1.1 (see Theorem 3.3 in [BG 3]) Suppose $\mathcal{M}$ is an Ω_u-invariant subspace of kxn matrices and W is a Hermitian kxk matrix. Associate the partial subspaccs $\mathcal{R}_{ij}$ $(1\le i\le n+1, 1\le j\le n)$ by (1.3), partial rank indices $r_{ij}=r_{ij}(\mathcal{M},W)$ by (1.4), partial signature indices $s_i^+=s_i^+(\mathcal{M},W)$ and $s_i^-=s_i^-(\mathcal{M},W)$ by (1.5) and (1.6) and the n-tuple $\underline{\nu}=\{\nu_1,\dots,\nu_n\}$ by

$$\nu_j=\dim \mathcal{M}E_j/\mathcal{M}E_{j-1}.$$

Let M be a $\underline{\nu}$-block Z-subpermutation matrix with this set of partial indices. Then there exists an injective $k\mathrm{x}N(\underline{\nu})$ matrix Θ such that

$$\mathcal{M}=\Theta\Omega_u(\underline{\nu})$$

and

$$\Theta^* W\Theta=M$$

Thus the pair $(\mathcal{M},W)$ is unitarily equivalent to the pair $(\Omega_u(\underline{\nu}),M)$.

Any such representing matrix Θ arises in the following way:

i) the columns of Θ can be organized as

$$\Theta=[\theta_{11},\theta_{21},\dots,\theta_{n1},\theta_{n+1,1};\theta_{12},\theta_{22},\dots,\theta_{n2},\theta_{n+1,2};\dots;$$
$$\theta_{1n},\theta_{2n},\dots,\theta_{nn},\theta_{n+1,n}]$$

where the images of the columns of θ_{ij} in $\mathbb{C}^k/(\mathcal{M}E_i^{<\perp>W} \cap \mathcal{M}E_j) + (\mathcal{M}E_{i-1}^{<\perp>W} \cap \mathcal{M}E_{j-1})$ form a basis for the partial subspace $\mathcal{R}_{ij}$,

ii) for $i>j$, $<W\theta_{ij}^{\alpha}, \theta_{ji}^{\beta}> = \delta_{\alpha\beta}$

where $\theta_{ij} = [\theta_{ij}^1, \theta_{ij}^2, \ldots, \theta_{ij}^{r_{ij}}]$,

and

iii) $<W\theta_{ii}^{+\alpha}, \theta_{ii}^{+\beta}> = \delta_{\alpha\beta}$

$<W\theta_{ii}^{-\alpha}, \theta_{ii}^{-\beta}> = -\delta_{\alpha\beta}$

and

$<W\theta_{ii}^{+\alpha}, \theta_{ii}^{-\beta}> = 0$

where $\theta_{ii} = [\theta_{ii}^{+1}, \ldots, \theta_{ii}^{+s_i^+}; \theta_{ii}^{-1}, \ldots, \theta_{ii}^{-s_i^-}]$ for $1 \le i \le n$.

2. Invariant subspaces with a Hermitian form: second format

For the study of unitary completions of partially defined lower triangular matrices, we need to have a convenient representation for the Ω_u-invariant maximal negative subspaces of a given Ω_u-invariant subspace $\mathcal{M}$ with Hermitian form $[\ ,\ ]_W$. To accomplish this, it is convenient to reorganize the presentation of the Ω_u-invariant subspace from [BG 3] given in §1. This involves sorting out basis vectors for $\mathcal{M}E_j/\mathcal{M}E_{j-1}$ into four separate categories and then listing by category first and then by block within each category. We do this first in §2.1 for the model $(\mathcal{M}=\Omega_u(\underline{\nu})$ with Hermitian form $[\ ,\ ]_M$ for a $\underline{\nu}$-block

Z-subpermutation M). This leads us to the definition of a $\underline{\nu}$-block S-subpermutation matrix. We then show in §2.2 how this leads to a reorganized general representation theorem. Then in §2.3 we develop the main application of this reorganized presentation, namely, the characterization of all Ω_u-invariant M-maximal W-negative subspaces of $\mathcal{M}$. Finally in §2.4 we give the new factorization theorems corresponding to these invariant subspace representations.

2.1 S-Permutation and S-Subpermutation Matrices

Let us suppose that we are given an n-tuple $\underline{\nu}=\{\nu_1,\dots,\nu_n\}$ of nonnegative integers and a $\underline{\nu}$-block Z-subpermutation matrix M as in §1. The standard basis vectors for $\mathbb{C}^{N(\underline{\nu})}$ we index in a way compatible with the n-tuple $\underline{\nu}$ as $\{e_{1,1},\dots,e_{1,\nu_1};e_{2,1},\dots,e_{2,\nu_2};\dots;$

$e_{n,1},\dots,e_{n,\nu_n}\}$. By $\underline{e}_i$ we denote the i-th block of standard basis vectors arranged as a $N(\underline{\nu})x\nu_i$ matrix

$$\underline{e}_i=[e_{i1},\dots,e_{i\nu_i}]$$

for $1\leq i\leq n$. Next consider $\mathbb{C}^{N(\underline{\nu})}$ in the indefinite inner product $[x,y]_M=\langle Mx,y\rangle$ induced by the $\underline{\nu}$-block Z-subpermutation matrix M. We assume that M is in canonical form, so M is the unique $\underline{\nu}$-block Z-subpermutation matrix obtained from its partial rank indices $\{r_{ij}:1\leq i,j\leq n\}$ and its partial signature induces $\{s_i^+,s_i^-:1\leq i\leq n\}$ by the "topmost-leftmost" and "1's before -1's" conventions as in §1 (see [BG 3]). Then one can easily check that each column $e_{ij}(1\leq j\leq\nu_i)$ of $\underline{e}_i$ satisfies exactly one of the following conditions:

i) $[e_{ij},e_{kl}]=0$ whenever $k\geq i$

but $[e_{ij}, e_{kl}] \neq 0$ for some $k<i$.

ii) $[e_{ij}, e_{ij}]=1$ and

$[e_{ij}, e_{kl}]=0$ if $(i,j)\neq(k,l)$.

iii) $[e_{ij}, e_{ij}]=-1$ and

$[e_{ij}, e_{kl}]=0$ if $(i,j)\neq(k,l)$.

and

iv) $[e_{ij}, e_{kl}]=0$ whenever $k\leq i$.

Moreover, due to the "topmost-leftmost" and "1's before −1's" conventions, the standard basis vectors of type i) are the first $\sum_{j'=1}^{i-1} r_{ij'}$ columns of $\underline{e}_i$, those satisfying ii) are the next s_i^+ columns, those satisfying iii) are the next s_i^- columns, and those satisfying iv) are the last $\nu_i - \sum_{j=1}^{i} r_{ij}$ columns. Thus we may arrange $\underline{e}_i$ as $\underline{e}_i=[\underline{e}_i^{x}, \underline{e}_i^{+}, \underline{e}_i^{-}, \underline{e}_i^{0}]$ where $\underline{e}_i^{x}$ consists of the first $\sum_{j=1}^{i-1} r_{ij}$ columns of $\underline{e}_i$ which satisfy i), e_i^+ consists of the next s_i^+ columns which satisfy ii), e_i^- consists of the next s_i^- columns which satisfy iii) and $\underline{e}_i^{0}$ consists of the remaining columns which satisfy iv). Now we wish to rearrange these standard basis vectors so that the category classification i), ii), iii), iv) supercedes the block decomposition from $\underline{\nu}$. To do this, let P be the permutation matrix

$$P=[\underline{e}^0, \underline{e}^+, \underline{e}^-, \underline{e}^x]$$

where

$$\underline{e}^0=[\underline{e}_1^0, \underline{e}_2^0, \dots, \underline{e}_n^0]$$

$$\underline{e}^{+}=[e_1^{+},\underline{e}_2^{+},\dots,\underline{e}_n^{+}]$$

$$\underline{e}^{-}=[\underline{e}_1^{-},\underline{e}_2^{-},\dots,\underline{e}_n^{-}]$$

and

$$\underline{e}^{x}=[\underline{e}_1^{x},\underline{e}_2^{x},\dots\underline{e}_n^{x}]$$

Introduce auxiliary n-tuples of nonnegative integers $\underline{\nu}^0=\{\nu_1^0,\dots,\nu_n^0\}$ where

$$\nu_i^0=\nu_i-\sum_{j=1}^{i} r_{ij} \tag{2.1}$$

$$\underline{\nu}^{+}=\{s_1^{+},\dots,s_n^{+}\} \tag{2.2}$$

$$\underline{\nu}^{-}=\{s_1^{-},\dots,s_n^{-}\} \tag{2.3}$$

and $\underline{\nu}^{x}=\{\nu_1^{x},\dots,\nu_n^{x}\}$ where

$$\nu_i^{x}=\sum_{j=1}^{i-1} r_{ij}. \tag{2.4}$$

Then one observes that

$$\Omega_u(\underline{\nu})=P\begin{bmatrix}\Omega_u(\underline{\nu}^0)\\ \Omega_u(\underline{\nu}^{+})\\ \Omega_u(\underline{\nu}^{-})\\ \Omega_u(\underline{\nu}^{x})\end{bmatrix}.$$

We see that the Ω_u-invariant subspace $\Omega_u(\underline{\nu})$ in the M-inner product is unitarily equivalent (via P) to the Ω_u-invariant subspace $\text{col.}[\Omega_u(\underline{\nu}^0),\Omega_u(\underline{\nu}^{+}),\Omega_u(\underline{\nu}^{-}),\Omega_u(\underline{\nu}^{x})]$ in the inner product induced by $N:=P^*MP$. It is a matter of bookkeeping to compute that N has the form

$$N=\begin{bmatrix}0 & 0 & \Pi\\ 0 & J & 0\\ \Pi^* & 0 & 0\end{bmatrix}$$

where J is the $(N(\underline{\nu}^+)+N(\underline{\nu}^-))x(N(\underline{\nu}^+)xN(\underline{\nu}^-))$ signature matrix

$J=\begin{bmatrix} I_{N(\underline{\nu}^+)} & 0 \\ 0 & -I_{N(\underline{\nu}^-)} \end{bmatrix}$, and Π is the $\underline{\nu}^0 x \underline{\nu}^x$-block subpermutation

matrix in canonical form with partial indices r^{Π}_{ij} given by

$$(2.5)\qquad r^{\Pi}_{ij}=\begin{cases} 0 & \text{if } i\geq j \\ r_{ji}\,(=r^{M}_{ji}) & \text{if } i<j \end{cases}.$$

In general by a $\underline{\nu}' x \underline{\nu}$ <u>block permutation matrix</u> L with partial indices r^{L}_{ij} we mean a matrix $L=[L_{ij}]_{1\leq i,j\leq n}$ of block class $\Omega(\underline{\nu}' x \underline{\nu})$ such that each row and column of L contains at most one nonzero entry, all nonzero entries are 1's, and rank $L_{ij}=r^{L}_{ij}$ for $1\leq i,j\leq n$. We say that L is in <u>canonical form</u> if L is the unique $\underline{\nu}' x \underline{\nu}$ block permutation matrix satisfying rank $L_{ij}=r^{L}_{ij}$ with nonzero entries assigned within each block according to the "top-most left-most" convention; for details see [BG 3]. For our situation here, the $\underline{\nu}^0 x \underline{\nu}^x$ block permutation matrix Π is in fact $\underline{\nu}^0 x \underline{\nu}^x$-inner (i.e. $\Pi\in\Omega_u(\underline{\nu}^0 x \underline{\nu}^x)$ and is isometric); moreover, the $\underline{\nu}^0 x \underline{\nu}^x$-block diagonal entries of Π are all zero, so in fact $\Pi\in\Omega^0_u(\underline{\nu}^0 x \underline{\nu}^x)$.

This motivates our definition of a $(\underline{\nu}^0,\underline{\nu}^+,\underline{\nu}^-,\underline{\nu}^x)$-block S-subpermutation matrix. Given four n-tuples $\underline{\nu}^0,\underline{\nu}^+,\underline{\nu}^-$ and $\underline{\nu}^x$, we say that a matrix N is a <u>$(\underline{\nu}^0,\underline{\nu}^+,\underline{\nu}^-,\underline{\nu}^x)$-block S-subpermutation</u>

matrix if N has the form $N=\begin{bmatrix} 0 & 0 & \Pi \\ 0 & J & 0 \\ \Pi^* & 0 & 0 \end{bmatrix}$ where J is the signature matrix $J=\begin{bmatrix} I_{N(\underline{\nu}^+)} & 0 \\ 0 & -I_{N(\underline{\nu}^-)} \end{bmatrix}$ and Π is an injective (and hence isometric) $\underline{\nu}^0 x \underline{\nu}^x$-block subpermutation matrix in canonical form which is also strictly $\underline{\nu}^0 x \underline{\nu}^x$-block upper triangular. If N is invertible (equivalently, if Π is a $\underline{\nu}^0 x \underline{\nu}^x$-block permutation matrix), then we say that N is a $(\underline{\nu}^0,\underline{\nu}^+,\underline{\nu}^-,\underline{\nu}^x)$-block S-permutation matrix. Given the n-tuples $\underline{\nu}^0,\underline{\nu}^+,\underline{\nu}^-$ and $\underline{\nu}^x$, such a matrix N is then uniquely determined by the partial indices r^{Π}_{ij} of Π for $i<j$. Our alternative model for an Ω_u-invariant subspace $\mathcal{M}$ with Hermitian form $[\ ,\ ]_W$ is then to take $\mathcal{M}$ to be equal to col. $[\Omega_u(\underline{\nu}^0),\ \Omega_u(\underline{\nu}^+),\ \Omega_u(\underline{\nu}^-),\ \Omega_u(\underline{\nu}^x)]$ with W equal to a $(\underline{\nu}^0,\underline{\nu}^+,\underline{\nu}^-,\underline{\nu}^x)$- block S-subpermutation matrix N.

Now suppose that we are given a $(\underline{\nu}^0,\underline{\nu}^+,\underline{\nu}^-,\underline{\nu}^x)$-block S-subpermutation matrix N. We now agrue that the Ω_u-invariant subspace col. $[\Omega_u(\underline{\nu}^0),\ \Omega_u(\underline{\nu}^+),\ \Omega_u(\underline{\nu}^-),\ \Omega_u(\underline{\nu}^x)]$ in the N-inner product is unitarily equivalent to the Ω_u-invariant subspace $\Omega_u(\underline{\nu})$ with the M-inner product for an appropriate n-tuple $\underline{\nu}$ and $\underline{\nu}$-block Z-subpermutation matrix M. Indeed, define $\underline{\nu}=\{\nu_1,\dots,\nu_n\}$ by $\nu_i=\nu^0_i+\nu^+_i+\nu^-_i+\nu^x_i$ for $1\le i\le n$. Define the partial signature indices $\{s^+_i,s^-_i:1\le i\le n\}$ by $s^+_i=\nu^+_i$ and $s^-_i=\nu^-_i$. Define partial rank indices

r_{ij} $(1\leq i,j\leq n)$ by $r_{ii}=s_i^+ + s_i^-, r_{ij}=r_{ji}^\Pi$ if $i>j$ and $r_{ij}=r_{ij}^\Pi$ if $i<j$. Let M be the unique $\underline{\nu}$-block Z-subpermutation matrix (in canonical form) with these partial indices and let P be the permutation martrix specified as above. Then we see that we recover N as $N=P^*MP$ and col.$[\Omega_u(\underline{\nu}^0)$, $\Omega_u(\underline{\nu}^+)$, $\Omega_u(\underline{\nu}^-)$, $\Omega_u(\underline{\nu}^x)]$ as equal to $P^*\Omega_u(\underline{\nu})$.

To specify an S-subpermutation matrix N, besides the four n-tuples $\underline{\nu}^0,\underline{\nu}^+,\underline{\nu}^-,\underline{\nu}^x$, we need only specify the rank indices r_{ij}^Π for $i<j$. Equivalently we need only specify the n-tuple $\underline{\nu}$ and the partial rank indices r_{ij}^M $(1\leq i<j\leq n)$ and the partial signature indices $s_i^+(M)$ and $s_i^-(M)$ for the equivalent $\underline{\nu}$-block Z-subpermutatin matrix M. When N is presented in this way we shall refer to N as a $\underline{\nu}$-block S-subpermutation matrix with partial rank indices r_{ij}^M $(1\leq i,j\leq n)$ and partial signature indices $s_i^+(M)$ and $s_i^-(M)(1\leq i\leq n)$.

2.2. S-Subpermutation Matrix Associated with an Invariant Subspace and Hermitian Form

Now suppose that $\mathcal{M}$ is an Ω_u-invariant subspace of kxn matrices with Hermitian form $[,]_W$. We may combine Theorem 1.1 with the discussion of §2.1 to see that we may represent $\mathcal{M}$ in the form $\mathcal{M}=\Theta$ col. $[\Omega_u(\underline{\nu}^0),\Omega_u(\underline{\nu}^+),\Omega_u(\underline{\nu}^-),\Omega_u(\underline{\nu}^x)]$ where $\Theta^*W\Theta=N$ and N is a S-subpermutation matrix uniquely determined by M. We state the result precisely as follows. The proof follows directly from Theorem 1.1 and §2.1. This is a more general form of Theorem 0.1 from the Introduction.

THEOREM 2.1. Suppose $\mathcal{M}$ is an Ω_u-invariant subspace of kxn matrices and W is a kxk Hermitian matrix. Then there exist unique n-tuples $\underline{\nu}^0, \underline{\nu}^+, \underline{\nu}^-$ and $\underline{\nu}^x$, a unique $(\underline{\nu}^0, \underline{\nu}^+, \underline{\nu}^-, \underline{\nu}^x)$-block S-subpermutation matrix N and an injective matrix Θ such that

$$\text{i)}\quad \mathcal{M} = \Theta \begin{bmatrix} \Omega_u(\underline{\nu}^0) \\ \Omega_u(\underline{\nu}^+) \\ \Omega_u(\underline{\nu}^-) \\ \Omega_u(\underline{\nu}^x) \end{bmatrix}$$

and

$$\text{ii)}\quad \Theta^* W \Theta = N.$$

Here $\nu_i^0 = \nu_i - \sum_{j=1}^{i} r_{ij}$, $\nu_i^+ = s_i^+$, $\nu_i^- = s_i^-$ and $\nu_i^x = \sum_{j=1}^{i-1} r_{ij}$ for $1 \le i \le n$, where $\{r_{ij}: 1 \le i,j \le n\}$ are the partial rank indices and $\{s_i^+, s_i^-: 1 \le i \le n\}$ are the partial signature indices of the Ω_u-invariant subspace $\mathcal{M}$ with respect to the Hermitian form $[,]_W$, as in Theorem 1.1. The matrix N has the form

$$N = \begin{bmatrix} 0 & 0 & \Pi \\ 0 & J & 0 \\ \Pi^* & 0 & 0 \end{bmatrix}$$

where $J = \begin{bmatrix} I_{N(\underline{\nu}^+)} & 0 \\ 0 & -I_{N(\underline{\nu}^-)} \end{bmatrix}$ and

Π is the unique $\underline{\nu}^0 x \underline{\nu}^x$-block subpermutation matrix in canonical form in $\Omega_u^0(\underline{\nu}^0 x \underline{\nu}^x)$ with partial indices $r_{ij}^{\Pi} = r_{ij}$ if $i<j$. Any representer Θ has the form $\Theta = [\Theta^0; \Theta^+; \Theta^-; \Theta^x]$

where

$$\Theta^0 = [\Theta_{21}, \ldots, \Theta_{n1}, \Theta_{n+1,1}; \Theta_{32}, \Theta_{42}, \ldots, \Theta_{n+1,2}; \ldots; \\ \Theta_{n,n-1}, \Theta_{n+1,n-1}; \Theta_{n+1,n}],$$

$$\Theta^{+}=[\Theta_{11}^{+}\Theta_{22}^{+}\ldots\Theta_{nn}^{+}],$$

$$\Theta^{-}=[\Theta_{11}^{-},\Theta_{22}^{-},\ldots,\Theta_{nn}^{-}]$$

and

$$\Theta^{x}=[\Theta_{12};\Theta_{13},\Theta_{23};\ldots;\Theta_{1n},\Theta_{2n},\ldots,\Theta_{n-1,n}].$$

Here the columns of Θ_{ij} $(1\leq i\leq n+1, 1\leq j\leq n)$ are as in Theorem 1.1.

2.3. Invariant maximal negative subspaces

Suppose $\mathcal{K}$ is an indefinite inner product space, that is K is a linear space with a Hermitian form [,]. We say that a subspace $\mathcal{G}\subset\mathcal{K}$ is negative if $[X,X]\leq 0$ for all $X\in\mathcal{G}$. We say that $\mathcal{G}$ is maximal negative if $\mathcal{G}$ is negative and is not contained in any larger negative subspace of $\mathcal{K}$. The isotropic subspace of $\mathcal{K}$ is by definition $\mathcal{K}\cap\mathcal{K}^{\perp}$, the set of all elements of $\mathcal{K}$ which are [,]-orthogonal to all of $\mathcal{K}$. A subspace $\mathcal{G}\subset\mathcal{K}$ is said to be neutral if $[X,X]=0$ for all $X\in\mathcal{G}$. In this case $[X,Y]=0$ for all $X,Y\in\mathcal{G}$ by the Schwarz inequality for positive semidefinite inner products.

For the matrix completion problems of §3 to come we shall need to characterize the Ω_u-invariant maximal negative subspaces of a given Ω_u-invariant subspace $\mathcal{M}$ with respect to a Hermitian form $[,]_W$. It is for this application that the representation of such a subspace in terms of a S-subpermutation matrix (rather than a Z-subpermuation matrix) is particularly useful, as we shall now see.

Now suppose that we are given a $(\underline{\nu}^0,\underline{\nu}^+,\underline{\nu}^-,\underline{\nu}^x)$-block S-subpermutation matrix N and that we consider the Ω_u-invariant subspace $\mathcal{M}=\mathrm{col.}[\Omega_u(\underline{\nu}^0),\Omega_u(\underline{\nu}^+),\Omega_u(\underline{\nu}^-),\Omega_u(\underline{\nu}^x)]$ as an indefinite inner produce space in the N-inner product $[X,Y]_N=\mathrm{tr}(Y^*NX)$. Our goal is to characterize all $\mathcal{M}$-maximal N-negative subspaces $\mathcal{G}$ of $\mathcal{M}$ which are also

Ω_u-invariant. Now it is easy to see that any $\mathcal{M}$-maximal negative subspace of $\mathcal{M}$ necessarily contains the isotropic subspace $\mathcal{M}\cap\mathcal{M}^{\langle\perp\rangle_N}$ of $\mathcal{M}$. Recall that the S-subpermutation matrix N has the form

$$N=\begin{bmatrix} 0 & 0 & \Pi \\ 0 & J & 0 \\ \Pi^* & 0 & 0 \end{bmatrix}$$ where Π is an injective $\underline{\nu}^0 x \underline{\nu}^x$-block subpermutation matrix in $\Omega_u^0(\underline{\nu}^0 x \underline{\nu}^x)$. Since the $\underline{\nu}^0 x \underline{\nu}^x$-block diagonal entries of Π are all zero, the subspace col. $[\Omega_d(\underline{\nu}^0),0,0,0]$ is contained in the isotropic subspace of $\mathcal{M}$, and thus is contained in any $\mathcal{M}$-maximal negative subspace $\mathcal{G}$. If $\mathcal{G}$ is simultaneously Ω_u-invariant, then necessarily {col. $[\Omega_d(\underline{\nu}^0),0,0,0]\}\Omega_u$=col. $[\Omega_u(\underline{\nu}^0),0,0,0]$ is contained in $\mathcal{G}$. (Note that since J is a nonsingular signature matrix and Π is injective, it is easy to see that $\mathcal{M}\cap\mathcal{M}^{\langle\perp\rangle_N}\subset$ col. $[\Omega_u(\underline{\nu}^0),0,0,0]$,

and thus

col. $[\Omega_d(\underline{\nu}^0),0,0,0]\subset\mathcal{M}\cap\mathcal{M}^{\langle\perp\rangle_N}\subset$ col. $[\Omega_u(\underline{\nu}^0),0,0,0]$; a more precise determination of $\mathcal{M}\cap\mathcal{M}^{\langle\perp\rangle_N}$ would involve the precise partial indices of Π but we shall not need this.) From the form of N, it is clear that col. $[\Omega_u(\underline{\nu}^0),0,0,0]$ is a neutral space. Suppose $X\in\mathcal{M}$ is such that $[X,Y]_N\neq 0$ for some Y in col. $[\Omega_u(\underline{\nu}^0),0,0,0]$. Then for k a scalar, we have

$$[X+kY,X+kY]_N$$

$$=[X,X]+2\mathrm{Re}\bar{k}[X,Y]$$

since $[Y,Y]_N=0$. Thus regardless of the value of $[X,X]$, we can find a scalar k so that $[X+kY,X+kY]>0$. Thus, if $\mathcal{G}$ is a N-negative space which contains col. $[\Omega_u(\underline{\nu}^0),0,0,0]$ (in particular if $\mathcal{G}$ is a $\mathcal{M}$-maximal N-negative space which is Ω_u-invariant), then necessarily $\mathcal{G}\subset\{\text{col. } [\Omega_u(\underline{\nu}^0),0,0,0]\}^{<\perp>N}$. Now from the form of N and the injectivity of Π, it is easy to see that $\{\text{col. } [\Omega_u(\underline{\nu}^0),0,0,0]\}^{<\perp>N}\cap\mathcal{M}$ = col. $[\Omega_u(\underline{\nu}^0),\Omega_u(\underline{\nu}^+),\Omega_u(\underline{\nu}^-),0]$. Conversely, if $\mathcal{G}$ is an N-negative subspace of $\widetilde{\mathcal{M}}:=$ col. $[\Omega_u(\underline{\nu}^0),\Omega_u(\underline{\nu}^+),\ \Omega_u(\underline{\nu}^-),0]$ which is $\widetilde{\mathcal{M}}$-maximal N-negative, then $\mathcal{G}$ must contain $\widetilde{\mathcal{M}}\cap\widetilde{\mathcal{M}}^{<\perp>N}$=col. $[\Omega_u(\underline{\nu}^0),0,0,0]$, and then by the argument above $\mathcal{G}$ is actually $\mathcal{M}$-maximal N-negative as well. Thus the Ω_u-invariant $\mathcal{M}$-maximal N-negative subspaces of $\mathcal{M}$ coincide with the Ω_u-invariant $\widetilde{\mathcal{M}}$-maximal N-negative subspaces of $\widetilde{\mathcal{M}}$.

Now an Ω_u-invariant $\widetilde{\mathcal{M}}$-maximal N-negative subspace necessarily has the form col. $[\Omega_u(\underline{\nu}^0),0,0,0]\dot{+}\mathcal{N}$, where $\mathcal{N}$ is an Ω_u-invariant col. $[0,\Omega_u(\underline{\nu}^+),\Omega_u(\underline{\nu}^-),0]$-maximal N-negative subspace. Temporarily identify col. $[0,\Omega_u(\underline{\nu}^+),\Omega_u(\underline{\nu}^-),0]$ with $\begin{bmatrix}\Omega_u(\underline{\nu}^+)\\ \Omega_u(\underline{\nu}^-)\end{bmatrix}$ in the obvious way. Note that N restricts to $J=\begin{bmatrix} I_{N(\underline{\nu}^+)} & 0 \\ 0 & -I_{N(\underline{\nu}^-)}\end{bmatrix}$ on this subspace. Now it is not hard to see that the Ω_u-invariant $\begin{bmatrix}\Omega_u(\underline{\nu}^+)\\ \Omega_u(\underline{\nu}^-)\end{bmatrix}$-maximal J-negative subspaces of $\begin{bmatrix}\Omega_u(\underline{\nu}^+)\\ \Omega_u(\underline{\nu}^-)\end{bmatrix}$ are precisely those of the form $\begin{bmatrix}G\\ I\end{bmatrix}\Omega_u(\underline{\nu}^-)$ where $G\in\Omega_u(\underline{\nu}^+ x \underline{\nu}^-)$ and $\|G\|\leq 1$; for greater detail on this step, see [BG 1]. Conclude that the Ω_u-invariant $\mathcal{M}$-maximal N-negative subspaces $\mathcal{G}$ are precisely those of the form

$$\mathcal{G}=\begin{bmatrix}I & 0\\ 0 & G\\ 0 & I\\ 0 & 0\end{bmatrix}\begin{bmatrix}\Omega_u(\underline{\nu}^0)\\ \Omega_u(\underline{\nu}^-)\end{bmatrix}$$

for a $G\in\Omega_u(\underline{\nu}^+ x \underline{\nu}^-)$ with $\|G\|\leq 1$.

Now suppose that $\mathcal{M}$ is <u>any</u> Ω_u-invariant subspace of kxn matrices with Hermitian form $[X,Y]_W=\mathrm{tr}(Y^*WX)$ for a kxk Hermitian matrix W. By Theorem 2.1, we may represent $\mathcal{M}$ as $\mathcal{M}=\Theta$ col. $[\Omega_u(\underline{\nu}^0),\Omega_u(\underline{\nu}^+),\Omega_u(\underline{\nu}^-),\Omega_u(\underline{\nu}^x)]$ where Θ is an injective matrix such that $\Theta^*W\Theta=N$ for a unique $(\underline{\nu}^0,\underline{\nu}^+,\underline{\nu}^-,\underline{\nu}^x)$-block S-subpermutation matrix N. Thus $(\mathcal{M}, [\ ,\ ]_W)$ is unitarily equivalent to (col. $[\Omega_u(\underline{\nu}^0),\Omega_u(\underline{\nu}^+),\Omega_u(\underline{\nu}^-),\Omega_u(\underline{\nu}^x)]$,N) under left multiplication by Θ. Moreover, the left multiplication by Θ commutes with right multiplication by a matrix in Ω_u. Thus an Ω_u-invariant $\mathcal{M}$-maximal W-negative subspace $\mathcal{G}$ necessarily has the form $\mathcal{G}=\Theta\mathcal{G}_1$ where $\mathcal{G}_1$ is an Ω_u-invariant col. $[\Omega_u(\underline{\nu}^0),\Omega_u(\underline{\nu}^+),\Omega_u(\underline{\nu}^-),\Omega_u(\underline{\nu}^x)]$-maximal N-negative subspace. We have arrived at the following.

THEOREM 2.2 Suppose $\mathcal{M}$ is an Ω_u-invariant subspace of kxn matrices and W is a kxk Hermitian matrix. Let the n-tuples $\underline{\nu}^0,\underline{\nu}^+,\underline{\nu}^-,\underline{\nu}^x$, the $(\underline{\nu}^0,\underline{\nu}^+,\underline{\nu}^-,\underline{\nu}^x)$-block S-subpermutation matrix N and the injective matrix Θ be as in Theorem 2.1, so $\mathcal{M}=\Theta$ col. $[\Omega_u(\underline{\nu}^0),\Omega_u(\underline{\nu}^+),\Omega_u(\underline{\nu}^-),\Omega_u(\underline{\nu}^x)]$ and $\Theta^*W\Theta=N$. Then the Ω_u-invariant $\mathcal{M}$-maximal W-negative subspaces $\mathcal{G}$ of $\mathcal{M}$ are precisely those of the form

$$\mathcal{G}=\Theta\begin{bmatrix} I & 0 \\ 0 & G \\ 0 & I \\ 0 & 0 \end{bmatrix}\begin{bmatrix} \Omega_u(\underline{\nu}^0) \\ \Omega_u(\underline{\nu}^-) \end{bmatrix}$$

for some $G\in\Omega_u(\underline{\nu}^+ x\underline{\nu}^-)$ with $\|G\|\leq 1$.

We shall also want to specify the Ω_u-invariant $\mathcal{M}$-maximal W-negative subspaces $\mathcal{G}$ of an Ω_u-invariant subspace $\mathcal{M}$ which are moreover W-neutral spaces, i.e. $[X,X]_W=0$ for all X in G. Note that for $\mathcal{G}_1=\begin{bmatrix} G \\ I \end{bmatrix}\Omega_u(\underline{\nu}^-)$ to be a J-neutral space ($J=\begin{bmatrix} I_{N(\underline{\nu}^+)} & 0 \\ 0 & -I_{N(\underline{\nu}^-)} \end{bmatrix}$), we must have that G is isometric. A matrix G in $\Omega_u(\underline{\nu}^+ x\underline{\nu}^-)$ which is also isometric we call $\underline{\nu}^+ x\underline{\nu}^-$-inner, in analogy with the function theory case (see [NF]). We therefore have the following.

COROLLARY 2.3 Suppose $\mathcal{M}$ is an Ω_u-invariant subspace of kxn matrices and W is a kxk Hermitian matrix. Let the notation be as in Theorem 2.2. Then the W-neutral Ω_u-invariant $\mathcal{M}$-maximal W-negative subspaces $\mathcal{G}$ of M are precisely those of the form

$$\mathcal{G}=\Theta \begin{bmatrix} I & 0 \\ 0 & G \\ 0 & I \\ 0 & 0 \end{bmatrix} \begin{bmatrix} \Omega_u(\underline{\nu}^0) \\ \Omega_u(\underline{\nu}^-) \end{bmatrix}$$

for a $\underline{\nu}^+ x \underline{\nu}^-$-inner matrix G.

2.4 Noncanonical Cholesky and QR Factorization: Second format

The invariant subspace representation of this section is merely a reorganization of that from [BG 3] given in §1. This implies that the factorization results from §4 of [BG 3] can also be presented in an alternative form. In fact, as the reader may well by now suspect, any invariant subspace representation result is roughly equivalent to a corresponding factorization theorem. Here we develop the analogues of the factorization results of §4 of [BG 3] based on the invariant subspace representations of this section. These results will not be needed in the sequel.

Suppose $\underline{\nu}, \underline{\nu}^0, \underline{\nu}^+, \underline{\nu}^-, \underline{\nu}^x$ are all n-tuples of nonnegative integers, and that A is a $(N(\underline{\nu}^0)+N(\underline{\nu}^+)+N(\underline{\nu}^-)+N(\underline{\nu}^x)) x\ N(\underline{\nu})$ matrix. We say that A is $(\underline{\nu}^0 \dot{+} \underline{\nu}^+ \dot{+} \underline{\nu}^- \dot{+} \underline{\nu}^x) x \underline{\nu}$-outer if $A\Omega_u(\underline{\nu})=$col. $[\Omega_u(\underline{\nu}^0)$, $\Omega_u(\underline{\nu}^+)$, $\Omega_u(\underline{\nu}^-)$, $\Omega_u(\underline{\nu}^x)]$. Similarly, a $N(\underline{\nu}) x (N(\underline{\nu}^0)+N(\underline{\nu}^+)+N(\underline{\nu}^-)+N(\underline{\nu}^x))$ matrix B we say is $\underline{\nu} x (\underline{\nu}^0 \dot{+} \underline{\nu}^+ \dot{+} \underline{\nu}^- \dot{+} \underline{\nu}^x)$-outer if B col. $[\Omega_u(\underline{\nu}^0)$, $\Omega_u(\underline{\nu}^+)$, $\Omega_u(\underline{\nu}^-)$, $\Omega_u(\underline{\nu}^x)]-\Omega_u(\underline{\nu})$. Define $\Omega_u(\underline{\nu} x (\underline{\nu}^0 \dot{+} \underline{\nu}^+ \dot{+} \underline{\nu}^- \dot{+} \underline{\nu}^x))$ as those matrices U which map col. $[\Omega_u(\underline{\nu}^0)$, $\Omega_u(\underline{\nu}^+)$, $\Omega_u(\underline{\nu}^-)$, $\Omega_u(\underline{\nu}^x)]$ into $\Omega_u(\underline{\nu})$. The following general factorization result is an alternative form of Theorem 4.1 from [BG 3] (generalized QR-factorization).

THEOREM 2.4 Suppose W is a Hermitian matrix of block class $\Omega(\underline{\nu}\times\underline{\nu})$ for some n-tuple $\underline{\nu}$ of nonnegative integers. Suppose F is a matrix of block type $\Omega_u(\underline{\nu}\times\underline{\nu}')$ for some second n-tuple $\underline{\nu}'$. Then there exist unique n-tuples $\underline{\nu}^0$, $\underline{\nu}^+$, $\underline{\nu}^-$, $\underline{\nu}^x$, a unique isometric $\underline{\nu}^0\times\underline{\nu}^x$-block subpermutation matrix Π in $\Omega_u^0(\underline{\nu}^0\times\underline{\nu}^x)$, a $(\underline{\nu}^0\dot{+}\underline{\nu}^+\dot{+}\underline{\nu}^-\dot{+}\underline{\nu}^x)\times\underline{\nu}'$-outer matrix G, and an injective matrix U in block class $\Omega_u(\underline{\nu}\times(\underline{\nu}^0\dot{+}\underline{\nu}^+\dot{+}\underline{\nu}^-\dot{+}\underline{\nu}^x))$ such that

$$F=UG$$

and

$$U^*WU=\begin{bmatrix} 0 & 0 & 0 & \Pi \\ 0 & I_{N(\underline{\nu}^+)} & 0 & 0 \\ 0 & 0 & -I_{N(\underline{\nu}^-)} & 0 \\ \Pi^* & 0 & 0 & 0 \end{bmatrix}.$$

PROOF. Apply Theorem 2.1 to the Ω_u-invariant subspace $\mathcal{M}=F\Omega_u(\underline{\nu}')$ with Hermitian form induced by W. Then $\mathcal{M}=\Theta\mathrm{col}.[\Omega_u(\underline{\nu}^0), \Omega_u(\underline{\nu}^+), \Omega_u(\underline{\nu}^-), \Omega_u(\underline{\nu}^x)]$ for uniquely determined n-tuples $\underline{\nu}^0$, $\underline{\nu}^+$, $\underline{\nu}^-$, $\underline{\nu}^x$ and an injective matrix Θ such that $\Theta^*W\Theta=N$ for a uniquely determined $(\underline{\nu}^0, \underline{\nu}^+, \underline{\nu}^-, \underline{\nu}^x)$-block S-subpermutation matrix N. One can now easily verify that F=UG where $U=\Theta$ and $G=\Theta^{-1}F$ (Θ^{-1}=any generalized inverse for Θ) is a factorization as asserted.

We remark that the technique in the proof above can be turned around; that is, we can base a proof of the invariant subspace representation Theorem 2.1 on the factorization Theorem 2.4. Indeed suppose $\mathcal{M}$ and W are as in the hypotheses of Theorem 5.1. By results from [BG 1] we know that we can represent $\mathcal{M}$ as $\mathcal{M}=F\Omega_u(\underline{\nu}')$ for a uniquely determined n-tuple $\underline{\nu}'$ and an injective matrix F. Choose $\underline{\nu}$ to be any convenient n-tuple with the property that $F \in\Omega_u(\underline{\nu}\times\underline{\nu}')$. Now apply Theorem 2.5 to the pair (W,F); thus F=UG and G is "outer" and U is "block upper triangular" and satisfies $U^*WU=N$. From all the properties of U and G in Theorem 2.4 and the identity $\mathcal{M}=F\Omega_u(\underline{\nu}')$, it is routine to check that $\Theta:=U$ is a representer for $(\mathcal{M},W)$ as in Theorem 5.1. The same remark applies to all our other combinations of invariant subspace representation and factorization theorems in [BG 3] as well.

If we take $\underline{\nu}'=\underline{\nu}$ and set $F=I_{N(\underline{\nu})}$ in Theorem 2.4, the noncanonical Cholesky factorization in Theorem 4.2 of [BG 3] takes on the following form.

THEOREM 2.5 Suppose W is a Hermitian matrix of block class $\Omega(\underline{\nu}\times\underline{\nu})$. Then there are unique n-tuples $\underline{\nu}^0$, $\underline{\nu}^+$, $\underline{\nu}^-$, $\underline{\nu}^x$, a unique isometric $\underline{\nu}^0\times\underline{\nu}^x$-block subpermutation matrix Π in $\Omega_u^0(\underline{\nu}^0\times\underline{\nu}^x)$ and a $(\underline{\nu}^0\dot{+}\underline{\nu}^+\dot{+}\underline{\nu}^-\dot{+}\underline{\nu}^x)\times\underline{\nu}$-outer matrix L^* such that

$$W=L\begin{bmatrix} 0 & 0 & 0 & \Pi \\ 0 & I_{N(\underline{\nu}^+)} & 0 & 0 \\ 0 & 0 & -I_{N(\underline{\nu}^-)} & 0 \\ \Pi^* & 0 & 0 & 0 \end{bmatrix} L^*.$$

3. Isometric completion problems

We consider in this section an interpolation (or completion) problem slightly more general than that considered in [BG1]. Let $\underline{\nu}$ and $\underline{\tilde{\nu}}$ be two n-tuples of nonnegative integers which we shall consider fixed in this section and the next. For $-n \le m \le n$ we let $\mathcal{A}_u(m)$ be the set of all block matrices $F=[F_{ij}]_{1\le i,j\le n}$ such that the (i,j) -th block F_{ij} has size $\tilde{\nu}_i \times \nu_j$ and such that $F_{ij}=0$ if $i>j-m$. Analogously $\mathcal{A}_\ell(m)$ denotes the class of block matrices $F=[F_{ij}]_{1\le i,j\le n}$ such that the (i,j)-th block F_{ij} has size $\tilde{\nu}_i \times \nu_j$ and such that $F_{ij}=0$ if $i<j-m$. Thus $\mathcal{A}_u(0)=\Omega_u(\underline{\tilde{\nu}}\times\underline{\nu})$, $\mathcal{A}_u(1)=\Omega_u^0(\underline{\tilde{\nu}}\times\underline{\nu})$, $\mathcal{A}_\ell(0)=\Omega_\ell(\underline{\tilde{\nu}}\times\underline{\nu})$ and $\mathcal{A}_\ell(-1) = \Omega_\ell^0(\underline{\tilde{\nu}}\times\underline{\nu})$ in terms of our old notation; we now want to consider $\underline{\tilde{\nu}}$ and $\underline{\nu}$ fixed and suppress them from the notation. We let $\mathcal{A}$ denote the full class $\Omega(\underline{\tilde{\nu}}\times\underline{\nu})$ of $\underline{\tilde{\nu}}\times\underline{\nu}$ -block matrices.

Now suppose K is a given block matrix in $\mathcal{A}$ and m is an integer satisfying $-n<m\le n$. For the problem to be considered, we may as well assume that $K\in\mathcal{A}_\ell(m-1)$, so all block entries on and above the m-th block diagonal are zero. The problem we wish to analyze in this section is the following.

PROBLEM A. Given $K\in\mathcal{A}_\ell(m-1)$, when is there an F in the coset $K+\mathcal{A}_u(m)$ such that i) $\|F\|\le 1$ or ii) F is isometric? If a solution F exists, describe the set of all solutions.

The problem has been analyzed before in [DG] and [BG1] for the case $\underline{\tilde{\nu}}=\underline{\nu}$; there is given the criterion for the existence of solutions. The analogous existence criterion for the general case can be obtained as an application of the lifting theorem in [BG 2]. From [BG1] and [BG2] the set of all solutions can be parametrized for a generic special case. It remains to parametrize the set of all solutions for the nongeneric degenerate situation.

We proceed much as in [BG 1]. We seek to characterize the graph spaces $\mathcal{G}=\begin{bmatrix} F \\ I \end{bmatrix} \Omega_u(\underline{\nu})$ associated with solutions F of Problem A rather than the solutions F themselves. We merely summarize the line of development; details are in [BG1]. Set $\widetilde{N}=N(\underline{\tilde{\nu}})$ and $N=N(\underline{\nu})$; thus such a subspace consists of $(\widetilde{N}+N)$ xn matrices. We let J now be the signature matrix $J=\begin{bmatrix} I_{\widetilde{N}} & 0 \\ 0 & -I_N \end{bmatrix}$ and let $\mathcal{M}$ be the subspace of $(\widetilde{N}+N)$xn matrices

$$\mathcal{M}(K;m)= \begin{bmatrix} K \\ I_N \end{bmatrix} \Omega_u(\underline{\nu})+\begin{bmatrix} \Omega_u(\underline{\tilde{\nu}};m) \\ 0 \end{bmatrix}$$

Here $\Omega_u(\underline{\tilde{\nu}};m)$ is the set of block matrices $[U_{ij}]_{1\leq i,j\leq n}$ such that U_{ij} has size $\tilde{\nu}_i$x1 and such that $U_{ij}=0$ if $i>j-m$. Introduce also the space $\mathcal{K}=\begin{bmatrix} M_{N'\,xn} \\ \Omega_u(\underline{\nu}) \end{bmatrix}$ of $(N'+N)$xn matrices. As in [BG1] one can check that the subspace $\mathcal{G}$ of $(N'+N)$xn matrices has the form $\mathcal{G}=\begin{bmatrix} F \\ I_n \end{bmatrix} \Omega_u(\underline{\nu})$ for a solution F of Problem A i) if and only if $\mathcal{G}$ satisfies the three conditions

i) $\mathcal{G} \subset \mathcal{M}$

ii) $\mathcal{G}$ is $\mathcal{K}$-maximal J-negative

and

iii) $\mathcal{G}$ is Ω_u-invariant.

Moreover $\mathcal{G} = \begin{bmatrix} F \\ I_N \end{bmatrix} \Omega_u(\underline{\nu})$ for a solution F of Problem A ii) if and only if, in addition to i) –iii), $\mathcal{G}$ satisfies

iv) $\mathcal{G}$ is J–neutral.

As in [BG1], we see that a subspace $\mathcal{G}$ can satisfy i) and ii) simultaneously if and only if the subspace $\chi := \mathcal{K} \cap \mathcal{M}^{<\perp>_J}$ is J–positive, i.e. $[X,X]_J \geq 0$ for all $X \epsilon \chi$.

One can give the condition that χ be a positive subspace more explicitly in terms of the original matrix $K \in \mathcal{A}_{\ell}(m-1)$. Indeed, let $K = [K_{ij}]_{1 \leq i,j \leq n}$ be the $\underline{\nu} x \underline{\tilde{\nu}}$ block decomposition of K. Then the condition is that each of the block submatrices $[K_{i'j'}]_{i \leq i' \leq n; 1 \leq j' \leq i-m-1}$ $(1 \leq i \leq n)$ have norm less than equal to 1.

With the assumption that χ is J–positive in force, condition ii) becomes equivalent to

ii′) $\mathcal{G}$ is $\mathcal{M}$–maximal J–negative.

Now solutions $\mathcal{G}$ of i) ii′) and iii) are characterized in Theorem 2.2, and solutions $\mathcal{G}$ of i), ii′), iii) and iv) are handled in Corollary 2.3. This leads to the following results which gives the desired parametrization of all solutions of Problem A. In the statement $\mathcal{P}_j$ denotes the subspace $\mathcal{P}^{\underline{\nu}}_j$ as defined in (1.1), $\widetilde{Q}_k$ is the orthogonal projection onto the subspace $\widetilde{\mathcal{Q}}_k := \mathcal{Q}^{\tilde{\underline{\nu}}}_k$ as defined in (1.2) and P_j is the orthogonal projection onto $\mathcal{P}_j$. The following is the general form of Theorem 0.2 from the Introduction.

THEOREM 3.1. Let $K = [K_{ij}]_{1 \leq i,j \leq n} \ \epsilon \ \mathcal{A}_{\ell}(m-1)$ be as in Problem A. Then Problem Ai) has a solution if and only if each of those block submatrices $[K_{i'j'}]_{i\leq i'\leq n;1\leq j'\leq i-m-1}$ which is nonempty has norm ≤ 1 for $1\leq i\leq n$.

In fact, in this case the set of all solutions of Problem Ai) can be represented as follows. Let $\mathcal{M}$ be the subspace

$$\mathcal{M}=\begin{bmatrix} K \\ I \end{bmatrix}\Omega_u(\underline{\nu}) + \begin{bmatrix} \Omega_u(\underline{\widetilde{\nu}};m) \\ \{0\} \end{bmatrix}$$

of $(\widetilde{N}+N)\mathrm{x}n$ matrices. Let J be the signature matrix $J=\begin{bmatrix} I_{\widetilde{N}} & 0 \\ 0 & -I_N \end{bmatrix}$ and let $\Theta = [\Theta^0;\Theta^+;\Theta^-;\Theta^\times]$ be a representer for $(\mathcal{M},J)$ as in Theorem 2.1. Thus there is a unique $(\underline{\nu}^0, \underline{\nu}^+,\underline{\nu}^-,\underline{\nu}^\times)$-block S-subpermutation matrix N such that

$$\mathcal{M} = \Theta \begin{bmatrix} \Omega_u(\underline{\nu}^0) \\ \Omega_u(\underline{\nu}^+) \\ \Omega_u(\underline{\nu}^-) \\ \Omega_u(\underline{\nu}^\times) \end{bmatrix}$$

and $\Theta^* J\Theta = N$. Further block-decompose Θ^0, Θ^+, Θ^-, $\Theta^\times$ as

$$\Theta^0=\begin{bmatrix} \psi \\ w \end{bmatrix}, \ \Theta^+=\begin{bmatrix} \alpha \\ k \end{bmatrix}, \ \Theta^- = \begin{bmatrix} \beta \\ \gamma \end{bmatrix}, \ \Theta^\times = \begin{bmatrix} \psi' \\ w' \end{bmatrix}$$

where ψ,α,β,ψ' consist of the first $\widetilde{N}$ rows and ω,k,γ,ω' consist of the last N rows of Θ^0, Θ^+, Θ^-, $\Theta^\times$, so $\Theta = \begin{bmatrix} \psi & \alpha & \beta & \psi' \\ w & k & \gamma & w' \end{bmatrix}$. Then a $\widetilde{N}\mathrm{x}N$ matrix F is a solution of Problem A i) if and only if

(3.1) $\quad F=[\psi, \ \alpha G+\beta] \ ([\omega, \ kG+\gamma])^{-1}$

for some $G\in\Omega_u(\underline{\nu}^+\mathrm{x}\underline{\nu}^-)$ with $\|G\|\leq 1$. Moreover the matrix F uniquely determines the matrix G.

Solutions to Problem Aii) exist if and only if in addition rank $\{(I-P_jK^*\tilde{Q}_{j-m+1}K)|P_j\} \le \sum_{j'=1}^{j-m} \tilde{\underline{\nu}}_{j'}$ for $1 \le j \le n$. This is the case precisely when $\sum_{j'=1}^{j} \nu'_{j'} \le \sum_{j'=1}^{j} \nu^+_{j'}$ for $1 \le j \le n$ where $\underline{\nu}^-$ and $\underline{\nu}^+$ are the n-tuples as specified above, or equivalently, when there exist $\underline{\nu}^+ x \underline{\nu}^-$-inner matrices. In this case a matrix F is a solution of Problem Aii) if and only if F is as in (3.1) for some $\underline{\nu}^+ x \underline{\nu}^-$-inner matrix G.

PROOF. As was mentioned above, the condition that all the submatrices $[K_{i'j'}]_{i \le i' \le n; 1 \le j' \le i-m-1}$ have norm ≤ 1 for $1 \le i \le n$ is equivalent to the subspace $\mathcal{K} \cap \mathcal{M}^{\langle \perp \rangle_J}$ being J-positive, a necessary condition for Problem Ai) to have solutions. By the discussion above and Theorem 2.3, in this case a subspace $\mathcal{G} \subset \mathcal{K}$ has the form $\begin{bmatrix} F \\ I \end{bmatrix} \Omega_u(\underline{\nu})$ for a solution F of Problem A i) if and only if $\mathcal{G}$ has the form

$$\mathcal{G} = \Theta \begin{bmatrix} I & 0 \\ 0 & G \\ 0 & I \\ 0 & 0 \end{bmatrix} \begin{bmatrix} \Omega_u(\underline{\nu}^0) \\ \Omega_u(\underline{\nu}^-) \end{bmatrix}$$

for a G in $\Omega_u(\underline{\nu}^+ x \underline{\nu}^-)$ with $\|G\| \le 1$ uniquely determined by $\mathcal{G}$. Inserting the block form for Θ described in the Theorem, we have $\begin{bmatrix} F \\ I \end{bmatrix}$

$$\Omega_u(\underline{\nu}) = \mathcal{G} = \begin{bmatrix} \psi & \alpha & \beta & \psi' \\ w & k & \gamma & w' \end{bmatrix} \begin{bmatrix} I & 0 \\ 0 & G \\ 0 & I \\ 0 & 0 \end{bmatrix} \begin{bmatrix} \Omega_u(\underline{\nu}^0) \\ \Omega_u(\underline{\nu}^-) \end{bmatrix}$$

$$= \begin{bmatrix} \psi & \alpha G+\beta \\ w & kG+\gamma \end{bmatrix} \begin{bmatrix} \Omega_u(\underline{\nu}^0) \\ \Omega_u(\underline{\nu}^-) \end{bmatrix}$$

From the last columns we read off

$$(**) \quad \begin{bmatrix} F \\ I \end{bmatrix} \mathbb{C}^N = \begin{bmatrix} \psi & \alpha G+\beta \\ \omega & kG+\gamma \end{bmatrix} \begin{bmatrix} \mathbb{C}^{N(\underline{\nu}^0)} \\ \mathbb{C}^{N(\underline{\nu}^-)} \end{bmatrix}.$$

From the fact that $\mathcal{K}\cap\mathcal{M}^{<\perp>_J}$ is J-positive, one can deduce that $N(\underline{\nu}^0) + N(\underline{\nu}^-) = N$. From the bottom row we read off $\mathbb{C}^N = [\omega \ \ kG+\gamma] \begin{bmatrix} \mathbb{C}^{N(\underline{\nu}^0)} \\ \mathbb{C}^{N(\underline{\nu}^-)} \end{bmatrix}$ so that the NxN matrix $[\omega \ \ kG+\gamma]$ is invertible. Then (**) may be rewritten as

$$\begin{bmatrix} F \\ I \end{bmatrix} \mathbb{C}^N = \begin{bmatrix} [\psi, \ \alpha G+\beta]([w, \ kG+\gamma])^{-1} \\ I_N \end{bmatrix} \mathbb{C}^N.$$

The identity of these graph subspaces implies the matrix identity

$$F=[\psi, \ \alpha G+\beta] \ ([\omega, \ kG+\gamma])^{-1}.$$

If F is to be a solution of Problem Aii), the same analysis yields that G must be $\underline{\nu}^+ x \underline{\nu}^-$-inner. One easily sees that $\underline{\nu}^+ x \underline{\nu}^-$-inner matrices exist if and only if $\sum_{j'=1}^{j} \nu_{j'}^- \le \sum_{j'=1}^{j} \nu_j^+$. To translate this condition to one more intrinsic to the data of the problem, compute the number of positive and negative squares in the subspace $\mathcal{M}E_j$ in the J inner product in two different ways as follows. From the representation $\mathcal{M}=\Theta$ col. $[\Omega_u(\underline{\nu}^0), \Omega_u(\underline{\nu}^+), \Omega_u(\underline{\nu}^-), \Omega_u(\underline{\nu}^x)]$ where

$$\Theta^* J\Theta = \begin{bmatrix} 0 & 0 & 0 & \pi \\ 0 & I & 0 & 0 \\ 0_* & 0 & -I & 0 \\ \pi & 0 & 0 & 0 \end{bmatrix},$$ we see that

$$\#\text{ of positive squares of } \mathcal{M}E_j = \sum_{j'=1}^{j} \nu^+_{j'} + \sum_{i=1}^{j}\sum_{j'=1}^{j} \operatorname{rank} \pi_{ij'}$$

$$\#\text{ of negative squares of } \mathcal{M}E_j = \sum_{j'=1}^{j} \nu^-_{j'} + \sum_{i=1}^{j}\sum_{j'=1}^{j} \operatorname{rank} \pi_{ij'}.$$

On the other hand, from the definition of $\mathcal{M}$ and the assumption that $\|Q_{j-m+1}K|\mathcal{P}_j\| \leq 1$ for all j, we see that

$$\#\text{ of positive squares of } \mathcal{M}E_j = \sum_{j'=1}^{j} \tilde{\nu}_{j-m}$$

and

$$\#\text{ of negative squares of } \mathcal{M}E_j = \operatorname{rank}\{(I-P_jK^*\tilde{Q}_{j-m+1}K)|\mathcal{P}_j\}.$$

Conclude that $\sum_{j'=1}^{j} \nu^-_{j'} \leq \sum_{j'=1}^{j} \nu^+_{j'}$ if and only if rank $\{I-P_jK^*\tilde{Q}_{j-m+1}K)|\mathcal{P}_j\}$

$\leq \sum_{j'=1}^{j} \tilde{\nu}_{j-m}$ for $1\leq j\leq n$. This concludes the proof of Theorem 3.1.

We should note that all the ingredients $\underline{\nu}^0$, $\underline{\nu}^+$, $\underline{\nu}^-$, $\underline{\nu}^x$,

$N=\begin{bmatrix} 0 & 0 & \pi \\ 0_* & J_1 & 0 \\ \pi & 0 & 0 \end{bmatrix}$ and $\Theta = \begin{bmatrix} \psi & \alpha & \beta & \psi' \\ w & k & \gamma & w' \end{bmatrix}$ in the formula for an arbitrary solution of Problem A can be given explicitly in terms of the original given matrix $K \in \mathcal{A}_\ell(m-1)$. Indeed all this data can be obtained from the partial subspaces $\mathcal{R}_{ij}$ $(1\le i,j\le n)$ of the subspace $\mathcal{M} = \begin{bmatrix} K \\ I \end{bmatrix} \Omega_u(\underline{\nu}) + \begin{bmatrix} \Omega_u(\underline{\widetilde{\nu}};m) \\ 0 \end{bmatrix}$

in the $J=\begin{bmatrix} I_{\widetilde{N}} & 0 \\ 0 & -I_N \end{bmatrix}$-inner product. We wish to delineate this information more explicitly. To do this, let $\mathcal{P}_i=\mathcal{P}_i^{\underline{\nu}}$ denote the span of the first $\sum_{i'=1}^{i} \nu_{i'}$ standard basis vectors in $\mathbb{C}^N$ $(N=N(\underline{\nu}))$, and let $\widetilde{\mathcal{P}}_i = \mathcal{P}_i^{\underline{\widetilde{\nu}}}$ denote the span of the first $\sum_{i'=1}^{i} \widetilde{\nu}_{i'}$ standard basis vectors in $\mathbb{C}^{\widetilde{N}}$ $(\widetilde{N} = N(\underline{\widetilde{\nu}}))$ as in (1.1). The orthogonal projections onto these subspaces we denote as P_i and $\widetilde{P}_i$ respectively. Similarly, $\mathcal{Q}_i=\mathcal{P}_{i-1}^{\perp} \subset \mathbb{C}^N$ and $\widetilde{\mathcal{Q}}_i = \widetilde{\mathcal{P}}_{i-1}^{\perp} \subset \mathbb{C}^{\widetilde{N}}$ with orthogonal projections $Q_i=I_N-P_{i-1}$ and $\widetilde{Q}_i = I_{\widetilde{N}} - \widetilde{P}_{i-1}$ respectively. Recall that E_i is the nx1 matrix with all entries 0 except for a 1 in the i-th row.

The partial subspaces $\mathcal{R}_{ij}$ for the pair $(\mathcal{M},J)$ are given by

$$\mathcal{R}_{ij} = \frac{\mathcal{M}E_{i-1}^{<\perp>J} \cap \mathcal{M}E_j}{(\mathcal{M}E_i^{<\perp>J} \cap \mathcal{M}E_j) + (\mathcal{M}E_{i-1}^{<\perp>J} \cap \mathcal{M}E_{j-1})}$$

The space $\mathcal{M}E_j$ has the form

$$\mathcal{M}E_j = \begin{bmatrix} \widetilde{Q}_{j-m+1}F \\ I \end{bmatrix} \mathcal{P}_j + \begin{bmatrix} \widetilde{\mathcal{P}}_{j-m} \\ 0 \end{bmatrix}$$

while

$$\mathcal{M}E_{i-1}^{<\perp>J} = \begin{bmatrix} I \\ \mathcal{P}_{i-1}F^* \end{bmatrix}\tilde{\mathcal{Q}}_{i-m} + \begin{bmatrix} 0 \\ \mathcal{Q}_i \end{bmatrix}$$

We wish to derive the following Lemma which gives the general form of formulas (0.1)–(0.8) from the Introduction for the scalar case.

LEMMA 3.2. Suppose $K = [K_{ij}]_{1\le i,j\le n} \in \mathcal{A}_\ell(m-1)$ is such that each of the block submatrices $[K_{i'j'}]_{i\le i'\le n; 1\le j'\le i-m-1}$ has norm ≤ 1 for $1\le i\le n$, and suppose $\mathcal{M}$ is the Ω_u-invariant subspace

$$\mathcal{M} = \begin{bmatrix} K \\ I \end{bmatrix} \Omega_u(\underline{\nu}) + \begin{bmatrix} \Omega_u(\underline{\tilde{\nu}};m) \\ 0 \end{bmatrix}.$$

Then for $1\le j<i\le n$,

$$\mathcal{M}E_{i-1}^{<\perp>J} \cap \mathcal{M}E_j = \begin{bmatrix} \tilde{Q}_{j+1-m}K \\ I \end{bmatrix} \operatorname{Ker}\{(I-P_jK^*\tilde{Q}_{i-m}K)\,|\,P_j\}.$$

Moreover, if we set

(3.2) $\varphi_{p,q} = \dim \operatorname{Ker}\{(I-P_qK^*\tilde{Q}_{p-m}K)\,|\,P_j\}$,

then the partial rank indices r_{pq} for the pair $(\mathcal{M},J)$ are given by

(3.3) $r_{p,q} = \varphi_{p,q} - \varphi_{p,q-1} - \varphi_{p+1,q} + \varphi_{p+1,q-1}$

as long as $p>q$. The invariant n-tuples $\underline{\nu}^0$, $\underline{\nu}^+$, $\underline{\nu}^-$, $\underline{\nu}^x$ and subpermutation matrix $\pi \in \Omega_u^0(\underline{\nu}^0 x \underline{\nu}^x)$ occuring as in Theorem 3.1 are then given by

(3.4) $\nu_p^0 = \sum_{j=p+1}^{n} r_{jp}$

(3.5) $\nu_p^+ = \tilde{\nu}_{p-m} - \sum_{i=1}^{p-1} r_{pi}$

(3.6) $\nu_p^- = \nu_p - \sum_{j=p+1}^{n} r_{jp}$

(3.7) $\nu_p^x = \sum_{i=1}^{p-1} r_{pi}$.

for $1 \le p \le n$, and $\pi = [\pi_{ij}]_{1 \le i,j \le n}$

where

(3.8) $\operatorname{rank} \pi_{ij} = \begin{cases} 0 & \text{if } i \ge j \\ r_{ji} & \text{if } i < j \end{cases}$.

PROOF. Suppose $X \in \mathcal{M}E_{i-1}^{<\perp>_J} \cap \mathcal{M}E_j$ and $i > j$. Since $X \in \mathcal{M}E_j$, X has the form

$$X = \begin{bmatrix} \tilde{Q}_{j+1-m} K \\ I \end{bmatrix} x + \begin{bmatrix} y \\ 0 \end{bmatrix}$$

where $x \in \mathcal{P}_j$ and $y \in \tilde{\mathcal{P}}_{j-m}$. Since X is J-orthogonal to $\mathcal{M}E_{i-1}$, in particular X is J-orthogonal to $\begin{bmatrix} \tilde{\mathcal{P}}_{i-1-m} \\ 0 \end{bmatrix}$. Since $i-1 \ge j$, $\tilde{\mathcal{P}}_{j-m} \subset \tilde{\mathcal{P}}_{i-1-m}$; from the form of X and its J-orthogonality to $\begin{bmatrix} \tilde{\mathcal{P}}_{i-1-m} \\ 0 \end{bmatrix}$, we get that $y=0$ and that $\tilde{Q}_{j+1-m} Kx$ is orthogonal to $\tilde{\mathcal{P}}_{i-1-m}$. This last condition can be expressed as

$$\tilde{Q}_{j+1-m} Kx = \tilde{Q}_{i-m} \tilde{Q}_{j+1-m} Kx.$$

Again since $i \ge j+1$, $\tilde{Q}_{i-m} \tilde{Q}_{j+1-m} = \tilde{Q}_{i-m}$ so this last condition becomes

$$\tilde{Q}_{j+1-m} Kx = \tilde{Q}_{i-m} Kx.$$

Now we use that X is also J-orthogonal to $\begin{bmatrix} \tilde{Q}_{i-m} K \\ I \end{bmatrix} \mathcal{P}_{i-1}$. Again since

$i-1 \geq j$, $\mathcal{P}_j \subset \mathcal{P}_{i-1}$ and thus $X = \begin{bmatrix} \widetilde{Q}_{i-m}K \\ I \end{bmatrix} x$ is J-orthogonal to itself. From this we get that $x \in \operatorname{Ker}\{(I - P_j K^* \widetilde{Q}_{i-m} K) \mid \mathcal{P}_j\}$, and that

$$X \in \begin{bmatrix} \widetilde{Q}_{j+1-m}K \\ I \end{bmatrix} \operatorname{Ker}\{(I - P_j K^* \widetilde{Q}_{i-m} K) \mid \mathcal{P}_j\}$$

Conversely, suppose that

$$X = \begin{bmatrix} \widetilde{Q}_{j+1-m}\ K \\ I \end{bmatrix} x \text{ where}$$

$x \in \operatorname{Ker}\ \{(I - P_j K^* \widetilde{Q}_{i-m} K) \mid \mathcal{P}_j\}$. From the form of X it is clear that $X \in \mathcal{M}E_j$. The condition on x can be restated as

$$\|\widetilde{Q}_{i-m}\ Kx\|^2 = \|x\|^2 \text{ and } x \in \mathcal{P}_j.$$

Since $i \geq j+1$, $\widetilde{Q}_{i-m} = \widetilde{Q}_{i-m}\widetilde{Q}_{j+1-m}$ and by hypothesis $Q_{j+1-m}\ K \mid \mathcal{P}_j$ has norm ≤ 1. Thus

$$\|x\| = \|\widetilde{Q}_{i-m}Kx\| = \|\widetilde{Q}_{i-m}\widetilde{Q}_{j+1-m}Kx\|$$

$$\leq \|\widetilde{Q}_{j+1-m}Kx\| \leq \|x\|$$

from which we get equality throughout. This forces

$$\widetilde{Q}_{i-m}Kx = \widetilde{Q}_{j+1-m}Kx$$

It is now easy to check directly that X is also J-orthogonal to $\mathcal{M}E_{i-1}$. We have thus established the claimed form for $\mathcal{M}E_{i-1}^{<\perp>_J} \cap \mathcal{M}E_j$. If φ_{pq} is given by (3.2) then we see that $\varphi_{pq} = \dim \mathcal{M}E_{p-1}^{<\perp>_J} \cap \mathcal{M}E_q$. By Formula a) in Remark 1.3 from [BG3], the set of partial rank indices r_{ij} for the pair $(\mathcal{M},J)$ (see §1 for definitions) is determined by

$$\sum_{i=p}^{n} \sum_{j=1}^{q} r_{ij} = \varphi_{pq}$$

If we solve for the r_{ij}'s in terms of the φ_{pq}'s, we get formula (3.3) for $r_{p,q}$. In this way, knowledge of φ_{pq} for $p>q$ determines r_{pq} for $p>q$. The formula (3.8) for π_{ij} and (3.7) for ν_p^x in terms of the r_{ij} have already been given in general by formula (2.5) and (2.4). For $\mathcal{M}$ of the special form we have here, $\mathcal{M}E_n$ is nondegenerate in the J-inner product, and hence $\sum_{j=1}^{n} r_{ij} = \tilde{\nu}_{i-m}+\nu_i$ (or ν_i in the notation from §2.1). Thus formula (2.1) collapses to formula (3.4) for ν_p^0. By viewing $\mathcal{M}$ as $\begin{bmatrix} K \\ I \end{bmatrix}$ $\Omega_u(\underline{\nu})$ + $\begin{bmatrix} \Omega_u(\underline{\tilde{\nu}};m) \\ 0 \end{bmatrix}$ in the J-inner product, we count that $\mathcal{M}E_p$ has $\sum_{j=1}^{p} \tilde{\nu}_{j-m}$ positive squares. On the other hand this must equal the number of positive squares had by col. $[\Omega_u(\underline{\nu}^0), \Omega_u(\underline{\nu}^+), \Omega_u(\underline{\nu}^-), \Omega_u(\underline{\nu}^x)]E_p$ in the N-inner product; one can compute that this quantity is given by $\sum_{j=1}^{p} \nu_j^+ + \sum_{j=2}^{p} \sum_{i=1}^{j-1} r_{ji}$. Equating these two expressions for $1\leq p\leq n$ and solving for ν_p^+ gives formula (3.5). Similarly one can compute two expressions for the number of negative squares had by $\mathcal{M}E_p$ in the J-inner product to deduce that

$$(*) \qquad \sum_{j=1}^{p} \nu_j^- + \sum_{j=2}^{p} \sum_{i=1}^{j-1} r_{ji} =$$

$$\operatorname{rank}\,(I-P_pK^*\widetilde{Q}_{p-m+1}K)\big|\mathcal{P}_p$$

Note that

$$\operatorname{rank}\,(I-P_pK^*\widetilde{Q}_{p-m+1}K)\big|\mathcal{P}_p$$

$$=\sum_{j=1}^{p}\nu_j-\dim\operatorname{Ker}\,\{(I-P_pK^*\widetilde{Q}_{p-m+1}K)\big|\mathcal{P}_p\}$$

$$=\sum_{j=1}^{p}\nu_j-\varphi_{p+1,p}$$

$$=\sum_{j=1}^{p}\nu_j-\sum_{i=p+1}^{n}\sum_{j=1}^{p}r_{ij}.$$

If we make this substitution for the right side of (*) and solve the resulting system of equations for ν_p^- we finally get formula (3.6).

One of the main themes of the article [BG3] was a classification of noncanonical LMU factorizations for the general block matrix setting. There it was shown that any block matrix $F=[F_{ij}]_{1\le i,j\le n}$ in $\Omega(\underline{\tilde{\nu}}x\underline{\nu})$ has a factorization F=LMU for a uniquely determined $\underline{\tilde{\nu}}x\underline{\nu}$ -block subpermutation matrix M in canonical form, where $L^{\pm1}\epsilon\Omega_{\ell}(\underline{\tilde{\nu}}x\underline{\tilde{\nu}})$ and $U^{\pm1}\epsilon\Omega_u(\underline{\nu}x\underline{\nu})$. The $(\underline{\tilde{\nu}}x\underline{\nu})$ block subpermutation matrix $M=[M_{ij}]_{1\le i,j\le n}$ is determined by the condition

$$\sum_{\substack{i'\ge i\\ j'\le j}} r^M_{i'j'}:=\operatorname{rank}\begin{bmatrix}M_{11}&\cdots&M_{1j}\\ \vdots&&\\ M_{i1}&\cdots&M_{ij}\end{bmatrix}=\operatorname{rank}\begin{bmatrix}F_{11}&\cdots&F_{1j}\\ \vdots&&\\ F_{i1}&\cdots&F_{ij}\end{bmatrix}$$

for $1\le i,j\le n$. For the case where F is isometric, by Formula (iii) in Remark 2.4 from [BG3] we have the alternative formula

$$\sum_{\substack{i'\ge i\\ j'\le j}} r^M_{i'j'}=\dim\operatorname{Ker}\{(I-P_jF^*\widetilde{Q}_iF)\big|\mathcal{P}_j\}\tag{3.9}$$

When we compare this with the formula in Lemma 3.2, we arrive immediately at the following.

COROLLARY 3.3. Suppose $K \in \mathcal{A}_{\ell}(m-1)$ and F is any isometric element of the coset $K+\mathcal{A}_u(m)$. Suppose F=LMU where $L^{\pm 1} \in \Omega_{\ell}(\underline{\tilde{\nu}} \times \underline{\tilde{\nu}})$, $U^{\pm 1} \in \Omega_u(\underline{\nu} \times \underline{\nu})$, and where M is the unique $\underline{\tilde{\nu}} \times \underline{\nu}$ block subpermutation matrix determined by

$$\sum_{\substack{i' \geq i \\ j' \leq j}} r^M_{i'j'} = \operatorname{rank} \begin{bmatrix} F_{11} & \cdots & F_{1j} \\ \vdots & & \\ F_{i1} & \cdots & F_{ij} \end{bmatrix}$$

Here $r^M_{ij} := \operatorname{rank} M_{ij}$ are the partial rank indices of M. Let $\mathcal{M}$ be the Ω_u-invariant subspace

$$\mathcal{M} = \begin{bmatrix} K \\ I \end{bmatrix} \Omega_u(\underline{\nu}) + \begin{bmatrix} \Omega_u(\underline{\tilde{\nu}};m) \\ 0 \end{bmatrix}$$

Then

$$\sum_{\substack{i' \geq i-m \\ j' \leq j}} r^M_{i'j'} = \dim\{\mathcal{M}E^{\langle\perp\rangle J}_{i-1} \cap \mathcal{M}E_j\}$$

for $1 \leq j < i \leq n$. In particular the partial indices r^M_{ij} are independent of the choice of isometric F in $K+\mathcal{A}_u(m)$ as long as $j<i-m$.

Thus we see that the blocks of the $\underline{\nu} \times \underline{\nu}$-block matrix M below the m-th diagonal are completely determined by the original data $K \in \mathcal{A}_{\ell}(m-1)$ of the interpolation problem, where M is the block subpermutation middle factor arising from a factorization F=LMU of <u>any</u> isometric element F of $K+\mathcal{A}_u(m)$. We shall analyze how the remaining blocks of M can be determined in the next section.

Finally we wish to make explicit one more connection for

factorization of isometric interpolants.

LEMMA 3.4. Suppose $K \in \mathcal{A}_\ell(m-1)$ and $N = \begin{bmatrix} 0 & 0 & \pi \\ 0 & J_1 & 0 \\ \pi^* & 0 & 0 \end{bmatrix}$ is the $(\underline{\nu}^0, \underline{\nu}^+, \underline{\nu}^-, \underline{\nu}^x)$-block S-subpermutation matrix associated with the pair $(\mathcal{M}, J)$ as in Theorem 2.1, where

$$\mathcal{M} = \begin{bmatrix} K \\ I \end{bmatrix} \Omega_u(\underline{\nu}) + \begin{bmatrix} \Omega_u(\tilde{\underline{\nu}}; m) \\ 0 \end{bmatrix} \text{ and}$$

$J = \begin{bmatrix} I_{\tilde{N}} & 0 \\ 0 & -I_N \end{bmatrix}$. We denote by r^π_{ij} $(1 \le i,j \le n)$ the partial indices of the $\underline{\nu}^0 x \underline{\nu}^x$-block subpermutation matrix π. Also suppose F is an isometric matrix in the coset $K + \mathcal{A}_u(m)$ with factorization $F = LMU$ as in Corollary 3.3. Let r^M_{ij} $(1 \le i,j \le n)$ be the partial indices of the $\tilde{\underline{\nu}} x \underline{\nu}$-block subpermutation matrix M. Then

$$r^\pi_{ji} = r^M_{i-m,j}$$

for $1 \le j \le n$.

PROOF. This follows immediately from the formula for how the partial indices r^π_{ij} are calculated from the partial rank indices $r_{ij}(\mathcal{M}, J)$ of the subspace $\mathcal{M}$ with Hermitian form $[\ ,\ ]_J$ as specified in Theorem 2.1, together with the connection between $r_{ij}(\mathcal{M}, J)$ and the partial indices r^M_{ij} observed in Corollary 3.3.

We close this section with a very simple example which illustrates the main ideas. We take $n=2$, $\underline{\nu} = \tilde{\underline{\nu}} = \{1,1\}$, $m=0$ and $K = \begin{bmatrix} 0 & 0 \\ 1 & 0 \end{bmatrix}$. By inspection one sees that a unitary matrix F is in the coset $K + \Omega_u$ if and only if F has the form $\begin{bmatrix} 0 & \epsilon \\ 1 & 0 \end{bmatrix}$ where $|\epsilon| = 1$. We now indicate how this follows from our machinery for the general problem. In this case the space $\mathcal{M}$ is given by

$$\mathcal{M} = \left\{ \begin{bmatrix} x' & y' \\ x & z' \\ x & y \\ 0 & z \end{bmatrix} : x,y,z,x',y',z' \in \mathbb{C} \right\}.$$

The representation from Theorem 2.2 for this case works out to be

$$\mathcal{M} = \Theta\mathcal{M}'$$

where

$$\mathcal{M}' = \begin{bmatrix} * & * \\ * & * \\ 0 & * \\ 0 & * \end{bmatrix} \quad (*'s = \text{arbitrary entries}),$$

$$\Theta = \begin{bmatrix} 0 & 1 & 0 & 0 \\ 1 & 0 & 0 & 1/2 \\ 1 & 0 & 0 & -1/2 \\ 0 & 0 & 1 & 0 \end{bmatrix}$$

and $J' := \Theta^* J \Theta = \begin{bmatrix} 0 & 0 & 0 & 1 \\ 0 & 1 & 0 & 0 \\ 0 & 0 & -1 & 0 \\ 1 & 0 & 0 & 0 \end{bmatrix}.$

Subspaces $\mathcal{G}'$ of $\mathcal{M}'$ which satisfy (i′)– (iv′) work out easily to be

$$\mathcal{G}' = \begin{bmatrix} 1 & 0 \\ 0 & \epsilon \\ 0 & 1 \\ 0 & 0 \end{bmatrix} \Omega_u$$

where $|\epsilon| = 1$. Then subspaces $\mathcal{G}$ of M satisfying (i)–(iv) must be

$$\mathcal{G} = \Theta\mathcal{G}' = \begin{bmatrix} 0 & \epsilon \\ 1 & 0 \\ 1 & 0 \\ 0 & 1 \end{bmatrix} \Omega_u.$$

On the other hand we know that any such subspace has the form $\mathcal{G}$ =

$\begin{bmatrix} F \\ I \end{bmatrix} \Omega_u$ where F is a unitary completion, and conversely. Thus we see that any unitary completion has the form $F = \begin{bmatrix} 0 & \epsilon \\ 1 & 0 \end{bmatrix}$. For this simple example, one sees by inspection that any such unitary completion F has LMU factorization

$$F = \begin{bmatrix} 1 & 0 \\ 0 & 1 \end{bmatrix} \begin{bmatrix} 0 & 1 \\ 1 & 0 \end{bmatrix} \begin{bmatrix} 1 & 0 \\ 0 & \epsilon \end{bmatrix}.$$

Thus the middle factor M is the permutation matrix $\begin{bmatrix} 0 & 1 \\ 1 & 0 \end{bmatrix}$ for all unitary completions. The content of Corollary 3.3 for this simple example is that any such middle factor must have the form $M = \begin{bmatrix} * & * \\ 1 & * \end{bmatrix}$; given that M must also be a permutation matrix, the only possibility is $M = \begin{bmatrix} 0 & 1 \\ 1 & 0 \end{bmatrix}$. Thus the uniqueness and form of the middle factor for all unitary completions in this simple example follows from Corollary 3.3.

4. Factorization of unitary completions

In this section we analyze Problem Ab) further for the case where $N(\underline{\nu}')=N(\underline{\nu})$; thus any isometric completion F is in fact unitary $(F^*F=FF^*=I)$. If $F\epsilon K+\Omega_u(m)$ is unitary, then F factors as F=LMU where M is a $\underline{\nu}'x\underline{\nu}$-block permutation matrix, $L^{\pm 1}\epsilon\Omega_u(\underline{\nu}'x\underline{\nu}')$ and $U^{\pm 1}\epsilon\Omega_u(\underline{\nu}x\underline{\nu})$. We have already seen how the block entries of the middle factor M_{ij} are completely determined by the data $K\epsilon\mathcal{A}_\ell(m-1)$ of the completion problem as long as i>j−m. In this section we show that if F is unitary this is the only constraint on the possibilities for middle factors M arising in this way. More precisely, we shall show that, given any $\underline{\nu}' x \underline{\nu}$-block permutation matrix M with rank M_{ij} = rank $[\pi^*]_{i+m,j}$ for i>j-m, then there is a unitary $F \epsilon K+\mathcal{A}_u(m)$ such that M is the middle factor for the $\underline{\nu}' x \underline{\nu}$ LU-factorization F=LMU of F. (Here π is the permutation matrix in $\Omega_u^0(\underline{\nu}^0 x \underline{\nu}^x)$ associated with the $(\underline{\nu}^0, \underline{\nu}^+, \underline{\nu}^-, \underline{\nu}^x)$-block S-subpermutation matrix as in Theorem 3.1). In fact we shall be able to specify which $\underline{\nu}^+ x \underline{\nu}^-$-inner matrices G to plug into the linear fractional map given in Theorem 3.1 give rise to a unitary completion $F \epsilon K+\mathcal{A}_u(m)$ with a prescribed middle factor M in its $\underline{\nu}' x \underline{\nu}$ LU factorization F=LMU. To handle this, we first must analyze how to complete partially defined matrices to block permutation matrices.

4.1 <u>Block permutation matrices with prescribed lower triangular entries</u>

In this section we consider the following variation on Problem A.

PROBLEM B. Given $P \epsilon \mathcal{A}_u(m-1)$, when is there an M in the coset $P+\mathcal{A}_u(m)$ such that i) M is a subpermutation matrix, or ii) M is an isometric subpermutation matrix? If a solution exists, describe the set of all solutions.

Note that Problem A reduces to Problem B if one substitutes "subpermutation matrix" for "contraction matrix" in i) and "isometric subpermutation matrix" for "isometry" in ii).

The existence criterion for Problem B is straightforward. We leave the details to the reader. The obvious necessary condition is also sufficient.

THEOREM 4.1. Let $P = [P_{ij}]_{1 \leq i,j \leq n} \epsilon \mathcal{A}_{\ell}(m-1)$. Then there exists a subpermutation matrix M in the coset $P+\mathcal{A}_u(m)$ if and only if each of the rectangular submatrices

$$\begin{bmatrix} P_{j-m+1,1} & \cdots & P_{j-m+1,j} \\ \vdots & & \\ P_{n,1} & & P_{n,j} \end{bmatrix}$$

is a subpermutation matrix for $1 \le j \le n$, or equivalently, the matrix P itself in $\mathcal{A}_\ell(m-1)$ is a subpermutation matrix. There exists an isometric subpermutation matrix M in the coset $P+\mathcal{A}_u(m)$ if and only if in addition

$$\sum_{j'=1}^{j} \nu_j - \operatorname{rank} \begin{bmatrix} P_{j-m+1,1} & \cdots & P_{j-m+1,j} \\ \vdots & & \\ P_{n,1} & & P_{n,j} \end{bmatrix} \le \sum_{j'=1}^{j} \tilde{\nu}_{j-m}.$$

The description of the set of all solutions of Problem Bi) or Bii) calls for further organization of the information. Introduce new n-tuples $\underline{\nu}^+=\{\nu_1^+,\ldots,\nu_n^+\}$ and $\underline{\nu}^- = \{\nu_1^-,\ldots,\nu_n^-\}$ of nonnegative integers by

$$\nu_j^+ = \tilde{\nu}_{j-m} - \operatorname{rank} [P_{j-m,1} \ldots P_{j-m,n}]$$

and

$$\nu_j^- = \nu_j - \operatorname{rank\ col.} [P_{1,j} \ldots P_{n,j}]$$

Note that then isometric subpermutation matrices exist in $P+\mathcal{A}_u(m)$ if and only if $\sum_{j'=1}^{j} \nu_{j'}^- \le \sum_{j'=1}^{j} \nu_{j'}^+$. Given any matrix M in $\mathcal{A} = \Omega(\tilde{\underline{\nu}} x \underline{\nu})$, there is induced a matrix M in $\tilde{\mathcal{A}} = \Omega(\underline{\nu}^+ x \underline{\nu}^-)$ by deleting all rows and columns in M corresponding to the locations of 1's in the subpermutation matrix $P \in \mathcal{A}_\ell(m-1)$. If M is a subpermutation matrix in the coset $P+\mathcal{A}_u(m)$, the induced matrix $\tilde{M}$ is a subpermutation matrix of upper triangular type $\Omega_u(\underline{\nu}^+ x \underline{\nu}^-)$. If M is in addition isometric, then $\tilde{M}$

is in addition isometric, i.e. $\widetilde{M}$ is a $\underline{\nu}^{+}x\underline{\nu}^{-}$-inner permutation matrix. Conversely, suppose $\widetilde{M}$ is a subpermutation matrix in $\Omega_u(\underline{\nu}^{+}x\underline{\nu}^{-})$. One may insert additional rows and columns in $\widetilde{M}$ to obtain a matrix M in $\mathcal{A}$ = $\Omega(\underline{\widetilde{\nu}}x\underline{\nu})$. Each row and column added to $\widetilde{M}$ to form M should agree with the corresponding row and column of P. Since the original matrix $\widetilde{M}$ was $\underline{\nu}^{+}x\underline{\nu}^{-}$ upper triangular, the resulting subpermutation matrix M built in this way will be in the coset $P+\mathcal{A}_u(m)$. Moreover the resulting M is in addition isometric if and only if the original $\widetilde{M}$ was $\underline{\nu}^{+}x\underline{\nu}^{-}$-inner. We summarize the situation as follows.

THEOREM 4.2. Suppose $P \in \mathcal{A}_{\ell}(m-1)$ is a subpermutation matrix. Then subpermutation matrices M in $P+\mathcal{A}_u(m)$ are in one-to-one correspondence with subpermutation matrices $\widetilde{M}$ in $\widetilde{\mathcal{A}}_u = \Omega_u(\underline{\nu}^{+}x\underline{\nu}^{-})$ under the correspondence $M \longrightarrow \widetilde{M}$ indicated above. Moreover $M \in P+\mathcal{A}_u(m)$ is isometric if and only if the associated $\widetilde{M}$ is $\underline{\nu}^{+}x\underline{\nu}^{-}$-inner.

4.2 Factorization partial indices for unitary completions

We now return to Problem Aii for a given $K \in \mathcal{A}_{\ell}(m-1)$. We suppose that the conditions of Theorem 3.1 hold so that isometric matrices F in the coset $K+\mathcal{A}_u(m)$ exist. We suppose now in addition that $N(\underline{\tilde{\nu}})$ = $N(\underline{\nu})$. This means that any such isometry F is in fact unitary so $F^{*}F$ =

$FF^* = I_k$ $(k=N(\underline{\nu})=N(\underline{\tilde{\nu}}))$. We know that any such unitary solution F has a $\underline{\tilde{\nu}} x \underline{\nu}$-block LU factorization

$$F=LMU$$

where $L^{\pm 1} \epsilon \Omega_u(\underline{\tilde{\nu}} x \underline{\tilde{\nu}})$, M is a $\underline{\tilde{\nu}} x \underline{\nu}$-block permutation (and hence unitary) matrix and $U^{\pm 1} \epsilon \Omega_u(\underline{\nu} x \underline{\nu})$. In ≷3 we saw how the partial rank indices r^M_{ij} for M were completely determined from the data of the completion problem as long as i>j−m; thus any such M has its block entries M_{ij} essentially tied down for i>j−m. In the previous ≷4.1 we analyzed all possible ways to complete a subpermutation matrix partially defined for indices (i,j) with i>j−m to a full permutation matrix. In this section we show that all such completions of the tied down entries of M can actually occur; that is, if M is any $\underline{\tilde{\nu}} x \underline{\nu}$-block permutation matrix whose partial indices r^M_{ij} are the appropriate values for all (i,j) such that i>j−m, then there is a unitary F in $K+\mathcal{A}_u(m)$ such that M occurs as the middle factor in the $\underline{\nu}' x \underline{\nu}$-block LU factorization F=LMU of F. Indeed we have the following general form of Theorem 0.3 from the Introduction.

THEOREM 4.3. Suppose $K \epsilon \mathcal{A}_\ell(m-1)$ satisfies the conditions of Theorem 3.1 for isometric F in $K+\mathcal{A}_u(m)$ to exist and suppose $N(\underline{\tilde{\nu}})=N(\underline{\nu})$. Let $P=[P_{ij}]_{1\leq i,j\leq n}$ be a subpermutation matrix in $\mathcal{A}_\ell(m-1)$ such that rank $P_{i-m,j}$ = rank π_{ji} for $1\leq j<i\leq n$, where π is the $\underline{\nu}^0 x \underline{\nu}^x$-inner matrix as in Theorem 3.1. Then the n-tuples $\underline{\nu}^+$ and $\underline{\nu}^-$ arising from K as in Theorem 3.1 are the same as the n-tuples $\underline{\nu}^+$ and$\underline{\nu}^-$ arising from

P as in Theorem 4.2. Suppose that the unitary $F \in K + \mathcal{A}_u(m)$ is related to the $\underline{\nu}^+ x \underline{\nu}^-$-inner matrix G as in formula (3.1). Let M be the middle factor for the $\tilde{\underline{\nu}} x \underline{\nu}$-block LU factorization F=LMU of F, and let the $\underline{\nu}^+ x \underline{\nu}^-$-inner permutation matrix $\widetilde{M}$ be the middle factor for the $\underline{\nu}^+ x \underline{\nu}^-$-block LU factorization $G = \widetilde{L}\widetilde{M}\widetilde{U}$ of G. Suppose $M' \in P + \mathcal{A}_u(m)$ is the augmentation of M as in Theorem 4.2. Then the partial indices r^M_{ij} of M agree with the partial indices $r^{M'}_{ij}$ of M′ for $1 \le i,j \le n$.

PROOF. First note that the two computations for $\underline{\nu}^+$ and $\underline{\nu}^-$ give the same result due to the strong connection between P and π in the definitions. Next recall that when F and G are related by formula (3.1) in Theorem 3.1, then we have the subspace equation.

$$\mathcal{G}_F := \begin{bmatrix} F \\ I \end{bmatrix} \Omega_u(\underline{\nu}) = \Theta \begin{bmatrix} I & 0 \\ 0 & G \\ 0 & I \\ 0 & 0 \end{bmatrix} \begin{bmatrix} \Omega_u(\underline{\nu}^0) \\ \Omega_u(\underline{\nu}^-) \end{bmatrix} = \Theta \mathcal{G}_G$$

Recall $\mathcal{M} = \begin{bmatrix} K \\ I \end{bmatrix} \Omega_u(\nu) + \begin{bmatrix} \Omega_u(\tilde{\nu};m) \\ \{0\} \end{bmatrix}$

is represented as $\mathcal{M} = \Theta\widetilde{\mathcal{M}}$ where $\widetilde{\mathcal{M}} = \text{col.}\ [\Omega_u(\underline{\nu}^0), \Omega_u(\underline{\nu}^+), \Omega_u(\underline{\nu}^-), \Omega_u(\underline{\nu}^x)]$ and $\Theta^* J \Theta = N$ where

$$J = \begin{bmatrix} I_{N(\tilde{\underline{\nu}})} & 0 \\ 0 & -I_{N(\underline{\nu})} \end{bmatrix} \text{ and } N = \begin{bmatrix} 0 & 0 & 0 & \pi \\ 0 & I_{N(\underline{\nu}^+)} & 0 & 0 \\ 0 & 0 & -I_{N(\underline{\nu}^-)} & 0 \\ \pi^* & 0 & 0 & 0 \end{bmatrix}$$

Conclude that

$$\mathcal{M}\cap\mathcal{G}_F^{<\perp>J} = \Theta(\widetilde{\mathcal{M}}\cap\mathcal{G}_G^{<\perp>N}).$$

Now direct computations give

$$\mathcal{M}\cap\mathcal{G}_F^{<\perp>J} = \mathcal{G}_F \dot{+} \begin{bmatrix} F\Omega_\ell^0(\underline{\tilde{\nu}})\cap\Omega_u(\underline{\tilde{\nu}};m) \\ \{0\} \end{bmatrix}$$

and that

$$\widetilde{\mathcal{M}}\cap\mathcal{G}_G^{<\perp>N} = \mathcal{G}_G \dot{+} \begin{bmatrix} \{0\} \\ G\Omega_\ell^0(\underline{\nu}^-)\cap\Omega_u(\underline{\nu}^+) \\ \{0\} \\ \{0\} \end{bmatrix}$$

Thus

$$(\mathcal{M}\cap\mathcal{G}_F^{<\perp>J})E_i = \begin{bmatrix} F \\ I \end{bmatrix}\mathcal{P}_i \dot{+} \begin{bmatrix} F\mathcal{Q}_{i+1}\cap\widetilde{\mathcal{P}}_{i-m} \\ \{0\} \end{bmatrix}$$

$$=\Theta(\widetilde{\mathcal{M}}\cap\mathcal{G}_G^{<\perp>N})E_i$$

where

$$(\widetilde{\mathcal{M}}\cap\mathcal{G}_G^{<\perp>N})E_i$$

$$= \begin{bmatrix} I & 0 \\ 0 & G \\ 0 & I \\ 0 & 0 \end{bmatrix} \begin{bmatrix} \mathcal{P}_i^0 \\ \mathcal{P}_i^- \end{bmatrix} \dot{+} \begin{bmatrix} G\mathcal{Q}_{i+1}^{\{0\}}\cap\widetilde{\mathcal{P}}_i^+ \\ \{0\} \\ \{0\} \end{bmatrix}.$$

Here $\mathcal{P}_i^0$ is the span of the first $\sum_{i'=1}^{i} \nu_i^0$ standard basis vectors, and the other spaces $\mathcal{P}_i^-$, $\mathcal{P}_i^+$ and $\mathcal{Q}_{i+1}^-$ are defined analogously. The spaces of particular interest to us are

$$\mathcal{K}_{ij} = \frac{(\mathcal{M}\cap\mathcal{G}_F^{<\perp>_J})E_i \cap (\mathcal{M}\cap\mathcal{G}_F^{<\perp>_J})E_j}{\mathcal{G}_F E_i \cap \mathcal{G}_F E_j}$$

and

$$\widetilde{\mathcal{K}}_{ij} = \frac{(\widetilde{\mathcal{M}}\cap\mathcal{G}_G^{<\perp>_N})E_i \cap (\widetilde{\mathcal{M}}\cap\mathcal{G}_G^{<\perp>_N})E_j}{\mathcal{G}_G E_i \cap \mathcal{G}_G E_j}$$

for $1 \le i,j \le n$.

By the above discussion we know that $\mathcal{K}_{ij} = \Theta\widetilde{\mathcal{K}}_{ij}$ for all i,j and hence

(4.1) $\quad \dim \mathcal{K}_{ij} = \dim \widetilde{\mathcal{K}}_{ij}.$

By the explicit formulas above for the various spaces involved, we see that for $i \ge j$

(4.2) $\quad \dim \mathcal{K}_{ij} = \dim F\mathcal{Q}_{i+1}\cap\widetilde{\mathcal{P}}_{j-m} = \dim\mathcal{Q}_{i+1}\cap F^*\widetilde{\mathcal{P}}_{j-m}$

and

(4.3) $\quad \dim \widetilde{\mathcal{K}}_{ij} = \dim G\, \mathcal{Q}^-_{i+1} \cap \mathcal{P}^+_j = \dim \mathcal{Q}^-_{i+1} \cap G^* \mathcal{P}^+_j.$

Here the last equality on each line follows from the fact that both F and G are unitary.

Now let us suppose that $F = LMU$ is the $\underline{\tilde{\nu}}\mathrm{x}\underline{\nu}$-block LU-factorization of F and that $G=\widetilde{L}\widetilde{M}\widetilde{U}$ is the $\underline{\nu}^+\mathrm{x}\underline{\nu}^-$-block LU factorization of G. Then $F^*=U^*M^*L^*$ is the $\underline{\nu}\mathrm{x}\underline{\tilde{\nu}}$-block factorization of F^* and $G^* = \widetilde{U}^*\widetilde{M}^*\widetilde{L}^*$ is the $\underline{\nu}^-\mathrm{x}\underline{\nu}^+$-block LU factorization of G^*. Now since both F^* and G^* are isometric, formula (3.9) applies with F^* and G^* in place of F. This gives

$$(4.4)\qquad \sum_{\substack{i'\geq i+1\\ j'\leq j-m}} \operatorname{rank}\,(M^*)_{i'j'} = \dim\,(\mathcal{Q}_{i+1}\cap F^*\widetilde{\mathcal{P}}_{j-m})$$

and

$$(4.5)\qquad \sum_{\substack{i'\geq i+1\\ j'\leq j}} \operatorname{rank}\,(\widetilde{M}^*)_{i'j'} = \dim\,(\mathcal{Q}^-_{i+1}\cap G^*\,\mathcal{P}^+_j).$$

Since we established above that the right hand sides in (4.4) and (4.5) are equal (by(4.1)–(4.3)), we get

$$\sum_{\substack{i'\geq i+1\\ j'\leq j-m}} \operatorname{rank}\, M_{j'i'} = \sum_{\substack{i'\geq i+1\\ j'\leq j}} \operatorname{rank}\, \widetilde{M}_{j'i'}.$$

whenever $i\geq j$. Conclude from this that

$$\operatorname{rank}\, M_{i-m,\, j} = \operatorname{rank}\, \widetilde{M}_{i,j} \text{ if } 1\leq i\leq j\leq n.$$

This shows how the partial indices of $\widetilde{M}$ determine the partial indices of M not in the lower triangular locations tied down as in Corollary 3.4. That moreover $\operatorname{rank}\, M_{i,j} = \operatorname{rank}\, M'_{i,j}$ for all i,j $(1\leq i,j\leq n)$, where M' is the augmentation of $\widetilde{M}$ as in Theorem 4.2, is now a matter of checking the bookkeeping. This completes the proof of Theorem 4.3.

5. The Abstract Framework: the Lifting Theorem

As was pointed out in [BG2], Problem A discussed above is but a special instance of a more abstract lifting problem, the main subject of [BG2]. To state the problem we need to first recall some terminology from [BG2]. Let Ω be the algebra of nxn matrices $U=[U_{ij}]_{1\leq i,j\leq n}$ over the complex numbers $\mathbb{C}$ and let Ω_u be the subalgebra of upper triangular matrices $(U=[U_{ij}]_{1\leq i,j\leq n}$ where $U_{ij}=0$ if $i>j)$. By a <u>contractive unital</u> (c.u.) <u>representation</u> of Ω_u we mean a mapping σ from Ω_u into the algebra $\mathcal{L}(\mathcal{H})$ of bounded linear operators on a Hilbert space $\mathcal{H}$ which is an algebra homomorphism carrying the nxn identity matrix I_n into the identity operator $I_{\mathcal{H}}$ on $\mathcal{H}$ and which has norm ≤ 1 ($\|\sigma(U)\|_{\mathcal{L}(\mathcal{H})} \leq \|U\|_{\mathcal{L}(\mathbb{C}^n)}$). As in [BG2] we consider only finite dimensional representations; thus the dimension of the Hilbert spaces to come in the sequel we assume to be finite. By a star-representation of Ω we mean a mapping τ from Ω into the bounded linear operators $\mathcal{L}(\mathcal{K})$ on some Hilbert space $\mathcal{K}$ which is a C^*-algebra homomorphism, i.e. $\tau(UV) = \tau(U)\tau(V)$, $\tau(U+V) = \tau(U)+\tau(V)$, $\tau(I_n) = I_{\mathcal{K}}$, and $\tau(U^*) = \tau(U)^*$. It then follows that $\|\tau(U)\| \leq \|U\|$ as well (see for example [Ar]). If τ: $\Omega \longrightarrow \mathcal{L}(\mathcal{K})$ is a star-representation and $\mathcal{M}$ is a subspace of $\mathcal{K}$ which is invariant under $\tau(\Omega_u)$

$$\tau(U)\,\mathcal{M} \subset \mathcal{M} \text{ whenever } U \in \Omega_u,$$

then we may define a c.u. representation ρ of Ω_u into $\mathcal{L}(\mathcal{M})$ by the formula

$$\rho(U) = \sigma(U)\,|\,\mathcal{M} \text{ for } U \in \Omega_u.$$

Any c.u. representation ρ of Ω_u arising in this way from a star-representation τ of Ω we shall call a substar-representation of Ω_u. Now suppose that $\rho:\Omega_u \longrightarrow \mathcal{L}(\mathcal{M})$ is a substar-representation of Ω_u as above and $\mathcal{N} \subset \mathcal{M}$ is an invariant subspace for all the operators in ρ (Ω_u):

$$\rho(U)\mathcal{N} \subset \mathcal{N} \text{ whenever } U \in \Omega_u.$$

Then we may define a maping σ: $\Omega_u \longrightarrow \mathcal{L}(\mathcal{H})$ where $\mathcal{H} = \mathcal{M} \ominus \mathcal{N}$ by

$$\sigma(U) = P_{\mathcal{H}}\,\rho(U)\Big|\mathcal{H}$$

where $P_{\mathcal{H}}$ is the orthogonal projection of $\mathcal{M}$ onto $\mathcal{H}$. The fact that $\mathcal{N}$

$= \mathcal{M} \ominus \mathcal{H}$ is invariant for $\rho(U)$ implies that σ defined as above is a c.u. representation of Ω_u. Whenever the c.u. representation σ: $\Omega_u \longrightarrow \mathcal{L}(\mathcal{H})$ arises in this way from some substar-representation $\rho:\Omega_u \longrightarrow \mathcal{L}(\mathcal{M})$ (where $\mathcal{H} \subset \mathcal{M}$), we say that ρ is a substar-dilation of σ, or that σ is a compression of ρ to the star-invariant subspace $\mathcal{H}$. Now it develops that any c.u. representation of Ω_u has a substar dilation; hence the above example is the model for any c.u. representation of Ω_u. Details and examples may be found in [BG2].

We can now state the lifting problem as follows.

PROBLEM C. Suppose $\sigma_j:\Omega_u \longrightarrow \mathcal{L}(\mathcal{H}_j)$ is a c.u. representation of Ω_u with substar-dilation $\rho_j:\Omega_u \longrightarrow \mathcal{L}(\mathcal{M}_j)$ (where $\mathcal{M}_j \supset \mathcal{H}_j$) for j=1,2, and $X \in \mathcal{L}(\mathcal{H}_1,\mathcal{H}_2)$ intertwines σ_1 with σ_2 (i.e. $X\sigma_1(U) = \sigma_2(U)X$ for all $U \in \Omega_u$). Find an operator $Y \in \mathcal{L}(\mathcal{M}_1,\mathcal{M}_2)$ such that

$$Y\rho_1(U) = \rho_2(U)Y \text{ for } U \in \Omega_u,$$

$$XP_{\mathcal{H}_1} = P_{\mathcal{H}_2}Y,$$

and either

i) $\|Y\| \le 1$

ii) Y is isometric.

If such a Y exists, describe the set of all such liftings.

One of the main results of [BG2] was that there exists a solution to Problem Ci) if and only if $\|X\| \le 1$. Also there was obtained a linear fractional map description for the set of all such Y's for the case where $\|X\| < 1$. In this section we indicate how the machinery of this paper can be used to obtain an analogous linear fractional map description for the general case $\|X\| \le 1$. Similar results hold for Problem Cii) as well.

Most of the work has already been done. Indeed, as was pointed out in [BG2], Problem C is just a more abstract version of Problem A in §3 of this paper. Thus all that needs to be done to handle Problem C is to translate the work already done above on Problem A to the more abstract framework of representation theory.

The reader should consult [BG2] for a listing of other instances of the abstract Problem C. Of course the results of this section apply to all these other examples as well.

The work of [BG2] also showed how to model any c.u. representation σ together with its substar-dilation in terms of Ω_u-invariant subspaces. This means that the invariant subspace representation theorems in this paper have direct analogues in the language of representations.

The following theorem, an essential tool for our analysis of Problem C, is just an abstract version of Theorem 2.1. We leave the proof to the reader.

THEOREM 5.1. Suppose $\sigma_j:\Omega_u \longrightarrow \mathcal{L}(\mathcal{H}_j)$, $\rho_j:\Omega_u \longrightarrow \mathcal{L}(\mathcal{M}_j)$ and $X \in \mathcal{L}(\mathcal{H}_1, \mathcal{H}_2)$ are as in the statement of Problem C where $j=1,2$. Then there are substar representations

$$\rho^0:\Omega_u \longrightarrow \mathcal{L}(\mathcal{M}^0)$$

$$\rho^+:\Omega_u \longrightarrow \mathcal{L}(\mathcal{M}^+)$$

$$\rho^-:\Omega_u \longrightarrow \mathcal{L}(\mathcal{M}^-)$$

and

$$\rho^x:\Omega_u \longrightarrow \mathcal{L}(\mathcal{M}^x),$$

an isometry $\pi \in \mathcal{L}(\mathcal{M}^x, \mathcal{M}^0)$ such that the image of π is contained in

$\rho^0(\Omega_u^0)$ $\mathcal{M}^0$ and such that π intertwines ρ^x with ρ^0 $(\pi\rho^x(U) = \rho^0(U)\pi$ for $U\in\Omega_u)$,

a mapping $\Theta\in\mathcal{L}(\mathcal{M}^0\oplus\mathcal{M}^+\oplus\mathcal{M}^-\oplus\mathcal{M}^x, \mathcal{M}_2\oplus\mathcal{M}_1)$ such that

i) Θ intertwines $\rho^0\oplus\rho^+\oplus\rho^-\oplus\rho^x$ with $\rho_2\oplus\rho_1$, i.e.

$$\Theta(\rho^0(U)\oplus\rho^+(U)\oplus\rho^-(U)\oplus\rho^x(U)) = (\rho_2(U)\oplus\rho_1(U))\Theta,$$

ii)

$$\Theta^*\begin{bmatrix} I_{\mathcal{M}_2} & 0 \\ 0 & -I_{\mathcal{M}_1} \end{bmatrix}\Theta = \begin{bmatrix} 0 & 0 & 0 & \pi \\ 0 & I_{\mathcal{M}^+} & 0 & 0 \\ 0 & 0 & -I_{\mathcal{M}^-} & 0 \\ \pi^* & 0 & 0 & 0 \end{bmatrix}$$

and

iii) $$\Theta \text{ col. } [\mathcal{M}^0, \mathcal{M}^+, \mathcal{M}^-, \mathcal{M}^x] = \begin{bmatrix} I_{\mathcal{M}_2} & XP_{\mathcal{H}_1} \\ 0 & I_{\mathcal{M}_1} \end{bmatrix}\begin{bmatrix} \mathcal{N}_2 \\ \mathcal{M}_1 \end{bmatrix}$$

(Here $\mathcal{N}_2 = \mathcal{M}_2 \ominus \mathcal{H}_2$.)

The next result is the extension of Theorem 3.1 to the abstract language of representations. The proof follows from the representation Theorem 5.1 in the same way as Theorem 3.1 follows from the representation Theorem 2.1 by using geometric ideas of indefinite inner product spaces. We leave the details to the interested reader.

THEOREM 5.2. Suppose that $\sigma_j:\Omega_u \longrightarrow \mathcal{L}(\mathcal{H}_j)$, $\rho_j:\Omega_u \longrightarrow \mathcal{L}(\mathcal{M}_j)$ and $X \in \mathcal{L}(\mathcal{H}_1, \mathcal{H}_2)$ are as in the statement of Problem C (where $j=1,2$), and that $\|X\|\le 1$. Let the substar-representations ρ^0, ρ^+, ρ^-, ρ^x, the mapping $\pi\in\mathcal{L}(\mathcal{M}^x,\mathcal{M}^0)$

and the mapping $\Theta \in \mathcal{L}(\mathcal{M}^0 \oplus \mathcal{M}^+ \oplus \mathcal{M}^- \oplus \mathcal{M}^x, \mathcal{M}_2 \oplus \mathcal{M}_1)$ be as in Theorem 5.1. Write Θ as a 2x4 block-operator matrix $\Theta = \begin{bmatrix} \psi & \alpha & \beta & \psi' \\ w & k & \gamma & w' \end{bmatrix}$ consistent with the indicated decompositions of its domain and range spaces. Then:

i) The map $Y \in \mathcal{L}(\mathcal{M}_1, \mathcal{M}_2)$ is a solution of Problem Ci) if and only if Y is of the form

(5.1) $Y = [\psi, \alpha Z + \beta]([w, kZ+\gamma])^{-1}$ for a uniquely determined $Z \in \mathcal{L}(\mathcal{M}^-, \mathcal{M}^+)$ such that

$$Z\rho^-(U) = \rho^+(U)Z \text{ for } U \in \Omega_u$$

and

(5.2) $\|Z\| \leq 1$

ii) The map $Y \in \mathcal{L}(\mathcal{M}_1, \mathcal{M}_2)$ is a solution of Problem Cii) if and only if Y is of the form (5.1) for a $Z \in \mathcal{L}(\mathcal{M}^-, \mathcal{M}^+)$ uniquely determined by Y as above, but with (5.2) replaced by the stronger constraint

(5.3) Z is isometric.

We remark that the map $\Theta = \begin{bmatrix} \psi & \alpha & \beta & \psi' \\ w & k & \gamma & w' \end{bmatrix}$ and the invariants ρ^0, ρ^+, ρ^-, ρ^x and π in Theorem 5.1 can be computed explicitly from the original data $X \in \mathcal{L}(\mathcal{H}_1, \mathcal{H}_2)$, $\sigma_j: \Omega_u \longrightarrow \mathcal{H}_j$ and $\rho_j: \Omega_u \longrightarrow \mathcal{M}_j$ (j=1,2) of the lifting problem. This follows ultimately from the explicit expression for the representer Θ in Theorem 2.1. Once ρ^+ and ρ^- are known explicitly, one can determine whether isometric intertwining maps $Z \in \mathcal{L}(\mathcal{M}^-, \mathcal{M}^+)$ exist, and hence whether Problem Cii) even has solutions. (Just as was the situation for Problem A, Problem Cii) may have no solutions even though Ci) has solutions.) Thus Theorem 5.2 represents a complete solution of Problem C. We do not pursue here however the development of explicit formulas at this level of generality.

REFERENCES

[Ar] W.B. Arveson, An Invitation to C^*-algebras, Springer-Verlag, New York-Heidelberg-Berlin (1976).

[B] J.A. Ball, Invariant subspace representations, unitary interpolants and factorization indices, in Topics in Operator Theory Systems and Networks, OT 12 Birkhauser Verlag, Basel-Boston-Stuttgart (1984).

[BG1] J.A. Ball and I. Gohberg, Shift invariant subspaces, factorization and interpolation for matrices I: The canonical case, Linear Algebra and Applications, to appear.

[BG2] J.A. Ball and I. Gohberg, A commutant lifting theorem for triangular matrices with diverse applications, Integral Equations and Operator Theory 8:205-267 (1985).

[BG3] J.A. Ball and I. Gohberg, Pairs of shift invariant subspaces of matrices and noncanonical factorization, to appear.

[BH1] J.A. Ball and J.W. Helton, A Beurling-Lax theorem for the Lie group U(m,n) which contains most classical interpolation, J. Operator Theory 9:107-142 (1983).

[BH2] J.A. Ball and J.W. Helton, Factorization results related to shifts in an indefinite metric, Integral Equations and Operator Theory, 5(1982), 632-658.

[DG1] H. Dym and I. Gohberg, Extensions of matrix valued functions and block matrices, Ind. U. Math. J. 31 (1982), 733-765.

[DG2] H. Dym and I. Gohberg, Unitary interpolants, factorization indices and infinite Hankel block matrices, J. Funct. Anal. 54 (1983), 229-289.

[DKW] C.Davis, W.M. Kahan and H.F. Weinberger, Norm preserving dilations and their applications to optimal error bounds, SIAM J.

Numer. Anal. 19 (1982), 445–469.

[GG1] I. Gohberg and S. Goldberg, Finite dimensional Wiener–Hopf equations and factorization of matrices, Linear Algebra Appl. 48 (1982), 216–226.

[GG2] I. Gohberg and S. Goldberg, Extensions of triangular Hilbert Schmidt operators, Integral Equations and Operator Theory 7 (1984), 743–790.

[NF] B. Sz.–Nagy and C. Foias, Harmonic Analysis of Operators on Hilbert Space, American Elsevier, New York (1970).

Joseph A. Ball
Department of Mathematics
Virginia Tech
Blacksburg, VA 24061–4097
U.S.A.

Israel Gohberg
School of Mathematical Sci.
Raymond and Beverly Sackler
Faculty of Exact Sciences
Tel Aviv University
Tel Aviv, Israel

Operator Theory:
Advances and Applications, Vol. 19

BALANCED REALISATIONS FOR INFINITE DIMENSIONAL SYSTEMS

Ruth F. Curtain and Keith Glover

ABSTRACT

Balanced realisations of finite-dimensional systems are those in which the controllability and observability gramians are both equal to the same symmetric, positive, diagonal matrix. For the class of infinite dimensional systems with a Hilbert Schmidt Hankel operator we obtain a characterization of a balanced realisation in terms of the singular values and the Schmidt pairs of the Hankel operator.

INTRODUCTION

Balanced realisations of finite dimensional systems are realisations whose controllability and observability gramians are both equal to the same positive diagonal matrix. The diagonal entries are the singular values of the Hankel operator defined by the system, and the balanced realisations can be expressed in terms of these singular values and the Schmidt pairs of the Hankel operator. Balanced realisations have special structural properties and their truncations usually provide good reduced order approximations of the original system (Moore [9], Pernebo and Silverman [10], Glover [8]). Consequently it is interesting to generalize the concept of balanced realisations to infinite dimensional systems.

We consider the class of stable infinite dimensional systems with finite dimensional input and output operators which are described by an impulse response function h such that

$$h \in L_1 \cap L_2(0,\infty;R^{k \times m}) \text{ and } t^{\frac{1}{2}}h(t) \in L_2(0,\infty;\ R^{k \times m})$$

In section 2 we define the induced Hankel operator Γ on $L_2(0,\infty)$ and its singular values and Schmidt pairs; the assumptions on h imply that Γ is compact and Hilbert Schmidt. In section 3 we find an explicit characterization of a balanced realisation in terms of the singular values and the Schmidt pairs of the Hankel operator. The realisation is characterized by a contraction semigroup and possibly unbounded input and output maps. However it does have bounded reachability and observability maps as defined in Curtain and Pritchard [2] and its controllability and observability gramians are both equal to the same positive diagonal operator. In section 4 we show that the truncations of the infinite dimensional balanced realisation generate a sequence of finite dimensional, stable, balanced realisations. The results are illustrated by a simple example in section 5 and in the conclusion section 6 we discuss the question of convergence of the finite dimensional balanced realisations which has been studied by Curtain and Glover in [5].

1. Hankel operators for stable infinite dimensional systems.

In this paper we consider the class of linear infinite dimensional systems defined by the following input-output map

$$y(t) = \int_0^t h(t-s)u(s)ds \tag{2.1}$$

where the outputs $y \in L_2(0,\infty;R^m)$, the inputs $u \in L_2(0,\infty;R^m)$ and the impulse function satisfies

$$h \in L_1 \cap L_2(0,\infty;R^{k\times m}) \text{ and } t^{\frac{1}{2}}h(t) \in L_2(0,\infty;\ R^{k\times m}) \tag{2.2}$$

The induces the following *Hankel operator* Γ: $L_2(0,\infty;R^m) \longrightarrow L_2(0,\infty;R^k)$ defined by

$$(\Gamma u)(t) = \int_0^\infty h(t+s)u(s)ds \tag{2.3}$$

(2.2) implies that Γ is compact and Hilbert Schmidt on $L_2(0,\infty)$ and is also a continuous operator on each of the spaces $L_p(0,\infty)$ $1 \le p < \infty$ and $C^1(0,\infty)$, which is the space of continuously differentiable functions with the norm

$$\|f\|_1 = \|f\|_\infty + \int_0^\infty |\dot{f}(s)|ds$$

This was used by AAK in [1] to deduce various properties of Γ for the case of a scalar kernel, h. The extension to the case of a matrix-valued kernel is obvious, so we summarize the properties of Γ deduced from [1].

$\Gamma^*\Gamma$ is compact and positive on $L_2(0,\infty)$ and so has countably many positive eigenvalues $\sigma_1^2 \geq \sigma_2^2 \geq \dots \sigma_i^2 \geq$. σ_i are the *singular values* of Γ. If v_i and w_i are the corresponding normalized eigenvectors of $\Gamma^*\Gamma$ and $\Gamma\Gamma^*$ respectively, then (v_i,w_i) are called the *Schmidt pairs* of Γ and we have

$$(2.4) \qquad \begin{cases} \Gamma v_i = \sigma_i w_i \\ \Gamma^* w_i = \sigma_i v_i \end{cases} ; \; i = 1,2,\dots.$$

Since w_i is the eigenvector of $\Gamma^*\Gamma$ which is compact in the spaces $L_2(0,\infty)$; $p > 1$ and $C^1(0,\infty)$, as proved in [1], we have

$$(2.5) \qquad w_i \in L_1 \cap L_2 \cap C^1(0,\infty;R^k)$$

and similarly

$$(2.6) \qquad v_i \in L_1 \cap L_2 \cap C^1(0,\infty;R^m)$$

Hankel kernels can be expressed in terms of their singular values and Schmidt pairs.

LEMMA 2.1

$$(2.7) \qquad (\Gamma u)(t) = \int_0^\infty h(t+s)u(s)ds = \sum_{i=1}^\infty w_i(t)\sigma_i \int_0^\infty v_i^*(s)u(s)ds$$

$$(2.8) \qquad (\Gamma^* y)(t) = \int_0^\infty h^*(t+s)y(s)ds = \sum_{i=1}^\infty v_i(t)\sigma_i \int_0^\infty w_i^*(s)y(s)ds$$

$$(2.9) \qquad h^*(t+s) = \sum_{i=0}^\infty v_i(t)\sigma_i w_i^*(s) \qquad \text{a.e. in } s \text{ for all } t \geq 0$$

$$(2.10) \qquad (\Gamma^\wedge y)(t) = \int_0^\infty h^*(t+s+r)y(r)dr = \sum_{i=1}^\infty v_i(t)\sigma_i \int_0^\infty w_i^*(s+r)y(r)dr$$

PROOF. (2.7) and (2.8) are just the polar decompositions of Γ and Γ^* respectively. Since $h \in L_2(0,\infty)$, so is its shifted function $h_t(.) = h(t+s)$

and its columns $\underline{h}_t^k$. Thus

$$\underline{h}_t^k(.) = \sum_{i=1}^{\infty} <\underline{h}_t^k, w_i> w_i^* + \sum_{i=1}^{\infty} <\underline{h}_t^k, \overline{w}_i> \overline{w}_i^*$$

where $<.,.>$ is the usual inner product in $L_2(0,\infty)$ and $\overline{w}_i$ extends $\overline{w}_i$ to form a complete orthonormal basis for $L_2(0,\infty)$. Now from (2.4) we have that

$$<\underline{h}_t^k, w> = \int_0^{\infty} \underline{h}_t^k(t+s) w_i(s) ds = \sigma_i v_i(t)$$

and so for $t \geq 0$

$$h^*(t+s) = \sum_{i=1}^{\infty} v_i(t)\sigma_i w_i^*(s) + \sum_{i=1}^{\infty} A_i \overline{w}_i^*$$

and substituting $y = \overline{w}_j$ in (2.8) shows that $A_i = 0$ for all i which proves (2.9). (2.10) follows from (2.9) and (2.8).
Notice that if the singular values are all distinct, then

(2.11) $$<v_i, v_j>_2 = \delta_{ij} = <w_i, w_j>_2$$

where $<.,.>_2$ denotes the inner product in $L_2(0,\infty;R^s)$ for $s = k$ or m. If σ_i has multiplicity r then there exists a D_o-chain of Schmidt pairs of length r and these form a basis for a set of Schmidt pairs corresponding to σ_i. ([1]). For simplicity of exposition we shall assume henceforth that the singular values are distinct, but the proofs can be extended to the case of multiple singular values in the usual way.

2. A Balanced Realisation.

We consider the class of linear infinite dimensional systems described by the Hankel operator Γ given by (2.3) where h satisfies (2.2).

We consider first the notion of a realisation (c.f. [7]).

DEFINITION 3.1. *A balanced realisation* for the impulse response h on the state space H is the triple (C,A,B), where A, B and C are linear operators: $C: H \longrightarrow R^k$, $B: R^m \longrightarrow H$ and A is the infinitesimal generator of a strongly continuous semigroup, T(t), on H such that $h(t) = CT(t)B$ a.e. in t.

If B and C are bounded operators and T(t) is exponentially stable then h satisfies our above assumptions and the following maps are bounded. The *reachability operator* $\underline{B}$: $L_2(0,\infty;R^m) \longrightarrow H$ given by

$$\underline{B}u = \int_0^\infty T(t)Bu(t)dt \tag{3.1}$$

The *observability operator* $\underline{C}$: $H \longrightarrow L_2(0,\infty;R^k)$

$$\underline{C}x = CT(t)x \tag{3.2}$$

$\underline{B}$ and $\underline{C}$ may also be well defined bounded maps for more general B and C (see [2], [4]). Whenever $\underline{B}$ and $\underline{C}$ are bounded we say that they realise the Hankel operator   if (c.f. [7])

$$\Gamma = \underline{C}\,\underline{B} \tag{3.3}$$

Furthermore the following controllability and observability gramians P and Q are well defined whenever the reachability and observability maps are by

$$P = \underline{B}\,\underline{B}^* = \int_0^\infty T(t)BB^*T^*(t)dt \tag{3.4}$$

$$Q = \underline{C}^*\underline{C} = \int_0^\infty T^*(t)C^*CT(t)dt \tag{3.5}$$

We are interested here in a special class of realisation.

DEFINITION 3.2. (C,T(t),B) is a *balanced realisation* for the Hankel operator Γ is it has bounded reachability and observability operators defined by (3.1) and (3.2), (3.3) holds and the observability and controllability gramians are both equal to the same positive diagonal operator.

Notice that in Definition 3.2 we do not necessarily require that B and C be bounded maps, and in fact C may be unbounded and the range of B may lie outside H. This is the usual situation for infinite dimensional continuous time realisations; as in [7] where a so-called rigged structure is introduced. We shall not bother with this as we only require that $\underline{B}$ and $\underline{C}$ be well defined bounded maps.

In the above sense we propose the realisation for h on ℓ_2 given by

$$T_{ij}(t) = (\sigma_j/\sigma_i)^{\frac{1}{2}} \int_0^\infty w_i^*(s) w_j(s+t)\,ds \tag{3.6}$$

$$C = (\sqrt{\sigma_1} w_1(0), \sqrt{\sigma_2} w_2(0), \ldots, \sqrt{\sigma_i} w_i(0), \ldots\ ) \tag{3.7}$$

$$B = (\sqrt{\sigma_1} v_1(0), \sqrt{\sigma_2} v_2(0), \ldots, \sqrt{\sigma_i} v_i(0), \ldots\ )^t \tag{3.8}$$

We proceed to prove some properties of T(t).

LEMMA 3.1.

$$\sigma_j \int_0^\infty w_i^*(s) w_j(t+s)\,ds = \sigma_i \int_0^\infty v_i^*(s+t) v_j(s)\,ds \tag{3.9}$$

PROOF

$$\sigma_j \int_0^\infty w_i^*(s) w_j(t+s)\,ds = \int_0^\infty w_i^*(s) \Gamma v_j(s+t)\,ds \qquad \text{from (2.4)}$$

$$= \int_0^\infty w_i^*(s) \int_0^\infty h(t+s+\beta) v_j(\beta)\,d\beta\,ds$$

from (2.7) and since $h \in L_2((0,\infty) \times (0,\infty))$ we may change the order of integration. Thus

$$\sqrt{\sigma_i \sigma_j}\, T_{ij}(t) = \int_0^\infty (\Gamma w_i(t+s))^* v_j(s)\,ds$$

$$= \sigma_i \int_0^\infty v_i^*(t+s) v_j(s)\,ds \qquad \text{from (2.4)}$$

LEMMA 3.2. $\Sigma^{\frac{1}{2}} T(t) \Sigma^{-\frac{1}{2}}$ *and* $\Sigma^{\frac{1}{2}} T^*(t) \Sigma^{-\frac{1}{2}}$ *are bounded contractions on* ℓ_2 *for all* $t \geq 0$, *where* $\Sigma = \text{diag}(\sigma_1, \sigma_2, \ldots, \sigma_i, \ldots)$.

PROOF. Let ℓ_2^0 be the subspace of ℓ_2 comprising all finite sequences. Then for $x \in \ell_2^0$, we calculate

$$\|\Sigma^{\frac{1}{2}}T^*(t)\Sigma^{-\frac{1}{2}}x\|^2$$

$$= \sum_{i=1}^{\infty} \sigma_i \Big(\sum_{j=1}^{N} (\sqrt{\sigma_i}/\sqrt{\sigma_j}) \int_0^{\infty} w_j^*(s) w_i(t+s)\,ds\,(1/\sqrt{\sigma_j})x_j \Big)^2$$

$$= \sum_{j,k=1}^{N}\ x_j x_k \sum_{i=1} \sigma_i/\sigma_j \int_0^{\infty} v_k^*(s+t) v_i(s)\,ds \int_0 w_j^*(s) w_i(s+t)\,ds$$

from (3.9)

$$= \sum_{j,k=1}^{N} x_k x_j < v_k(t+.), v_j(t+.) >$$

since expanding $v_j(t+.)$ in terms of $v_i(.)$, we have

$$v_j(t+\beta) = \sum_{i=1}^{\infty} v_i(\beta) \int_0^{\infty} v_j^*(t+s) v_i(s)\,ds$$

$$= \sum_{i=1}^{\infty} (\sigma_i/\sigma_j) v_i(\beta) \int_0^{\infty} w_j^*(s) w_i(t+s)\,ds$$

and so from (3.9)

(3.10) $$< v_k(t+.), v_j(t+.) > = 1/\sigma_j \sum_{i=1}^{\infty} \sigma_i \int_0^{\infty} v_k^*(\beta+t) v_i(\beta)\,d\beta \int_0^{\infty} w_j^*(s) w_i(s+t)\,ds$$

Notice that the fact that $v_j(t+.)$ has an expansion in terms of $v_i(.)$ follows from (2.4) and (2.8). Now

$$\|\Sigma^{\frac{1}{2}}T^*(t)\Sigma^{-\frac{1}{2}}x\|^2 = \sum_{j,k=1}^{N} x_k x_j < v_k(t+.), v_j(t+.) >$$

$$\sum_{j,k=1}^{N} x_k x_j \int_0^{\infty} v_j^*(s+t) v_k(s+t)\,ds$$

$$= \|x\|^2 - \int_0^t \Big| \sum_{j=1}^{N} x_j v_j(s) \Big|^2 ds$$

$$\leq \| x \|^2$$

Since ℓ_2^0 is dense in ℓ_2, we have proved that $\Sigma^{\frac{1}{2}} T^*(t) \Sigma^{-\frac{1}{2}}$ is a contraction on ℓ_2. That $\Sigma^{\frac{1}{2}} T(t) \Sigma^{-\frac{1}{2}}$ is a contraction can be proved similarly. (see [5]).

LEMMA 3.3. $T(t)$ *is a strongly continuous contraction semigroup on* ℓ_2.

PROOF.

(a) We first show that $T(t)$ is a bounded contraction on ℓ_2 by appealing to the crucial lemma 3.1 in [12], which was developed for the discrete time balanced realisation problem. Notice that we already have shown in lemma 3.2 that $Z_1 = \Sigma^{\frac{1}{2}} T(t) \Sigma^{-\frac{1}{2}}$ and $Z_2 = \Sigma^{\frac{1}{2}} T^*(t) \Sigma^{-\frac{1}{2}}$ are bounded contractions and writing

$$(3.11) \qquad T(t) = \Sigma^{\frac{1}{2}} Z_2^* \Sigma^{-\frac{1}{2}}, \quad T^*(t) = \Sigma^{\frac{1}{2}} Z_1^* \Sigma^{-\frac{1}{2}}$$

it is clear that $D(T(t)) \supset \Sigma^{\frac{1}{2}} \ell_2$ and $D(T^*(t)) \supset \Sigma^{\frac{1}{2}} \ell_2$ and $T^*(t) : \Sigma^{\frac{1}{2}} \ell_2 \longrightarrow \Sigma^{\frac{1}{2}} \ell_2$. So $M = T(t) T^*(t) : \Sigma^{\frac{1}{2}} \ell_2 \longrightarrow \ell_2$ is a densely defined, self adjoint, positive operator such that $\Sigma^{\frac{1}{2}} M \Sigma^{-\frac{1}{2}} = Z_1 Z_2$ is a contraction (3.11) and from lemma 3.1, [12], we may conclude that $T(t)$ is a contraction on ℓ_2 for all $t \geq 0$.

(b) Since $T(t)$ is a contraction, if suffices to prove the semigroup property for $x \in \ell_2^0$.

$$\{T(t)T(s)x\}_i = \Sigma \sum_{k,j=1}^{N} (\sqrt{\sigma_i}/\sqrt{\sigma_j}) \int_0^\infty v_i^*(r+t) v_k(r) dr \int_0^\infty v_k^*(r+s) v_j(r) dr x_j$$

from (3.6), (3.9)

$$= \sum_{j=1}^{N} (\sqrt{\sigma_i}/\sqrt{\sigma_j}) \int_s^\infty v_i^*(r+t) v_j(r-s) dr x_j$$

expanding $v_i(t+.)$ and $v_j(.-s)$ as infinite series in $v_k(.)$ as the in proof of lemma 3.2.

$$= \sum_{j=1}^{N} (\sqrt{\sigma_i}/\sqrt{\sigma_j}) \int_0^\infty v_i^*(t+s+r)v_j(r)dr\, x_j$$

$$= \{T(t+s)x\}_i \qquad \text{by (3.6), (3.9).}$$

(c) By virtue of the semigroup property proved in (b) and $T(t) = I$ from the definition, it is sufficient to establish the weak continuity of $T(t)$ at the origin, ([6], p.15). Furthermore since $T(t)$ is a contraction we need only prove this for $x,y \in \ell_2^0$. From (3.6) and (3.9) we have

$$< (T(t) - I)x,y > =$$

$$= \sum_{i=1}^{N} \sum_{j=1}^{p} x_i y_j \big(\int_0^\infty (\sigma_i/\sigma_j)^{\frac{1}{2}} x_i x_j v_i^*(t+s)v_j(s)ds - < x,y >$$

$$= \sum_i^N \sum_j^p x_i y_j \big(\int_0^\infty v_i^*(s)v_j(s)ds - (\sigma_i/\sigma_j)^{\frac{1}{2}} \int_0^\infty v_i^*(t+s)v_j(s)ds$$

$$\text{using (2.11)}$$

$$\leq \sum_{i=1}^{N} x_i \big(\sum_{j=1}^{p} y_j^2 \big)^{\frac{1}{2}} \big(\sum_{j=1}^{p} f_{ij}^2(t) \big)^{\frac{1}{2}}$$

$$\leq \|x\| \, \|y\| \big(\sum_{i=1}^{N} \sum_{j=1}^{p} f_{ij}^2(t) \big)^{\frac{1}{2}}$$

$$\longrightarrow 0 \text{ as } n \longrightarrow \infty$$

where

$$f_{ij}(t) = \int_0^\infty v_i^*(s)v_j(s)ds - (\sigma_i/\sigma_j)^{\frac{1}{2}} \int_0^\infty v_i^*(t+s)v_j(s)ds$$

$$\longrightarrow 0 \text{ as } n \longrightarrow \infty \text{ by (2.11).}$$

So $T(t)$ has the strong continuity property and (a), (b) and (c) show that

$T(t)$ is a strongly continuous contraction semigroup ([6]).

It is of interest to have a matrix representation for the infinitesimal generator A of $T(t)$.

LEMMA 3.4. *The infinitesimal generator of* $T(t)$, A, *satisfies*

$$A_{ij} = (\sigma_j/\sigma_i)^{\frac{1}{2}} \int_0^\infty w_i^*(s)\dot{w}_j(s)ds \tag{3.12}$$

PROOF. First we notice that since $w_i \in C^1(0,\infty) \cap L_1$, the integral (3.12) is well defined and furthermore

$$\lim_{t \to 0} (1/t) \int_0^\infty w_i^*(s)w_j(s+t)ds = \int_0^\infty w_i^*(s)\dot{w}_j(s)ds$$

for all i and j. So componentwise we have if $e_j = (0,0,..,1,0,..)$, then

$$\lim_{t \to 0} [(T(t)e_j - e_j)/t]_i = A_{ij} \text{ given by (3.12)} \tag{3.13}$$

This shows that the infinitesimal generator of $T(t)$ satisfies (3.12).

Although B and C need not be bounded (see the example in section 5), we always obtain bounded reachability and observability maps defined by (3.1) and (3.2).

LEMMA 3.5. $(T(t),B)$ *has a bounded reachability map and* $(C,T(t))$ *has a bounded observability map given by*

$$(\mathcal{B}u)_i = \sqrt{\sigma_i} \langle v_i, u \rangle \tag{3.14}$$

$$(\mathcal{C}x)(t) = \sum_i \sqrt{\sigma_i} x_i w_i(t) \tag{3.16}$$

PROOF.

$$(\mathcal{B}u)_i = \int_0^\infty \sum_j (\sigma_j/\sigma_i)^{\frac{1}{2}} \int_0^\infty w_i^*(s)w_j(s+t)ds \sqrt{\sigma_j} v_j^*(0)u(t)dt$$

from (3.4), (3.6)

$$= \int_0^\infty 1/\sqrt{\sigma_i}[\sum_j \sigma_j \int_0^\infty w_i^*(s)w_j(s+t)ds v_j^*(0)]u(t)dt$$

$$= 1/\sqrt{\sigma_i} \int_0^\infty \int_0^\infty w_i^*(s)h(s+t)ds u(t)dt \qquad \text{from (2.8)}$$

$$= \sqrt{\sigma_i} \int_0^\infty v_i^*(t)u(t)dt \qquad \text{from (2.4), (2.8)}$$

So $$\|B\|^2 = \sum_i \sigma_i (\int_0^\infty v_i^*(t)u(t)dt)$$

$$\leq \sigma_1 \|u\|^2$$

and B is bounded. Similarly we can show that C is bounded and

$$(C^* f)_i = \sqrt{\sigma_i} \int_0^\infty w_i^*(t)f(t)dt$$

and this proves (3.15)

LEMMA 3.6. (C,T(t),B) *defines a realisation for* Γ *on* ℓ_2 *which is approximately controllable and initially observable with controllability and observability gramians* $P = Q = \Sigma$. *It is a balanced realisation.*

PROOF.

(a) From lemma 3.5 we have that

$$(\underline{C}\,\underline{B}u)(t) = \sum_i \sigma_i w_i(t) \int_0^\infty v_i^*(s)u(s)ds$$

$$= \int_0^\infty h(t+s)u(s)ds \qquad \text{from (2.7)}$$

$$= (\Gamma u)(t) \qquad \text{from (2.3)}$$

(b) Now from lemma 3.5 $\underline{C}$ and $\underline{C}^*$ are bounded operators and so

$\underline{C}^*\underline{C} = \int_0^\infty T^*(t)C^*CT(t)dt = Q$ is bounded and from (3.15)

$$\{Q\}_{ij} = \int_0^\infty \sum^\infty \sum_{i,j=1}^\infty \sqrt{\sigma_i}\sqrt{\sigma_j} w_i^*(t) w_j(t)dt = \sigma_i \delta_{ij}$$

Thus

$$Q = \Sigma = \mathrm{diag}(\sigma_1, \sigma_2, \ldots\ldots)$$

Similarly $P = \underline{B}\,\underline{B}^* = \Sigma$ and so the realisation is balanced.

(c) Since P and Q are positive, the realisation is approximately controllable and initially observable.

4. Truncations of the balanced realisation.

In finite dimensions it is known that truncations of balanced realisations are also balanced and provide a good reduced-order approximation. ([9], [10], [8]) This motivates us to define the following sequence of n th order truncations of $(C, T(t), B)$

$$(4.1) \qquad \{A_n\}_{ij} = \{(\sigma_j/\sigma_i)^{\frac{1}{2}} \int_0^\infty w_i^*(s)\dot{w}_j(s)ds\} \ ; \ i,j = 1,..n$$

$$= \{(\sigma_i/\sigma_j)^{\frac{1}{2}} \int_0^\infty v_i^*(s)\dot{v}_j(s)ds\}$$

$$(4.2) \qquad C_n = (\sigma_1 w_1(0), \ldots\ldots, \sigma_n w_n(0))$$

$$(4.3) \qquad B_n = (\sigma_1 v_i(0), \ldots\ldots, \sigma_n v_n(0))^t$$

(A_n, B_n, C_n) is a minimal system with state space R^n, and as one would expect, it is balanced. However this takes a few lemmas to establish.

LEMMA 4.1. *The truncated system* (A_n, B_n, C_n) *satisfies the following Lyapunov equations*

$$(4.4) \qquad A_n\Sigma_n + \Sigma_n A_n^* + B_n B_n^* = 0$$

$$(4.5) \qquad \Sigma_n A_n + A_n^*\Sigma_n + C_n^* C_n = 0$$

where $\Sigma_n = \mathrm{diag}(\sigma_1,\sigma_2,\ldots,\sigma_n)$.

PROOF.

$$(A_n\Sigma_n + \Sigma_n A_n^*)_{ij} = (\sigma_i\sigma_j)^{\frac{1}{2}}\{\int_0^\infty \dot{v}_i^*(s)v_j(s)ds + \int_0^\infty v_i^*(s)\dot{v}_j(s)ds\}$$

$$= -(\sigma_i\sigma_j)^{\frac{1}{2}}v_i^*(0)v_j(0) \qquad \text{since } v_i \in C^1(0,\infty)$$

$$= -(B_nB_n^*)_{ij}$$

and so (4.4) is satisfied and similarly for (4.5).

To prove that A_n is stable we need the following finite dimensional results.

LEMMA 4.2. *Suppose that* (C,A,B) *are finite dimensional complex matrices satisfying the Lyapunov equations*

(4.6) $$AP + PA^* + BB^* = 0$$

(4.7) $$A^*P + PA + C^*C = 0$$

where $P = \mathrm{diag}(p_1,p_2,\ldots,p_n)$ *and* $p_1 > p_2 > \ldots > p_n > 0$.
Then the following asertions are equivalent.

(a) $\sigma(A) \subset \{z: \mathrm{Re}\, z < 0\}$.
(b) $e_i^*B \neq 0$ *for all* i.
(c) $Ce_i \neq 0$ *for all* i.
(d) (A,B) *is controllable.*
(e) (C,A) *is observable.*
where $e_i = (0,0,..,1,..)$.

PROOF. (b) ⇔ (c): $e_i^*(4.6)e_i$ and $e_i^*(4.7)e_i$ give

(4.8) $$p_i(a_{ii} + \bar{a}_{ii}) = -e_i^*BB^*e_i = -e_i^*C^*Ce_i$$

(d) ⇒ (d): Suppose that $e_i^*B = 0$ for some i and hence from (4.8) $Ce_i = 0$.
Then $e_i^*(4.6)$ and $e_i^*(4.7)$ give the following

(4.9) $$e_i^*AP + p_ie_i^*A^* = 0$$

(4.10) $\quad e_i^* A^* P + p_i e_i^* A = 0$

Now (4.9) implies that $e_i^* AP = -p_i e_i^* A^* = p_i^2 e_i^* AP^{-1}$ from (4.10). So $e_i^* AP^2 = p_i^2 e_i^* A$ and since the eigenvectors of P^2 are unique, and (A,B) is not controllable.

(e) ⇒ (c): This follows by a dual argument to the above.

(a) ⇒ (d): Suppose that (A,B) is not controllable, then there exist x and k such that $x^* A = kx^*$ and $x^* B = 0$. Then from (4.6) we obtain

$$kx^* P + x^* PA^* = 0$$

and hence $-\bar{k}$ is an eigenvalue of A and this contadicts (a).

(a) ⇒ (e): This follows by a dual argument to the above.

(c) or (b) ⇒ (a): (4.6) implies that $\sigma(A) \subset \{z: \text{Re}\, z \leq 0\}$. If (a) is false then there exists an eigenvalue z of A such that $z + \bar{z} = 0$. Let X and Y^* denote the null spaces of $(A - zI)$ and $(A^* - \bar{z}I)$, respectively, both of dimension r. Then composing $Y(4.6)Y^*$ we obtain

$$YAPY^* + YPA^*Y^* + YBB^*Y^* = 0$$

which implies

$$zYPY^* + \bar{z}YPY^* + YBB^*Y^* = 0$$

and so $YB = 0$ and similarly $X^*(4.7)X$ implies that $CX = 0$. Now $(4.6)Y^*$ implies

$$APY^* = -\bar{z}PY^* = zPY^*$$

In other words, PY^* is spanned by the eigenvectors of A associated with z, which is Y^* itself. So Y^* is invariant under P and since P is diagonal, $Y = \text{span}\{e_{i1}^*, \ldots, e_{ir}^*\}$. $YB = 0$ implies that $e_{iq}^* B = 0$ for q = i,..,r and this contradicts (b). Similarly one can show that X is invariant under P and so $X = \text{span}\{e_{jk}, \ldots, e_{jr}\}$ and $Ce_{jq} = 0$; q = 1,..,r, contradicting (c).

We now prove that for the truncations of infinite dimensional balanced realisations obtained in section 3, (b) and (d) of lemma 2.1 always hold.

LEMMA 4.3. *Suppose that* Γ *of* (2.2) *has distinct singular values, then* $w_i(0) \neq 0$ *and* $v_i(0) \neq 0$ *for all* i.

PROOF. This proof follows ideas in Theorem 7.2, [1]. Suppose that $v_i(0) = 0$ and differentiate the identity

$$\sigma_i w_i(t) = (\Gamma v_i)(t) = \int_t^\infty h(s) v_i(s-t)\,ds$$

to obtain

$$\sigma_i \dot{w}_i(t) = -h(t) v_i(0) - \int_t^\infty h(s) \dot{v}_i(s-t)\,ds$$

$$= -(\Gamma \dot{v}_i)(t)$$

Similarly one can show that

$$\sigma_i \dot{v}_i(t) = -(\Gamma^* \dot{w}_i)(t)$$

and hence $(\dot{v}_i, -\dot{w}_i)$ form a Schmidt pair corresponding to σ_i, but since σ_i are assumed distinct, the Schmidt pairs are unique and so we must have

$$\dot{v}_i = k v_i; \; v_i(0) = 0$$

which leads to the contradiction $v_i(t) = 0$.

Our final result follows from the last three lemmas.

THEOREM 4.1. The truncated system (A_n, B_n, C_n) is a balanced realisation with A_n asymptotically stable and the controllability and observability gramians are both equal to Σ_n, the unique solution of the Lyapunov equations (4.4) and (4.5).

5. An Example

Consider the system with impulse response function

$$h(t) = \begin{cases} 1 & ; \; 0 \leq t \leq 1 \\ 0 & ; \; t > 1 \end{cases} \tag{5.1}$$

Then h satisfies all our assumptions and we can calculate its singular values and Schmidt vectors explicitly to be

(5.2) $\sigma_i = 2/(\pi(2i-1)) \quad ; \quad i = 1,2,\ldots$

(5.3) $v_i(t) = \begin{cases} \sqrt{2}\cos\pi(i-\frac{1}{2})t & ; \quad 0 \le t \le 1 \\ 0 & ; \quad t > 1 \end{cases}$

(5.4) $w_i(t) = (-1)^{i+1} v_i(t)$

Its balanced realisation is then

(5.4) $C = \{\sqrt{2\sigma_i}, -\sqrt{2\sigma_2}, \ldots, (-1)^{i+1}\sqrt{2\sigma_i}, \ldots\}$

(5.5) $B = \{\sqrt{2\sigma_1}, \sqrt{2\sigma_2}, \ldots\}^t$

(5.6) $A_{ij} = \begin{cases} -[(2i-1)(2j-1)]^{\frac{1}{2}}/[i+j-i] & \text{for } i+j \text{ even} \\ [(2i-1)(2j-1)]^{\frac{1}{2}}/\ j-1 & \text{for } i+j \text{ odd} \end{cases}$

(5.7) $T(t)_{ii} = (1-t)\cos\pi t(i-\frac{1}{2}) \quad ; \quad 0 \le t \le 1$

$T(t)_{ij} = 1/\pi[\sin\pi t(i-\frac{1}{2}) - (-1)^{i+j}\sin t(j-\frac{1}{2})]\alpha_{ij} \quad ; \quad i \ne j$

$T(t)_{ij} = 0 \quad ; \quad t > 0$

where $\alpha_{ij} = [(2i-1)(2j-1)]^{\frac{1}{2}}/(j-i)(i+j-1)$.

Notice that in this example C is unbounded and B has its range outside ℓ_2, however B and C are bounded maps.

6. Conclusions

Given a system defined by its Hankel operator whose kernel h satisfies $h \in L_1 \cap L_2(0,\infty;R^{k \times m})$ and $h(t+s) \in L_2((0,\infty) \times (0,\infty);R^{k \times m})$, we have shown that it always has a well defined balanced realisation on a Hilbert space. This realisation is defined explicitly in terms of the singular values and Schmidt vectors of Γ and its truncations are also balanced realisations of some finite dimensional system. It is natural to ask whether these truncations approximate the original system as in the finite dimensional case ([9], [10] and [8]). This question is investigated in [5], where using a different class of realisations, it is shown that if Γ is nuclear, then the truncated transfer functions do converge to the original one in the L_∞-norm. So balanced realisations do provide one with useful finite dimensional approximations of infinite dimensional systems. An exactly analagous analysis for the discrete time case is given in [3], that is, discrete time systems with an L_2 kernel have a balanced realisations and if Γ is nuclear the finite dimensional truncated transfer functions converge. In [12] it is shown that balanced realisations exist for a much wider class of discrete time systems, but the important question of convergence of truncations was not considered there.

ACKNOWLEDGEMENTS. At the original presentation of these results at the Workshop on Operator Theory and its Applications in Amsterdam, June, 1985, we had not proved the contraction property of the semigroup in the balanced realisation. We are indebted to Nicholas Young for his lemma in [12], which has made this improvement possible.

REFERENCES

1. V.M. Adamjan, D.Z. Arov and M.G. Krein, Analytic properties of Schmidt pairs for a Hankel Operator and the Generalized Schur-Takagi problem, Math. USSR Sbornik, 15, p. 31-73, 1971.

2. R.F. Curtain and A.J. Pritchard, "Infinite Dimensional Linear Systems Theory", Lecture Notes in Control and Information Sciences, Vol. 8, 1978, Springer Verlag.

3. R.F. Curtain, Balanced Realisations for Discrete time Infinite-dimensional Systems. (submitted)

4. R.F. Curtain, Sufficient conditions for Infinite-rank Hankel Operators to be Nuclear, J. Math. Control and Info. (to appear).

5. R.F. Curtain and K. Glover, Realisation and Approximation for infinite dimensional systems. (in preparation).

6. E.B. Davies, One Parameter Semigroups, 1980, Academic Press Inc., London.

7. P.A. Fuhrmann, Linear Systems and Operators in Hilbert Space, 1981, McGraw Hill.

8. K. Glover, All Optimal Hankel-norm approximations of linear multivariable systems and their L -error bounds. Int.J. Control, 1984, 39, pp. 1115-1193.

9. B.C. Moore, Principal Component ANalysis in Linear Systems: controllability, observability and model reduction. IEEE. Trans. Aut. Control, Vol. AC-26, 1981, p. 17-32.

10. L. Pernebo and L.M. Silverman, Model Reduction via Balanced State-space Representiation, IEEE. Trans. Aut. Control, Vol. AC-27, 1982, p. 382-387.

11. J. Weidmann, Linear Operators in Hilbert Spaces, 1980, Springer Verlag.

12. N. Young, Balanced Realisations in Infinite Dimensions, Proc. Workshop on Operator Theory and its Applications, Amsterdam, June 4-7, 1985.

Operator Theory:
Advances and Applications, Vol. 19

UNITARY LINEAR SYSTEMS WHOSE TRANSFER FUNCTIONS ARE RIEMANN MAPPING FUNCTIONS

Louis de Branges*

The aim of the present work is to place the proof of the Bieberbach conjecture [4] in the context of contemporary ideas in functional analysis and to formulate a new coefficient conjecture.

The importance of invariant subspace theory has been so universally recognized in the last twenty-five years that it is no longer necessary to apologize when the subject is introduced. Engineers and applied mathematicians have experienced the value of its concepts through the theory of linear systems. Almost everyone knows that the important properties of a linear system are described by its transfer function. And most people are aware that the effectiveness of the theory of linear systems is due to the good relationship between factorization of transfer functions and invariant subspaces of the underlying transformations [2], [9].

Assume that a bounded linear transformation A of a Hilbert space $\mathcal{H}$ into itself is given. The problem is to determine or make use of the invariant subspaces of A. Some sort of mechanism needs to be found for observing what is happening in $\mathcal{H}$. Computational experience shows that it is advantageous to construct an associated linear system.

The present discussion is concerned with the important scalar case of the theory of linear systems. The external Hilbert space $\mathcal{C}$ is assumed to be the complex numbers, considered

* Research supported by the National Science Foundation

with absolute value as norm. The Cartesian product $\mathcal{H} \oplus \mathcal{C}$ is realized as a Hilbert space of column vectors, the upper entry in $\mathcal{H}$ and the lower entry in $\mathcal{C}$.

A linear system is a matrix

$$\begin{vmatrix} A & B \\ C & D \end{vmatrix}$$

acting on $\mathcal{H} \oplus \mathcal{C}$, where the input B is a bounded linear transformation of $\mathcal{C}$ into $\mathcal{H}$, the output C is a bounded linear transformation of $\mathcal{H}$ into $\mathcal{C}$, and the external operator D is a bounded linear transformation of $\mathcal{C}$ into $\mathcal{C}$. Since $\mathcal{C}$ is now assumed to be the complex numbers, the range of B is spanned by an element of $\mathcal{H}$, C is a continuous linear functional on $\mathcal{H}$, and D is multiplication by a number.

A linear system is said to be unitary if the defining matrix is unitary. M. S. Brodskii [8] has shown that the unitary hypothesis characterizes the well-known Nagy-Foias model theory for contractive transformations [20].

The unitary condition on the matrix is the ideal hypothesis for constructing a model theory. Some twenty-five years ago, James Rovnyak and the author introduced a model theory under a weaker hypothesis. This theory is characterized by the fact that the defining matrix is conjugate-isometric [7]. That is, it has an isometric adjoint.

There are at least two reasons why a more general model theory might be constructed. One is that you cannot always have what you want and have to make do with less. The other is that you may not always have what you want at the outset and need to create some mechanism by which to achieve the ideal situation.

The present work is concerned principally with conjugate-isometric and unitary linear systems. It is also interesting to consider linear systems without any further hypothesis and to introduce hypotheses as needed [5].

A characteristic feature of the model theory of James Rovnyak and the author is flexibility in the use of norms. There is provision in the theory for norms to change in a way which is rare in other work on invariant subspaces. A construction which has since been called complementation is used to make these changes. The advantage of complementation is that it allows the construction of Hilbert spaces without proceeding through tedious quadratic inequalities for reproducing kernel functions.

If $\mathcal{P}$ is a Hilbert space which is contained contractively in a Hilbert space $\mathcal{H}$, then a unique Hilbert space $\mathcal{Q}$ exists, which is contained contractively in $\mathcal{H}$ and which has these properties: The inequality

$$\|h\|^2 \leq \|f\|_{\mathcal{P}}^2 + \|g\|_{\mathcal{Q}}^2$$

holds whenever f is in $\mathcal{P}$, g is in $\mathcal{Q}$, and $h = f + g$. Every element h of $\mathcal{H}$ admits a minimal decomposition for which equality holds.

The space $\mathcal{Q}$ is called the complementary space to $\mathcal{P}$ in $\mathcal{H}$. The space $\mathcal{P}$ is recovered as the complementary space to $\mathcal{Q}$ in $\mathcal{H}$. The minimal decomposition of an element h of $\mathcal{H}$ with f in $\mathcal{P}$ and g in $\mathcal{Q}$ is unique.

Define $\mathcal{C}(z)$ to be the Hilbert space of square summable power series $f(z) = \Sigma\, a_n z^n$ with complex coefficients,

$$\|f(z)\|_{\mathcal{C}(z)}^2 = \Sigma\, |a_n|^2 .$$

Note that a power series B(z) with complex coefficients represents a function which is bounded by one in the unit disk if, and only if, multiplication by B(z) is contractive in $\mathcal{C}(z)$. Assume that B(z) does not vanish identically. The range $\mathcal{M}(B)$ of multiplication by B(z) in $\mathcal{C}(z)$ is a Hilbert space in the unique norm such that multiplication by B(z) is an isometric transformation of $\mathcal{C}(z)$ onto the space. The space is contained contractively in $\mathcal{C}(z)$. Multiplication by z is isometric in the space.

Define $\mathcal{H}(B)$ to be the complementary space to $\mathcal{M}(B)$ in $\mathcal{C}(z)$. The series $[B(z) - B(0)]/z$ is an element of norm at most one in the space. The series $[f(z) - f(0)]/z$ belongs to the space whenever $f(z)$ belongs to the space. The inequality for difference-quotients

$$\|[f(z) - f(0)]/z\|^2_{\mathcal{H}(B)} \leq \|f(z)\|^2_{\mathcal{H}(B)} - |f(0)|^2$$

is always satisfied. The space is said to satisfy the identity for difference-quotients if equality always holds.

The reason why a space $\mathcal{H}(B)$ is interesting is that a conjugate-isometric linear system is associated with the space. The main transformation is the difference-quotient transformation, which takes every element $f(z)$ of $\mathcal{H}(B)$ into $[f(z) - f(0)]/z$. The input is the transformation which takes every complex number c into $[B(z) - B(0)]\, c/z$. The output is the transformation which takes $f(z)$ into $f(0)$. The external operator is multiplication by the complex number $B(0)$.

The example so constructed is essentially the most general conjugate-isometric linear system with the complex numbers as external space. For assume given an arbitrary conjugate-isometric linear system

$$\begin{vmatrix} A & B \\ C & D \end{vmatrix}$$

where A maps the Hilbert space $\mathcal{H}$ into itself, B maps $\mathcal{C}$ into $\mathcal{H}$, C maps $\mathcal{H}$ into $\mathcal{C}$, and D maps $\mathcal{C}$ into $\mathcal{C}$, and $\mathcal{C}$ is the complex numbers with absolute value as norm. Then the transformation which takes f into

$$\sum_{n=0}^{\infty} C A^n f z^n$$

is a partial isometry of $\mathcal{H}$ onto a space $\mathcal{H}(W)$ with $W(z)$ specified by the conditions $W(0) = D$ and

$$[W(z) - W(0)]\, c/z = \sum_{n=0}^{\infty} C A^n B c z^n$$

for every number c. The power series $W(z)$ so obtained is called the transfer function of the given linear system.

A further construction is needed if the natural linear system associated with a given space $\mathcal{H}(B)$ is not unitary. The construction is also of interest for the study of adjoint transformations when the linear system is unitary.

If a bounded linear transformation of a Hilbert space into itself has an isometric adjoint but is not unitary, then a unitary transformation exists in a larger Hilbert space which extends the given transformation. The construction of the larger Hilbert space is made by an elementary inductive procedure. When this construction is applied to a conjugate-isometric linear system, it produces a unitary linear system.

Assume that $\mathcal{H}(B)$ is a given space. Define the extension space $\mathcal{D}(B)$ of $\mathcal{H}(B)$ to be the space of pairs $(f(z),g(z))$ consisting of an element $f(z)$ of $\mathcal{H}(B)$ and an element $g(z) = \Sigma\, a_n z^n$ of $\mathcal{C}(z)$ such that

$$z^{n+1} f(z) - B(z)(a_0 z^n + \cdots + a_n)$$

belongs to $\mathcal{H}(B)$ for every nonnegative integer n and such that the sequence of numbers

$$\|z^{n+1} f(z) - B(z)(a_0 z^n + \cdots + a_n)\|^2_{\mathcal{H}(B)} + |a_0|^2 + \cdots + |a_n|^2$$

is bounded. Since the sequence is nondecreasing, it has a limit, which is taken as the definition of the square of the norm of $(f(z),g(z))$ as an element of $\mathcal{D}(B)$. The space $\mathcal{D}(B)$ is a Hilbert space which contains

$$(z f(z) - B(z) g(0), [g(z) - g(0)]/z)$$

whenever it contains $(f(z),g(z))$, and the identity

$$\|(z f(z) - B(z) g(0), [g(z) - g(0)]/z)\|^2_{\mathcal{D}(B)} = \|(f(z),g(z))\|^2_{\mathcal{D}(B)} - |g(0)|^2$$

is satisfied. The transformation which takes $(f(z),g(z))$ into $f(z)$ is a partial isometry of $\mathcal{D}(B)$ onto $\mathcal{H}(B)$.

Define $B^*(z) = \Sigma \overline{B}_n z^n$ if $B(z) = \Sigma B_n z^n$. Then multiplication by $B^*(z)$ is contractive in $\mathcal{C}(z)$ and the transformation which takes $(f(z), g(z))$ into $(g(z), f(z))$ is an isometry of the extension space $\mathcal{D}(B)$ of $\mathcal{H}(B)$ onto the extension space $\mathcal{D}(B^*)$ of $\mathcal{H}(B^*)$.

A unitary linear system is associated with the space $\mathcal{D}(B)$. The main transformation maps the space $\mathcal{D}(B)$ into itself and its action on $(f(z), g(z))$ is

$$([f(z) - f(0)]/z, \; z\, g(z) - B^*(z)\, f(0)) .$$

The input maps the complex numbers into $\mathcal{D}(B)$ and its action on a number c is

$$([B(z) - B(0)]\, c/z, \; [1 - B^*(z)\, B(0)]\, c) .$$

The output maps the space $\mathcal{D}(B)$ into the complex numbers and its action on $(f(z), g(z))$ is $f(0)$. The external operator maps complex numbers into complex numbers and coincides with multiplication by $B(0)$.

In summary there are two kinds of linear systems which naturally appear in the invariant subspace theory for contractive transformations in Hilbert space. One is the kind of linear system which is associated with a space $\mathcal{D}(B)$ and is unitary. The other is the kind of linear system which is associated with a space $\mathcal{H}(B)$ and is conjugate-isometric. These linear systems are now constructed only in the case of the complex numbers as external space, but the issues which arise are typical of the general situation in which an arbitrary coefficient Hilbert space is used.

The question now arises as to when the linear system associated with a space $\mathcal{H}(B)$ is unitary. The answer in the scalar case is that it is if, and only if, the identity for difference-quotients holds in the space and $B(z)$ is not identically zero. An equivalent condition is that the series $B(z)$ is not an element of $\mathcal{H}(B)$.

Another equivalent condition has recently been obtained by Donald Sarason [18]. It is that $B(z)$ is an extreme point of the convex set of power series which represent functions which are bounded by one in the unit disk. Related results outside the scalar theory have also been obtained independently by Ball and Kriete [1] and by Nikol'skii and Vasyunin [15]. Unitary linear systems always determine extreme points. Extreme points determine unitary linear systems under the existence of a scalar majorant for the transfer function. There do exist extreme points which do not correspond to unitary linear systems.

These results outside of the scalar theory are applications of a construction originating with David Lowdenslager [12]. Lowdenslager's fundamental contribution to factorization has been well appreciated by Rosenblum and Rovnyak [16].

Some remarks on reproducing kernel functions need to be made before proceeding with the issues at hand. The series

$$\frac{1 - B(z)\,\overline{B}(w)}{1 - z\,\overline{w}}$$

belongs to the space $\mathcal{H}(B)$ for every point w of the unit disk. The identity

$$f(w) = \langle f(z), \frac{1 - B(z)\,\overline{B}(w)}{1 - z\,\overline{w}} \rangle_{\mathcal{H}(B)}$$

holds for every element f(z) of the space. The pair

$$\left(\frac{1 - B(z)\,\overline{B}(w)}{1 - z\,\overline{w}}, \frac{B^*(z) - \overline{B}(w)}{z - \overline{w}}\right)$$

belongs to the space $\mathcal{D}(B)$ for every point w of the unit disk. The identity

$$f(w) = \langle (f(z), g(z)), \left(\frac{1 - B(z)\,\overline{B}(w)}{1 - z\,\overline{w}}, \frac{B^*(z) - \overline{B}(w)}{z - \overline{w}}\right) \rangle_{\mathcal{D}(B)}$$

holds for every element (f(z),g(z)) of the space.

Now consider the question of what the theory of linear systems has to do with the Bieberbach conjecture.

A Riemann mapping function is a power series which represents a function which has distinct values at distinct points of the unit disk. A Riemann mapping function is said to be normalized if it has constant coefficient zero and has coefficient of z positive. Most of the Riemann mapping functions which now appear will be so normalized.

If B(z) is a normalized Riemann mapping function for a subregion of the unit disk, then B(z) represents a function which is bounded by one in the disk. So the spaces $\mathcal{H}(B)$ and $\mathcal{D}(B)$ are defined. The question now is to determine what influence this choice of B(z) has on the properties of these spaces.

The answer to this question involves a remarkable generalization of the previously stated theory of square summable power series. The underlying space in the new theory is the space of power series

$$f(z) = \sum_{n=1}^{\infty} a_n z^n$$

with constant coefficient zero which have finite Dirichlet norm,

$$\|f(z)\|_{\mathcal{G}}^2 = \sum_{n=1}^{\infty} n\,|a_n|^2 \quad .$$

The notation $\mathcal{G}$ is used to suggest the names of Grunsky and Goluzin, who discovered the present use of the space.

Some elementary observations will be made by way of motivation before getting down to substance. The elements of $\mathcal{G}$ are convergent power series in the unit disk. The Dirichlet norm is computable as an area integral,

$$\|f(z)\|_{\mathcal{G}}^2 = \frac{1}{\pi} \iint_{\Delta} |f'(z)|^2 \, dxdy$$

where Δ denotes the unit disk. Such integrals are of interest because they are well-behaved under conformal mapping.

Assume that B(z) is a normalized Riemann mapping function for a subregion of the unit disk. If f(z) belongs to $\mathcal{G}$, then so does f(B(z)) and the inequality

$$\|f(B(z))\|_{\mathcal{G}}^2 \leq \|f(z)\|_{\mathcal{G}}^2$$

is satisfied. This occurs because the desired inequality can be written

$$\frac{1}{\pi} \iint_{\Delta} |f'(B(z))|^2 |B'(z)|^2 dxdy \le \frac{1}{\pi} \iint_{\Delta} |f'(z)|^2 dxdy$$

where the integral on the left is equal to

$$\frac{1}{\pi} \iint_{B(\Delta)} |f'(z)|^2 dxdy$$

by the change of variable theorem for multiple integrals. The change of variable is possible because the mapping is one-to-one and because $|B'(z)|^2$ is equal to the Wronskian determinant by the Cauchy-Riemann equations. The desired inequality holds because the image $B(\Delta)$ of the unit disk is contained in Δ by hypothesis.

Consider the range $\mathcal{M}(B)$ of $B(z)$-substitution as a Hilbert space in the unique norm such that the transformation which takes $f(z)$ into $f(B(z))$ is an isometry of $\mathcal{G}$ onto $\mathcal{M}(B)$. Since $B(z)$-substitution is contractive in the metric of $\mathcal{G}$, the space $\mathcal{M}(B)$ is contained contractively in $\mathcal{G}$. The Goluzin space $\mathcal{G}(B)$ is defined to be the complementary space to $\mathcal{M}(B)$ in $\mathcal{G}$.

Goluzin [10] arrived at the existence of such a space from entirely different considerations which will now be explained. They have to do with a computation of adjoints of substitution transformations which is due to Grunsky [11]. The outcome of Grunsky's calculation demonstrates the importance of the space $\mathcal{G}$ for studying the properties of substitution transformations.

Grunsky obtained his result as a property of Faber polynomials. But the only significant aspect of these polynomials is that they are essentially the images of the powers of z under the adjoint of $B(z)$-substitution, considered in the space $\mathcal{G}$. The definition of the Faber polynomials is complicated by the fact that they are given nontrivial constant coefficients. But that feature of the polynomials is not relevant to the computation of adjoints.

In order to formulate the Grunsky result, observe that $B(z)$-substitution in $\mathcal{G}$ has invariant subspaces. An invariant subspace is formed by the elements of $\mathcal{G}$ which have a zero of order at least $r+1$ for a given positive integer r. The orthogonal complement of the invariant subspace is an invariant subspace for the adjoint transformation. So the adjoint of $B(z)$-substitution takes polynomials into polynomials, and it does not raise the degree of a polynomial.

Since the polynomials are dense in $\mathcal{G}$, it is sufficient to compute the adjoint transformation on polynomials. Assume that $f(z)$ and $g(z)$ are given polynomial elements of $\mathcal{G}$. Then the adjoint of $B(z)$-substitution takes $f(z)$ into $g(z)$ if, and only if,

$$f\left(\frac{1}{z}\right) - g\left(\frac{1}{B^*(z)}\right)$$

is a power series.

Thus the adjoint of $B(z)$-substitution is something like $B^*(z)$-substitution, but in a different space, a space of functions in which the independent variable z has been replaced by $1/z$. Also the adjoint of $B(z)$-substitution involves truncation.

Grunsky was the first to realize that the part which is lopped off can be estimated. But Grunsky made an estimate of remainders which applies to arbitrary normalized Riemann mapping functions. Goluzin gave a more precise estimate of the remainder which applies when the Riemann mapping function is for a subregion of the unit disk.

Assume that $B(z)$ is a normalized Riemann mapping function for a subregion of the unit disk. The Grunsky transformation of the Goluzin space $\mathcal{G}(B)$ into the Goluzin space $\mathcal{G}(B^*)$ is a contractive transformation of $\mathcal{G}(B)$ into $\mathcal{G}(B^*)$ which will now be described. Assume that $f(z)$ is a polynomial element of $\mathcal{G}$ and that $g(z)$ is the element of $\mathcal{G}$ obtained from $f(z)$ under the adjoint of $B(z)$-substitution in $\mathcal{G}$. Then $f(z) - g(B(z))$ is the element of $\mathcal{G}(B)$ obtained from $f(z)$ under the adjoint of the inclusion of $\mathcal{G}(B)$ in $\mathcal{G}$. A dense set of elements of $\mathcal{G}(B)$ are of this form. The Grunsky transformation of $\mathcal{G}(B)$ into $\mathcal{G}(B^*)$ takes

$f(z) - g(B(z))$ into

$$f\left(\frac{1}{z}\right) - g\left(\frac{1}{B^*(z)}\right) + \text{constant} .$$

The stated expression is a power series by Grunsky's computation of the adjoint. The constant is determined so that the resulting power series has constant coefficient zero.

The adjoint of the Grunsky transformation of the space $\mathcal{G}(B)$ into the space $\mathcal{G}(B^*)$ is the Grunsky transformation of the space $\mathcal{G}(B^*)$ into the space $\mathcal{G}(B)$. Some remarks will be made later on the reason why these transformation are contractive. It will first be explained why they are interesting.

The best way to do this is to construct the extension space $\mathcal{J}(B)$ of the space $\mathcal{G}(B)$. The space $\mathcal{J}(B)$ is a Hilbert space whose elements are pairs $(f(z),g(z))$ of power series, $f(z)$ with constant coefficient zero. The transformation which takes $(f(z),g(z))$ into $f(z)$ is a partial isometry of $\mathcal{J}(B)$ onto $\mathcal{G}(B)$. The transformation which takes $(f(z),g(z))$ into $(g(z) - g(0),\ f(z) + g(0))$ is an isometry of $\mathcal{J}(B)$ onto $\mathcal{J}(B^*)$. If $f(z)$ belongs to $\mathcal{G}(B)$ and if $g(z)$ is the unique power series such that $(f(z),g(z))$ belongs to $\mathcal{J}(B)$ and has the same norm in $\mathcal{J}(B)$ as $f(z)$ has in $\mathcal{G}(B)$, then $g(z) - g(0)$ is the image of $f(z)$ under the Grunsky transformation of $\mathcal{G}(B)$ into $\mathcal{G}(B^*)$.

So the extension space $\mathcal{J}(B)$ of $\mathcal{G}(B)$ is a construction which specifies the graph of the Grunsky transformation of $\mathcal{G}(B)$ into $\mathcal{G}(B^*)$ and its adjoint.

The crucial properties of the spaces $\mathcal{G}(B)$ and $\mathcal{J}(B)$ are seen from a computation of reproducing kernel functions. For every point w of the unit disk, the series

$$\log \frac{1 - B(z)\,\bar{B}(w)}{1 - z\,\bar{w}}$$

belongs to $\mathcal{G}(B)$ and the identity

$$f(w) = \left\langle f(z),\ \log \frac{1 - B(z)\,\bar{B}(w)}{1 - z\,\bar{w}} \right\rangle_{\mathcal{G}(B)}$$

holds for every element $f(z)$ of $\mathcal{G}(B)$. For every point w of the unit disk, the series

$$\left(\log \frac{1 - B(z)\, \bar{B}(w)}{1 - z\bar{w}},\ \log \frac{1 - \bar{B}(w)/B^*(z)}{1 - \bar{w}/z}\right)$$

belongs to $\mathcal{J}(B)$, and the identity

$$f(w) = \left\langle (f(z), g(z)),\ \left(\log \frac{1 - B(z)\, \bar{B}(w)}{1 - z\bar{w}},\ \log \frac{1 - \bar{B}(w)/B^*(z)}{1 - \bar{w}/z}\right)\right\rangle_{\mathcal{J}(B)}$$

holds for every element $(f(z), g(z))$ of $\mathcal{J}(B)$.

The reader will notice a remarkable resemblance between these expressions and the reproducing kernel functions for the spaces $\mathcal{H}(B)$ and $\mathcal{D}(B)$. The new formulas are essentially obtained from the old ones by a passage to logarithms.

A similar phenomenon is seen in a more elementary context with the reproducing kernel functions of the spaces $\mathcal{C}(z)$ and $\mathcal{G}$. If w is a point in the unit disk, then

$$\frac{1}{1 - z\bar{w}}$$

belongs to $\mathcal{C}(z)$, and the identity

$$f(w) = \left\langle f(z), \frac{1}{1 - z\bar{w}}\right\rangle_{\mathcal{C}(z)}$$

holds for every element $f(z)$ of $\mathcal{C}(z)$. If w is a point of the unit disk, then

$$\log \frac{1}{1 - z\bar{w}}$$

belongs to $\mathcal{G}$, and the identity

$$f(w) = \left\langle f(z), \log \frac{1}{1 - z\bar{w}}\right\rangle_{\mathcal{G}}$$

holds for every element $f(z)$ of $\mathcal{G}$.

A theorem of Lebedev and Milin states that exp f(z) belongs to $\mathcal{C}(z)$ whenever f(z) belongs to $\mathcal{G}$ and that the inequality

$$\|\exp f(z)\|^2_{\mathcal{C}(z)} \le \exp \|f(z)\|^2_{\mathcal{G}}$$

is always satisfied. It should be added that the identity

$$\langle \exp f(z), \exp g(z)\rangle_{\mathcal{C}(z)} = \exp \langle f(z), g(z)\rangle_{\mathcal{G}}$$

holds for every element g(z) of $\mathcal{G}$ if f(z) is an element of $\mathcal{G}$ for which equality holds.

Lebedev and Milin obtain the inequality by elementary calculations with coefficients. These results appear in a generalized form in Milin's book [13]. But the inequality is also a consequence of the stated logarithmic relationship between reproducing kernel functions. It is an application of the theory of iterated tensor products for Hilbert spaces.

The same general considerations show that exp f(z) belongs to $\mathcal{K}(B)$ whenever f(z) belongs to $\mathcal{G}(B)$ and that the inequality

$$\|\exp f(z)\|^2_{\mathcal{K}(B)} \le \exp \|f(z)\|^2_{\mathcal{G}(B)}$$

is satisfied. The identity

$$\langle \exp f(z), \exp g(z)\rangle_{\mathcal{K}(B)} = \exp \langle f(z), g(z)\rangle_{\mathcal{G}(B)}$$

holds for every element g(z) of $\mathcal{G}(B)$ if equality holds.

For the same reasons, the pair

$$(\exp f(z),\ B^*(z)/z \exp g(z))$$

belongs to $\mathcal{D}(B)$ whenever (f(z),g(z)) belongs to $\mathcal{J}(B)$, and the inequality

$$\|(\exp f(z),\ B^*(z)/z \exp g(z))\|^2_{\mathcal{D}(B)} \le \exp \|(f(z),g(z))\|^2_{\mathcal{J}(B)}$$

is satisfied. The identity

$$\langle(\exp f(z),\ B^*(z)/z \exp g(z)),\ (\exp u(z),\ B^*(z)/z \exp v(z))\rangle_{\mathcal{D}(B)}$$

$$= \exp \langle(f(z),g(z)),\ (u(z),v(z))\rangle_{\mathcal{J}(B)}$$

holds for every element $(u(z), v(z))$ of $\mathcal{D}(B)$ if equality holds.

It is now clear that the hypothesis that $B(z)$ is a normalized Riemann mapping function does have an influence on the structure of the unitary linear system whose transfer function is $B(z)$. Some consequences of the exponential relations will be noted.

If $f(z)$ belongs to $\mathcal{G}(B)$, then $\exp f(z)$ is an element of $\mathcal{C}(z)$ such that $B(z)/z \exp f(z)$ belongs to $\mathcal{H}(B)$. Since $B(z)/z$ belongs to $\mathcal{H}(B)$, the series

$$g(z) = \frac{\exp f(z) - 1}{z}$$

is an element of $\mathcal{C}(z)$ such that $B(z)\, g(z)$ belongs to $\mathcal{H}(B)$.

Now if $\mathcal{H}(B)$ is contained isometrically in $\mathcal{C}(z)$, it is orthogonal to the range of multiplication by $B(z)$ in $\mathcal{C}(z)$. In this case it follows that $g(z)$, and hence $f(z)$, vanishes identically. Since the space $\mathcal{G}(B)$ contains no nonzero element, an inspection of reproducing kernel functions shows that $B(z) = z$.

There has been recent interest, for instance by Donald Sarason [19], in the issue of whether the identity for difference-quotients holds in a space $\mathcal{H}(B)$. The spaces which do not satisfy the identity for difference-quotients have fascinating and unexpected properties. The identity for difference-quotients can hold in a nontrivial space $\mathcal{H}(B)$ where $B(z)$ is a normalized Riemann mapping function for a subregion of the unit disk. This is always the case when the region has an arc of the unit circle as part of its boundary.

A central issue in the theory of unitary linear systems is interpolation. The problem precedes this theory and its solutions are often formulated outside of the theory. But the methods of the theory have been important ever since Nagy and Foias obtained their commutant lifting theorem [21], generalizing a theorem of Sarason [17]. Applications of the theorem to interpolation are given, for example, by Rosenblum and Rovnyak [16].

Interpolation has also been viewed outside of the theory of unitary linear systems by Beurling [3] in a generalization of the celebrated solution which Lennart Carleson gave to the corona problem.

Interpolation is viewed in still another way by Nikol'skii [14], who uses methods both from and outside of the theory of unitary linear systems.

An interpolation problem which is relevant to the present work was stated and solved by Carathéodory and Fejér in 1911. If r is a nonnegative integer, define the $\mathcal{C}_r(z)$-norm of a power series $f(z) = \Sigma\, a_n z^n$ by

$$\|f(z)\|^2_{\mathcal{C}_r(z)} = \sum_{n=0}^{r} |a_n|^2 .$$

Consider two power series f(z) and g(z) to be r-equivalent if the coefficient of z^n in f(z) is equal to the coefficient of z^n in g(z) for $n = 0,\ldots,r$. Then a power series B(z) is r-equivalent to a power series which represents a function which is bounded by one in the unit disk if, and only if, the inequality

$$\|B(z)\, f(z)\|^2_{\mathcal{C}_r(z)} \le \|f(z)\|^2_{\mathcal{C}_r(z)}$$

holds for every power series f(z).

The problem is now considered of characterizing the power series B(z) which are r-equivalent to a normalized Riemann mapping function for a subregion of the unit disk.

At this point it becomes necessary to give the promised explanation of why the Grunsky transformation of a Goluzin space $\mathcal{G}(B)$ into the Goluzin space $\mathcal{G}(B^*)$ is contractive.

Define $\mathcal{G}^\nu$ to be the set of generalized power series

$$f(z) = \sum_{n=1}^{\infty} a_n z^{\nu+n}$$

such that

$$\|f(z)\|^2_{\mathcal{G}^\nu} = \sum_{n=1}^{\infty} (\nu+n)\, |a_n|^2$$

is finite. A finite number of negative or zero coefficients will appear if $1+\nu$ is not positive. So the space obtained is in general not a Hilbert space. But the usual Hilbert space conventions have an obvious generalization to these spaces.

Assume that $B(z)$ is a normalized Riemann mapping function for a subregion of the unit disk. Since $B(z)/z$ is a power series which represents a function without zeros in the unit disk,

$$\log \frac{B(z)}{z}$$

is meaningful as a power series. The arbitrary constant in the definition of the logarithmic power series is chosen so that the constant coefficient is real. The expression

$$\left(\frac{B(z)}{z}\right)^{\nu} = \exp\left(\nu \log \frac{B(z)}{z}\right)$$

is meaningful as a power series.

If a generalized power series

$$f(z) = \sum_{n=1}^{\infty} a_n z^{\nu+n}$$

belongs to $\mathscr{G}^{\nu}$, then the expression

$$g(z) = f(B(z))$$

is meaningful as a generalized power series

$$g(z) = \sum_{n=1}^{\infty} b_n z^{\nu+n} .$$

The interpretation is that

$$\sum_{n=1}^{\infty} b_n z^n = \left(\frac{B(z)}{z}\right)^{\nu} \sum_{n=1}^{\infty} a_n B(z)^n .$$

A fundamental theorem states that $f(B(z))$ belongs to $\mathscr{G}^{\nu}$ whenever $f(z)$ belongs to $\mathscr{G}^{\nu}$ and that the inequality

$$\|f(B(z))\|^2_{\mathscr{G}^{\nu}} \leq \|f(z)\|^2_{\mathscr{G}^{\nu}}$$

is satisfied. The proof of the inequality is a generalization, due to Grunsky [11] and Goluzin [10], of the area theorem. The contractive property of the Grunsky transformation of the

Goluzin space $\mathcal{G}(B)$ into the Goluzin space $\mathcal{G}(B^*)$ is a consequence of the inequality for nonpositive integers ν.

Indeed it is a consequence of the theory of that transformation that the inequality characterizes the coefficients of a Riemann mapping function for a subregion of the unit disk. For assume that $B(z)$ is a power series with constant coefficient zero and coefficient of z nonzero. If the last inequality holds for every element $f(z)$ of $\mathcal{G}^{\nu}$ whenever ν is a nonpositive integer, then $B(z)$ is a Riemann mapping function for a subregion of the unit disk.

This observation about coefficients raises the question of interpolation within the class of Riemann mapping functions for subregions of the unit disk. Both Nevanlinna-Pick and Carathéodory-Fejér interpolation deserve consideration in this context. The present work is restricted for definiteness to Carathéodory-Fejér interpolation.

If r is a positive integer, define $\mathcal{G}_r^{\nu}$ to be the set of r-equivalence classes of generalized power series

$$f(z) = \sum_{n=1}^{\infty} a_n z^{\nu+n}$$

considered with

$$\|f(z)\|^2_{\mathcal{G}_r^{\nu}} = \sum_{n=1}^{r} (\nu+n) |a_n|^2 .$$

Then the inequality

$$\|f(B(z))\|^2_{\mathcal{G}_r^{\nu}} \le \|f(z)\|^2_{\mathcal{G}_r^{\nu}}$$

holds for every power series $f(z)$ if $B(z)$ is a power series which is r-equivalent to a normalized Riemann mapping function for a subregion of the unit disk and if $r+1+\nu \ge 0$.

It is natural to conjecture that the inequality, for nonpositive integers ν, characterizes the power series which are r-equivalent to a Riemann mapping function for a subregion of the unit disk. That is, if $B(z)$ is a power series with constant coefficient zero and coefficient of z nonzero such that the last inequality holds for every nonpositive integer ν, then

B(z) is r-equivalent to a Riemann mapping function for a subregion of the unit disk.

The conjecture seems to be new. The author first stated it at a lecture in memory of Stefan Bergman given at Stanford University on April 4, 1985. The conjecture, and the considerations on which it is based, were new to an audience containing experts on interpolation, such as Tom Kailath and Donald Sarason, and experts on Riemann mapping functions, such as Max Schiffer.

The conjecture was also presented at the conference on function-theoretic operator theory at Indiana University, November 18 and 19, 1985.

REFERENCES

1. J. A. Ball and T. L. Kriete: Operator-valued Nevanlinna-Pick kernels and the functional models for contraction operators, unpublished manuscript, 1985.
2. H. Bart, I. Gohberg, and M. A. Kaashoek: "Minimal Factorization of Matrix and Operator Functions," Birkhäuser Verlag, Basel, 1979.
3. A. Beurling: On interpolation, Blaschke products, and balayage of measures, "Proceedings of the Symposium on the Occasion of the Proof of the Bieberbach Conjecture," Mathematical Surveys, American Mathematical Society, Providence, 1986.
4. L. de Branges: A proof of the Bieberbach conjecture, Acta Math. 154:1-2 (1985), 137-152.
5. L. de Branges: Nodal Hilbert spaces of analytic functions, J. Math. Anal. Appl. 108 (1985), 447-465.
6. L. de Branges: Powers of Riemann mapping functions, "Proceedings of the Symposium on the Occasion of the Proof of the Bieberbach Conjecture," Mathematical Surveys, American Mathematical Society, Providence, 1986.

7. L. de Branges: "Square Summable Power Series," Springer-Verlag, Heidelberg, 1986.

8. M. S. Brodskii: Unitary linear systems and their characteristic functions, Uspehi Mat. Nauk 33:4 (1978), 141-168 (Russian). Russian Mathematical Surveys 33:4 (1978), 159-191.

9. P. A. Fuhrmann: "Linear Systems and Operators in Hilbert Space," Mc Graw-Hill, New York, 1981.

10. G. M. Goluzin: "Geometric Theory of Functions of a Complex Variable," second edition, Nauka, Moscow, 1966. Translations of Mathematical Monographs, Volume 26, American Mathematical Society, Providence, 1969.

11. H. Grunsky: Koeffizienten Bedingungen für schlichte abbildende meromorphe Funktionen, Math. Z. 45 (1939), 26-66.

12. D. B. Lowdenslager: On factoring matrix-valued functions, Annals of Math. 78 (1960), 450-454.

13. I. M. Milin: "Univalent Functions and Orthonormal Systems," Nauka, Moscow, 1971 (Russian). Translations of Mathematical Monographs, Volume 49, American Mathematical Society, Providence, 1977.

14. N. K. Nikol'skii: "A Treatise on Shift Operators," Nauka, Moscow, 1980 (Russian). Springer-Verlag, Heidelberg, 1985.

15. N. K. Nikol'skii and V. I. Vasyunin: Notes on two function models, "Proceedings of the Symposium on the Occasion of the Proof of the Bieberbach Conjecture," Mathematical Surveys, American Mathematical Society, Providence, 1986.

16. M. Rosenblum and J. Rovnyak: "Hardy Classes and Operator Theory," Clarendon Press, Oxford, 1986.

17. D. Sarason: Generalized interpolation in H^∞, Trans. Amer. Math. Soc. 127 (1967), 179-203.

18. D. Sarason: Invariant subspaces from the Brangesian point of view, "Proceedings of the Symposium on the Occasion of the Proof of the Bieberbach Conjecture," Mathematical Surveys, American Mathematical Society, Providence, 1986.

19. D. Sarason: Doubly shift-invariant spaces in H^2, unpublished manuscript, 1985.

20. B. Sz.-Nagy and C. Foias: "Analyse Harmonique des Opérateurs de l'Espace de Hilbert," Masson & Cie, Paris, 1967.

21. B. Sz.-Nagy and C. Foias: On the structure of intertwining operators, Acta Sci. Math. 35 (1978), 225-254.

Department of Mathematics
Purdue University,
Lafayette,
Indiana 47907,
U.S.A.

Operator Theory:
Advances and Applications, Vol. 19

CHARACTERISTIC FUNCTIONS OF UNITARY OPERATOR COLLIGATIONS IN Π_κ - SPACES

A. Dijksma, H. Langer and H.S.V. de Snoo*)

1. INTRODUCTION

By an operator colligation or colligation, for short, we shall mean a quadruple $\Delta = (\hat{\mathbf{H}},\mathbf{F},\mathbf{G};U)$, where $\hat{\mathbf{H}}$ is a Pontryagin space of index κ or π_κ - space (see [5],[11]), the inner space, $\mathbf{F}$ and $\mathbf{G}$ are Hilbert spaces, the left outer space and the right outer space, respectively, and U is a mapping from $\binom{\hat{\mathbf{H}}}{\mathbf{F}}$ into $\binom{\hat{\mathbf{H}}}{\mathbf{G}}$, orthogonal sums. In terms of these orthogonal sums, the connecting operator U has a matrix representation

$$U = \begin{pmatrix} T & F \\ G & H \end{pmatrix} : \binom{\hat{\mathbf{H}}}{\mathbf{F}} \to \binom{\hat{\mathbf{H}}}{\mathbf{G}}.$$

For such a colligation Δ we define, following M.G. Krein (who allowed the inner and outer spaces to be indefinite inner product spaces), the characteristic functies Θ_Δ by

$$\Theta_\Delta(z) = H + zG(1-zT)^{-1}F,\ z^{-1} \in \rho(T).$$

Here $\rho(T)$ denotes the resolvent set of the operator T.

In this paper we will give a systematic exposition of isometric and unitary colligations, i.e., colligations with isometric or unitary connecting operator. In some parts the results are parallel to some results in the expository paper [6] by M.S. Brodskii, who restricted himself to the case where the inner space is a Hilbert space.

The need for a study of unitary and isometric colligations in π_κ-spaces arose in connection with our attempt to generalize the main results of A.V. Štraus [21] - [25] about selfadjoint extensions of a

*) This work was supported by the Netherlands Organization for the Advancement of Pure Research (Z.W.O.).

symmetric linear operator S in a Hilbert space to the case where the selfadjoint extensions are operators or relations in some π_κ-space (and also S is allowed to be a symmetric relation instead of an operator). We mention that this extension theory gives a framework for the study of a large class of differential eigenvalue problems where the eigenvalue parameter appears in the boundary conditions. This class contains, in particular, all problems satisfying some symmetry condition and with eigenvalue parameter entering the boundary conditions polynomially. Special problems of this kind have received a lot of attention recently.

It was shown in [8] how generalizations of the main results of A.V. Štraus to the situation described above can easily be obtained from statements about the boundary behaviour of the characteristic function of a colligation which corresponds to the considered extension. So the original aim of this paper was to give the complete proofs of these statements. Now, however, also some other related results are included.

We shall briefly describe the contents of the paper; some historical remarks are given at the end of each section. In Section 2 we collect results concerning contractions in π_κ-spaces and, especially, their minimal unitary dilations. Section 3 gives a survey of isometric and unitary colligations. Such colligations $\Delta=(\hat{\mathbf{H}},\mathbf{F},\mathbf{G};T,F,G,H)$ are called closely connected if the operator T is minimal in a sense to be explained below. The notion "closely connected" is characterized in several ways. In Section 4 we present general factorizations for an isometric or unitary colligation, relative to the invariant subspaces of the operator T. Especially the existence of a κ-dimensional nonpositive invariant subspace of a contraction gives rise to an important factorization. In Section 5 we introduce characteristic functions of isometric and unitary colligations. The class of these functions coincides with the (generalized) Schur class $S_\kappa(\mathbf{F},\mathbf{G})$, consisting of all functions θ with values in $L(\mathbf{F},\mathbf{G})$ which are meromorphic in the open unit disc $\mathbb{D}$, holomorphic at z=0 and such that the kernel S_θ:

$$S_\theta(z,w) = \frac{I-\theta(w)^*\theta(z)}{1-z\bar{w}} \quad , \; z,w \in \mathbf{D}_\theta,$$

where $\mathbf{D}_\theta$ is the domain of holomorphy of θ, has κ negative squares. In Section 6 up till 9 we study functions in the Schur class $S_\kappa(\mathbf{F},\mathbf{G})$ by means of unitary colligations. Section 6 shows that functions of this class may be characterized by means of different kernels. In this way we obtain a π_κ-version of an inequality due to Ginzburg, see [6]. In Section 7 we study a factorization of functions of class S_κ in terms of an inverse Blaschke-Potapov product and a holomorphic contraction. These results are closely related to some statements of [13] and of [16]; this is explained at the end of Section 7. In Section 8 we formulate a statement about functions of class $S_\kappa(\mathbf{F},\mathbf{G})$, which for $\kappa=0$ reduces to the maximum principle for holomorphic contractions. Section 9 contains results about the behaviour on the boundary $\partial\mathbb{D}$ of functions of class $S_\kappa(\mathbf{F},\mathbf{G})$. Here we meet a condition involving a derivative in the sense of Carathéodory, see also Štraus [22], [23]. If the operator T in the matrix representation of a unitary colligation is a partial isometry, then the corresponding characteristic function θ has the property $\theta(0)=0$. Such characteristic functions are considered in Section 10; if $\mathbf{F}=\mathbf{G}$ they have been studied in [13]. Finally, we mention that some of the results of this paper have also some points in common with results of [1], [2], [4] and [9].

2. AUXILIARY STATEMENTS CONCERNING CONTRACTIONS IN Π_κ-SPACES

Let $\hat{\mathbf{H}}$ be a π_κ-space and let $T : \hat{\mathbf{H}} \to \hat{\mathbf{H}}$ be a contraction. Then according to [11, Lemma 11.3] T^* is a contraction too. By $\mathbb{D}$ we denote the open unit disc in the complex plane $\mathbb{C}$, by $\partial\mathbb{D}$ its boundary and by $\mathbb{D}^c$ its closure. The spectrum $\sigma(T)$ of T consists of points from $\mathbb{D}^c$ and of at most κ points from $\mathbb{C}\setminus\mathbb{D}^c$, which are normal eigenvalues of T. Thus with the exception of at most κ points all points of $\mathbb{C}\setminus\mathbb{D}^c$ are contained in

$\rho(T)$, the resolvent set of T. The same is true for T^*. In [11,Lemma 11.6] it is shown that the nullspaces $\nu(T-z)$ and $\nu(T^*-\bar{z})$ coincide for $z \in \partial\mathbb{D}$ and hence we have that

$$\sigma_p(T) \cap \partial\mathbb{D} = \overline{\sigma_p(T^*) \cap \partial\mathbb{D}} \;,$$

where $\sigma_p(T)$ denotes the point spectrum of T.

By $\mathbf{J}_-$ we denote a κ-dimensional nonpositive subspace of $\hat{\mathbf{H}}$ such that

(2.1) $$T\mathbf{J}_- \subset \mathbf{J}_-, \quad |\sigma(T|_{\mathbf{J}_-})| \geqq 1$$

and by $\mathbf{J}_-^*$ a κ-dimensional nonpositive subspace of $\hat{\mathbf{H}}$ such that

(2.2) $$T^*\mathbf{J}_-^* \subset \mathbf{J}_-^*, \quad |\sigma(T^*|_{\mathbf{J}_-^*})| \geqq 1.$$

For the existence of $\mathbf{J}_-$ and $\mathbf{J}_-^*$ see [11, Theorem 11.2]. Note that if $\sigma_p(T) \cap \partial\mathbb{D} = \emptyset$, then $|\sigma(T|_{\mathbf{J}_-})| > 1$ and $|\sigma(T^*|_{\mathbf{J}_-^*})| > 1$.

Lemma 2.1. <u>If a nonpositive subspace $\mathbf{L} \subset \hat{\mathbf{H}}$ is invariant under the contraction T then T maps $\mathbf{L}$ onto itself. In particular, the isotropic part $\mathbf{J}_-^o = \mathbf{J}_- \cap \mathbf{J}_-^\perp$ is mapped onto itself.</u>

Proof. The first statement follows from the fact that $\dim \mathbf{L} < \infty$ and that $\nu(T)$ is positive (see [11]). The second statement will follow if we show that $\mathbf{J}_-^o$ is invariant under T. To this end we denote $T|_{\mathbf{J}_-}$ by T_-. Then we have for arbitrary $x \in \mathbf{J}_-^o$:

$$0 = [x_1 x] \leq [T_-^{-1}x, T_-^{-1}x] \leqq 0,$$

hence T_-^{-1} maps $\mathbf{J}_-^o$ onto itself and T has the same property.

Lemma 2.2. <u>If T is a contraction in $\hat{\mathbf{H}}$ and $\mathbf{H}$ is a subspace of $\hat{\mathbf{H}}$ such that $T\mathbf{H}=\mathbf{H}$ and $T|_{\mathbf{H}}$ is isometric on $\mathbf{H}$, then $T^*\mathbf{H}=\mathbf{H}$ and $T^*|_{\mathbf{H}}$ is isometric on $\mathbf{H}$. In fact,</u>

$$(T^*|_{\mathbf{H}})(T|_{\mathbf{H}}) = (T|_{\mathbf{H}})(T^*|_{\mathbf{H}}) = I,$$

<u>the identity on $\mathbf{H}$.</u>

This is the content of [11, Lemma 11.5].

Corollary 2.3. If there is no nontrivial neutral subspace $\mathbf{J} \subset \hat{\mathbf{H}}$ such that $T\mathbf{J}=\mathbf{J}$, then $\mathbf{J}_-$ and $\mathbf{J}_-^*$ are both negative and uniquely determined by their properties (2.1) and (2.2), respectively.

Proof. From Lemma 2.2 it follows that also T^* does not have a neutral subspace $\mathbf{J}^*$ such that $T^*\mathbf{J}^*=\mathbf{J}^*$, and Lemma 2.1 implies that $\mathbf{J}_-$ and $\mathbf{J}_-^*$ have no isotropic parts. Hence they are negative and their uniqueness follows from [11, Theorem 11.3].

We recall that under the conditions of Corollary 2.3 we can write

$$\hat{\mathbf{H}}=\mathbf{J}_+ \dotplus \mathbf{J}_- = \mathbf{J}_+^* \dotplus \mathbf{J}_-^*, \text{ direct sum,}$$

where $\mathbf{J}_+ = (\mathbf{J}_-^*)^{\perp}$, $\mathbf{J}_+^* = (\mathbf{J}_-)^{\perp}$. In this decomposition all the subspaces $\mathbf{J}_\pm$, $\mathbf{J}_\pm^*$ are definite and besides (2.1) and (2.2) we have

$$T\mathbf{J}_+ \subset \mathbf{J}_+, \quad |\sigma(T|_{\mathbf{J}_+})| \leqq 1 ,$$

$$T^*\mathbf{J}_+^* \subset \mathbf{J}_+^*, \quad |\sigma(T^*|_{\mathbf{J}_+^*})| \leqq 1.$$

In Section 4 we shall make use of the decomposition

(2.3) $\quad \hat{\mathbf{H}}=\mathbf{J}_+ \oplus \mathbf{J}_-^* = \mathbf{J}_+^* \oplus \mathbf{J}_-$,orthogonal direct sums.

Let T be an arbitrary contraction in the π_κ-space $\hat{\mathbf{H}}$, and let $\mathbf{L}_0$ be a maximal neutral subspace of $\hat{\mathbf{H}}$ which is invariant under T. We choose a skewly linked subspace $\mathbf{L}_0'$ of $\mathbf{L}_0$ and decompose the space $\hat{\mathbf{H}}$:

(2.4) $\quad \hat{\mathbf{H}}=(\mathbf{L}_0 \dotplus \mathbf{L}_0') \oplus \hat{\mathbf{H}}_0.$

If we observe the invariance of $\mathbf{L}_0$ under T and Lemma 2.2 it follows that with respect to this decomposition T has the matrix representation

$$T = \begin{pmatrix} T_{11} & T_{12} & T_{13} \\ 0 & (T_{11}^*)^{-1} & 0 \\ 0 & T_{32} & T_0 \end{pmatrix}.$$

It is easy to see that T_0 and hence also T_0^* does not have any nontrivial neutral invariant subspace, hence their invariant subspaces $\mathbf{J}_\pm, \mathbf{J}^*_\pm$ are uniquely determined. In $\hat{\mathbf{H}}_o$ there can still exist a (necessarily nondegenerated) subspace $\mathbf{L}_{o1}$, on which T_o (and hence also T_o^*) is unitary: It contains, in particular, all the negative eigenspaces of T_o corresponding to eigenvalues on $\partial\mathbb{D}$. Then the subspace $\hat{\mathbf{H}}_1 = \mathbf{L}_{o1}^\perp$ of $\hat{\mathbf{H}}_o$ is again invariant under T_o and T_o^*. If we denote the restriction $T_o|\hat{\mathbf{H}}_1$ by T_1, then T_1 is "completely nonunitary" in the sense that it has no nontrivial invariant subspace where it is unitary, and $T_1^*(=T_o^*|\hat{\mathbf{H}}_1)$ has the same property. This is equivalent to the fact that the space $\hat{\mathbf{H}}_1$ can be uniquely decomposed as

$$\hat{\mathbf{H}}_1 = \mathbf{J}_{+,1} \dot{+} \mathbf{J}_{-,1} = \mathbf{J}^*_{+,1} \dot{+} \mathbf{J}^*_{-,1} \quad ,$$

where $\mathbf{J}_{+,1} = (\mathbf{J}^*_{-,1})^\perp$, $\mathbf{J}^*_{+,1} = \mathbf{J}^\perp_{-,1}$; all these subspaces are definite and we have

$$T_1\mathbf{J}_{\pm,1} \subset \mathbf{J}_{\pm,1},\ T_1^*\mathbf{J}^*_{\pm,1} \subset \mathbf{J}^*_{\pm,1},$$

$$|\sigma(T_1|\mathbf{J}_{+,1})| \leqq 1,\ |\sigma(T_1|\mathbf{J}_{-,1})| > 1,$$

$$|\sigma(T_1^*|\mathbf{J}^*_{+,1})| \leqq 1,\ |\sigma(T_1^*|\mathbf{J}^*_{-,1})| > 1.$$

It is contractions with these properties which in later sections will be associated with, closely connected unitary colligations or with functions of class $S_\kappa(\mathbf{F},\mathbf{G})$.

We mention that the (nontrivial) maximal neutral subspace $\mathbf{L}_o$ which gave rise to decomposition (2.4) is in general not uniquely determined, and the same is always true for the chosen skewly linked subspace $\mathbf{L}'_o$. We shall not study here the relations between different decompositions (2.4) and the corresponding decompositions of T.

In order to formulate the next lemma we shall first give the following definition. Let $\mathbf{H}$ be a Hilbert space, Z a nonempty set and K a mapping from a subset of $\mathbf{D}_K$ of ZxZ into $L(\mathbf{H})$. The "kernel" K is said to have κ negative squares (on $\mathbf{D}_K$), if for any positive integer n, any $z_1,..,z_n \in \mathbf{D}_K$ and $x_1,...,x_n \in \mathbf{H}$ the nxn matrix

$$((K(z_i,z_j)x_i,x_j)_{\mathbf{H}})_{i,j=1}^{n}$$

has at most κ negative eigenvalues and for some choice of $n, z_1,\dots,z_n$ and $x_1,\dots,x_n$ it has exactly κ negative eigenvalues. If it is clear from the context where the kernel is being defined, then "on $\mathbf{D}_K$" will be omitted. Let T be again a contraction in the π_κ-space $\hat{\mathbf{H}}$. Define

$$\mathbf{D}_T := \{z : z^{-1} \in \rho(T),\ z \in \mathbb{D}\} \cup \{0\},$$

$$F_T(z) := (I+zT)(I-zT)^{-1}, \quad z \in \mathbf{D}_T,$$

and

$$C_T(z,w) := (1-z\bar{w})^{-1}(F_T(z)+F_T(w)^*), \quad z,w \in \mathbf{D}_T.$$

Lemma 2.4. <u>The kernel C_T has κ negative squares.</u>

<u>Proof</u>. An easy calculation yields for $z,w \in \mathbf{D}_T$:

$$C_T(z,w) = 2(1-\bar{w}T^*)^{-1}(1-zT)^{-1} +$$

(2.5)
$$+\ 2z\bar{w}(1-z\bar{w})^{-1}(1-\bar{w}T^*)^{-1}(1-T^*T)(1-zT)^{-1}.$$

The first term on the right hand side defines a kernel with κ negative squares. The kernel

$$(z,w) \to z\bar{w}(1-z\bar{w})^{-1}, \quad z,w \in \mathbb{D},$$

is positive definite as can be seen from the Cauchy formula and, as T is a contraction, we have for arbitrary $n \in \mathbb{N}$, $z_1,\dots,z_n \in \mathbf{D}_T, x_1\dots,x_n \in \hat{\mathbf{H}}$:

$$\sum_{i,j=1}^{n} [(1-\bar{z}_jT^*)^{-1}(1-T^*T)(1-z_iT)^{-1}x_i,x_j] =$$

$$= [\sum_{i=1}^{n}(1-z_iT)^{-1}x_i, \sum_{i=1}^{n}(1-z_iT)^{-1}x_i] -$$

$$-\ [T\sum_{i=1}^{n}(1-z_iT)^{-1}x_i, T\sum_{i=1}^{n}(1-z_iT)^{-1}x_i] \geq 0.$$

Hence the second term on the right hand side of (2.5) defines a positive definite kernel, and C_T has at most κ negative squares. On the other hand, choosing $z_1=\dots=z_\kappa=0$ and $x_1,\dots,x_\kappa$ such that they span a negative subspace of $\mathbf{H}$ of dimension κ it follows that C_T has at least κ negative squares. The lemma is proved.

Proposition 2.5. Let T be a contraction in the π_κ-space $\hat{\mathbf{H}}$. Then there exists a π_κ-space $\tilde{\mathbf{H}} \supset \hat{\mathbf{H}}$ and a unitary operator $\tilde{U}$ in $\tilde{\mathbf{H}}$ such that $\tilde{U}$ is a dilation of T:

$$T^n = P\tilde{U}^n|_{\hat{\mathbf{H}}}, \quad T^{*n} = P\tilde{U}^{*n}|_{\hat{\mathbf{H}}}, \quad n \in \mathbb{N} \cup \{0\}, \tag{2.6}$$

where P denotes the orthogonal projection from $\tilde{\mathbf{H}}$ onto $\hat{\mathbf{H}}$. The space $\tilde{\mathbf{H}}$ can be chosen such that it is minimal, that is,

$$\begin{aligned} \tilde{\mathbf{H}} &= \text{c.l.s.}\{(\tilde{U}-zI)^{-1}\hat{\mathbf{H}} : z \in \rho(\tilde{U})\} = \\ &= \text{c.l.s.}\{\tilde{U}^n\hat{\mathbf{H}} : n=0,\pm1,\ldots\}, \end{aligned} \tag{2.7}$$

Then $\tilde{\mathbf{H}}$ and $\tilde{U}$ are uniquely determined up to an isomorphism.

Proof. According to lemma 2.4 and [13, Satz 2.4,b)] there exist a π_κ-space $\tilde{\mathbf{H}}$, a mapping Γ from $\hat{\mathbf{H}}$ into $\tilde{\mathbf{H}}$ and a unitary $\tilde{U}$ in $\tilde{\mathbf{H}}$ such that

$$F_T(z) = \Gamma^*(1+z\tilde{U})(1-z\tilde{U})^{-1}\Gamma, \quad z \in \mathbf{D}_T. \tag{2.8}$$

The relation $F_T(0)=I$ implies $\Gamma^*\Gamma=I$, hence Γ is an isometry from $\hat{\mathbf{H}}$ into $\tilde{\mathbf{H}}$ and $\hat{\mathbf{H}}$ can be identified with the range of Γ in $\tilde{\mathbf{H}}$. Then Γ^* is the orthogonal projection P from $\tilde{\mathbf{H}}$ onto $\hat{\mathbf{H}}$, and the minimality relations (2.7) are equivalent to the closely connectedness of $\tilde{U}$ and Γ, see [13]. Further, the definition of F_T and (2.8) imply the relation

$$I+2\sum_{j=1}^{\infty} z^j T^j = P(I+2\sum_{j=1}^{\infty} z^j \tilde{U}^j)|_{\hat{\mathbf{H}}},$$

for sufficiently small $|z|$, which yields (2.6). The proof of the uniqueness will be left to the reader.

Proposition 2.6. Let T be a contraction in a π_κ-space $\hat{\mathbf{H}}$ and let $\tilde{U}$ be its minimal unitary dilation in $\tilde{\mathbf{H}} \supset \hat{\mathbf{H}}$. Then we have

(i) if $x \in \hat{\mathbf{H}}$ and $[Tx,Tx]=[x,x]$, then $\tilde{U}x=Tx$,

(ii) in particular, for $\zeta \in \partial\mathbb{D}$, $\nu(T-\zeta) \subset \nu(\tilde{U}-\zeta)$,

(iii) if $\mathbf{L}$ is a nonpositive subspace of $\hat{\mathbf{H}}$ such that $T\mathbf{L} \subset \mathbf{L}$ and $|\sigma(T|_{\mathbf{L}})|=1$, then $\tilde{U}x=Tx$, $x \in \mathbf{L}$.

Proof. Item (i) follows from

$$[x,x]=[Tx,Tx]=[P\tilde{U}x,P\tilde{U}x] \leqq [\tilde{U}x,\tilde{U}x]=[x,x],$$

so that $Tx=P\tilde{U}x=\tilde{U}x$, while (ii) is an obvious consequence of (i). For the proof of (iii) we refer to [16, Lemma 1.3].

For $z \in \sigma_p(T)(z \in \sigma_p(\tilde{U}))$, we define $\mathbf{L}_z(\tilde{\mathbf{L}}_z)$ to be the algebraic eigenspace of T ($\tilde{U}$, respectively), with corresponding Riesz projection P_z from $\hat{\mathbf{H}}$ onto $\mathbf{L}_z$ ($\tilde{P}_z$ from $\tilde{\mathbf{H}}$ onto $\tilde{\mathbf{L}}_z$, respectively). We recall that if $z \in \sigma_p(\tilde{U})$, $|z|>1$, then $\tilde{\mathbf{L}}_z$ is a neutral subspace of $\hat{\mathbf{H}}$. We denote by $\sigma_0(T)(\sigma_0(\tilde{U}))$ the set of all $z \in \mathbb{C}, |z| \geqq 1$, such that $\nu(T-z)(\nu(\tilde{U}-z)$, respectively) contains a nonpositive, nonzero element. Note that $\sigma_0(\tilde{U})=\sigma(\tilde{U}|_{\tilde{\mathbf{J}}_-})$, where $\tilde{\mathbf{J}}_-$ is an arbitrary κ-dimensional nonpositive subspace of $\tilde{\mathbf{H}}$ invariant under $\tilde{U}$.

Proposition 2.7. Let T be a contraction in a π_κ-space $\hat{\mathbf{H}}$ and let $\tilde{U}$ be its minimal unitary dilation in $\tilde{\mathbf{H}} \supset \hat{\mathbf{H}}$ and corresponding orthogonal projection P. Then the following statements are true:

(i) If $z \in \sigma_p(T), |z|>1$, then $z \in \sigma_p(\tilde{U})$ and

$$P\tilde{\mathbf{L}}_z = \mathbf{L}_z, \quad \dim \tilde{\mathbf{L}}_z = \dim \mathbf{L}_z.$$

(ii) $\sigma_0(T) = \sigma_0(\tilde{U})$.

(iii) Let $\zeta \in \partial\mathbb{D}$. Then $\nu(T-\zeta)$ is (positive or negative) definite if and only if $\nu(\tilde{U}-\zeta)$ is (positive or negative) definite. In this case we have $P_\zeta = \tilde{P}_\zeta|_{\hat{\mathbf{H}}}$ and

$$\mathbf{L}_\zeta = \nu(T-\zeta) = \nu(\tilde{U}-\zeta) = \tilde{\mathbf{L}}_\zeta.$$

Proof. Item (i) follows from the last part of the proof of Proposition 1.2 of [16]. We repeat the arguments for the convenience of the reader. We have

$$P_z = -\frac{1}{2\pi i}\oint(T-w)^{-1}dw=$$

$$=-P\frac{1}{2\pi i}\oint(\tilde{U}-w)^{-1}dw|_{\hat{\mathbf{H}}}=P\tilde{P}_z|_{\hat{\mathbf{H}}},$$

where the contour of integration is a small circle around z. Hence

(2.9) $\dim \tilde{\mathbf{L}}_z \geqq \dim \mathbf{L}_z$.

Let $\mathbf{J}_-$ be a κ-dimensional nonpositive subspace of $\hat{\mathbf{H}}$ which is invariant under T and such that $|\sigma(T|_{\mathbf{J}_-})| \geqq 1$. Decompose $\mathbf{J}_-=\mathbf{L}+\mathbf{L}'$ with subspaces $\mathbf{L},\mathbf{L}'$ invariant under T such that

$$|\sigma(T|_{\mathbf{L}})|=1, \quad |\sigma(T|_{\mathbf{L}'})|>1.$$

According to Proposition 2.6 (iii) $\mathbf{L}$ is also invariant under $\tilde{U}$, $|\sigma(\tilde{U}|_{\mathbf{L}})|=1$. Then $\mathbf{L}$ can be extended to a κ-dimensional nonpositive invariant subspace $\tilde{\mathbf{J}}_-$ of $\tilde{U}$ (see [5, Theorem VIII.2.1] which contains

$$\tilde{\mathbf{L}}' := \text{l.s.}\{\tilde{\mathbf{L}}_z : z \in \sigma(\tilde{U}), |z|>1\}.$$

If we decompose $\tilde{\mathbf{J}}_-$ as $\tilde{\mathbf{J}}_-=\tilde{\mathbf{L}}\dot{+}\tilde{\mathbf{L}}'$, where $|\sigma(\tilde{U}|_{\tilde{\mathbf{L}}})|=1$, it follows from $\mathbf{L} \subset \tilde{\mathbf{L}}$ and (2.9) that

$$\kappa=\dim \mathbf{L}+\dim \mathbf{L}' \leqq \dim \tilde{\mathbf{L}}+\dim \tilde{\mathbf{L}}'=\kappa.$$

Hence $\mathbf{L}=\tilde{\mathbf{L}}$ and $\dim \mathbf{L}'=\dim \tilde{\mathbf{L}}'$ which yields (i).

These considerations also imply that (ii) is valid.

Now we shall prove (iii). We assume that $\nu(\tilde{U}-\zeta)$ is definite, then we have immediately $\nu(\tilde{U}-\zeta)=\tilde{\mathbf{L}}_\zeta$. By Proposition 2.6 (ii) it is clear that $\nu(T-\zeta) \subset \nu(\tilde{U}-\zeta)$. So we shall prove $\nu(\tilde{U}-\zeta) \subset \nu(T-\zeta)$. For the proof we suppose that $\zeta=1$. Further we have that if $\tilde{P}_1$ denotes the orthogonal projection onto $\tilde{\mathbf{L}}_1$, then

(2.10) $\tilde{P}_1 \hat{\mathbf{H}} = \tilde{\mathbf{L}}_1$.

Indeed, the inclusion $\tilde{P}_1\hat{\mathbf{H}} \subset \mathbf{L}_1$ is obvious. If there exists an $\tilde{y} \in \tilde{\mathbf{L}}_1$, $[\tilde{y},\tilde{P}_1\hat{\mathbf{H}}]=\{0\}$, it follows easily that $[\tilde{y},\tilde{U}^n\hat{\mathbf{H}}]=\{0\}$, $n=0,\pm1,\dots$, hence $\tilde{y}=0$, and (2.10) is proved. Using the spectral function of the unitary operator $\tilde{U}$ in $\tilde{\mathbf{H}}$ (see [17],[18]) it follows as in the Hilbert space situation, that for the eigenvalue 1 with $\nu(\tilde{U}-\tilde{I})$ nondegenerated we have

(2.11) $$\tilde{P}_1= \lim_{z \hat{\to} 1} (1-z)(\tilde{I}-z\tilde{U})^{-1}, \text{ strongly}, \ z \in \mathbb{D},$$

where by $\lim\limits_{z \hat{\to} \zeta}$ we denote the sectorial limit as $z \in \mathbb{D}$ tends to $\zeta \in \partial\mathbb{D}$.

If $\tilde{x}=\tilde{P}_1 y$ with $y \in \hat{\mathbf{H}}$ then $\tilde{x}= \lim\limits_{z \hat{\to} 1} (1-z)(\tilde{I}-z\tilde{U})^{-1}y$. Applying P we find

$$P\tilde{x} = \lim_{z \hat{\to} 1} (1-z)(I-zT)^{-1}y$$

and

$$T P\tilde{x} = \lim_{z \uparrow 1} \frac{(1-z)}{z} [-I+(I-zT)^{-1}]y = P\tilde{x}.$$

According to Proposition 2.6 (ii) this implies $\tilde{U}P\tilde{x}=P\tilde{x}$, hence $(\tilde{x}-P\tilde{x},\tilde{U}^{n}\hat{\mathbf{H}})=(\tilde{x}-P\tilde{x},\hat{\mathbf{H}})=\{0\}, n=0,\pm 1,\ldots$ and $\tilde{x}=P\tilde{x}$ follows. Thus we have proved $\tilde{P}_1\hat{\mathbf{H}} \subset \nu(T-I) \subset \hat{\mathbf{H}}$ and now it follows from (2.10) that $\nu(\tilde{U}-\tilde{I}) \subset \nu(T-I)$. So under the condition $\nu(\tilde{U}-\zeta)$ is definite, we have shown

$$\nu(T-\zeta) = \nu(\tilde{U}-\zeta) = \tilde{\mathbf{L}}_{\zeta}. \tag{2.12}$$

Now we assume that $\nu(T-\zeta)$ is definite, and we shall show that $\nu(T-\zeta)=\mathbf{L}_{\zeta}=\tilde{\mathbf{L}}_{\zeta}$ and $P_{\zeta}=\tilde{P}_{\zeta}|_{\hat{\mathbf{H}}}$, which then completes the proof of (iii). As $\nu(T-\zeta)=\nu(T^{*}-\bar{\zeta})$, both subspaces in the decomposition $\hat{\mathbf{H}}=\nu(T-\zeta) \oplus (\nu(T-\zeta))^{\perp}$ are invariant under T, $\nu(T-\zeta)=\mathbf{L}_{\zeta}$ and $\zeta \notin \sigma_p(T|_{\mathbf{L}_{\zeta}^{\perp}})$. As $T|_{\mathbf{L}_{\zeta}}$ is unitary in $\mathbf{L}_{\zeta}$, the minimal unitary dilation of T is the orthogonal sum of $T|_{\mathbf{L}_{\zeta}}$ and the minimal unitary dilation $\tilde{U}_1$ of $T|_{\mathbf{L}_{\zeta}^{\perp}}$. Now $\zeta \notin \sigma_p(T|_{\mathbf{L}_{\zeta}^{\perp}})$, and the arguments given in (2.12), and (ii) imply that $\zeta \notin \sigma_p(\tilde{U}_1)$, hence $\tilde{\mathbf{L}}_{\zeta}=\mathbf{L}_{\zeta}$. The fact that $P_{\zeta}=\tilde{P}_{\zeta}|_{\hat{\mathbf{H}}}$ follows from $\mathbf{L}_{\zeta}=\tilde{\mathbf{L}}_{\zeta}$. This completes the proof of the lemma.

Proposition 2.8. Let T be a contraction in the π_κ-space $\hat{\mathbf{H}}$, $\zeta \in \partial\mathbb{D}$ and suppose that $\nu(T-\zeta I)$ is (positive or negative) definite. Then for the orthogonal projection P_ζ onto $\nu(T-\zeta)$ we have

$$P_{\zeta} = \lim_{z \hat{\to} \bar{\zeta}} (1-\zeta z)(I-zT)^{-1}, \text{ strongly.}$$

Proof. Proposition 2.7 (ii) and (2.11) imply

$$P_{\zeta} = \lim_{z \hat{\to} \bar{\zeta}} (1-z\zeta)(\tilde{I}-z\tilde{U})^{-1}|_{\hat{\mathbf{H}}} = \lim_{z \hat{\to} \bar{\zeta}} (1-z\zeta)P(I-zT)^{-1}|_{\hat{\mathbf{H}}} = \lim_{z \hat{\to} \bar{\zeta}} (1-z\zeta)(I-zT)^{-1}.$$

Corollary 2.9. Let T be a contraction in the π_κ-space $\hat{\mathbf{H}}$. If $\zeta \in \partial\mathbb{D}$ is not an eigenvalue of T, then

(i) $\lim_{z \hat{\to} \zeta} (z-\zeta)(I-zT)^{-1}=0,$

(ii) $\lim_{z \hat{\to} \zeta}(I-zT)^{-1}\psi=(I-\zeta T)^{-1}\psi, \quad \psi \in \mathbf{R}(I-\zeta T).$

Proof. Proposition 2.8 implies (i), and (ii) follows from

$$z(I-zT)^{-1}\psi-\zeta(I-\zeta T)^{-1}\psi=(z-\zeta)(I-zT)^{-1}(I-\zeta T)^{-1}\psi$$

which tends to 0 as $z \hat{\to} \zeta$, according to (i).

The first statements of this section are easy consequences of results of [11]. They lead in a natural way to the definition of the completely nonunitary part of the contraction T which is, apparently, new. (cf. [13] for the case of an isometric operator). The existence of a minimal unitary dilation (Proposition 2.5) was first proved in a different way in [20]. The above proof shows that after the simple Lemma 2.4 it is a direct consequence of the operator representation of functions of class C_κ given in [13]. Proposition 2.6- 2.8 and Corollary 2.9, which will be used in Section 9, are, according to our knowledge, published here for the first time.

3. ISOMETRIC AND UNITARY COLLIGATIONS

Recall that the kind of colligation Δ we shall consider in this paper is a quadruple $\Delta = (\hat{\mathbf{H}},\mathbf{F},\mathbf{G};U)$ consisting of a Pontryagin space $\hat{\mathbf{H}}$, the inner space, two Hilbert spaced $\mathbf{F}$ and $\mathbf{G}$, the left outer and right outer space, respectively, and a bounded operator

$$U : \begin{pmatrix}\hat{\mathbf{H}}\\ \mathbf{F}\end{pmatrix} \to \begin{pmatrix}\hat{\mathbf{H}}\\ \mathbf{G}\end{pmatrix},$$

the connecting operator. Here, e.g., $\begin{pmatrix}\hat{\mathbf{H}}\\ \mathbf{F}\end{pmatrix}$ denotes the orthogonal sum of $\hat{\mathbf{H}}$ and $\mathbf{F}$. We shall use [,] as the notation for the scalar or inner product of all these and other spaces; it should be clear from the context to which space it refers. If $\hat{\mathbf{H}}$ is a π_κ-space we shall say that Δ has index κ. The operator U can be represented by the matrix

$$(3.1) \qquad U = \begin{pmatrix}T & F\\ G & H\end{pmatrix},$$

where $T \in L(\hat{\mathbf{H}})$, $F \in L(\mathbf{F},\hat{\mathbf{H}})$, $G \in L(\hat{\mathbf{H}},\mathbf{G})$ and $H \in L(\mathbf{F},\mathbf{G})$, where e.g., $L(\mathbf{F},\hat{\mathbf{H}})$ denotes the set of all bounded linear operators from $\mathbf{F}$ into $\hat{\mathbf{H}}$ and $L(\hat{\mathbf{H}}) = L(\hat{\mathbf{H}},\hat{\mathbf{H}})$, and we shall sometimes write $\Delta = (\hat{\mathbf{H}},\mathbf{F},\mathbf{G};\ T,F,G,H)$.

The operator T is called the basic operator. The adjoint Δ^* of the colligation Δ is the colligation $\Delta^* = (\hat{\mathbf{H}},\mathbf{G},\mathbf{F};\ T^*,G^*,F^*,H^*)$.

The colligation is called isometric if U is isometric, i.e., $U^*U = I$ or, equivalently,

$$(3.2)\qquad T^*T + G^*G = I,\ F^*F + H^*H = I,\ F^*T + H^*G = 0.$$

Here we use I and 0, to denote the identity and zero operator. It should be clear from the context in which spaces they act. If we want to indicate that they act in the space $\mathbf{F}$, say, we write $I_{\mathbf{F}}$ and $0_{\mathbf{F}}$. The adjoint Δ^* of Δ is isometric if and only if U^* is isometric, i.e., $UU^* = I$ or, equivalently,

$$(3.3)\qquad TT^* + FF^* = I,\ GG^* + HH^* = I,\ TG^* + FH^* = 0.$$

The colligation is called unitary if U is unitary, i.e., $U^*U = I$ and $UU^* = I$ or, equivalently, if (3.2) and (3.3) are valid.

If a colligation $\Delta = (\hat{\mathbf{H}},\mathbf{F},\mathbf{G};\ T,F,G,H)$ is isometric, then its basic operator $T \in L(\hat{\mathbf{H}})$ is a contraction. It follows from Proposition 2.5 that any contraction $T \in L(\hat{\mathbf{H}})$ is the basic operator of a unitary colligation. In fact, let U be a unitary dilation of T, then U has a matrix representation

$$\begin{pmatrix} T & F \\ G & H \end{pmatrix} : \begin{pmatrix} \hat{\mathbf{H}} \\ \mathbf{F} \end{pmatrix} \to \begin{pmatrix} \hat{\mathbf{H}} \\ \mathbf{F} \end{pmatrix},$$

where $\mathbf{F}$ is a Hilbert space. Note that T acts now as the basic operator of the corresponding unitary colligation $(\hat{\mathbf{H}},\mathbf{F},\mathbf{F};U)$. We remark that for any unitary colligation $(\hat{\mathbf{H}},\mathbf{F},\mathbf{F};U)$ with $U = \begin{pmatrix} T & F \\ G & H \end{pmatrix}$, the connecting operator U is a dilation of the basic operator T if and only if $FH^nG = 0$, $n \in \mathbb{N} \cup \{0\}$. In a similar way it is clear that the connecting operator U of a unitary colligation $\Delta = (\hat{\mathbf{H}},\mathbf{F},\mathbf{F};U)$ with matrix representation

$$\begin{pmatrix} T & F \\ G & H \end{pmatrix} : \begin{pmatrix} \hat{\mathbf{H}} \\ \mathbf{F} \end{pmatrix} \to \begin{pmatrix} \hat{\mathbf{H}} \\ \mathbf{F} \end{pmatrix}$$

is a dilation of the contraction $H \in L(\mathbf{F})$ if and only if $GT^nF = 0$, $n \in \mathbb{N} \cup \{0\}$.

A colligation Δ is called closely innerconnected if $\hat{\mathbf{H}}$ is the closed linear span of elements of the form T^nFf, $n = 0,1,2,\ldots$ and $f \in \mathbf{F}$, and closely outerconnected if $\hat{H}$ is the closed linear span of elements of the form $T^{*n}G^{*}g$, n = 0,1,2,... and g ∈ **G**. It is called closely connected if $\hat{H}$ is the closed linear span of all elements of the afore mentioned types. Equivalent definitions are formulated in the following propositions. Note that Δ is closely inner- or outerconnected if and only if Δ^* is closely outer- or innerconnected, respectively.

Proposition 3.1. Let $\Delta = (\hat{H},F,G;U)$ be a colligation. Then Δ is closely innerconnected (outerconnected) if and only if there exists no nonzero subspace **H** of $\hat{H}$ which is invariant under U^* (U, respectively).

Proof. Let U be represented as in (3.1). If $H \subset \hat{H}$ is a subspace invariant under U^* then $F^*H = \{0\}$ and $T^*H \subset H$, and hence

(3.4) $[T^nF\mathbf{F},\mathbf{H}] = 0, \; n = 0,1,2,...$

The closely innerconnectedness of Δ implies that **H** = {0}. Conversely, if Δ is not closely innerconnected, let **H** be the orthogonal complement of the linear span of all elements of the form T^nFf, n = 0,1,2,... and f ∈ **F**. Then **H** ≠ {0} and (3.4) holds. It readily follows that $F^*H = \{0\}$ and that T^*H is also orthogonal to this linear span, which implies that $T^*H \subset H$. Hence $U^*H \subset H$. The statement concerning the outerconnectedness can be proved in a similar way.

Note that the subspace **H** in Proposition 3.1 may be degenerated.

Proposition 3.2. Let $\Delta = (\hat{H},F,G;U)$ be a unitary colligation, where U has the matrix representation (3.1). Then the following three statements are equivalent:

(i) Δ is closely connected.

(ii) There exists no nonzero subspace **H** of $\hat{H}$ which is invariant under both U and U^*.

(iii) There exists no nonzero subspace **H** of $\hat{H}$ such that TH = **H** and $T|_H$ is isometric.

Proof. (i) ⇒ (ii): If $H \subset \hat{H}$ is invariant under U and U^* then $GH = \{0\}$, $F^*H = \{0\}$, $TH \subset H$ and $T^*H \subset H$ and so

$$(3.5)\qquad [T^n\mathbf{FF},\mathbf{H}] = 0,\ [T^{*n}G^*\mathbf{G},\mathbf{H}] = 0,\ n = 0,1,2,\ldots$$

The closely connectedness of Δ implies $\mathbf{H} = \{0\}$.
(ii) ⇒ (iii): Let $\mathbf{H}$ be a subspace of $\hat{\mathbf{H}}$ such that $T\mathbf{H} = \mathbf{H}$ and $T|_{\mathbf{H}}$ is isometric. We note that

$$(3.6)\qquad U\binom{\phi}{0} = \binom{T\phi}{G\phi},\ U^*\binom{\phi}{0} = \binom{T^*\phi}{F^*\phi},\ \phi \in \mathbf{H}.$$

As U is isometric we find that $G\phi = 0$ and so $G\mathbf{H} = \{0\}$. From the relation $F^*T + H^*G = 0$ and the fact that $T\mathbf{H} = \mathbf{H}$ we obtain $F^*\mathbf{H} = \{0\}$. Hence we have $T|_{\mathbf{H}} = U|_{\mathbf{H}}$ and $T^*|_{\mathbf{H}} = U^*|_{\mathbf{H}}$. Now (ii) implies that $\mathbf{H} = \{0\}$.
(iii) ⇒ (ii): Let $\mathbf{H}$ be a subspace of $\hat{\mathbf{H}}$ such that $U\mathbf{H} \subset \mathbf{H}$ and $U^*\mathbf{H} \subset \mathbf{H}$. Then (3.6) implies that $G\mathbf{H} = \{0\}$ and hence $T|_{\mathbf{H}} = U|_{\mathbf{H}}$. Therefore $T\mathbf{H} \subset \mathbf{H}$ and $T|_{\mathbf{H}}$ is isometric. To show $T\mathbf{H} = \mathbf{H}$, we let $\phi \in \mathbf{H}$. Then $U^*\phi \in \mathbf{H}$ and $TU^*\phi = UU^*\phi = \phi$, which shows that $T\mathbf{H} \supset \mathbf{H}$. From (iii) it follows that $\mathbf{H} = \{0\}$.
(ii) ⇒ (i): Assume that Δ is not closely connected and let $\mathbf{H}$ be the orthogonal complement in $\hat{\mathbf{H}}$ of the linear span of all elements of the form T^nFf, $T^{*m}G^*g$, $n,m = 0,1,2,\ldots,f \in \mathbf{F}$ and $g \in \mathbf{G}$. Then $\mathbf{H} \neq \{0\}$ and (3.5) is valid. It follows that $G\mathbf{H} = \{0\}$, $F^*\mathbf{H} = \{0\}$ and hence also that

$$\begin{aligned}
[\mathbf{FF},T\mathbf{H}] &= [\mathbf{F},F^*T\mathbf{H}] = -[\mathbf{F},H^*G\mathbf{H}] = 0,\\
[T^n\mathbf{FF},T\mathbf{H}] &= [T^{n-1}\mathbf{FF},T^*T\mathbf{H}] = [T^{n-1}\mathbf{FF},(I-G^*G)\mathbf{H}]\\
&= [T^{n-1}\mathbf{FF},\mathbf{H}] = 0,\ n \geqq 1,\\
[T^{*m}G^*\mathbf{G},T\mathbf{H}] &= [T^{*(m+1)}G^*\mathbf{G},\mathbf{H}] = 0,\ m \geqq 0,\\
[G^*\mathbf{G},T^*\mathbf{H}] &= [\mathbf{G},GT^*\mathbf{H}] = -[\mathbf{G},HF^*\mathbf{H}] = 0,\\
[T^{*m}G^*\mathbf{G},T^*\mathbf{H}] &= [T^{*(m-1)}G^*\mathbf{G},TT^*\mathbf{H}] = [T^{*(m-1)}G^*\mathbf{G},(I-FF^*)\mathbf{H}]\\
&= [T^{*(m-1)}G^*\mathbf{G},\mathbf{H}] = 0,\ m \geqq 1,\\
[T^n\mathbf{FF},T^*\mathbf{H}] &= [T^{n+1}\mathbf{FF},\mathbf{H}] = 0,\ n \geqq 0,
\end{aligned}$$

which shows that $T\mathbf{H} \subset \mathbf{H}$ and $T^*\mathbf{H} \subset \mathbf{H}$. Hence $U\mathbf{H} \subset \mathbf{H}$ and $U^*\mathbf{H} \subset \mathbf{H}$, which contradicts (ii). This completes the proof of the lemma.

Using Proposition 3.1 and arguments like in the proof of the implication (ii) ⇒ (iii) of Proposition 3.2, one can prove the following result.

Proposition 3.3. *Let* $\Delta = (\hat{\mathbf{H}},\mathbf{F},\mathbf{G};U)$ *be a colligation, where* U *has the matrix representation* (3.1). *If*

Δ *is isometric and closely innerconnected,*

or, if

Δ^* *is isometric and* Δ *is closely outerconnected,*

then there exists no nonzero subspace $\mathbf{H}$ *of* $\hat{\mathbf{H}}$ *such that* $T\mathbf{H} = \mathbf{H}$ *and* $T|_{\mathbf{H}}$ *is isometric.*

Corollary 3.4. *Let* $\Delta = (\hat{\mathbf{H}},\mathbf{F},\mathbf{G};U)$ *be a colligation, where* U *has the matrix representation* (3.1). *If one of the following conditions*

(i) Δ *is isometric and closely innerconnected*

(ii) Δ^* *is isometric and* Δ *is closely outerconnected*

(iii) Δ *is unitary and closely connected*

is valid, then

$$\partial\mathbb{D} \cap \sigma_p(T) = \partial\mathbb{D} \cap \overline{\sigma_p(T^*)} = \emptyset.$$

A colligation $\Delta = (\hat{\mathbf{H}},\mathbf{F},\mathbf{G};U)$ is said to be unitarily equivalent to a colligation $\Delta = (\hat{\mathbf{H}}',\mathbf{F},\mathbf{G};U')$ with the same outer spaces, if there exists a unitary operator $Z \in L(\hat{\mathbf{H}},\hat{\mathbf{H}}')$ such that

$$U' = \begin{pmatrix} Z & 0 \\ 0 & I_{\mathbf{G}} \end{pmatrix} U \begin{pmatrix} Z^{-1} & 0 \\ 0 & I_{\mathbf{F}} \end{pmatrix}.$$

Let $\Delta = (\hat{\mathbf{H}},\mathbf{F},\mathbf{G};\ T,F,G,H)$ be an isometric colligation. Then the identity $T^*T + G^*G = I$ implies that $T|_{\nu(G)}$ is an isometric operator and the identity $F^*T + H^*G = 0$ implies that T maps $\nu(G)$ into $\nu(F^*)$. If, in addition, Δ is unitary then the identity $TG^* + FH^* = 0$ implies that T maps $\nu(G)$ onto $\nu(F^*)$. For an isometric colligation $\Delta = (\hat{\mathbf{H}},\mathbf{F},\mathbf{G};\ T,F,G,H)$ we define the isometric part of the basic contraction T by the restriction $T|_{\nu(G)}$. If isometric colligations are unitarily equivalent, then not only the basic contractions but also their isometric parts are unitarily equivalent.

In this section we have followed the lines of Brodskii [6], but we have considered the more general situation where the inner space is a π_κ-space. Instead of the term closely connected, the term prime has been used in [4]. In Systems theory the terms controllable and observable are used instead of closely innerconnected and closely outerconnected, respectively.

4. A FACTORIZATION OF ISOMETRIC AND UNITARY COLLIGATIONS

We introduce the product $\Delta = \Delta_1\Delta_2$ of two colligations Δ_1 and Δ_2 with the property that the left outer space of Δ_1 coincides with the right outer space of Δ_2: If

$$(4.1)\qquad \Delta_1 = (\hat{\mathbf{H}}_1,\mathbf{H},\mathbf{G};\ U_1),\ \text{where } U_1 = \begin{pmatrix} T_1 & F_1 \\ G_1 & H_1 \end{pmatrix} : \begin{pmatrix} \hat{\mathbf{H}}_1 \\ \mathbf{H} \end{pmatrix} \to \begin{pmatrix} \hat{\mathbf{H}}_1 \\ \mathbf{G} \end{pmatrix}$$

and

$$(4.2)\qquad \Delta_2 = (\hat{\mathbf{H}}_2,\mathbf{F},\mathbf{H};\ U_2),\ \text{where } U_2 = \begin{pmatrix} T_2 & F_2 \\ G_2 & H_2 \end{pmatrix} : \begin{pmatrix} \hat{\mathbf{H}}_2 \\ \mathbf{F} \end{pmatrix} \to \begin{pmatrix} \hat{\mathbf{H}}_2 \\ \mathbf{H} \end{pmatrix},$$

then $\Delta = \Delta_1\Delta_2$ is defined as the colligation

$$(4.3)\qquad \Delta = (\hat{\mathbf{H}},\mathbf{F},\mathbf{G};\ U),$$

where

$$\hat{\mathbf{H}} = \begin{pmatrix} \hat{\mathbf{H}}_1 \\ \hat{\mathbf{H}}_2 \end{pmatrix},\quad U = (U_1 \oplus I_{\hat{\mathbf{H}}_2})(U_2 \oplus I_{\hat{\mathbf{H}}_1}),$$

that is,

$$U = \left(\begin{array}{cc|c} T_1 & 0 & F_1 \\ 0 & I & 0 \\ \hline G_1 & 0 & H_1 \end{array}\right)\left(\begin{array}{cc|c} I & 0 & 0 \\ 0 & T_2 & F_2 \\ \hline 0 & G_2 & H_2 \end{array}\right) = \left(\begin{array}{cc|c} T_1 & F_1G_2 & F_1H_2 \\ 0 & T_2 & F_2 \\ \hline G_1 & H_1G_2 & H_1H_2 \end{array}\right) : \begin{pmatrix} \hat{\mathbf{H}} \\ \mathbf{F} \end{pmatrix} \to \begin{pmatrix} \hat{\mathbf{H}} \\ \mathbf{G} \end{pmatrix}.$$

We note that $\hat{\mathbf{H}}_1$ is a nondegenerated subspace of $\hat{\mathbf{H}}$ and that $U\hat{\mathbf{H}}_1 \subset \begin{pmatrix} \hat{\mathbf{H}}_1 \\ \mathbf{G} \end{pmatrix}$. It is easy to see that $(\Delta_1\Delta_2)^* = \Delta_2^*\Delta_1^*$, and that if Δ_1 and Δ_2 are either both isometric or both unitary, then $\Delta_1\Delta_2$ is isometric or unitary, respectively. Furthermore, if $\Delta = \Delta_1\Delta_2$ is closely innerconnected (outerconnected, connected) then each of the factors Δ_1,Δ_2 is closely innerconnected (outerconnected, connected, respectively). In general, the converse is not true, see Brodskii [6].

Proposition 4.1. *Let $\Delta = (\hat{\mathbf{H}},\mathbf{F},\mathbf{G};\ T,F,G,H)$ be a colligation such that Δ is isometric (unitary). Let $\hat{\mathbf{H}}_1$ be a nondegenerated subspace of $\hat{\mathbf{H}}$ and put $\hat{\mathbf{H}}_2 = \hat{\mathbf{H}} \ominus \hat{\mathbf{H}}_1$. Then Δ admits a factorization $\Delta = \Delta_1\Delta_2$, where Δ_1,Δ_2, are isometric (unitary) colligations given by (4.1) and (4.2) for some Hilbert space $\mathbf{H}$ if and only if $T\hat{\mathbf{H}}_1 \subset \hat{\mathbf{H}}_1$. Here, Δ_1 can be constructed such that it is unitary, even when Δ is isometric.*

The proof is similar to the proof of [6, Theorem 2.6] and we omit the details.

If $\Delta = (\hat{\mathbf{H}},\mathbf{F},\mathbf{G};\ T,F,G,H)$ is a colligation, we have defined the characteristic function θ_Δ by

$$\theta_\Delta(z)=H+zG(I-zT)^{-1}F, \qquad \frac{1}{z}\in\rho(T).$$

Factorization of a colligation implies factorization of its characteristic function:

Proposition 4.2. *If $\Delta = \Delta_1\Delta_2$ where Δ_1,Δ_2 and Δ are given by (4.1), (4.2) and (4.3), then $\theta_\Delta(z) = \theta_{\Delta_1}(z)\theta_{\Delta_2}(z)$ for $z^{-1}\in\rho(T_1)\cap\rho(T_2)$.*

For a proof of Proposition 4.2 we refer to Brodskii [6].

The next result gives sufficient conditions, such that in a closely connected unitary colligation $\Delta = (\hat{\mathbf{H}},\mathbf{F},\mathbf{G};\ T,F,G,H)$ the spaces $\mathbf{F}$ and $\mathbf{G}$ may be identified in a sense to be explained below.

Proposition 4.3. *Let $\Delta = (\hat{\mathbf{H}},\mathbf{F},\mathbf{G};\ T,F,G,H)$ be a closely connected unitary colligation, such that T or, equivalenty, T^* is invertible. Then $\mathbf{R}(H)^c = \mathbf{G}$, $\mathbf{R}(H^*)^c = \mathbf{F}$.*

(i) *If $\mathbf{R}(F^*)$ is closed, we let $A = (I-F^*F)|_{\mathbf{R}(F^*)}$. Then $A\in L(\mathbf{R}(F^*))$, $A\geqq 0$, $\nu(A) = \{0\}$ or, equivalently, $\mathbf{R}(A)^c = \mathbf{R}(F^*)$ and*

$$\Delta_A = (\hat{\mathbf{H}},\mathbf{F},\mathbf{F};\ T,F,-A^{1/2}F^*(T^*)^{-1},\ I-P_{\mathbf{R}(F^*)}+A^{1/2}P_{\mathbf{R}(F^*)})$$

is a closely connected unitary colligation. There exists a unitary operator V_A from $\mathbf{F}$ onto $\mathbf{G}$ such that

$$\begin{pmatrix} T & F \\ G & H\end{pmatrix} = \begin{pmatrix} I & 0 \\ 0 & V_A\end{pmatrix}\begin{pmatrix} T & F \\ -A^{1/2}F^*(T^*)^{-1} & I-P_{\mathbf{R}(F^*)}+A^{1/2}P_{\mathbf{R}(F^*)}\end{pmatrix}.$$

Hence we also have $\theta_\Delta(z) = V_A\Theta_{\Delta_A}(z)$, $\frac{1}{z}\in\rho(T)$.

(ii) *If $\mathbf{R}(G)$ is closed, we let $B = (I-GG^*)|_{\mathbf{R}(G)}$. Then $B\in L(\mathbf{R}(G))$, $B\geqq 0$, $\nu(B) = \{0\}$ or, equivalently, $\mathbf{R}(B)^c = \mathbf{R}(G)$ and*

$$\Delta_B=(\hat{\mathbf{H}},\mathbf{G},\mathbf{G};\ T,-(T^*)^{-1}G^*B^{1/2}P_{\mathbf{R}(G)},\ G,I-P_{\mathbf{R}(G)}+B^{1/2}P_{\mathbf{R}(G)})$$

is a closely connected unitary colligation. There exists a unitary

<u>operator</u> V_B <u>from</u> **F** <u>onto</u> **G** <u>such that</u>

$$\begin{pmatrix} T & F \\ G & H \end{pmatrix} = \begin{pmatrix} T & -(T^*)^{-1}G^*B^{1/2}P_{\mathbf{R}(G)} \\ G & I - P_{\mathbf{R}(G)} + B^{1/2}P_{\mathbf{R}(G)} \end{pmatrix} \begin{pmatrix} I & 0 \\ 0 & V_B \end{pmatrix}.$$

<u>Hence we also have</u> $\theta_\Delta(z) = \theta_{\Delta_B}(z)V_B, \quad \frac{1}{z} \in \rho(T)$.

<u>Proof</u>. If T or T^* is invertible, then we find

$$F = -(T^*)^{-1}G^*H, \; G = -HF^*(T^*)^{-1}.$$

Also we have

$$\mathbf{R}(H)^c = \mathbf{G}, \; \mathbf{R}(H^*)^c = \mathbf{F}.$$

Let us show the first relation or, equivalenty, $\nu(H^*) = \{0\}$. Suppose $g \in \mathbf{G}$ is such that $H^*g = 0$. Then also $TG^*g = 0$. By assumption it follows that $G^*g = 0$, but then $g = HH^*g = 0$. The second relation is shown similarly.

Next we assume that $\mathbf{R}(G)$ is closed. Then $GG^*|_{\mathbf{R}(G)}$ is selfadjoint and we have

$$GG^*|_{\mathbf{R}(G)} \leq I_{\mathbf{R}(G)}.$$

With $B=(I-GG^*)|_{\mathbf{R}(G)}$ we obtain $B \in L(\mathbf{R}(G))$, $B \geq 0$ and

$$GG^* = GG^*P_{\mathbf{R}(G)} = (I-B)P_{\mathbf{R}(G)}.$$

Therefore we have

$$HH^* = I - P_{\mathbf{R}(G)} + BP_{\mathbf{R}(G)},$$

and, in particular,

$$(HH^*)^{1/2} = I - P_{\mathbf{R}(G)} + B^{1/2}P_{\mathbf{R}(G)}.$$

Now we consider the polar decomposition of H^* as a mapping from **G** to **F**, see [12]. There exists a unique partial isometry W from **G** to **F**, with initial set $\mathbf{R}((HH^*)^{1/2})^c = \mathbf{R}(H)^c$ and final set $\mathbf{R}(H^*)^c$, such that

$$H^* = W(HH^*)^{1/2},$$

and therefore

$$H = (I - P_{\mathbf{R}(G)} + B^{1/2}P_{\mathbf{R}(G)})W^*.$$

It is clear that $F = -(T^*)^{-1}G^*H = -(T^*)^{-1}G^*B^{\frac{1}{2}}P_{\mathbf{R}(G)}W^*$. Above we have shown that $\mathbf{R}(H)^c = \mathbf{G}$ and $\mathbf{R}(H^*)^c = \mathbf{F}$, so that the partial isometry W is in fact a unitary operator from $\mathbf{G}$ onto $\mathbf{F}$. With $V_B = W^*$ we have a unitary operator from $\mathbf{F}$ onto $\mathbf{G}$, for which the stated product decomposition in the proposition holds. But then the colligation Δ_B is automatically unitary and closely connected. Finally suppose $Bg = 0$, then $g \in \mathbf{R}(G)$ and $0 = g - GG^*g = HH^*g$, so that $H^*g = 0$. This implies $TG^*g = 0$ and by the invertibility of T we obtain $G^*g = 0$ and hence $g = 0$. This proves $\nu(B) = \{0\}$, or, equivalently, $\mathbf{R}(B)^c = \mathbf{R}(G)$. This completes the proof of (ii). The proof of (i) is completely similar and uses the polar decomposition of H.

Let $\Delta = (\hat{\mathbf{H}},\mathbf{F},\mathbf{G};U)$ be a colligation of index κ satisfying one of the conditions (i)-(iii) of Corollary 3.4 and let U have the matrix representation (3.1). Then $T,T^* \in L(\hat{\mathbf{H}})$ are contractions, $\partial\mathbb{D} \cap \sigma_p(T) = \partial\mathbb{D} \cap \sigma_p(T^*) = \emptyset$, and, on account of Propositions 3.2 and 3.3, the condition of Corollary 2.3 is satisfied. Hence $\hat{\mathbf{H}}$ can be decomposed as in (2.3), and this leads to the following results.

(i) If Δ is isometric and closely innerconnected or if Δ is unitary and closely conneted, then $\hat{\mathbf{H}} = \mathbf{J}_+ \oplus \mathbf{J}_-^*$ and U can be represented by the matrix

$$U = \begin{pmatrix} T_{11} & T_{12} & F_1 \\ 0 & T_{22} & F_2 \\ G_1 & G_2 & H \end{pmatrix} : \begin{pmatrix} \mathbf{J}_+ \\ \mathbf{J}_-^* \\ \mathbf{F} \end{pmatrix} \to \begin{pmatrix} \mathbf{J}_+ \\ \mathbf{J}_-^* \\ \mathbf{G} \end{pmatrix},$$

where $T = \begin{pmatrix} T_{11} & T_{12} \\ 0 & T_{22} \end{pmatrix}$, $F = \begin{pmatrix} F_1 \\ F_2 \end{pmatrix}$, and $G = (G_1 \; G_2)$. Note that $T_{22} = (T^*|_{\mathbf{J}_-^*})^* \in L(\mathbf{J}_-^*)$ is an invertible operator in the κ-dimensional negative subspace $\mathbf{J}_-^*$ with $|\sigma_p(T_{22})| > 1$. We apply Proposition 4.1: There exists a Hilbert space $\mathbf{H}$ such that $\Delta = \Delta_1\Delta_2$ where

$$\Delta_1(\mathbf{J}_+,\mathbf{H},\mathbf{G};U_1) \text{ is unitary,}$$

and

$$\Delta_2(\mathbf{J}_-^*,\mathbf{F},\mathbf{H};U_2) \text{ is isometric.}$$

(ii) If Δ^* is isometric and Δ is closely outerconnected, or if Δ is unitary and closely connected, then $\hat{\mathbf{H}} = \mathbf{J}_+^* \oplus \mathbf{J}_-$ and U can be represented by the matrix

$$U = \begin{pmatrix} \tilde{T}_{11} & 0 & \tilde{F}_1 \\ \tilde{T}_{21} & \tilde{T}_{22} & \tilde{F}_2 \\ \tilde{G}_1 & \tilde{G}_2 & H \end{pmatrix} : \begin{pmatrix} \mathbf{J}_+^* \\ \mathbf{J}_- \\ \mathbf{F} \end{pmatrix} \to \begin{pmatrix} \mathbf{J}_+^* \\ \mathbf{J}_- \\ \mathbf{G} \end{pmatrix},$$

where $T = \begin{pmatrix} \tilde{T}_{11} & 0 \\ \tilde{T}_{21} & \tilde{T}_{22} \end{pmatrix}$, $F = \begin{pmatrix} \tilde{F}_1 \\ \tilde{F}_2 \end{pmatrix}$ and $G = (\tilde{G}_1 \ \tilde{G}_2)$. Note that $\tilde{T}_{22} = T|_{\mathbf{J}_-} \in L(\mathbf{J}_-)$ is also an invertible operator in the κ-dimensional negative subspace $\mathbf{J}_-$ with $|\sigma_p(\tilde{T}_{22})| > 1$. Again we apply Proposition 4.1: There exists a Hilbert space $\mathbf{K}$ such that $\Delta = \Delta'_1\Delta'_2$ where

$$\Delta'_1 = (\mathbf{J}_-, \mathbf{K}, \mathbf{G}; U'_1) \text{ and } (\Delta'_1)^* \text{ is isometric,}$$

and

$$\Delta'_2 = (\mathbf{J}_+^*, \mathbf{F}, \mathbf{K}; U'_2) \text{ is unitary.}$$

The spaces $\mathbf{J}_-^*$ and $\mathbf{J}_-$ are finite dimensional, and the operators T_{22} and $\tilde{T}_{22}$ are invertible in $\mathbf{J}_-^*$ and $\mathbf{J}_-$, respectvely. If the colligation Δ is unitary, then by Proposition 4.1 the colligations Δ_2 and Δ'_1 are unitary. Hence we may apply Proposition 4.3 to obtain the following result.

Theorem 4.4. <u>Let</u> $\Delta = (\hat{\mathbf{H}}, \mathbf{F}, \mathbf{G}; U)$ <u>be a unitary closely connected colligation of index</u> κ <u>and assume that</u> U <u>is given by</u> (3.1). <u>Then</u> Δ <u>admits the factorizations</u> $\Delta = \Delta_1\Delta_2 = \Delta'_1\Delta'_2$, <u>where the factors are unitary, closely connected colligations of the form</u>

$$\Delta_1 = ((\mathbf{J}_-^*)^{\perp}, \mathbf{F}, \mathbf{G}; U_1), \quad \Delta_2 = (\mathbf{J}_-^*, \mathbf{F}, \mathbf{F}; U_2)$$

<u>and</u>

$$\Delta'_1 = (\mathbf{J}_-, \mathbf{G}, \mathbf{G}; U'_1), \quad \Delta'_2 = (\mathbf{J}_-^{\perp}, \mathbf{F}, \mathbf{G}; U'_2),$$

<u>in which</u> $\mathbf{J}_-(\mathbf{J}_-^*)$ <u>is the uniquely determined</u> κ<u>-dimensional negative subspace invariant under</u> $T(T^*$, <u>respectively)</u> <u>such that</u> $|\sigma(T|_{\mathbf{J}_-})| > 1$ $(|\sigma(T^*|_{\mathbf{J}_-^*})| > 1$, <u>respectively</u>).

In this section we have followed the theory as presented by Brodskii [6]. In general invariant subspaces of the basic operator give rise to factorizations of the characteristic function. This is in particular the case in our situation, where the basic operator is a contraction in a π_κ-space and hence it possesses a κ-dimensional invariant nonpositive (here even negative) subspace. Hence Theorem 4.4 is a direct application of the general factorization, with in each factorization one of the factors being a unitary colligation in a κ-dimensional space.

5. CHARACTERISTIC FUNCTIONS OF COLLIGATIONS

The characteristic function $\theta = \theta_\Delta$ of a colligation $\Delta = (\hat{\mathbf{H}},\mathbf{F},\mathbf{G};T,F,G,H)$ of index κ is defined by

$$\theta(z) = H+zG(I-zT)^{-1}F, \qquad z^{-1} \in \rho(T).$$

It is a locally holomorphic function with values in $L(\mathbf{F},\mathbf{G})$ and satisfies

$$(5.1) \qquad \theta(z)-\theta(w)=(z-w)G(I-wT)^{-1}(I-zT)^{-1}F, \qquad z^{-1},w^{-1} \in \rho(T).$$

If $\Delta(\Delta^*)$ is isometric, then the domain of holomorphy $\mathbf{D}_\theta$ of Θ in $\mathbb{D}$ coincides with $\mathbb{D}$ except for at most κ points and, if Δ is isometric,

$$(5.2) \qquad I-\theta(w)^*\theta(z)=(1-z\bar{w})F^*(I-\bar{w}T^*)^{-1}(I-zT)^{-1}F, \qquad z,w \in \mathbf{D}_\theta,$$

whereas, if Δ^* is isometric,

$$(5.3) \qquad I-\theta(z)\theta(w)^*=(1-z\bar{w})G(I-zT)^{-1}(I-\bar{w}T^*)^{-1}G^*, \qquad z,w \in \mathbf{D}_\theta.$$

The last equality follows immediately from the first one by making use of the relation

$$(5.4) \qquad \theta_\Delta(z)^* = \theta_{\Delta^*}(\bar{z}), \qquad z \in \mathbf{D}_\theta.$$

We denote by $S_\kappa^i(\mathbf{F},\mathbf{G})$ the set of all functions θ with the following properties:

(a) θ is defined and meromorphic on $\mathbb{D}$ with $0 \in \mathbf{D}_\theta$, the domain of holomorphy of θ in $\mathbb{D}$, and has values in $L(\mathbf{F},\mathbf{G})$, and

(b) The kernel

$$(5.5) \qquad S_\theta(z,w) = \frac{I-\theta(w)^*\theta(z)}{1-z\bar{w}}, \qquad z,w \in \mathbf{D}_\theta,$$

has κ negative squares on $\mathbf{D}_\theta$. (Note that this kernel has values in $L(\mathbf{F})$.)

Proposition 5.1. Let $\Delta = (\hat{\mathbf{H}},\mathbf{F},\mathbf{G};T,F,G,H)$ be an isometric colligation of index κ. Then its characteristic function θ belongs to class $S^i_{\kappa'}(\mathbf{F},\mathbf{G})$ for some κ' with $0 \leqq \kappa' \leqq \kappa$. If Δ is closely innerconnected then $\kappa' = \kappa$.

The proof follows easily from (5.2), as it implies that

$$[S_\theta(z,w)f,g] = [(I-zT)^{-1}Ff,\ (I-wT)^{-1}Fg],\ z,w \in \mathbf{D}_\theta,\ f,g \in \mathbf{F}.$$

The converse of Proposition 5.1 is also true. We formulate it in more general terms:

Theorem 5.2. Let $\mathbf{D} \subset \mathbb{D}$ be a set which contains 0 as an accumulation point and let $\theta\colon \mathbf{D} \to L(\mathbf{F},\mathbf{G})$ be continuous at 0 and such that S_θ defined as in (5.5) has κ negative squares on $\mathbf{D}$. Then there exists an isometric colligation $\Delta = (\hat{\mathbf{H}},\mathbf{F},\mathbf{G};U)$ of index κ such that $\theta = \theta_\Delta$ on $\mathbf{D} \cap \mathbf{D}_{\theta_\Delta}$. The colligation can be constructed so that it is closely innerconnected, in which case it is uniquely determined up to unitary equivalence.

Proof. Let $\mathbf{L}$ be the linear space of all formal finite sums

$$\sum_z \varepsilon_z x_z, \qquad x_z \in \mathbf{F}, \quad z \in \mathbf{D},$$

in which ε_z is a symbol associated with $z \in \mathbf{D}$, and provide $\mathbf{L}$ with the scalar product

$$[\sum_z \varepsilon_z x_z, \sum_w \varepsilon_w y_w] = \sum_{z,w} [S_\theta(z,w)x_z, y_w].$$

Define the linear operators

$$T_o\varepsilon_z x = \frac{1}{z}(\varepsilon_z x - \varepsilon_o x) \quad \in \mathbf{L}, \qquad z \in \mathbf{D}\backslash\{0\},\ x \in \mathbf{F},$$

$$F_o x = \varepsilon_o x \quad \in \mathbf{L}, \qquad x \in \mathbf{F},$$

$$G_o\varepsilon_z x = \frac{1}{z}(\theta(z) - \theta(0))x \in \mathbf{G}, \qquad z \in \mathbf{D}\backslash\{0\},\ x \in \mathbf{F},$$

$$Hx = \theta(0)x \quad \in \mathbf{G}, \qquad x \in \mathbf{F}.$$

The continuity of θ at 0 implies that T_o and G_o are defined on a dense linear manifold $\tilde{\mathbf{L}} \subset \mathbf{L}$. Straightforward calculations show that

$$U_o = \begin{pmatrix} T_o & F_o \\ G_o & H \end{pmatrix}$$

maps the dense linear manifold $\binom{\tilde{\mathbf{L}}}{\mathbf{F}}$ in $\binom{\mathbf{L}}{\mathbf{F}}$ isometrically in $\binom{\mathbf{L}}{\mathbf{G}}$. Let $\mathbf{L}_o$ be the isotropic part of $\mathbf{L}$, i.e., $\mathbf{L}_o = \mathbf{L} \cap \mathbf{L}^{\perp}$. Then it can be verified that U_o maps $\begin{pmatrix} \tilde{\mathbf{L}} \cap \mathbf{L}_o \\ 0 \end{pmatrix}$ in $\begin{pmatrix} \mathbf{L}_o \\ 0 \end{pmatrix}$, and hence U_o induces an isometric operator $\tilde{U}_o$ from a dense linear manifold in $\begin{pmatrix} \mathbf{L}/\mathbf{L}_o \\ \mathbf{F} \end{pmatrix}$ into $\begin{pmatrix} \mathbf{L}/\mathbf{L}_o \\ \mathbf{G} \end{pmatrix}$. The space $\mathbf{L}/\mathbf{L}_o$ can be imbedded in a canonical way in a π_κ-space $\hat{\mathbf{H}}$, and $\tilde{U}_o$ can be extended by continuity to an isometric closed operator U from all of $\binom{\hat{\mathbf{H}}}{\mathbf{F}}$ in $\binom{\hat{\mathbf{H}}}{\mathbf{G}}$. Put $\Delta = (\hat{\mathbf{H}},\mathbf{F},\mathbf{G};U)$. Then Δ is isometric. From

$$(H+zG_o(I-zT_o)^{-1}F_o)x=\theta(z)x, \qquad z \in \mathbf{D},\ x \in \mathbf{F},$$

it follows that $\theta = \theta_\Delta$ on $\mathbf{D} \cap \mathbf{D}_{\theta_\Delta}$ and from

$$(I - zT_o)^{-1}F_o x = \varepsilon_z x, \qquad z \in \mathbf{D},\ x \in \mathbf{F},$$

it follows that Δ is closely innerconnected. Write $\Delta = (\hat{\mathbf{H}},\mathbf{F},\mathbf{G};T,F,G,H)$ and let $\Delta_1 = (\hat{\mathbf{H}}_1,\mathbf{F},\mathbf{G};T_1,F_1,G_1,H_1)$ be another isometric closely innerconnected colligation such that $\theta = \theta_{\Delta_1}$ on $\mathbf{D} \cap \mathbf{D}_{\theta_{\Delta_1}}$. Then $H = H_1$,

$$G(I-zT)^{-1}F=G_1(I-zT_1)^{-1}F_1, \qquad z \in \mathbf{D} \cap \mathbf{D}_{\theta_\Delta} \cap \mathbf{D}_{\theta_{\Delta_1}},$$

and, on account of (5.2),

$$F^*(I-\bar{w}T^*)^{-1}(I-zT)^{-1}F=F_1^*(I-\bar{w}T_1^*)(I-zT_1)^{-1}F_1, \qquad z,w \in \mathbf{D} \cap \mathbf{D}_{\theta_\Delta} \cap \mathbf{D}_{\theta_{\Delta_1}}.$$

The last two equalities imply that

$$GT^nF = G_1T_1^nF_1,\ n = 0,1,2,\dots,$$
$$F^*(T^*)^nT^mF = F_1^*(T_1^*)^nT_1^mF_1,\ n,m = 0,1,2,\dots$$

From these relations and the closely connectedness of Δ and Δ_1 it readily follows that there exists a unitary map $Z \in L(\hat{\mathbf{H}}_1,\hat{\mathbf{H}})$ such that $F = ZF_1$, $G = G_1Z^{-1}$ and $T = ZT_1Z^{-1}$, cf. [6, pp. 169, 170].

We denote by $S_\kappa^o(\mathbf{F},\mathbf{G})$ the set of all functions θ with the properties (a) and

(c) The kernel

$$(5.6) \qquad S_\theta^{(*)}(z,w) = \frac{I-\theta(\bar{w})\theta(\bar{z})^*}{1-z\bar{w}}, \qquad z,w \in \mathbf{D}_\theta,$$

has κ negative squares on $\mathbf{D}_\theta$. (Note that this kernel has values in $L(\mathbf{G})$.)

From (5.4) it follows that $S^{(*)}_{\theta_\Delta} = S_{\theta_{\Delta^*}}$ and hence Proposition 5.1 and Theorem 5.2 imply the following corollaries.

Corollary 5.3. Let $\Delta = (\hat{\mathbf{H}},\mathbf{F},\mathbf{G};U)$ be a colligation of index κ such that Δ^* is isometric. Then its characteristic function θ belongs to class $S^0_{\kappa'}(\mathbf{F},\mathbf{G})$ for some κ' with $0 \leqq \kappa' \leqq \kappa$. If Δ is closely outerconnected then $\kappa' = \kappa$.

Corollary 5.4. Let $\mathbf{D} \subset \mathbb{D}$ be a set which contains 0 as an accumulation point and let θ: $\mathbf{D} \to L(\mathbf{F},\mathbf{G})$ be continuous at 0 and such that $S^{(*)}_\theta$ defined as in (5.6) has κ negative squares on $\mathbf{D}$. Then there exists a colligation $\Delta = (\hat{\mathbf{H}},\mathbf{F},\mathbf{G};U)$ of index κ such that Δ^* is isometric and $\theta = \theta_\Delta$ on $\mathbf{D} \cap \mathbf{D}_{\theta_\Delta}$. The colligation can be constructed so that it is closely outerconnected, in which case it is uniquely determined up to unitary equivalence.

We now consider the case where $\Delta = (\hat{\mathbf{H}},\mathbf{F},\mathbf{G};U)$ with

$$U = \begin{pmatrix} T & F \\ G & H \end{pmatrix}$$

is a unitary colligation. If $\mathbf{D} \subset \mathbb{C}$ is a set then we denote by $\mathbf{D}^*$ the set $\mathbf{D}^* = \{\frac{1}{\bar{z}} \mid z \in \mathbf{D} \setminus \{0\}\}$. The characteristic function $\theta = \theta_\Delta$ of Δ defined on $\mathbf{D}_\theta$ can be extended to a map $\tilde{\theta}$ defined on $\mathbf{D}_\theta \cup \mathbf{D}^*_\theta$

$$(5.7) \qquad \tilde{\theta}(z) = \begin{cases} \theta(z), & z \in \mathbf{D}_\theta, \\ \theta(\frac{1}{\bar{z}})^*, & z \in \mathbf{D}^*_\theta. \end{cases}$$

We note that $\tilde{\theta}(z) \in L(\mathbf{F},\mathbf{G})$, $z \in \mathbf{D}_\theta$ and $\tilde{\theta}(z) \in L(\mathbf{G},\mathbf{F})$, $z \in \mathbf{D}^*_\theta$. The kernel S_θ given by (5.5) can also be extended:

$$(5.8) \qquad \tilde{S}_\theta(z,w) = \begin{cases} \dfrac{I - \tilde{\theta}(w)^*\tilde{\theta}(z)}{1 - z\bar{w}}, & z,w \in \mathbf{D}_\theta, \\ -\bar{w}\,\dfrac{\tilde{\theta}(z) - \tilde{\theta}(w)^*}{1 - z\bar{w}}, & z \in \mathbf{D}_\theta, w \in \mathbf{D}^*_\theta, \\ z\,\dfrac{\tilde{\theta}(z) - \tilde{\theta}(w)^*}{1 - z\bar{w}}, & z \in \mathbf{D}^*_\theta, w \in \mathbf{D}_\theta, \\ -z\bar{w}\,\dfrac{I - \tilde{\theta}(w)^*\tilde{\theta}(z)}{1 - z\bar{w}}, & z,w \in \mathbf{D}^*_\theta. \end{cases}$$

Using (5.1), (5.2) and (5.3) we obtain that

$$\tilde{S}_\Theta(z,w) = \begin{cases} F^*(I-\bar{w}T^*)^{-1}(I-zT)^{-1}F & \in L(\mathbf{F}), \quad z,w \in \mathbf{D}_\theta, \\ \bar{w}G(\bar{w}-T)^{-1}(I-zT)^{-1}F & \in L(\mathbf{F},\mathbf{G}), \quad z \in \mathbf{D}_\theta, w \in \mathbf{D}_\theta^*, \\ zF^*(I-\bar{w}T^*)^{-1}(z-T^*)^{-1}G^* & \in L(\mathbf{G},\mathbf{F}), \quad z \in \mathbf{D}_\theta^*, w \in \mathbf{D}_\theta, \\ z\bar{w}G(\bar{w}-T)^{-1}(z-T^*)^{-1}G^* & \in L(\mathbf{G}), \quad z,w \in \mathbf{D}_\theta^*. \end{cases} \tag{5.9}$$

We denote by $S_\kappa(\mathbf{F},\mathbf{G})$ the set of all functions θ with the following properties: (a) and

(d) The kernel $\tilde{S}_\theta$ defined by (5.8) where the extension $\tilde{\theta}$ of θ is defined by (5.7) has κ negative squares on $\mathbf{D}_\theta \cup \mathbf{D}_\theta^*$.

The analog of Proposition 5.1 reads as follows:

Proposition 5.5. Let $\Delta = (\hat{\mathbf{H}},\mathbf{F},\mathbf{G};T,F,G,H)$ be a unitary colligation of index κ. Then its characteristic function θ belongs to class $S_{\kappa'}(\mathbf{F},\mathbf{G})$ for some κ' with $0 \leqq \kappa' \leqq \kappa$. If Δ is closely connected, then $\kappa' = \kappa$.

The proof follows immediately from (5.9). The converse of Proposition 5.5 is contained in the following result, which is the analog of Theorem 5.2.

Theorem 5.6. Let $\mathbf{D} \subset \mathbb{D}$ be a set which contains 0 as an accumulation point and let θ: $\mathbf{D} \to L(\mathbf{F},\mathbf{G})$ be continuous at 0 and such that $\tilde{S}_\theta$ defined as in (5.8) where $\tilde{\theta}$ is defined on $\mathbf{D} \cup \mathbf{D}^*$ as in (5.7), has κ negative squares on $\mathbf{D} \cup \mathbf{D}^*$. Then there exists a unitary colligation $\Delta = (\hat{\mathbf{H}},\mathbf{F},\mathbf{G};U)$ of index κ such that $\theta = \theta_\Delta$ on $\mathbf{D} \cap \mathbf{D}_{\theta_\Delta}$. The colligation can be constructed so that it is closely connected, in which case it is uniquely determined up to unitary equivalence.

Proof. Let $\mathbf{L}$ be the linear space of all formal finite sums

$$\sum_z \varepsilon_z x_z, \qquad x_z \in \mathbf{F} \text{ if } z \in \mathbf{D},\ x_z \in \mathbf{G} \text{ if } z \in \mathbf{D}^*,$$

in which ε_z is a symbol associated with $z \in \mathbf{D} \cup \mathbf{D}^*$, and provide $\mathbf{L}$ with the scalar product

$$[\sum_z \varepsilon_z x_z, \sum_w \varepsilon_w y_w] = \sum_{z,w} [\tilde{S}_\theta(z,w)x_z, y_w].$$

Define the linear operators

$$T_o\varepsilon_z x = \begin{cases} \frac{1}{z}(\varepsilon_z x - \varepsilon_o x) & , z \in \mathbf{D},\ x \in \mathbf{F}, \\ \frac{1}{z}\varepsilon_z x - \varepsilon_o\tilde{\theta}(z)x, & z \in \mathbf{D}^*,\ x \in \mathbf{G}, \end{cases}$$

$$F_o x = \varepsilon_o x \qquad , x \in \mathbf{F},$$

$$G_o\varepsilon_z x = \begin{cases} \frac{1}{z}(\tilde{\theta}(z) - \tilde{\theta}(0))x, & z \in \mathbf{D},\ x \in \mathbf{F}, \\ (I - \tilde{\theta}(0)\tilde{\theta}(z)x & , z \in \mathbf{D}^*,\ x \in \mathbf{G}, \end{cases}$$

$$Hx = \theta(0)x \qquad , x \in \mathbf{F},$$

and let

$$U_o = \begin{pmatrix} T_o & F_o \\ G_o & H \end{pmatrix}.$$

Then the continuity of θ at 0 implies that U_o is an operator with dense domain and range in $\binom{\mathbf{L}}{\mathbf{F}}$ and $\binom{\mathbf{L}}{\mathbf{G}}$, respectively. Moreover, straightforward calculations show that U_o is isometric. The remaining arguments to complete the proof are similar to the ones in the proof of Theorem 5.2, where the operator U now turns out to be unitary, and we therefore omit further details.

Remark. Let θ be given as in Theorem 5.6. Then we know, using Theorem 5.2 that the limit

$$\lim_{\substack{z \to 0 \\ z \in \mathbf{D}}} \frac{\theta(z) - \theta(0)}{z}$$

exists. In the proof of Theorem 5.6 we may enlarge **L** be adding the element $\varepsilon_\infty y$, $y \in \mathbf{G}$, and extend the definition of the scalar product by

$$[\varepsilon_o x, \varepsilon_\infty y] = \lim_{z\to 0} \left[\frac{\theta(z) - \theta(0)}{z} x, y\right], \qquad x \in \mathbf{F},\ y \in \mathbf{G},$$

$$[\varepsilon_z x, \varepsilon_\infty y] = \lim_{w\to 0} [\tilde{S}_\theta(z,w)x,y],\ y \in \mathbf{G},\ x \in \mathbf{F} \text{ if } z \in \mathbf{D}, \text{ and } x \in \mathbf{G} \text{ if } z \in \mathbf{D}^*, \text{ respectively,}$$

$$[c_\infty x, c_\infty y] = [(I\ \ \theta(0)\theta(0)^*)x,y], \qquad x,y \in \mathbf{G}.$$

It is then possible to give a description of the formal "adjoints" of T_o, G_o, F_o and H:

$$T_o^* \varepsilon_z x = \begin{cases} z\varepsilon_z x - \varepsilon_\infty \theta(z)x, & z \in \mathbf{D},\ x \in \mathbf{F}, \\ z(\varepsilon_z x - \varepsilon_\infty x), & z \in \mathbf{D}^*,\ x \in \mathbf{G}, \end{cases}$$

$$G_o^* x = \varepsilon_\infty x, \qquad x \in \mathbf{G},$$

$$F_o^* \varepsilon_z x = \begin{cases} (I-\theta(0)^*\theta(z))x, & z \in \mathbf{D},\ x \in \mathbf{F}, \\ z(\tilde{\theta}(z)-\theta(0)^*)x, & z \in \mathbf{D}^*,\ x \in \mathbf{G}, \end{cases}$$

$$H^* x = \theta(0)^* x, \qquad x \in \mathbf{G}$$

If $\Delta = (\hat{\mathbf{H}},\mathbf{F},\mathbf{G};T,F,G,H)$ is an isometric colligation then (5.2) implies that $\theta(\frac{1}{\bar{z}})^*\theta(z) = I$ for $z \in \mathbb{C}$ with $\bar{z},\frac{1}{z} \in \rho(T)$. If Δ is unitary, then on account of (5.3) we also have that $\theta(z)\theta(\frac{1}{\bar{z}})^* = I$ for $z \in \mathbb{C}$ with $\bar{z},\frac{1}{z} \in \rho(T)$. This implies the following result.

Proposition 5.7. Let $\Delta = (\hat{\mathbf{H}},\mathbf{F},\mathbf{G};T,F,G,H)$ be a unitary colligation such that $\rho(T) \cap \rho(T)^* \neq \emptyset$. Then

$$\theta(z)^{-1} = \theta(\tfrac{1}{\bar{z}})^*, \qquad z \in \rho(T)^* \cap \rho(T).$$

Corollary 5.8. Let $\Delta = (\hat{\mathbf{H}},\mathbf{F},\mathbf{G};T,F,G,H)$ be a unitary colligation such that $\rho(T) \cap \partial\mathbb{D} \neq \emptyset$. Then $\theta(z)$ is a unitary operator for all $z \in \rho(T) \cap \partial\mathbb{D}$. In particular, if the π_κ-space $\hat{\mathbf{H}}$ is finite dimensional, then $\rho(T) \cap \rho(T)^* \neq \emptyset$, and if, in addition, the unitary colligation Δ is closely connected, then $\partial\mathbb{D} \subset \rho(T)$, so that $\theta(z)$ is unitary for all $z \in \partial\mathbb{D}$.

We note that the constructions in the previous proofs imply the following result.

Proposition 5.9. Let $\Delta = (\hat{\mathbf{H}},\mathbf{F},\mathbf{G};T,F,G,H)$ be a closely connected unitary colligation. Then $\dim \hat{\mathbf{H}}$ is equal to the sum of the number of negative squares and the number of positive squares of the kernel $\tilde{S}_\theta$.

Some of the theorems in this section are similar to results, previously obtained by Krein and Langer [13], [15], [16]. However, here they are obtained in a direct way, without use of the Cayley transform. Even in the case $\kappa = 0$ our arguments are straightforward without making use of the Naimark extension results, as was done for instance by Brodskii [6].

In special cases simlar results were obtained by Ball and Helton [3]. Related work was done by Arov and Grossmann [1] and Filimonov [9], while a more general situation was studied by Azizov [2]. We note that for $\kappa = 0$ the class $S^i_\kappa(\mathbf{F},\mathbf{G})$ coincides with the class of all holomorphic contractions from **F** to **G**. For a proof of this we refer to [13] or [26]. In Systems theory results such as Theorem 5.2 and Theorem 5.6 are sometimes called realization theorems.

6. DIFFERENT KERNELS CHARACTERIZING THE SCHUR CLASS $S_\kappa(\mathbf{F},\mathbf{G})$

For functions θ satisfying condition (a) of the preceding section, i.e., defined and meromorhic on $\mathbb{D}$ with $0 \in \mathbf{D}_\theta$ and with values in $L(\mathbf{F},\mathbf{G})$, we have considered three kernels

$$S_\theta,\ S_\theta^{(*)} \text{and } \tilde{S}_\theta$$

(see (5.5), (5.6) and (5.8), respectively). We define a fourth kernel S_θ^m by the matrix

$$S_\theta^m(z,w) = \begin{pmatrix} S_\theta(z,w) & \dfrac{\theta(\bar{z})^* - \theta(w)^*}{z-\bar{w}} \\ \dfrac{\theta(z)-\theta(\bar{w})}{z-\bar{w}} & S_\theta^{(*)}(z,w) \end{pmatrix}, \quad z,w \in \mathbf{D}_\theta.$$

We note that this kernel has values in $L(\binom{\mathbf{F}}{\mathbf{G}})$. If $\Delta = (\hat{\mathbf{H}},\mathbf{F},\mathbf{G};T,F,G,H)$ is unitary, then it follows from (5.1), (5.2) and (5.3) that

$$(6.1)\quad [S^m_{\theta_\Delta}(z,w)\begin{pmatrix} f_1 \\ g_1 \end{pmatrix},\begin{pmatrix} f_2 \\ g_2 \end{pmatrix}] = [(I-zT)^{-1}Ff_1+(I-zT^*)^{-1}G^*g_1,(I-wT)^{-1}Ff_2+(I-wT^*)^{-1}G^*g_2],$$
$$z,w \in \mathbf{D}_{\theta_\Delta},\ f_1,f_2 \in \mathbf{F},\ g_1,g_2 \in \mathbf{G}.$$

Theorem 6.1. Let θ be a function which is defined and meromorphic on $\mathbb{D}$, holomorphic at 0 and has values in $L(\mathbf{F},\mathbf{G})$. If one of the kernels S_θ, $S_\theta^{(*)}$, $\tilde{S}_\theta$ and S_θ^m has κ negative squares, then each of the other three kernels also has κ negative squares.

We first state and prove a lemma.

Lemma 6.2. Let $\Delta = (\hat{\mathbf{H}},\mathbf{F},\mathbf{G};V)$ be an isometric closely innerconnected colligation of index κ. Then there exists a unitary closely connected colligation $\Delta_1 = (\hat{\mathbf{K}},\mathbf{F},\mathbf{G};U)$ also of index κ such that $\hat{\mathbf{H}} \subset \hat{\mathbf{K}}$, $V \subset U$ and $\theta_\Delta = \theta_{\Delta_1}$.

Proof. As the domain $\mathbf{D}(V)$ of V is all of $\binom{\hat{\mathbf{H}}}{\mathbf{F}}$, the range $\mathbf{R}(V)$ of V is a π_κ-space and hence $\mathbf{R}(V)^\perp$ is a Hilbert space. Let $\mathbf{K}_o$ be a Hilbert space and let $V_o: \mathbf{D}(V_o) \subset \mathbf{K}_o \to \mathbf{K}_o$ be an isometry such that $\dim(\mathbf{K}_o \ominus \mathbf{D}(V_o)) = \dim\mathbf{R}(V)^\perp$ and $\mathbf{R}(V_o) = \mathbf{K}_o$. (E.g., $\mathbf{K}_o$ is the Hilbert space of all sequences (f_j), $f_j \in \mathbf{R}(V)^\perp$, $j = 1,2,\ldots$, with $\sum \|f_j\|^2 < \infty$, provided with the inner product $[(f_j),(g_j)] = \sum[f_j,g_j]$ and $V_o(0,f_1,f_2,\ldots) = (f_1,f_2,\ldots)$.) The isometry

$$\tilde{V} = V_o \oplus V: \mathbf{D}(\tilde{V}) \subset \begin{pmatrix} \hat{\mathbf{K}}_o \\ \mathbf{F} \end{pmatrix} \to \begin{pmatrix} \hat{\mathbf{K}}_o \\ \mathbf{G} \end{pmatrix},$$

where $\hat{\mathbf{K}}_o = \begin{pmatrix} \mathbf{K}_o \\ \hat{\mathbf{H}} \end{pmatrix}$, has equal defect indices $(\dim\mathbf{D}(\tilde{V})^\perp = \dim\mathbf{R}(\tilde{V})^\perp = \dim\mathbf{R}(V)^\perp)$ and can be extended to a linear operator U_o:

$$U_o h = \begin{cases} \tilde{V}h, & h \in \mathbf{D}(\tilde{V}), \\ h, & h \in \mathbf{D}(\tilde{V})^\perp. \end{cases}$$

Write $V = \binom{T\ F}{G\ H}$, then, as $V \subset U_o$, U_o has a matrix representation of the form

$$U_o = \begin{pmatrix} A_o & 0 & 0 \\ B_o & T & F \\ C_o & G & H \end{pmatrix} : \begin{pmatrix} \hat{\mathbf{K}}_o \\ \mathbf{F} \end{pmatrix} \to \begin{pmatrix} \hat{\mathbf{K}}_o \\ \mathbf{G} \end{pmatrix}.$$

Let $\hat{\mathbf{K}}$ be the closure in $\hat{\mathbf{K}}_o$ of the linear space of all elements of the form

$$\begin{pmatrix} A_o & 0 \\ B_o & T \end{pmatrix}^n \begin{pmatrix} 0 \\ F \end{pmatrix} f = \begin{pmatrix} 0 \\ T^n Ff \end{pmatrix}, \begin{pmatrix} A_o^* & B_o^* \\ 0 & T \end{pmatrix}^m \begin{pmatrix} C_o^* \\ G^* \end{pmatrix} g,\ n,m = 0,1,2,\ldots,\ f \in \mathbf{F},\ g \in \mathbf{G}.$$

Then, as Δ is closely innerconnected, $\hat{\mathbf{H}} \subset \hat{\mathbf{K}}$, so that $\hat{\mathbf{K}}$ is a π_κ-space. The Hilbert space $\mathbf{H}_o = \hat{\mathbf{K}}_o \ominus \hat{\mathbf{K}}$ is invariant under U_o and U_o^* (see the proof of Proposition 3.2). Hence $\hat{\mathbf{K}}$ is invariant under U_o and U_o^*. Put $U = U_o|_{\binom{\hat{\mathbf{K}}}{\mathbf{F}}}$ and $\mathbf{H} = \hat{\mathbf{K}} \ominus \hat{\mathbf{H}}$. Then U has a matrix representation of the form

$$U = \begin{pmatrix} A & 0 & 0 \\ B & T & F \\ C & G & H \end{pmatrix} : \begin{pmatrix} \mathbf{H} \\ \hat{\mathbf{H}} \\ \mathbf{F} \end{pmatrix} \to \begin{pmatrix} \mathbf{H} \\ \hat{\mathbf{H}} \\ \mathbf{G} \end{pmatrix}$$

and one easily checks that $\Delta_1 = (\hat{\mathbf{K}},\mathbf{F},\mathbf{G};U)$ has the desired properties.

Proof of Theorem 6.1. If S stands for one of the four kernels, we denote by n(S) the number of negative squares. From the definitions it is clear that

(i) $n(S_\theta) \leqq n(\tilde{S}_\theta)$, (ii) $n(S_\theta^{(*)}) \leqq n(\tilde{S}_\theta)$,

(iii) $n(S_\theta) \leqq n(S_\theta^m)$, (iv) $n(S_\theta^{(*)}) \leqq n(S_\theta^m)$.

If $n(\tilde{S}_\theta) = \kappa$, then also $n(S_\theta^m) = \kappa$, on account of Theorem 5.6 and (6.1). This and (i)-(iv) imply that we only have to prove that if $n(S_\theta) = \kappa$ or if $n(S_\theta^{(*)}) = \kappa$ then $n(\tilde{S}_\theta) = \kappa$. Assume $n(S_\theta) = \kappa$. Then Theorem 5.2 and Lemma 6.2 imply that $\theta = \theta_\Delta = \theta_{\Delta_1}$ where Δ and Δ_1 are as in Lemma 6.2, and now Proposition 5.5 implies that $n(\tilde{S}_\theta) = \kappa$. To prove the other implication, assume that $n(S_\theta^{(*)}) = \kappa$. Then Corollary 5.4 implies that $\theta = \theta_\Delta$ where $\Delta = (\hat{\mathbf{H}},\mathbf{F},\mathbf{G};V)$ is of index κ and Δ^* is isometric and closely innerconnected. On account of Lemma 6.2 there exists a unitary closely connected colligation $\Delta_1 = (\hat{\mathbf{K}},\mathbf{F},\mathbf{G};U)$ of index κ such that $\theta_{\Delta^*} = \theta_{\Delta_1}$. It is easy to see that $\theta = \theta_{\Delta_1^*}$. Again Proposition 5.5 implies that $n(\tilde{S}_\theta) = \kappa$. This completes the proof.

Corollary 6.3. We have $S_\kappa(\mathbf{F},\mathbf{G}) = S_\kappa^i(\mathbf{F},\mathbf{G}) = S_\kappa^o(\mathbf{F},\mathbf{G})$.

It follows from Theorem 6.1 that if a function θ, which is meromorphic on $\mathbb{D}$, holomorphic at 0, with values in $L(\mathbf{F},\mathbf{G})$ is such that the kernel S_θ has κ negative squares then also the matrix kernel S_θ^m has κ negative squares. This fact may be viewed as a generalization of Ginzburg's inequality: for $\theta \in S_0(\mathbf{F},\mathbf{G})$, $f \in \mathbf{F}, g \in \mathbf{G}$, and $z,w \in \mathbb{D}$ we have

$$\left|\left[\frac{\theta(z)-\theta(\bar{w})}{z-\bar{w}} f,g\right]\right|^2 \leqq \left[\frac{I-\theta(z)^*\theta(z)}{1-|z|^2} f,f\right]\left[\frac{I-\theta(\bar{w})\theta(\bar{w})^*}{1-|w|^2} g,g\right].$$

In [13] Krein and Langer gave a generalization of Ginzburg's inequality in anther direction, stated in the next proposition. We shall present a simple proof, based on Schwarz' inequality.

Proposition 6.4. Let $\theta \in S_0(\mathbf{F},\mathbf{G})$, $n \in \mathbb{N}$ and let $z_1,\dots,z_n,w_1,\dots,w_n \in \mathbb{D}, f_1,\dots,f_n \in \mathbf{F}$ and $g_1,\dots,g_n \in \mathbf{G}$ be arbitrary. Then

$$\left|\sum_{i,j=1}^{n}\left[\frac{\theta(z_i)-\theta(\bar{w}_j)}{z_i-\bar{w}_j}f_i,g_j\right]\right|^2 \leq$$

$$\leq\left(\sum_{i,j=1}^{n}\left[\frac{I-\theta(z_j)^*\theta(z_i)}{1-z_i\bar{z}_j}f_i,f_j\right]\right)\left(\sum_{i,j=1}^{n}\left[\frac{I-\theta(\bar{w}_j)\theta(\bar{w}_i)^*}{1-w_i\bar{w}_j}g_i,g_j\right]\right).$$

Proof. We choose a unitary colligation Δ, such that $\theta = \theta_\Delta$. Then we obtain

$$\frac{\theta(z_i)-\theta(\bar{w}_j)}{z_i-\bar{w}_j} = G(I-\bar{w}_jT)^{-1}(I-z_iT)^{-1}F.$$

From this it follows that

$$\left|\sum_{i,j=1}^{n}\left[\frac{\theta(z_i)-\theta(\bar{w}_j)}{z_i-\bar{w}_j}f_i,g_j\right]\right|^2 = \left|\left[\sum_{i=1}^{n}(I-z_iT)^{-1}Ff_i,\ \sum_{j=1}^{n}(I-w_jT^*)^{-1}G^*g_j\right]\right|^2$$

$$\leq \left[\sum_{i=1}^{n}(I-z_iT)^{-1}Ff_i,\ \sum_{j=1}^{n}(I-z_jT)^{-1}Ff_j\right]\left[\sum_{i=1}^{n}(I-w_iT^*)^{-1}G^*g_i,\ \sum_{j=1}^{n}(I-w_jT^*)^{-1}G^*g_j\right]$$

$$= \left(\sum_{i,j=1}^{n}[F^*(I-\bar{z}_jT^*)^{-1}(I-z_iT)^{-1}Ff_i,f_j]\right)\left(\sum_{i,j=1}^{n}[G(I-\bar{w}_jT)^{-1}(I-w_iT^*)^{-1}G^*g_i,g_j]\right)$$

$$= \left(\sum_{i,j=1}^{n}\left[\frac{I-\theta(z_j)^*\theta(z_i)}{1-z_i\bar{z}_j}f_i,f_j\right]\right)\left(\sum_{i,j=1}^{n}\left[\frac{I-\theta(\bar{w}_j)\theta(\bar{w}_i)^*}{1-w_i\bar{w}_j}g_i,g_j\right]\right),$$

which completes the proof.

The introduction of the matrix kernel S_θ^m is apparently new. We obtain Ginzburg's inequality as a result of the representation of θ as a characteristic function of a unitary colligation. In the exposition of Brodskii [6] Ginzburg's inequality is an essential step to show that θ is a characteristic function.

7. STRONGLY REGULAR FACTORIZATION OF FUNCTIONS IN $S_\kappa(\mathbf{F},\mathbf{G})$.

Let $\mathbf{F},\mathbf{G}$ be Hilbert spaces and let A be a function with values in $L(\mathbf{F},\mathbf{G})$ which is defined and holomorphic in some deleted neighbourhood $U\backslash\{\gamma\}$ of the point $\gamma \in \mathbb{C}$ and has a pole at γ. A pole function of A at γ is a holomorphic function u with values in $\mathbf{G}$, such that $u(\gamma) \neq 0$ and $u(z) = A(z)g(z)$ with some function g with values in $\mathbf{F}$ which is holomorphic at γ. Now the pole multiplicity of A at γ, denoted by $\pi_A(\gamma)$, is defined in the usual way (see [10], [14] for the case $\mathbf{F} = \mathbf{G}$).

In the following we are only interested in poles of finite pole multiplicity. Recall that if the Laurent expansion of A at γ is given:

$$(7.1) \qquad A(z) = A_n(z-\gamma)^{-n}+A_{n-1}(z-\gamma)^{n+1}+\ldots+A_1(z-\gamma)^{-1}+A_o(z)$$

with A_o holomorphic at γ, $A_n \neq 0$, then the pole multiplicity of A at γ equals the rank of the (n+n)-matrix

$$N: = \begin{pmatrix} A_n & 0 & . & . & . & 0 \\ A_{n-1} & A_n & . & . & . & 0 \\ . & . & & & & . \\ . & . & & & & . \\ . & . & & & & . \\ A_2 & A_2 & . & . & . & 0 \\ A_1 & A_2 & . & . & . & A_n \end{pmatrix},$$

(see [14]). If, in particular, $\mathbf{G}=\mathbf{F}$ and $A(z):=(T-z)^{-1}$ for some linear operator T and γ is an isolated normal eigenvalue of T of finite algebraic mulitplicity ν then γ is pole of A with $\pi_A(\gamma) = \nu$. In this case, the order n of the pole γ of A in (7.1) equals the maximal length of the Jordan chaines of T corresponding to the eigenvalue γ.

Using the definition of a pole function, it can easily be seen that γ is a pole of A if and only if γ^{-1} is a pole of the function $B:B(z):=A(\frac{1}{z})$, and that $\pi_A(\gamma) = \pi_B(\gamma^{-1})$ if only one of these pole multiplicities is finite.

A special triangularization of an operator T with eigenvalue α of finite algebraic multiplicity is essential in the following.

Lemma 7.1. Let T be a contraction in the π_κ-space $\hat{\mathbf{H}}$ and let $\alpha, |\alpha| > 1$, be an eigenvalue of T of order m such that the corresponding algebraic eigenspace is negative. Then $\hat{\mathbf{H}}$ can be decomposed in a unique way as

(7.3) $$\hat{\mathbf{H}} = \hat{\mathbf{H}}_1 \oplus \hat{\mathbf{H}}_2 \oplus \ldots \oplus \hat{\mathbf{H}}_m \oplus \hat{\mathbf{K}},$$

such that with respect to this decomposition the operator T has the triangular matrix representation

$$T = \begin{pmatrix} \alpha I_1 & T_{12} & \cdots & T_{1m} & T_{1,m+1} \\ 0 & \alpha I_2 & \cdots & T_{2m} & T_{2,m+1} \\ \vdots & \vdots & & \vdots & \vdots \\ 0 & 0 & & \alpha I_m & T_{m,m+1} \\ 0 & 0 & & 0 & T_o \end{pmatrix}$$

with the properties

(7.5) $$\mathbf{R}(T_{j,j+1}) = \hat{\mathbf{H}}_j, \quad j=1,2,\ldots,m-1,$$

(7.6) $$\alpha \notin \sigma(T_o).$$

Evidently, as the algebraic eigenspace of T at α is negative by hypothesis, the spaces $\hat{\mathbf{H}}_j$, $j=1,2,\ldots,m$ are negative; the space $\hat{\mathbf{K}}$ is a Pontryagin space of index $\kappa - \sum_{j=1}^{m} \dim \hat{\mathbf{H}}_j$. Moreover, property (7.5) implies that

$$\dim \hat{\mathbf{H}}_j \leqq \dim \hat{\mathbf{H}}_{j+1}, \quad j=1,\ldots,m-1.$$

Proof of Lemma 7.1. We consider the Laurent expansion of $(T-z)^{-1}$ at α:

(7.7) $$(T-z)^{-1} = (\alpha-z)^{-m}E_m+(\alpha-z)^{-m+1}E_{m-1}+\ldots+(\alpha-z)^{-1}E_1+S(z)$$

and put $\hat{\mathbf{H}}_1 := \mathbf{R}(E_m)$. Then the nonzero elements of $\hat{\mathbf{H}}_1$ are eigenelements of T, hence $\hat{\mathbf{H}}_1$ is a negative subspace of $\hat{\mathbf{H}}$ and we can write $\hat{\mathbf{H}} = \hat{\mathbf{H}}_1 \oplus \hat{\mathbf{H}}'_2$. With respect to this decomposition T has the representation

$$T = \begin{pmatrix} \alpha I_1 & T'_{12} \\ 0 & T'_2 \end{pmatrix}.$$

The space $\hat{\mathbf{H}}'_2$ is a Pontryagin space of index κ-dim $\hat{\mathbf{H}}_1$, and T'_2 is a contraction in $\hat{\mathbf{H}}'_2$. In order to see this we observe that with T also T^* is a contraction, this yields immediately that $(T'_2)^*$ is a contraction. Hence also T'_2 has this property. If we denote the orthogonal projection onto $\hat{\mathbf{H}}'_2$ by Q'_2 we have

$$Q'_2 T = Q'_2 T Q'_2.$$

Now we suppose $m > 1$, then the resolvent of T'_2 admits near α the Laurent expansion

$$(T'_2 - z)^{-1} = \alpha - z)^{m-1} Q'_2 E_{m-1} Q'_2 + \ldots + (\alpha - z)^{-1} Q'_2 E_1 Q'_2 + Q'_2 S(z) Q'_2.$$

In this case we put $\hat{\mathbf{H}}_2 := \mathbf{R}(Q'_2 E_{m-1} Q'_2)$. The nonzero elements of $\hat{\mathbf{H}}_2$ are eigenelements of T'_2 corresponding to the eigenvalue α. Thus, with respect to the decomposition $\hat{\mathbf{H}} = \hat{\mathbf{H}}_1 \oplus \hat{\mathbf{H}}_2 \oplus \hat{\mathbf{H}}'_3$ the operator T has the representation

$$T = \begin{pmatrix} \alpha I_1 & T_{12} & T'_{13} \\ 0 & \alpha I_2 & T'_{23} \\ 0 & 0 & T'_3 \end{pmatrix}.$$

Continuing this procedure we find the decomposition (7.3) and the corresponding representation (7.4) of T. Starting with this representation of T, it can easily be checked that E_m in (7.7) admits a matrix representation with

$$0 \quad 0 \ldots 0 \quad (-1)^{m-1} T_{12} T_{23} \cdots T_{m-1,m} \quad (-1)^m T_{12} T_{23} \cdots T_{m-1,m} T_{m,m+1} (T_0 - \alpha)^{-1}$$

as the first row and zeros elsewhere. Hence we have

$$\hat{\mathbf{H}}_1 = \mathbf{R}(E_m) \subset \mathbf{R}(T_{12}) \subset \hat{\mathbf{H}}_1,$$

and the relation (7.5) for j=1 follows. In the same way we find that the first row of the matrix of $Q'_2 E_{m-1} Q'_2$ in the decomposition $\hat{\mathbf{H}}_2 \oplus \ldots \oplus \hat{\mathbf{H}}_m \oplus \mathbf{K}$ equals

$$0 \quad 0 \ldots 0 \quad (-1)^m T_{23} T_{34} \cdots T_{m-1,m} \quad (-1)^{m+1} T_{23} T_{34} \cdots T_{m-1,m} T_{m,m+1} (T_0 - \alpha)^{-1}$$

and we find $\hat{\mathbf{H}}_2 = \mathbf{R}(T_{23})$, etc. Finally, as the resolvent of T_0 is given by

$$(T_0 - z)^{-1} = Q'_{m+1} S(z) Q'_{m+1},$$

where Q'_{m+1} is the orthogonal projection onto $\mathbf{K}$, and this function is holomorphic at α, the relation (7.6), $\alpha \notin \sigma(T_0)$, follows.

Conversely, let a decomposition

$$\hat{\mathbf{H}} = \hat{\mathbf{H}}_1^o \oplus \hat{\mathbf{H}}_2^o \oplus \ldots + \hat{\mathbf{H}}_{m^o}^o \oplus \hat{\mathbf{K}}_o^o$$

be given such that T is of the form (7.4) with T_{jk}, T_o replaced by T_{jk}^o, T_o^o and with the properties corresponding to (7.5), (7.6). Then in the Laurent expansion of $(T-z)^{-1}$ at α the first term is $(\alpha-z)^{-m^o} E_{m^o}^o$ with $E_{m^o}^o$ of the same matrix structure as E_m. Hence $m^o = m$ and we find

$$\hat{\mathbf{H}}_1^o = \mathbf{R}(T_{12}^o) = \mathbf{R}(E_m^o) = \mathbf{R}(E_m) = \hat{\mathbf{H}}_1 .$$

In the same way the equality of the other corresponding components in (7.3) follows. The lemma is proved.

We mention that the Jordan chain structure of the eigenspace of T at α follows immediately from the decomposition (7.4) (with the properties (7.5), (7.6)). Namely, it has dim $\hat{\mathbf{H}}_1$ chains of lenght m, dim $\hat{\mathbf{H}}_2$-dim $\hat{\mathbf{H}}_1$ additional chains of length m-1, etc.

Lemma 7.2. Let $\Delta = (\hat{\mathbf{H}}, \mathbf{F}, \mathbf{G}; U)$ be a unitary colligation such that

$$U = \begin{pmatrix} T_1 & T_{12} & F' \\ 0 & T_2 & F'' \\ G' & G'' & H \end{pmatrix}.$$

with respect to a decomposition $\hat{\mathbf{H}} = \mathbf{H}' \oplus \mathbf{H}''$ of the inner space, in terms of non degenerated subspaces. We assume that

(i) $T_1^{-1} \in L(\mathbf{H}')$,

(ii) $I - G'G'^* \geq c > 0$ as an operator in $L(\mathbf{G})$.

Let $\tilde{H}$ be an operator in $L(\mathbf{G})$ with $G'G'^* + \tilde{H}\tilde{H}^* = I$, $\nu(\tilde{H}) = \{0\}$, and let $\tilde{F} = -(T_1^*)^{-1} G'^* \tilde{H}$. Then $\tilde{H}^{-1} \in L(\mathbf{G})$ and $\tilde{F} \in L(\mathbf{G}, \mathbf{H}')$.

We define

$$U' = \begin{pmatrix} T_1 & \tilde{F} \\ G' & \tilde{H} \end{pmatrix}, \quad U'' = \begin{pmatrix} T_2 & F'' \\ \tilde{H}^*(G'' - G'T_1^{-1}T_{12}) & \tilde{H}^*(H - G'T_1^{-1}F') \end{pmatrix}.$$

Then

$$U'' = \begin{pmatrix} T_2 & F'' \\ \tilde{H}^{-1}G'' & \tilde{H}^{-1}H \end{pmatrix}$$

and $\Delta' = (\mathbf{H}', \mathbf{G}, \mathbf{G}; U')$, $\Delta'' = (\mathbf{H}'', \mathbf{F}, \mathbf{G}; U'')$ are unitary colligations such that $\Delta = \Delta'\Delta''$.

Proof. Note that operators $\tilde{H} \in L(\mathbf{G})$ with $G'G'^* + \tilde{H}\tilde{H}^* = I$, $\nu(\tilde{H}) = \{0\}$ exist. For instance we may take $\tilde{H} = (I - G'G'^*)^{1/2}$. Since $\tilde{H}\tilde{H}^* \geq c$, we observe that $\nu(\tilde{H}^*) = \{0\}$ and $\mathbf{R}(\tilde{H}^*)$ is closed. Hence we see that $\mathbf{R}(\tilde{H})$ is closed, so that $\mathbf{R}(\tilde{H}) = \mathbf{G}$. The condition $\nu(\tilde{H}) = \{0\}$ now shows that $\tilde{H}$ is an invertible operator. Similarly, $\tilde{H}^*$ is an invertible operator. Also, it is clear that $\tilde{F} \in L(\mathbf{G}, \mathbf{H'})$. Next we show that U' is unitary. It is clear that $T_1^* T_1 + G'^* G' = I$, $G'G'^* + \tilde{H}\tilde{H}^* = I$, and

$$\tilde{F}^* T_1 + \tilde{H}^* G' = -\tilde{H}^* G' T_1^{-1} T_1 + \tilde{H}^* G' = 0.$$

Note that

$$G'T_1^* + \tilde{H}\tilde{F}^* = (G'T_1^* T_1 - \tilde{H}\tilde{H}^* G')T_1^{-1} = (G'(I - G'^* G') - (I - G'G'^*)G')T_1^{-1} = 0$$

and that

$$\begin{aligned} \tilde{F}^*\tilde{F} + \tilde{H}^*\tilde{H} &= \tilde{H}^*(G'T_1^{-1}T_1^{*-1}G'^* + I)\tilde{H} \\ &= \tilde{H}^*(G'(I - G'^*G')^{-1}G'^* + I)\tilde{H} \\ &= \tilde{H}^*(I - G'G'^*)^{-1}\tilde{H} = I, \end{aligned}$$

where we have used the identity

$$G'(I - G'^*G')^{-1}G'^* + I = (I - G'G'^*)^{-1}.$$

Finally we see that

$$\begin{aligned} T_1 T_1^* + \tilde{F}\tilde{F}^* &= (T_1^*)^{-1}((T_1^* T_1)^2 + G'^*\tilde{H}\tilde{H}^* G')T_1^{-1} \\ &= (T_1^*)^{-1}((I - G'^*G')^2 + G'^*(I - G'G'^*)G')T_1^{-1} \\ &= (T_1^*)^{-1}(I - G'^*G')T_1^{-1} = I. \end{aligned}$$

This shows that $\Delta' = (\mathbf{H'}, \mathbf{G}, \mathbf{G}; U')$ is a unitary colligation.
Next we observe that

$$\begin{pmatrix} T_1^* & 0 & G'^* \\ 0 & I & 0 \\ \tilde{F}^* & 0 & \tilde{H}^* \end{pmatrix} \begin{pmatrix} T_1 & T_{12} & F' \\ 0 & T_{12} & F'' \\ G' & G'' & H \end{pmatrix} = \begin{pmatrix} I & 0 & 0 \\ 0 & T_2 & F'' \\ 0 & \tilde{H}^*(G'' - G'T_1^{-1}T_{12}) & \tilde{H}^*(H - G'T_1^{-1}F') \end{pmatrix}.$$

This implies that U'' is a unitary operator, and thus $\Delta'' = (\mathbf{H''}, \mathbf{F}, \mathbf{G}; U'')$ is a unitary colligation. In addition we observe that $\Delta = \Delta'\Delta''$. The above identity is equivalent to

$$U = (U' \oplus I_{\mathbf{H''}})(U'' \oplus I_{\mathbf{H'}}),$$

which implies that we also have

$$U''=\begin{pmatrix} T_2 & F'' \\ \tilde{H}^{-1}G'' & \tilde{H}^{-1}H \end{pmatrix}.$$

Remark. If $\mathbf{R}(G')$ is closed, we may choose $\tilde{H}=I-P_{\mathbf{R}(G')}+AP_{\mathbf{R}(G')}$ where $A\in L(\mathbf{R}(G'))$ is an invertible operator such that

$$AA^*=(I-G'G'^*)|_{\mathbf{R}(G')}.$$

In this case we may write

$$U'=\begin{pmatrix} T_1 & -(T_1^*)^{-1}G'^*AP_{\mathbf{R}(G')} \\ G' & I-P_{\mathbf{R}(G')}+AP_{\mathbf{R}(G')} \end{pmatrix}.$$

Corollary 7.3. Let $\mathbf{H}'$ be a negative (and hence finite dimensional) space and let T_1 be multiplication by $\alpha\in\mathbb{C}$ with $|\alpha|>1$. Then we may take

$$U'=\begin{pmatrix} \alpha I & \frac{\alpha}{\bar{\alpha}}G'^* \\ G' & I-P_{\mathbf{R}(G')}-\alpha P_{\mathbf{R}(G')} \end{pmatrix}$$

and for the characteristic function $\theta_{\Delta'}$ we have with $\gamma=\frac{1}{\alpha}$

$$\theta_{\Delta'}(z)=I-P_{\mathbf{R}(G')}+\frac{1-z\bar{\gamma}}{z-\gamma}P_{\mathbf{R}(G')},\qquad z\neq\gamma.$$

Proof. It is clear that $\mathbf{R}(G')$ is closed and that $G'^*G'=(1-|\alpha|^2)I_{\mathbf{H}'}$, but then we also have

$$G'G'^*=(1-|\alpha|^2)P_{\mathbf{R}(G')}.$$

This shows that $I-G'G'^*=I-P_{\mathbf{R}(G')}+|\alpha|^2P_{\mathbf{R}(G')}\geq I$ and by the remark we may choose $A=-\alpha I|_{\mathbf{R}(G')}$. It is clear that

$$G'^*AP_{\mathbf{R}(G')}=-\alpha G'^*P_{\mathbf{R}(G')}=-\frac{\alpha}{1-|\alpha|^2}G'^*G'G'^*=-\alpha G'^*,$$

which completes the proof.

Note that the inverse of $\theta_{\Delta'}(z)$, $z\neq\bar{\gamma}^{-1}$, exists in $L(\mathbf{G})$ and is given by

$$I-P_{\mathbf{R}(G')}+\frac{z-\gamma}{1-z\bar{\gamma}}P_{\mathbf{R}(G')}.$$

This last function is called a Blaschke-Potapov factor. It is evident that for $z\in\partial\mathbb{D}$ such a factor is unitary.

Lemma 7.4. Let T be a contraction in $\mathbf{H}$ and let $\Delta=(\hat{\mathbf{H}},\mathbf{F},\mathbf{G};T,F,G,H)$ be a unitary colligation. If $Tx=\alpha x$ for some $\alpha\in\mathbb{C}\setminus\partial\mathbb{D}$, $[x,x]\neq 0$, then $Gx\neq 0$. Similarly, if $T^*y=\bar{\alpha}y$ for some $\alpha\in\mathbb{C}\setminus\partial\mathbb{D}$, $[y,y]\neq 0$, then $F^*y\neq 0$.

Proof. Assume $Gx=0$. Then the relation (3.2) implies $T^*Tx=x$, hence $[Tx,Tx]=[x,x]$, which is impossible as $[Tx,Tx]=|\alpha|^2[x,x]\neq[x,x]$. The proof for T^* is similar.

Corollary 7.5. Any definite (geometric) eigenspace of $T(T^*)$ corresponding to an eigenvalue not belonging to $\partial\mathbb{D}$ is of dimension $\leqq$ dim $\mathbf{G}$ (dim $\mathbf{F}$, respectively).

Proposition 7.6. Let $\theta\in S_\kappa(\mathbf{F},\mathbf{G})$ and let $\Delta=\Delta(\hat{\mathbf{H}},\mathbf{F},\mathbf{G};T,F,G,H)$ be a closely connected unitary colligation with $\theta=\theta_\Delta$. Then $\gamma\in\mathbb{D}$ is a pole of θ of pole multiplicity ν if and only if $\alpha=\gamma^{-1}$ is an eigenvalue of the basic operator T of algebraic multiplicity ν.

Proof. According to the remarks at the beginning of this section the function θ has a pole of pole multiplicity ν at γ if and only if the function Q_1: $Q_1(z):=H-G(T-z)^{-1}F$ has a pole of multiplicity ν at α. Hence it remains to show that the matrices

$$A:=\begin{pmatrix} E_m & 0 & \dots & 0 & 0\\ E_{m-1} & E_m & \dots & 0 & 0\\ \cdot & \cdot & & \cdot & \cdot\\ \cdot & \cdot & & \cdot & \cdot\\ \cdot & \cdot & & \cdot & \cdot\\ E_2 & E_3 & \dots & E_m & 0\\ E_1 & E_2 & \dots & E_{m-1} & E_m \end{pmatrix} \quad\text{and}\quad A_1:=\begin{pmatrix} GE_mF & 0 & \dots & 0 & 0\\ EE_{m-1}F & GE_mF & \dots & 0 & 0\\ \cdot & \cdot & & \cdot & \cdot\\ \cdot & \cdot & & \cdot & \cdot\\ \cdot & \cdot & & \cdot & \cdot\\ GE_2F & GE_3F & \dots & GE_mF & 0\\ GE_1F & GE_2F & \dots & GE_{m-1} & GE_mF \end{pmatrix}$$

have the same rank, where E_j are given according to (7.7). In order to do this we observe that because of the closely connectedness of the colligation Δ the operator T has a unique κ-dimensional negative invariant subspace $\mathbf{J}_{-1}$ and it has the property $|\sigma(T|_{\mathbf{J}_-})|>1$; in the same way, T^* has a unique κ-dimensional negative invariant subspace $\mathbf{J}_-^*$ and $|\sigma(T^*|_{\mathbf{J}_-^*})|>1$. It is easy to see that the operators $T_2',\dots,T_0$ in the proof of Lemma 7.1 have the same properties with κ changed correspondingly.

In order to prove that A and A_1 have the same ranks, it is sufficient to assume that $\alpha \in \sigma_p(T)$, $|\alpha| > 1$ i.e., that T has an expansion of the form (7.7) around α. For, if $\gamma \in \mathbf{D}$ is a pole of θ, then $\alpha = \frac{1}{\gamma} \in \sigma_p(T)$. Consider the decomposition (7.3) of $\hat{\mathbf{H}}$ and the corresponding representation for U:

$$(7.9) \qquad U = \begin{pmatrix} \alpha I_1 & T_{12} & \cdots & T_{1m} & T_{1m+1} & F_1 \\ 0 & \alpha I_2 & \cdots & T_{2m} & T_{2m+1} & F_2 \\ \cdot & \cdot & & \cdot & \cdot & \cdot \\ \cdot & \cdot & & \cdot & \cdot & \cdot \\ \cdot & \cdot & & \cdot & \cdot & \cdot \\ 0 & 0 & & \alpha I_m & T_{m\,m+1} & F_m \\ 0 & 0 & \cdots & 0 & T_o & F_{m+1} \\ G_1 & G_2 & \cdots & G_m & G_{m+1} & H \end{pmatrix}$$

In what follows we use the operators used in the proof of Lemma 7.1 and denote by Q_j the projections of $\hat{\mathbf{H}}$ onto $\hat{\mathbf{H}}_j$, $j=1,\ldots,m$.
The operator E_m is of the form $\sum_{k=1}^{\dim \hat{\mathbf{H}}_1} [.,f_k]e_k$ with eigenvectors e_k, f_k of T and T^*, respectively, corresponding to the eigenvalue α and $\bar{\alpha}$. Hence

$$GE_mF = \sum_{k=1}^{\dim \hat{\mathbf{H}}_1} [.,F^*f_k]Ge_k,$$

and Lemma 7.4 implies

$$\text{rank } GE_mF = \dim \hat{\mathbf{H}}_1.$$

Moreover,

$$GE_{m-1}F = G(Q_1+Q_2')E_{m-1}F = G_1Q_1E_{m-1}F + GQ_2'E_{m-1}F.$$

The range of $G_1Q_1E_{m-1}F$ is contained in the range of GE_mF, and hence in the range of the element in the right lower corner of the matrix A_1. Therefore

$$\text{rank } A_1 \geqq \dim \hat{\mathbf{H}}_1 + \text{rank } GQ_2'E_{m-1}Q_2'F.$$

If we represent $Q_2'E_{m-1}Q_2'$ by a matrix with respect to the decomposition $\hat{\mathbf{H}}_2 \oplus \ldots \oplus \hat{\mathbf{H}}_m \oplus \mathbf{K}$, the first row is given by (7.8) and it has zeros elsewhere. This yields the relation

$$(7.10) \qquad \mathbf{R}(GQ_2'E_{m-1}Q_2'F) = \mathbf{R}(G_2Q_2E_{m-1}Q_mF_m).$$

On the other hand, using Lemma 7.2 we find

$$
(7.11)\quad \begin{pmatrix}
\alpha I_1 & T_{12} & \dots & T_{1m} & T_{1m+1} & F_1 \\
0 & \alpha I_2 & \dots & T_{2m} & T_{2m+1} & F_2 \\
\cdot & \cdot & & \cdot & \cdot & \cdot \\
\cdot & \cdot & & \cdot & \cdot & \cdot \\
\cdot & \cdot & & \cdot & \cdot & \cdot \\
0 & 0 & \dots & \alpha I_m & T_{mm+1} & F_m \\
0 & 0 & \dots & 0 & T_o & F_{m+1} \\
G_1 & G_2 & \dots & G_m & G_{m+1} & H
\end{pmatrix} =
$$

$$
\begin{pmatrix}
\alpha I_1 & 0 & \dots & 0 & F'_1 \\
0 & I & \dots & 0 & 0 \\
\cdot & \cdot & & \cdot & \cdot \\
\cdot & \cdot & & \cdot & \cdot \\
\cdot & \cdot & & \cdot & \cdot \\
0 & 0 & \dots & I & 0 \\
G_1 & 0 & \dots & 0 & H_1
\end{pmatrix}
\begin{pmatrix}
I_1 & 0 & \dots & 0 \\
0 & \alpha I_2 & \dots & F_2 \\
\cdot & \cdot & & \cdot \\
\cdot & \cdot & & \cdot \\
\cdot & \cdot & & \cdot \\
0 & 0 & & F_m \\
0 & H_1^{-1}G_2 & \dots & H_1^{-1}H
\end{pmatrix},
$$

where the matrices on the right hand side define closely connected colligations. As $Q'_2E_{m-1}Q'_2$ can be expressed by eigenvectors of T'_2 and $(T'_2)^*$, Lemma 7.4 yields

$$\text{rank } H_1^{-1}G_2Q_2E_{m-1}Q_mF_m=\dim \hat{\mathbf{H}}_2,$$

which, together with (7.10), gives

$$\text{rank } GQ'_2E_{m-1}Q'_2F=\dim \hat{\mathbf{H}}_2,$$

hence

$$\text{rank } A_1 \geqq \dim \hat{\mathbf{H}}_1+\dim \hat{\mathbf{H}}_2.$$

Observe, that we obtained this estimation by considering only the last two columns of A_1. Continuing this procedure we find

$$\text{rank } A_1 \geqq \dim \hat{\mathbf{H}}_1+\dim \hat{\mathbf{H}}_2+\dots+\dim \hat{\mathbf{H}}_m= \text{rank } A.$$

On the other hand, the inequality rank $A_1 \leqq$ rank A is evidently true, and Proposition 7.6 is proved.

We mention that from this proof it follows that all the operators $G_1, G_2, \dots, G_m$ in (7.9) are of full rank.

We have to introduce some more notations. If mutually different points $\gamma_1, \dots, \gamma_\ell \in \mathbf{D}\setminus\{0\}$ are given and with each such point γ_j there are associated orthogonal projections $P_{jk}, k=1,\dots,m_j$ in $\mathbf{G}$, we denote by B_{jk} the Blaschke-Potapov factor

$$B_{jk}(z)=I-P_{jk}+\frac{z-\gamma_j}{1-z\bar{\gamma}_j}P_{jk}.$$

Further, we put

$$B_j(z):=\overleftarrow{\prod_{k=1}^{m_j}} B_{jk}(z),\ j=1,2,\dots,\ell,\ B(z):=\overleftarrow{\prod_{j=1}^{\ell}} B_j(z).$$

Let $\theta_o \in S_o(\mathbf{F},\mathbf{G})$. It is easy to see that the function θ:

$$\theta(z)=B(z)^{-1}\theta_o(z) \tag{7.12}$$

belongs to $S_\kappa(\mathbf{F},\mathbf{G})$ for some κ. Besides Θ we also consider the funcions θ_{jk}:

$$\theta_{jk}(z)=B_{jk}(z)^{-1}B_{jk+1}(z)^{-1}\dots B_{jm_j}(z)^{-1}B_{j+1}(z)^{-1}\dots B_\ell(z)^{-1}\theta_o(z)$$

$k=1,\dots,m_j$, $j=1,2,\dots,\ell$. The representation (7.12) of the function θ or the product on the right hand side of (7.12) is said to be strongly regular if the following conditions are satisfied:

$$\begin{aligned}&\mathbf{R}(P_{jk}P_{jk+1}) = \mathbf{R}(P_{jk}),\ k=1,2,\dots,m_j-1,\ j=1,2,\dots,\ell,\\ &\mathbf{R}(P_{jm_j}\theta_{j+1,1}(\gamma_j)) = \mathbf{R}(P_{jm_j}),\ j=1,2,\dots,\ell;\end{aligned} \tag{7.13}$$

here we put $\theta_{\ell+1,1}:=\theta_o$. These conditions can also be written as

$$\mathbf{R}(P_{jk}P_{jk+1}\dots P_{jm_j}\theta_{j+1,1}(\gamma_j))=\mathbf{R}(P_{jk}),$$

$j=1,2,\dots,m_j$, $j=1,2,\dots,\ell$. Evidently, the relations (7.13) imply

$$\operatorname{rank} P_{j1} \lneqq \operatorname{rank} P_{j2} \lneqq \dots \lneqq \operatorname{rank} P_{jm_j}.$$

If the representation (7.12) is strongly regular, the number $n:=\sum_{j=1}^{\ell}\sum_{k=1}^{m_j}\operatorname{rank} P_{jk}$ is called the index of the representation.

Theorem 7.7. Each function $\theta \in S_\kappa(\mathbf{F},\mathbf{G})$ admits a strongly regular representation (7.12) of index κ. If the ordering of the poles $\gamma_1,\dots,\gamma_\ell \in \mathbb{D}\setminus\{0\}$ of θ is fixed, this representation is uniquely determined. Conversely, if $\theta_0 \in S_0(\mathbf{F},\mathbf{G})$, mutually different numbers $\gamma_1,\dots,\gamma_\ell \in \mathbb{D}\setminus\{0\}$ and orthogonal projections P_{jk} in $\mathbf{G}$ are given, such that the product on the right hand side of (7.12) is strongly regular, then $\theta \in S_\kappa(\mathbf{F},\mathbf{G})$ where κ is the index of the representation (7.12).

Proof. Let $\theta \in S_\kappa(\mathbf{F},\mathbf{G})$ be given. We consider a closely connected unitary colligation $\Delta=(\hat{\mathbf{H}},\mathbf{F},\mathbf{G};U)$ with $\theta=\theta_\Delta$. Let $\gamma_1,\gamma_2,\dots,\gamma_\ell$ be the poles of θ in $\mathbb{D}$. Then the points $\alpha_j := \gamma_j^{-1}$, $j=1,2,\dots,\ell$, are the eigenvalues of the basic operator T of Δ in $\mathbb{C}\setminus\mathbb{D}^c$. We apply the representation (7.4) and the factorization (7.11) for $\alpha=\alpha_1$ and denote the corresponding order m of the pole γ_1 by m_1. Then Lemma 7.3 and Corollary 7.4 imply the factorization

$$\theta(z)=B_{11}(z)^{-1}\theta_1(z),$$

where P_{11} is the orthogonal projection onto $\mathbf{R}(G_1)=\hat{\mathbf{H}}_1$ and θ_1 is the characteristic function of the unitary colligation given by the second matrix on the right hand side of (7.11).

If $m_1 > 1$, we can apply the same procedure to the function θ_1 and its unitary colligation and find

$$\theta_1(z) = B_{12}(z)^{-1}\theta_2(z),$$

where P_{12} is the orthogonal projection onto $\mathbf{R}(H_1^{-1}G_2)$. The relation (7.11) yields $T_{12}=F_1'H_1^{-1}G_2$ and by Corollary 7.3 we have that $F_1' = \frac{\alpha_1}{\bar{\alpha}_1}G_1^*$, hence

$$T_{12} = \frac{\alpha_1}{\bar{\alpha}_1}G_1^*H_1^{-1}G_2. \tag{7.14}$$

We shall show that

$$\mathbf{R}(P_{11}) = \mathbf{R}(P_{11}P_{12}).$$

Indeed, $\mathbf{R}(P_{11}) \supset \mathbf{R}(P_{11}P_{12})$ is obvious. Assume $x=P_{11}x$, such that $[x,P_{11}P_{12}\mathbf{G}]=\{0\}$. As $x=G_1\hat{x}$ for some $\hat{x} \in \hat{\mathbf{H}}_1$ it follows that

$$\{0\} = [G_1\hat{x},P_{11}P_{12}\mathbf{G}] = [G_1\hat{x},P_{12}\mathbf{G}] =$$
$$= [\hat{x},G_1^*H_1^{-1}G_2\hat{\mathbf{H}}_2] =$$
$$= \frac{\alpha_1}{\bar{\alpha}_1}[\hat{x},T_{12}\hat{\mathbf{H}}_2],$$

which implies $\hat{x}=0$ as $\mathbf{R}(T_{12})=\hat{\mathbf{H}}_1$ (see Lemma 7.1).

Continuing this procedure we are led to a factorization

$$\theta(z)=B_{11}(z)^{-1}B_{12}(z)^{-1}\dots B_{1m_1}(z)^{-1}\theta_{m_1+1}(z)$$

with

$$\mathbf{R}(P_{1k})=\mathbf{R}(P_{1k}P_{1\,k+1}),\ k=1,2,\dots,m_1-1,$$

and a function θ_{m_1+1} which is holomorphic at γ_1.

We shall show that

$$(7.15)\qquad \mathbf{R}(P_{1m_1}\theta_{m_1+1}(\gamma_1)) = \mathbf{R}(P_{1m_1}).$$

To this and we consider the function

$$\theta_{1m_1}(z): = B_{1m_1}(z)^{-1}\theta_{m_1+1}(z),$$

which corresponds to the closely connected unitary colligation with basic operator

$$\begin{pmatrix} \alpha_1 I_{m_1} & T_{m_1\,m_1+1} \\ 0 & T_o \end{pmatrix}$$

in the Pontryagin space $\hat{\mathbf{H}}_{m_1} \oplus \hat{\mathbf{K}}$ of index $\kappa-\sum_{k=1}^{m_1-1}$ rank P_{1k}. Proposition 7.6 implies that θ_{1m_1} has at $\gamma_1=\alpha_1^{-1}$ a pole of pole multiplicity dim $\hat{\mathbf{H}}_{m_1}$=rank P_{1m_1}. On the other hand, the relation

$$\theta_{1m_1}(z) = B_{1m_1}(z)^{-1}(\theta_{m_1+1}(z)-\theta_{m_1+1}(\gamma_1))+B_{1m_1}(z)^{-1}\theta_{m_1+1}(\gamma_1)$$

implies that this pole multiplicity equals rank $P_{1m_1}\theta_{m_1+1}(\gamma_1)$, and the statement (7.15) follows. Thus we have shown that any function $\theta \in S_\kappa(\mathbf{F},\mathbf{G})$ admits a strongly regular representation (7.11). Before we continue the proof of Theorem 7.7. we state the following lemma.

Lemma 7.8. Let $A(z) = (1-P + \frac{1-z\bar{\gamma}}{z-\gamma}P)D(z)$, where

$$D(z) = \frac{D_n}{(z-\gamma)^n} + \dots + \frac{D_1}{z-\gamma} + D_o + \hat{D}(z),$$

<u>and where</u> $\hat{D}(z)$ <u>is holomorphic at</u> γ, $\hat{D}(\gamma)=0$. <u>If</u> $\mathbf{R}(PD_n) = \mathbf{R}(P)$, <u>then</u> $\pi_A(\gamma) = \pi_D(\gamma)+\text{rank } P$.

<u>Proof</u>. If we write $I-P + \frac{1-z\bar{\gamma}}{z-\gamma}P = a\frac{P}{z-\gamma} + bP+I$, with $a=1-\gamma\bar{\gamma}$, $b=-(1+\bar{\gamma})$, then the matrix (7.2) corresponding to the function $A(z)$ is

$$A = \begin{pmatrix} aPD_n & 0 & \cdot & \cdot & 0 \\ aPD_{n-1}+(1+bP)D_n & APD_n & \cdot & \cdot & 0 \\ \cdot & & & & \cdot \\ \cdot & & & & \cdot \\ \cdot & & & & \cdot \\ aPD_1 \quad +(1+bP)D_2 & \cdot & \cdot & \cdot & 0 \\ aPD_o \quad +(1+bP)D_1 & \cdot & \cdot & \cdot & aPD_n \end{pmatrix}.$$

As $R(PD_n)=\mathbf{R}(P)$, the range of A does not change if the last row in A is replaced by

$$(1+bP)D_1 \quad (1+bP)D_2 \dots (1+bP)D_n \quad aPD_n.$$

Denote this new matrix by $\tilde{A}$. Then we have

$$\text{(7.16)} \quad \mathbf{R}(\tilde{A})=\mathbf{R}\left(\begin{pmatrix} aP & 0 & \cdot & \cdot & \cdot & 0 \\ 1+bP & aP & \cdot & \cdot & \cdot & 0 \\ 0 & 1+bP & \cdot & \cdot & \cdot & 0 \\ \cdot & \cdot & & & & \cdot \\ \cdot & \cdot & & & & \cdot \\ \cdot & \cdot & & & & \cdot \\ 0 & 0 & & & & aP \\ 0 & 0 & & & & 1+bP \end{pmatrix} D\right)+\mathbf{R}\left(\begin{pmatrix} 0 \\ 0 \\ 0 \\ \cdot \\ \cdot \\ \cdot \\ 0 \\ P \end{pmatrix}\right),$$

where the first matrix on the right hand side is of size $(n+1)\times n$, D is the $n\times n$-matrix corresponding to the Laurent expansion of $D(z)$ at γ and the last column denotes the range of an $(n+1)\times 1$-matrix having P at the bottom corner and zeros elsewhere. It is easy to check that the $(n+1)\times n$-matrix in (7.16) does not annihilate any nonzero vector. Hence the dimension of the first term on the right hand side of (7.16) is $\pi_D(\gamma)$

and the statement follows, as the two ranges on this right hand side are linearly independent.

Continuing the proof of Theorem 7.7, let a function θ be given by a strongly regular representation of index κ. The hypothesis $\mathbf{R}(P_{\ell m_\ell})=\mathbf{R}(P_{\ell m_\ell}\theta_o(\gamma_\ell))$ implies

$$\pi_{\theta_{\ell m_\ell}}(\gamma_\ell) = \text{rank } P_{\ell m_\ell}.$$

Assume that we have shown the relation

$$(7.7) \qquad \pi_{\theta_{\ell n}}(\gamma_\ell)=\sum_{k=n}^{m_\ell} \text{rank } P_{\ell k},\ 1 < n \leqq m_\ell.$$

The Laurent expansion of $\theta_{\ell n}$ at γ_ℓ starts with the coefficient $P_{\ell n}P_{\ell n+1} \cdots P_{\ell m_\ell}\theta_o(\gamma_\ell)$, and the assumption of strong regularity implies

$$\mathbf{R}(P_{\ell n-1}P_{\ell n} \cdots P_{\ell m_\ell}\theta_o(\gamma_\ell)) = \mathbf{R}(P_{\ell n-1}).$$

Hence Lemma 7.8 yields the relation (7.17) with n replaced by n+1. Repeating these arguments for the other poles $\gamma_{\ell-1},\dots,\gamma_1$ of θ in $\mathbf{D}$, we find that the total pole multiplicity of θ equals the index of the representation (7.12). Now it is easy to see that the function θ belongs to some class $S_\kappa(\mathbf{F},\mathbf{G})$ with

$$(7.18) \qquad \kappa \leqq \sum_{j=1}^{\ell} \sum_{k=1}^{m_j} \text{rank } P_{jk}.$$

Hence it is the characteristic funcion of a closely connected unitary colligaton with inner space a π_κ-space. Proposition 7.6 implies that κ is the total pole multiplicity of θ in $\mathbf{D}$. Hence in (7.18) the equality holds and the last statement of the theorem is proved.

The uniqueness of the strongly regular representation (7.12) follows easily from the fact that the projections P_{jk} can be found independently by considering the colligation as follows. Let

$$\theta(z)=(z-\gamma_1)^{-m_1}D_{m_1}+(z-\gamma_1)^{-m_1+1}D_{m_1-1}+\dots$$

be the Laurent expansion of θ at γ_1. Then we have

$$D_{m_1}=(1-|\gamma_1|^2)^{m_1}P_{11}\cdots P_{1m_1}\theta_{21}(\gamma_1)$$

and the assumptions (7.13) imply $\mathbf{R}(D_{m_1})=\mathbf{R}(P_{11})$, hence P_{11} is the orthogonal projection onto $\mathbf{R}(D_{m_1})$. Now we can multiply $\theta(z)$ from the left by $B_{11}(z)$ and apply the same argument, etc. Thus Theorem 7.7 is proved.

The procedure in the last paragraph of this proof can be applied in order to find the strongly regular representation (7.12) of a given function $\theta \in S_\kappa(\mathbf{F},\mathbf{G})$.

We have shown in Proposition 7.6, that the pole multiplicity $\pi_\theta(\gamma)$ at a pole $\gamma \in \mathbb{D}$ of θ equals the algebraic multiplicity of the eigenvalue $\alpha=\gamma^{-1}$ of the basic operator of the closely connected unitary colligation Δ with $\theta=\theta_\Delta$. It also coincides with the index of the corresponding Blaschke-Potapov factor B_j if $\gamma=\gamma_j$, this index being defined by $\sum_{j=1}^{m_j} P_{jk}$. Moreover, with the pole $\gamma \in \mathbb{D}$ of θ a local index of negativity can be associated as follows.

Let A_γ be the set of all functions f which are defined and holomorphic in some neighbourhood of γ (which may depend on f) and with values in $\mathbf{F}$. On A_γ we define the kernel $S_{\theta;\gamma}$ as follows:

$$S_{\theta;\gamma}(f,g):=\frac{1}{4\pi^2}\oint_{\bar{C}}\oint_{C}\left[\frac{1-\theta(w)^*\theta(z)}{1-\bar{w}z}f(z),g(w)\right]dz\,\overline{dw},$$

where the circle C around γ is chosen such that f,g are holomorphic on the disc surrounded by C and that γ is the only pole of θ in this disc. Evidently, $S_{\theta;\gamma}(f,g)$ does not depend on the special choice of C.

Proposition 7.9. <u>If</u> $\gamma \in \mathbb{D}$ <u>is a pole of the function</u> $\theta \in S_\kappa(\mathbf{F},\mathbf{G})$ <u>of pole multiplicity</u> $\pi_\theta(\gamma)$, <u>then the kernel</u> $S_{\theta;\gamma}$ <u>on</u> A_γ <u>is negative definite with</u> $\pi_\theta(\gamma)$ <u>negative squares.</u>

<u>Proof</u>. Using

$$(7.19)\qquad \frac{1}{1-z\bar{w}}=\frac{1}{2\pi i}\oint_{|u|=1}\frac{1}{u-z}\,\frac{1}{1-u\bar{w}}\,du=\frac{1}{2\pi}\int_0^{2\pi}\left(\frac{1}{e^{i\tau}-z}\right)\overline{\left(\frac{1}{e^{i\tau}-w}\right)}d\tau,$$

we find that

$$S_{\theta;\gamma}(f,g)=-\frac{1}{2\pi}\int_0^{2\pi}[\frac{1}{2\pi i}\oint_C \frac{\theta(z)f(z)}{e^{i\tau}-z}\,dz,\ \frac{1}{2\pi i}\oint_C \frac{\theta(z)g(z)}{e^{i\tau}-z}\,dz]d\tau.$$

If the Laurent expansion of θ at γ is given by

$$\theta(z)=(z-\gamma)^{-n}A_n+(z-\gamma)^{n+1}A_{n-1}+\dots+(z-\gamma)^{-1}A_1+\theta_0(z)$$

with $A_n \neq 0$ and θ_0 holomorphic at γ, then it is not difficult to verify that

$$\frac{1}{2\pi i}\oint\frac{\theta(z)f(z)}{e^{i\tau}-z}\,dz=((e^{i\tau}-\gamma)^{-n},\ (e^{i\tau}-\gamma)^{-n+1},\ \dots,\ (e^{i\tau}-\gamma)^{-1})A\tilde{f}(\gamma),$$

where A is the $n \times n$ matrix (7.2) and

$$\tilde{f}(z)=\begin{pmatrix} f(z) \\ \frac{1}{1!}f'(z) \\ \cdot \\ \cdot \\ \cdot \\ \frac{1}{(n-1)!}f^{(n-1)}(z)\end{pmatrix}.$$

If $E=(E_{jk})_{j,k=1,\dots,n}$ is the $n \times n$ matrix defined by

$$E_{jk}=\frac{1}{2\pi}\int_0^{2\pi}(e^{i\tau}-\gamma)^{-k}\overline{(e^{i\tau}-\gamma)^{-j}}\,d\tau,$$

then

$$S_{\theta;\gamma}(f,g)=-[EA\tilde{f}(\gamma),A\tilde{g}(\gamma)]$$

and the proposition now follows from the fact that E is positive definite.

The conditions (7.13) about the inverse Blaschke-Potapov factors $B(z)^{-1}$ in the strongly regular representation (7.12) of θ are needed in order to assure that the multiplication of $B(z)^{-1}$ by $\theta_0(z)$ does not cancel or reduce the order of any of the poles of $B(z)^{-1}$. The following proposition shows that the total multiplicity of the poles of any inverse Blaschke-Potapov factor $B(z)^{-1}$ and also the index κ for which $B^{-1} \in S_\kappa(\mathbf{G},\mathbf{G})$ can easily be found.

Proposition 7.10. Let (not necessarily mutually different) points $\gamma_j \in \mathbb{D}\setminus\{0\}$ and orthogonal projections $P_j \in L(\mathbf{G})$ with rank $P_j < \infty$, $j=1,2,\dots,n$, be given. Then the function θ:

$$\theta(z) := \overrightarrow{\prod_{j=1}^{n}} \left(1-P_j + \frac{1-z\bar{\gamma}_j}{z-\gamma_j} P_j\right),\ z \in D, z \neq g_j,$$

belongs to the class $S_\kappa(\mathbf{G},\mathbf{G})$ with $\kappa = \sum_{j=1}^{n} \text{rank } P_j$.

Proof. For $j=1,2,\dots,n$, let $\hat{\mathbf{H}}_j$ be a negative space with $\dim \hat{\mathbf{H}}_j = \text{rank } P_j$, choose $G_j \in L(\hat{\mathbf{H}}_j, \mathbf{G})$ such that

$$G_j^* G_j = (1-|\alpha_j|^2) I_{\hat{\mathbf{H}}_j},\quad G_j G_j^* = (1-|\alpha_j|^2) P_j,$$

where $\alpha_j = \gamma_j^{-1}$ and define the unitary colligation $\Delta_j = (\hat{\mathbf{H}}_j, \mathbf{G}, \mathbf{G}; U_j)$ by

$$U_j = \begin{pmatrix} \alpha_j I_{\hat{\mathbf{H}}_j} & \dfrac{\alpha_j}{\bar{\alpha}_j} G_j^* \\ G_j & I-P_j-\alpha_j P_j \end{pmatrix}.$$

Then $\Delta = \Delta_1 \Delta_2 \dots \Delta_n$ is a unitary colligation and $\theta = \theta_\Delta$. The inner space $\hat{\mathbf{H}}$ of Δ is the negative space $\hat{\mathbf{H}} = \hat{\mathbf{H}}_1 \oplus \hat{\mathbf{H}}_2 \oplus \dots \oplus \hat{\mathbf{H}}_n$. Clearly, $\dim \hat{\mathbf{H}} = \sum_{j=1}^{n} \text{rank } P_j = \kappa$. Let T be the basic operator of Δ. Let $\mathbf{H}$ be a subspace of $\hat{\mathbf{H}}$ such that $T\mathbf{H}=\mathbf{H}$ and $T|_{\mathbf{H}}$ is isometric. Then either $\mathbf{H}=\{0\}$ or $\mathbf{H}$ is a negative subspace, in which case $T|_{\mathbf{H}}$ is unitary in $\mathbf{H}$ and so $|\sigma(T|_{\mathbf{H}})|=1$. This last case cannot hold as $\sigma(T)=\{\alpha_1,\dots,\alpha_n\}$, so that $|\sigma(T)| > 1$. Hence $\mathbf{H}=\{0\}$. Proposition 3.2 now implies that Δ is closely connected and, on account of Proposition 5.5, it follows that $\theta=\theta_\Delta \in S_\kappa(\mathbf{G},\mathbf{G})$.

If $\mathbf{F}=\mathbf{G}$ the factorization of a function $\theta \in S_\kappa(\mathbf{G},\mathbf{G})$ was essentially proved in [11, 3.4]. There, however, in the definition of regularity the ordering (3.15) of the projections has to be replaced by the ordering of their dimensions (see formula (7.13) above), and then Lemma 3.5 of [13] is not sufficiently general. Instead of extending this lemma, or simply replacing it by Lemma 7.8 above, we have preferred to prove the factorization (in the slightly more general situation where $\mathbf{F} \neq \mathbf{G}$ is allowed) by applying the factorization results for unitary colligations. This seems to be more lengthly than the direct analytical procedure of [13]. However, it gives some insight into the structure of the colligation corresponding to θ which (if $\theta(0) \neq 0$) was not considered in [13]. As the notion of a regular factorization was also used in a different meaning by M.S. Brodskii [6], we call the factorization considered here (and essentially in [13], [16]) strongly regular. We mention that any strongly regular representation is regular in Brodskii's sense. There are, however, in general many more regular factorizations then strongly regular ones. E.g., starting with any triangularization of T, we are led to an regular factorization of θ in Brodskii's sense. It seems to be difficult to characterize this regularity directly in terms of the projections P_j. For results such as given in Proposition 7.11 we refer to Chamfy [7].

Let again B(z) be an arbitrary Blaschke-Potapov product, and consider the closely connected unitary colligation Δ_1 such that $B(z)^{-1}=\theta_{\Delta_1}(z)$. Then the algebraic multiplicity of an eigenvalue $\alpha=1/\gamma$, $|\alpha| > 1$, of the basic operator T_1 of Δ_1 equals $\sum_{j:\gamma_j=\gamma} \operatorname{rank} P_j$. If we multiply $B(z)^{-1}$ from the right by some function $\theta_o \in S_o(\mathbf{F},\mathbf{G})$: $\theta(z)=B(z)^{-1}\theta_o(z)$, the algebraic multiplicity of this eigenvalue α of the corresponding basic operator T is $\leqq \sum_{j:\gamma_j=\gamma} \operatorname{rank} P_j$. If, however the product $B(z)^{-1}\theta_o(z)$ is strongly regular, then equality holds, according to the above statement. Moreover, in this case the structure of the algebraic eigenspace of T at α can be read off from the projections P_{jk}, $k=1,2,\dots,m_j$. Indeed, the proof of Theorem 7.7 implies that rank P_{j1} is the number of Jordan chains of length m_j, rank P_{j2}-rank P_{j1} is the additional number of Jordan chains of length m_1-1, etc.

We finally mention, that also an arbitrary triangularization of T in the κ-dimensional negative invariant subspace gives a representation of θ in the form (7.12) with some inverse Blaschke-Potapov product $B(z)^{-1}$ of the form

$$(7.20)\qquad B(z)^{-1}=\overset{\curvearrowright}{\prod_{k=1}^{\kappa}}\left(1-P_k+\frac{1-z\bar{\delta}_k}{z-\delta_k}P_k\right),\qquad \operatorname{rank} P_k=1,$$

where the P_k's are orthogonal projections in $\mathbf{G}$. Here the number of δ_k's which are equal to γ_j equals the algebraic multiplicity of the eigenvalue $\alpha_j=\gamma_j^{-1}$ of T or the pole multiplicity of the pole γ_j of θ. If, conversely, a function θ:

$$\theta(z):=B(z)^{-1}\theta_o(z)$$

with $\theta_o \in S_o(\mathbf{F},\mathbf{G})$ and $B(z)^{-1}$ as in (7.20) is given, then $\theta \in S_{\kappa'}(\mathbf{F},\mathbf{G})$ for some κ', $0 \leqq \kappa' \leqq \kappa$. Lemma 7.8 can be used in order to give a sufficient condition for $\kappa'=\kappa$: This equality holds if

$$\mathbf{R}(P_k\hat{B}_{k+1}(\delta_k)^{-1}\ldots\hat{B}_\kappa(\delta_k)^{-1}\theta_o(\delta_k))=\mathbf{R}(P_k)$$

for each $k=1,\ldots,\kappa$, where

$$\hat{B}_k(\delta_j)=\begin{cases} B_k(\delta_j) & \text{if } \delta_k \neq \delta_j,\\ P_k & \text{if } \delta_k = \delta_j,\end{cases}$$

and B_k denotes the k-th factor on the right hand side of (7.20).

An immediate consequence of the factorization (7.12) is the following result. Let $\theta \in S_\kappa(\mathbf{F},\mathbf{G})$, then

$$(7.21)\qquad \lim_{r\to 1}\ \sup_{|z|=r}\|\theta(z)\| \leqq 1.$$

In order to show this, we note that $\theta=B^{-1}\theta_o$, with an inverse Blaschke-Potapov product B^{-1} and a holomorphic contraction θ_o. Hence $\|\theta(z)\| \leqq \|B(z)^{-1}\|$, $z \in D_\theta$ and the result follows from the fact that $B(z)^{-1}$ is unitary if $z \in \partial\mathbf{D}$. The next proposition shows that in a sense the converse statement is also true.

Proposition 7.11. <u>Let</u> θ <u>be a meromorphic function on</u> $\mathbf{D}$, <u>holomorphic at</u> 0, <u>with values in</u> $L(\mathbf{F},\mathbf{G})$ <u>and with finite pole multiplicity</u> κ, <u>such that</u> (7.21) <u>holds. Then</u> $\theta \in S_\kappa(\mathbf{F},\mathbf{G})$.

<u>Proof</u>. There exists a Blaschke-Potapov product B, such that $\theta_o=B\theta$ is holomorphic on $\mathbf{D}$. As B is unitary on $\partial\mathbf{D}$, this implies that

$$\lim_{r\to 1} \sup_{|z|=r} \|\theta_o(z)\| \leqq 1$$

and hence θ_o is a holomorphic contraction. Now $\theta = B^{-1}\theta_o$ implies that $\theta \in S_{\kappa'}(\mathbf{F},\mathbf{G})$ for some κ' and Proposition 7.6 shows that $\kappa' = \kappa$.

8. A MAXIMUM PRINCIPLE.

Before stating our main results we recall (see (7.19)) that for $z, w \in \mathbb{D}$ we have the multiplicative representation

$$\frac{1}{1-z\bar{w}} = \frac{1}{2\pi} \int_0^{2\pi} \left(\frac{1}{e^{i\tau}-z}\right)\left(\frac{1}{e^{i\tau}-w}\right) d\tau.$$

This representation readily implies that for any distinct points $z_1, \ldots, z_n \in \mathbb{D}$ the $n \times n$ matrix

$$(8.1) \qquad \left(\frac{1}{1-z_i\bar{z}_j}\right)_{i,j=1,\ldots,n}$$

is strictly positive.

Proposition 8.1 (maximum principle). Let $\theta \in S_\kappa(\mathbf{F},\mathbf{G})$, $f \in \mathbf{F}$, $g \in \mathbf{G}$ and assume $g = \theta(z)f$ for more than κ points $z \in \mathbf{D}_\theta$. Then we have $\|g\| \leqq \|f\|$, and the equality $\|g\| = \|f\|$ implies that $g = \theta(z)f$, $z \in \mathbf{D}_\theta$.

Proof. Suppose we have $g = \theta(z_i)f$ for $z_i \in \mathbf{D}_\theta$, $i = 1, \ldots, \kappa+1$. Then it follows that

$$(8.2) \qquad [S_\theta(z_i,z_j)f,f] = \frac{[(I-\theta(z_j)^*\theta(z_i))f,f]}{1-z_i\bar{z}_j}$$

$$= \frac{[f,f]-[g,g]}{1-z_i\bar{z}_j}$$

Hence if $\|f\| < \|g\|$, by (8.1) it follows that the $(\kappa+1)\mathrm{x}(\kappa+1)$ matrix $([S_\theta(z_i,z_j)f,f])_{i,j=1,\ldots,n}$ is strictly negative which contradicts the fact that $\theta \in S_\kappa(\mathbf{F},\mathbf{G})$ and thus we obtain $\|g\| \leqq \|f\|$. Now assume $\|g\| = \|f\|$. Let $\Delta = (\hat{\mathbf{H}}, \mathbf{F}, \mathbf{G}; T, F, G, H)$ be a closely connected unitary colligation such that

$\theta_\Delta=\theta$. Then using (8.2) and (5.1) we obtain

$$[(I-z_iT)^{-1}Ff,(I-z_jT)^{-1}Ff]=0,\quad i,j=1,\dots,\kappa+1.$$

This shows that the $\kappa+1$ elements $(I-z_iT)^{-1}Ff$ form a neutral subspace in $\hat{\mathbf{H}}$. Since any neutral subspace in $\hat{\mathbf{H}}$ has dimension at most κ, there exist $\alpha_i \in \mathbb{C}$, $i=1,\dots,\kappa+1$, not all zero, such that

$$\sum_{i=1}^{\kappa+1} \alpha_i(I-z_iT)^{-1}Ff=0.$$

Then we have for all $z \in \mathbf{D}_\theta$ and $h \in \mathbf{F}$

$$[\sum_{i=1}^{\kappa+1} \frac{\alpha_i}{1-z_i\bar{z}} (f-\theta(z)^*g),h]$$

$$=\sum_{i=1}^{\kappa+1} \frac{\alpha_i}{1-z_i\bar{z}} [(I-\theta(z)^*\theta(z_i)f,h]$$

$$=[\sum_{i=1}^{\kappa+1} \alpha_i(I-z_iT)^{-1}Ff,(I-zT)^{-1}Fh]=0,$$

Hence we obtain for all $z \in \mathbf{D}_\theta$

$$\sum_{i=1}^{\kappa+1} \frac{\alpha_i}{1-z_i\bar{z}} (f-\theta(z)^*g)=0.$$

Since the factor

$$\sum_{i=1}^{\kappa+1} \frac{\alpha_i}{1-z_i\bar{z}}$$

is not identally 0, we find $f=\theta(z)^*g$, $z \in \mathbf{D}_\theta$. Hence we have shown for $f \in \mathbf{F}, g \in \mathbf{G}$, $\|f\|=\|g\|$ with $g=\theta(z)f$ for more than κ points $z \in \mathbf{D}_\theta$, that $f=\theta(z)^*g$ for all $z \in \mathbf{D}_\theta$. As we have seen before $\theta_\Delta(z)^*=\theta_\Delta{}^*(\bar{z})$, $z \in \mathbf{D}_\theta$, so we finally conclude $g=\theta(z)f$ for all $z \in \mathbf{D}_\theta$.

Let $\Delta=(\hat{\mathbf{H}},\mathbf{F},\mathbf{G};T,F,G,H)$ be a unitary colligation with characteristic function $\theta=\theta_\Delta$:

$$\theta(z)=H+zG(I-zT)^{-1}F,\quad \frac{1}{z} \in \rho(T).$$

If $f \in \nu(F)$, then $\theta(z)f=Hf$ for all $z \in \mathbf{D}_\theta$ and $\|Hf\|=\|f\|$. A converse of this is contained in the following result.

Proposition 8.2. Let $\theta \in S_\kappa(\mathbf{F},\mathbf{G})$, $f \in \mathbf{F}, g \in \mathbf{G}$, $\|g\|=\|f\|$ and assume $g=\theta(z)f$, $z \in \mathbf{D}_\theta$. If $\Delta=(\hat{\mathbf{H}},\mathbf{F},\mathbf{G};T,F,G,H)$ is a closely connected unitary colligation, such that $\theta_\Delta=\theta$, then $f \in \nu(F)$, $g \in \nu(G^*)$ and $g=Hf$.

Proof. As in the proof of Proposition 8.1 we have $g=\theta(z)f$ and $f=\theta(z)^*g$, $z \in \mathbf{D}_\theta$. Therefore we have for all $h \in \mathbf{F}$

$$[Ff,(I-zT)^{-1}Fh]=[(I-\theta(z)^*\theta(0))f,h]=0, \quad z \in \mathbf{D}_\theta.$$

By our assumption we have $\theta(z)f=\theta(0)F$, $z \in \mathbf{D}_\theta$, and hence $G(I-zT)^{-1}Ff=0$, $z \in \mathbf{D}_\theta$. Hence we have for all $k \in \mathbf{G}$

$$[Ff,(I-\bar{z}T^*)^{-1}G^*k]=[G(I-zT)^{-1}Ff,k]=0, \ z \in \mathbf{D}_\theta.$$

Since Δ is closely connected, we obtain $Ff=0$. The rest of the statements in the proposition is clear.

We require the colligation in Proposition 8.2 to be closely connected in order to exclude the situation that $Ff \in \mathbf{N}$, where $\mathbf{N} \subset \hat{\mathbf{H}}$ is a neutral subspace in $\hat{\mathbf{H}}$ which is invariant under T. Any such subspace satisfies $\mathbf{N} \subset \nu(G)$, since $T^*T+G^*G=I$, and thus $G(I-zT)^{-1}|_\mathbf{N}=0$ for all $z \in \mathbb{D}$ with $\frac{1}{z} \in \rho(T)$. Therefore we would have $zG(I-zT)^{-1}Ff=0$, without $Ff=0$.

The propositions in this section show that functions $\theta \in S_\kappa(\mathbf{F},\mathbf{G})$ behave in a certain sense like holomorphic contractions on $\mathbb{D}$. These results are apparently new.

9. BOUNDARY BEHAVIOUR OF FUNCTIONS IN $S_\kappa(\mathbf{F},\mathbf{G})$.

Any function $\theta \in S_\kappa(\mathbf{F},\mathbf{G})$ is meromorphic on $\mathbb{D}$, with at most κ poles in $\mathbb{D}$. In particular, if $\theta=\theta_\Delta$, where $\Delta=(\hat{\mathbf{H}},\mathbf{F},\mathbf{G};T,F,G,H)$ is a closely connected unitary colligation, then the eigenvalues of T outside $\mathbb{D}^c$ give rise to poles of θ inside $\mathbb{D}$. Also we mention that then the contraction T does not have any eigenvalues on $\partial\mathbb{D}$. In this section we study the behaviour of θ on $\partial\mathbb{D}$. The following result depends on the boundary behaviour near $\partial\mathbb{D}$ of the resolvent of a contraction in a π_κ-space, as given in Proposition 2.8 and its Corollary 2.9.

Theorem 9.1. *Let* $\theta \in S_\kappa(\mathbf{F},\mathbf{G})$, $\zeta \in \partial\mathbb{D}$ *and* $\psi \in \mathbf{F}$. *Then the following statements are equivalent:*

(i) *there exists an element* $\phi \in \mathbf{G}$, *such that* $\|\phi\|=\|\psi\|$ *and*

$$\lim_{z \hat{\to} \zeta} \frac{[\psi,\psi]-[\theta(z)\psi,\phi]}{1-z\bar{\zeta}} \text{ exists,}$$

(ii) $$\lim_{z \hat{\to} \zeta} \frac{\|\psi\|-\|\theta(z)\psi\|}{1-|z|} \text{ exists,}$$

(iii) *for some sequence (and hence for all sequences)* (z_n) *in* $\mathbf{D}_\theta$ *with* $z_n \hat{\to} \zeta$ *we have that*

$$\sup_n \left| \frac{\|\psi\|-\|\theta(z_n)\psi\|}{1-|z_n|} \right| < \infty.$$

If one of these statements is valid, then the element ϕ *in* (i) *is uniquely determined and*

$$\phi=\lim_{z \hat{\to} \zeta} \theta(z)\psi.$$

If $\Delta=(\hat{\mathbf{H}},\mathbf{F},\mathbf{G};T,F,G,H)$ *is a closely connected unitary colligation such that* $\theta=\theta_\Delta$, *then equivalent to* (i), (ii) *or* (iii) *is*

(iv) $F\psi \in \mathbf{R}(I-\zeta T)$

and then (iv) *implies*

$$\phi=(H+\zeta G(I-\zeta T)^{-1}F)\psi.$$

Proof. We shall show the following implications: (iv)⇒(i) and (ii), (i)⇒(iii), (ii)⇒(iii) and (iii)⇒(iv).
Note that the implication (ii) ⇒ (iii) is trivial. We assume (iv). Since the colligation Δ is closely connected T does not have any eigenvalues on $\partial\mathbb{D}$ and we may use Corollary 2.9. Now let $\psi \in \mathbf{F}$ such that $F\psi \in \mathbf{R}(I-\zeta T)$. First we show that (i) is satisfied with $\phi=(H+\zeta G(I-\zeta T)^{-1}F)\psi$. We observe that for $z \in \mathbf{D}_\theta$

$$(I-\theta(z)^*[H+\zeta G(I-\zeta T)^{-1}F])\psi=(I-\bar{z}\zeta)F^*(I-\bar{z}T^*)^{-1}(I-\zeta T)^{-1}F$$

and hence for $z \in \mathbf{D}_\theta$

$$\frac{[\psi,\psi]-[\theta(z)\psi,\phi]}{1-z\bar{\zeta}} = [(I-zT)^{-1}F\psi,(I-\zeta T)^{-1}F\psi].$$

Thus by Corollary 2.9 the right hand side tends to $[(I-\xi T)^{-1}F\psi,(I-\zeta T)^{-1}F\psi]$. Next we show that (ii) is satisfied. Therefore we observe that

$$\frac{[\psi,\psi]-[\theta(z)\psi,\theta(w)\psi]}{1-z\bar{w}} = [(I-zT)^{-1}F\psi,(I-wT)^{-1}F\psi] \to [(I-\zeta T)^{-1}F\psi,(I-\zeta T)^{-1}F\psi]$$

as $z,w \hat{\to} \zeta$ by Corollary 2.9. In particular,

$$\lim_{z\hat{\to}\zeta} \frac{\|\psi\|^2-\|\theta(z)\psi\|^2}{1-|z|^2} \text{ exists}$$

and

$$\lim_{z\hat{\to}z} \|\theta(z)\psi\| = \|\psi\|.$$

Therefore we obtain for $\psi \neq 0$, that

$$\frac{\|\psi\|-\|\theta(z)\psi\|}{1-|z|} = \frac{\|\psi\|^2-\|\theta(z)\psi\|^2}{1-|z|^2} \cdot \frac{1+|z|}{\|\theta(z)\psi\|+\|\psi\|}$$

has a limit as $z\hat{\to}\zeta$.

Now we assume (i). Then we have that

$$\frac{[\psi,\psi]-[\theta(z)\psi,\phi]}{1-|z|^2} = \frac{[\phi-\theta(z)\psi,\phi]}{1-|z|^2}$$

is bounded for $z\hat{\to}\zeta$, and hence also that its real part is bounded as $z\hat{\to}\zeta$. For any α,β in the Hilbert space $\mathbf{G}$ we apply the simple inequality $2\mathrm{Re}[\alpha-\beta,\alpha] \geqq \|\alpha\|^2-\|\beta\|^2$, which was used by Štraus [22], [23], in order to obtain

$$2\mathrm{Re}\, \frac{[\phi-\theta(z)\psi,\phi]}{1-|z|^2} \geqq \frac{\|\phi\|^2-\|\theta(z)\psi\|^2}{1-|z|^2} = \frac{[\psi,\psi]-[\theta(z)\psi,\theta(z)\psi]}{1-|z|^2}$$

By Theorem 4.4 we can factor θ as follows: $\theta=\theta_1\theta_2$, where in the notation of that theorem $\theta_1=\theta_{\Delta_1}$, $\theta_2=\theta_{\Delta_2}$ and the colligations Δ_1 and Δ_2 are of the form

$$\Delta_1=((\mathbf{J}_-^*)^\perp,\mathbf{F},\mathbf{G};T_1,F_1,G_1,H_1),\quad \Delta_2=(\mathbf{J}_-^*,\mathbf{F},\mathbf{F};T_2,F_2,G_2,H_2).$$

It follows that the righthand side of the above inequality is equal to

$$(9.1)\qquad \frac{[\psi,\psi]-[\theta_2(z)\psi,\theta_2(z)\psi]}{1-|z|^2}+\frac{[\theta_2(z)\psi,\theta_2(z)\psi]-[\theta(z)\psi,\theta(z)\psi]}{1-|z|^2}.$$

By Corollary 5.8 the first term is convergent as $z \hat{\to} \zeta$ since $\frac{1}{\zeta}\in\rho(T_2)$, and the second term is non-negative since $\mathbf{J}_-^*$ is a Hilbert space. Therefore we conclude that for any sequence $z_n \in \mathbf{D}_\theta$ with $z_n \hat{\to} \zeta$

$$\frac{\|\psi\|^2-\|\theta(z_n)\psi\|^2}{1-|z_n|^2}$$

and hence also

$$\frac{\|\psi\|-\|\theta(z_n)\psi\|}{1-|z_n|}$$

is bounded, which proves (iii).

Next we assume the assertion in (iii) for some sequence (z_n) in $\mathbf{D}_\theta$ with $z_n \hat{\to} \zeta$. Then it is clear that also

$$\frac{\|\psi\|^2-\|\theta(z_n)\psi\|^2}{1-|z_n|^2}$$

is bounded for that sequence. Factoring θ in the above manner we see that the expression in (9.1) with $z=z_n$ is bounded. The first term in (9.1) with $z=z_n$ converges as $z_n \hat{\to} \zeta$ by Corollary 5.8, so that the second term in (9.1) with $z=z_n$, which is equal to

$$[(I-z_nT_1)^{-1}F_1\theta_2(z_n)\psi,(I-z_nT_1)^{-1}F_1\theta_2(z_n)\psi],$$

is bounded as $z_n \hat{\to} \zeta$. Hence there exists a subsequence of (z_n), which we again denote by (z_n), such that

$$\psi_n=(I-z_nT_1)^{-1}F_1\theta_2(z_n)\psi$$

converges weakly in the Hilbert space $(J_-^*)^\perp$ to, say, $\eta \in (J_-^*)^\perp$. Since

$$(I-zT)^{-1}F=(I-zT_1)^{-1}F_1\theta_2(z)+(I-zT_2)^{-1}F_2, \qquad z \in \mathbf{D}_\theta,$$

(see [6, p.169, formula (3.8)]) we obtain

$$F\psi=(I-z_nT)(\psi_n+(I-z_nT_2)^{-1}F_2\psi).$$

Therefore we see that

$$F\psi=(I-\zeta T)(\eta+(I-\zeta T)^{-1}F_2\psi)$$

and thus $F\psi \in \mathbf{R}(I-\zeta T)$, which implies (iv). It remains to show that the element ϕ in (i) is uniquely determined and that $\phi=\lim_{z\hat{\to}\zeta} \theta(z)\psi$. Therefore we consider

$$[\theta(z)\psi-\phi,\theta(z)\psi-\phi]=[\theta(z)\psi,\theta(z)\psi]-[\theta(z)\psi,\phi]$$
$$-[\phi,\theta(z)\psi]+[\phi,\phi],$$

so that by (i) and (ii) the righthand side tends to 0 as $z\hat{\to}\zeta$.

Let $\theta \in S_\kappa(\mathbf{F},\mathbf{G})$ and $\zeta \in \partial\mathbb{D}$. By Theorem 9.1 we may extend the definition of θ to ζ, by putting

$$\mathbf{D}_o(\theta(\zeta))=\{\psi \in F \mid \lim_{z\hat{\to}\zeta} \frac{\|\psi\|-\|\theta(z)\psi\|}{1-|z|} \text{ exists}\}$$

and defining

$$\theta(\zeta)\psi=\lim_{z\hat{\to}\zeta} \theta(z)\psi, \text{ strongly}, \qquad \psi \in \mathbf{D}_o(\theta(\zeta)).$$

We note that for $\zeta \in \partial\mathbb{D}$ $\theta(\zeta)$ is an isometry on $\mathbf{D}_o(\theta(\zeta))$. In general, the set $\mathbf{D}_o(\theta(\zeta))$ is contained in the set

$$\{\psi \in \mathbf{F} \mid \lim_{z\hat{\to}\zeta} \theta(z) \text{ exists strongly}\}$$

and the set $\mathbf{D}_o(\theta(\zeta))$ need not be closed. If $\Delta=(\hat{\mathbf{H}},\mathbf{F},\mathbf{G};T,F,G,H)$ is a closely connected unitary colligation with $\theta=\theta_\Delta$, then it is easy to see that

$$\mathbf{D}_o(\theta(\zeta))=\{\psi \in F \mid F\psi \in \mathbf{R}(I-\zeta T)\}$$

and

$$\theta(\zeta)\psi=(H+\zeta G(I-\zeta T)^{-1}F)\psi, \qquad \psi \in \mathbf{D}_o(\theta(\zeta)).$$

We observe the following identities for $z \in \mathbf{D}_\theta$ and $\zeta \in \partial\mathbb{D}$

$$\frac{\theta(z)-\theta(\zeta)}{z-\zeta}\psi=G(I-zT)^{-1}(I-\zeta T)^{-1}F\psi, \qquad \psi \in \mathbf{D}_0(\theta(\zeta)),$$

and

$$\frac{I-\theta(z)^*\theta(\zeta)}{1-\bar{z}\zeta}\psi=F^*(I-\bar{z}T^*)^{-1}(I-\zeta T)^{-1}F\psi, \qquad \psi \in \mathbf{D}_0(\theta(\zeta)).$$

Proposition 9.2. <u>If $\theta \in S_\kappa(\mathbf{F},\mathbf{G})$ is holomorphic at $\zeta \in \partial\mathbf{D}$, and $\theta(\zeta)$ is isometric, then $\mathbf{D}_0(\theta(\zeta))=\mathbf{F}$ and the two notions of $\theta(\zeta)$ coincide</u>

<u>Proof</u>. Let $\psi \in \mathbf{F}$ and define $\phi=\theta(\zeta)\psi$ (which is defined since θ is holomorphic at ζ). Now we may apply Theorem 9.1, because $\|\phi\|=\|\psi\|$ and

$$\frac{[\psi,\psi]-[\theta(z)\psi,\phi]}{1-z\bar{\zeta}} = (-\zeta)\frac{[\theta(\zeta)\psi-\theta(z)\psi,\phi]}{\zeta-z}$$

has a limit as $z\hat{\to}\zeta$.

In particular, if $\zeta \in \partial\mathbf{D}$ and $\frac{1}{\zeta} \in \rho(T)$, then θ is holomorphic at ζ and $\theta(\zeta)$ is unitary.

Proposition 9.3. <u>Let $\theta \in S_\kappa(\mathbf{F},\mathbf{G})$ and assume that θ admits a continuous extension (in the uniform topology), which is unitary on $\partial\mathbf{D}$. Then θ is holomorphic on $\partial\mathbf{D}$.</u>

The proof of this proposition is based on the Schwarz reflection principle, see McKelvey [19] or Štraus [22], cf. Proposition 5.7.

Our Theorem 9.1 has many points of contact with results of Štraus. In fact, for $\kappa=0$ the analytic statements (i), (ii) and (iii) reduce to Štraus' results [22], [23], [25]. For $\kappa > 0$ these results could also be proved using Theorem 4.4 and by applying Štraus' results for the corresponding Hilbert space situation. Statement (iv) in Theorem 9.1 is apparently new.

10. CHARACTERISTIC FUNCTIONS OF PARTIAL ISOMETRIES

In this section we consider functions $\theta \in S_\kappa(\mathbf{F},\mathbf{G})$ with the additional property $\theta(0)=0$. Then the basic operator in a corresponding unitary colligation is a partial isometry and the characteristic function can be interpreted in terms of a parallel projection.

Proposition 10.1 (parallel projection). *Let $\hat{\mathbf{H}}$ be a π_κ-space and let V be a closed isometric operator in $\hat{\mathbf{H}}$ with domain $\mathbf{D}(V)$ and range $\mathbf{R}(V)$, such that $\mathbf{D}(V)^\perp$ is a Hilbert space. Let $\tilde{V}$ be the trivial contraction extension of V: $\tilde{V}f=Vf$, $f \in \mathbf{D}(V)$, $\tilde{V}f=0$, $f \in \mathbf{D}(V)^\perp$. Then we have*

$$\hat{\mathbf{H}}=\mathbf{R}(I-zV)+\mathbf{D}(V)^\perp, \qquad \frac{1}{z} \in \rho\,(\tilde{V}), \text{ direct sum,}$$

and the corresponding parallel projection P_z from $\hat{\mathbf{H}}$ onto $\mathbf{D}(V)^\perp$ is given by

$$P_z=P_{\mathbf{D}(V)^\perp}(I-z\tilde{V})^{-1}, \qquad \frac{1}{z} \in \rho(\tilde{V}),$$

where $P_{\mathbf{D}(V)^\perp}$ denotes the orthogonal projection from $\hat{\mathbf{H}}$ onto $\mathbf{D}(V)^\perp$. In addition we have that

$$(\hat{\mathbf{H}},\mathbf{R}(V)^\perp,\mathbf{D}(V)^\perp;\ \tilde{V},I|_{\mathbf{R}(V)^\perp},P_{\mathbf{D}(V)^\perp},0)$$

is a unitary colligation, with corresponding characteristic function

$$zPz|_{\mathbf{R}(V)^\perp} = zP_{\mathbf{D}(V)^\perp}(I-z\tilde{V})^{-1}I|_{\mathbf{R}(V)^\perp}, \qquad \frac{1}{z} \in \rho(\tilde{V}).$$

Proof. The partial isometry $\tilde{V}$ is a contraction, since we assume that $\mathbf{D}(V)^\perp$ is a Hilbert space. It is clear that $I-z\tilde{V}$ maps $\mathbf{D}(V)$ onto $\mathbf{R}(I-zV)$ and acts as the identity on $\mathbf{D}(V)^\perp$, when $\frac{1}{z} \in \rho(\tilde{V})$. Hence we obtain

$$\begin{aligned}\hat{\mathbf{H}}&=(I-z\tilde{V})\mathbf{H}\\ &=(I-z\tilde{V})(\mathbf{D}(V)+D(V)^\perp)\\ &=(I-z\tilde{V})\mathbf{D}(V)+(I-z\tilde{V})\mathbf{D}(V)^\perp\\ &=\mathbf{R}(I-zV)+D(V)^\perp, \qquad \frac{1}{z} \in \rho(\tilde{V}), \text{ direct sum.}\end{aligned}$$

From the relation

$$I-z\tilde{V}=(I-z\tilde{V})(I-P_{D(V)^{\perp}})+(I-z\tilde{V})P_{D(V)^{\perp}}$$
$$=(I-zV)(I-P_{D(V)^{\perp}})+P_{D(V)^{\perp}},$$

we obtain $P_z(I-z\tilde{V})=P_{D(\tilde{V})^{\perp}}$ or $P_z=P_{D(\tilde{V})^{\perp}}(I-z\tilde{V})^{-1}$, $\frac{1}{z}\in\rho(\tilde{V})$.

It is clear that the colligation is isometric. We show that its connecting operator is surjective. So let $\binom{\alpha}{\beta}\in\begin{pmatrix}\hat{H}\\ D(V)^{\perp}\end{pmatrix}$, then we have to find $\binom{\phi}{\psi}\in\begin{pmatrix}\hat{H}\\ R(V)^{\perp}\end{pmatrix}$ so that $\tilde{V}\phi+\psi=\alpha$ and $P_{D(V)^{\perp}}\phi=\beta$. Hence we must have $\phi=\beta+\chi$ with $\chi\in D(V)$. Hence we obtain $\alpha=\tilde{V}\phi+\psi=V\chi+\psi$, and therefore $\chi\in D(V)$ and $\psi\subset R(V)^{\perp}$ are uniquely determined.

In a similar way we can prove the following result.

Proposition 10.2. Let the conditions of Proposition 10.1 be satisfied and assume in addition that $\dim D(V)^{\perp}=\dim R(V)^{\perp}$. If U is a unitary operator extension of V to all of $\hat{H}$, then

$$(\hat{H},R(V)^{\perp},R(V)^{\perp};\ \tilde{V},I|_{R(V)^{\perp}},\ UP_{D(V)^{\perp}},0)$$

is a unitary colligation with characteristic function

$$zUP_z|_{R(V)^{\perp}}=zUP_{D(V)^{\perp}}(I-z\tilde{V})^{-1}I|_{R(V)^{\perp}},\qquad \frac{1}{z}\in\rho(\tilde{V}).$$

We shall show that the characteristic function in Proposition 10.1 characterizes -up to isometries- all functions $\theta\in S_\kappa(F,G)$ with $\theta(0)=0$.

Proposition 10.3. Let $\theta\in S_\kappa(F,G)$ and let $\theta(0)=0$. If $\theta=\theta_\Delta$, where

$$\Delta=(\hat{H},F,G;T,F,G,0)$$

is a closely connected unitary colligation of index κ, then T is a partial isometry with isometric part V, F is an isometry mapping F onto $R(V)^{\perp}=\nu(T^*)$ and G^* is an isometry mapping G onto $D(V)^{\perp}=\nu(T)$. We have

$$\theta(z)=G[zP_{D(V)^{\perp}}(I-zT)^{-1}I|_{R(V)^{\perp}}]F,\quad \frac{1}{z}\in\rho(T).$$

Proof. We have $T^*T+G^*G=I$, $F^*F=I$, $F^*T=0$, $TT^*+FF^*=0$, $GG^*=I$ and $TG^*=0$. This shows that **FF** and G^***G** are Hilbert spaces and that T is a partial isometry with isometric part V with $\mathbf{D}(V)^{\perp}=G^*\mathbf{G}$ and $\mathbf{R}(V)^{\perp}=F\mathbf{F}$. It is clear that $G^*G=P_{\mathbf{D}(V)^{\perp}}$, and therefore we may rewrite $\theta(z)$:

$$\theta(z)=zG(I-zT)^{-1}F$$

$$=zGG^*G(I-zT)^{-1}F$$

$$=G[zP_{\mathbf{D}(V)^{\perp}}(I-zT)^{-1}I|_{\mathbf{R}(V)^{\perp}}]F, \qquad \frac{1}{z}\in\rho(T).$$

Remark. Let $(\hat{H},\mathbf{F},\mathbf{G};T,F,G,H)$ be a unitary colligation. Then it is clear that $T|_{\mathbf{R}(G^*)}=0$ if and only if $H|_{\mathbf{R}(F^*)}=0$. We have defined the isometric part of T by $V=T|_{\nu(G)}$ and hence we have a necessary and sufficient condition for T to be a partial isometry, i.e., $Tf=Vf$, $f\in\mathbf{D}(V)=\nu(G)$ and $Tf=0$, $f\in\mathbf{D}(V)^{\perp}=\mathbf{R}(G^*)^c$. We have seen that $\theta(z)|_{\nu(F)}=H|_{\nu(F)}$ is an isometry from $\nu(F)$ onto $\nu(G^*)$. By considering $\theta(z)$ as a mapping from $\mathbf{R}(F^*)^c$ into $\mathbf{R}(G^*)^c$, we could delete this isometric part without loss of generality. Completely similar, the characteristic function in Proposition 10.2 characterizes - up to isometries - all functions $\theta\in S_\kappa(\mathbf{F},\mathbf{F})$ with $\theta(0)=0$.

Proposition 10.4. Let $\theta\in S_\kappa(\mathbf{F},\mathbf{F})$ with $\theta(0)=0$. Then there exist a π_κ-space $\hat{H}$, an isometry F from **F** into $\hat{H}$, an isometry V in $\hat{H}$ with trivial extension $\tilde{V}$ and a unitary extension $U\in L(\hat{H})$ of V, such that

$$\theta(z)=F^*[zUP_{\mathbf{D}(V)^{\perp}}(I-z\tilde{V})^{-1}I|_{\mathbf{R}(V)^{\perp}}]F, \qquad \frac{1}{z}\in\rho(\tilde{V}).$$

Proof. In terms of the colligation in Proposition 10.3, we have

$$\theta(z)=F^*[zFG(I-z\tilde{V})^{-1}I|_{\mathbf{R}(V)^{\perp}}]F, \quad \frac{1}{z}\in\rho(\tilde{V}).$$

We show that FG is a partial isometry with kernel $\nu(G)$ and range $\mathbf{R}(F)$. This follows from $FGG^*f=Ff$, $f\in\mathbf{G}$, and hence

$$[FGG^*f,FGG^*f]=[Ff,Ff]=[f,f]=[G^*f,G^*f].$$

Thus FG maps $\mathbf{R}(G^*)=\mathbf{D}(V)^\perp$ isometrically onto $\mathbf{R}(F)=\mathbf{R}(V)^\perp$. Therefore U, defined by $Uf=Vf$, $f \in \mathbf{D}(V)$, $Uf=FGf$, $f \in \mathbf{D}(V)^\perp$, is a unitary operator extension of V, and it is clear that

$$FG=FGP_{\mathbf{D}(V)^\perp}=UP_{\mathbf{D}(V)^\perp}.$$

With regard to boundary values we shall state a proposition, which is a direct consequence of Theorem 9.1. In view of Proposition 10.3 it is sufficient to deal with the situation as described in Proposition 10.1.

Proposition 10.5. Let V be a closed isometric operator in a π_κ-space $\hat{\mathbf{H}}$ with domain $\mathbf{D}(V)$ and range $\mathbf{R}(V)$ such that $\mathbf{D}(V)^\perp$ is a Hilbert space. Let $\tilde{V}$ be the trivial extension of V, and suppose that the unitary colligation

$$(\hat{\mathbf{H}},\mathbf{R}(V)^\perp,\mathbf{D}(V)^\perp;\tilde{V},I|_{\mathbf{R}(V)^\perp},P_{\mathbf{D}(V)^\perp},0)$$

is closely connected. Let θ be its characteristic function. Then, for $\zeta \in \partial\mathbb{D}$, $\mathbf{D}_0(\theta(\zeta))$ is the set of all $\psi \in \mathbf{R}(V)^\perp$ for which there exists a unique element $\phi \in \mathbf{D}(V)^\perp$ such that $\psi-\phi \in \mathbf{R}(I-\zeta V)$, and then

$$\theta(\zeta)\psi=\zeta\phi, \qquad \psi \in \mathbf{D}_0(\theta(\zeta)).$$

In other words, $\mathbf{D}_0(\theta(\zeta))$ is the set of all elements $\psi \in \mathbf{R}(V)^\perp$, that have a projection $P_\zeta\psi$ onto $\mathbf{D}(V)^\perp$, parallel to $\mathbf{R}(I-\zeta V)$, and on those elements $\theta(\zeta)$ acts like ζP_ζ.

Proof. Let $\psi \in \mathbf{D}_0(\theta(\zeta))$, then $\psi \in \mathbf{R}(V)^\perp$ and $\psi=(I-\zeta\tilde{V})\alpha$, for some unique $\alpha \in \hat{\mathbf{H}}$. Now $\alpha=\phi+\chi$, $\phi \in \mathbf{D}(V)^\perp$, $\chi \in \mathbf{D}(V)$, so that $\psi=\phi+(I-\zeta V)\chi$ or $\psi-\phi \in \mathbf{R}(I-\zeta V)$. We may also reverse the argument. Next we observe $\theta(\zeta)\psi=P_{\mathbf{D}(V)^\perp}(I-\zeta\tilde{V})^{-1}\psi = P_{\mathbf{D}(V)^\perp}\alpha = \phi$.

Proposition 10.6 (Schwarz' lemma). Let $\theta \in S_\kappa(\mathbf{F},\mathbf{G})$ and $\theta(0)=0$. We define θ_1 by $\theta_1(z)=\frac{\theta(z)}{z}$, $z \in \mathbf{D}_\theta\backslash\{0\}$, $\theta_1(0)=\theta'(0)$. Then $\theta_1 \in S_\kappa(\mathbf{F},\mathbf{G})$. More specifically, if $\theta=\theta_\Delta$, where $\Delta=(\hat{\mathbf{H}},\mathbf{F},\mathbf{G};T,F,G,0)$ is a closely connected unitary colligation, then

$$\Delta_1=(\nu(G),\mathbf{F},\mathbf{G};P_{\nu(G)}TP_{\nu(G)},P_{\nu(G)}F,GTP_{\nu(G)},GF)$$

<u>is a closely connected unitary colligation of index</u> κ, <u>such that</u> $\theta_1=\theta_{\Delta_1}$. <u>Here</u> $P_{\nu(G)}$ <u>stands for the orthogonal projection from</u> $\hat{\mathbf{H}}$ <u>onto</u> $\nu(G)$.

<u>Proof</u>. Since $\mathbf{R}(G^*)$ is a Hilbert space, it follows that $\nu(G)\subset\hat{\mathbf{H}}$ is a π_κ-space. We shall use the abbreviation P for $P_{\nu(G)}$. It is clear that $P=I-G^*G$. Let us check that the connecting operator

$$U_1=\begin{pmatrix} PTP & PF \\ GTP & GF \end{pmatrix} : \begin{pmatrix} \nu(G) \\ \mathbf{F} \end{pmatrix} \to \begin{pmatrix} \nu(G) \\ \mathbf{G} \end{pmatrix}$$

is unitary. We note that

$$U_1^*U_1=\begin{pmatrix} PT^*PTP+PT^*G^*GTP \\ F^*PTP+F^*G^*GTP \end{pmatrix}\begin{pmatrix} PT^*PF+PT^*G^*GF \\ F^*PF+F^*G^*GF \end{pmatrix}$$

$$=\begin{pmatrix} PT^*TP & PT^*F \\ F^*TP & F^*F \end{pmatrix}=\begin{pmatrix} I_{\nu(G)} & 0 \\ 0 & I_{\mathbf{F}} \end{pmatrix},$$

since $F^*T=0$ and $F^*F=I$. Hence V_1 is isometric and an analogous calculation shows that U_1^* is isometric:

$$U_1U_1^*=\begin{pmatrix} P(TT^*+FF^*)P & P(TT^*+FF^*)G^* \\ G(TT^*+FF^*)P & G(TT^*+FF^*)G^* \end{pmatrix}$$

$$=\begin{pmatrix} P & 0 \\ 0 & GG^* \end{pmatrix}=\begin{pmatrix} I_{\nu(G)} & 0 \\ 0 & I_{\mathbf{G}} \end{pmatrix},$$

since $GT^*=0$, $TT^*+FF^*=I$ and $GG^*=I$. Therefore Δ_1 is a unitary colligation and we claim that Δ_1 is closely connected. Suppose Δ_1 is not closely connected, then there exists by Proposition 3.3 a nontrivial subspace

$$U_1^*U_1=\begin{pmatrix} PT^*PTP+PT^*G^*GTP & PT^*PF+PT^*G^*GF \\ F^*PTP+F^*G^*GTP & F^*PF+F^*G^*GF \end{pmatrix}$$

$$=\begin{pmatrix} PT^*TP & PT^*F \\ F^*TP & F^*F \end{pmatrix}=\begin{pmatrix} I_{\nu(G)} & 0 \\ 0 & I_{\mathbf{F}} \end{pmatrix},$$

since $F^*T=0$ and $F^*F=I$. Hence V_1 is isometric and an analogous calculation shows that U_1^* is isometric:

$$U_1U_1^* = \begin{pmatrix} P(TT^*+FF^*)P & P(TT^*+FF^*)G^* \\ G(TT^*+FF^*)P & G(TT^*+FF^*)G^* \end{pmatrix}$$

$$= \begin{pmatrix} P & 0 \\ 0 & GG^* \end{pmatrix} = \begin{pmatrix} I_{\nu(G)} & 0 \\ 0 & I_{\mathbf{G}} \end{pmatrix},$$

since $GT^*=0$, $TT^*+FF^*=I$ and $GG^*=I$. Therefore Δ_1 is a unitary colligation and we claim that Δ_1 is closely connected. Suppose Δ_1 is not closely connected, then there exists by Proposition 3.3 a nontrivial subspace $\mathbf{H} \subset \nu(G)$, such that $\mathbf{H}$ is mapped isometrically onto itself by PTP. Suppose $PTP\psi=\phi$ for $\psi,\phi \in \mathbf{H}$ with $[\psi,\psi]=[\phi,\phi]$, then $T\psi-\phi=G^*g$ for some $g \in \mathbf{G}$. From the inequality

$$[\psi,\psi]+[G^*g,G^*g]=[T\psi,T\psi] \leqq [\psi,\psi],$$

it follows that $G^*g=0$, so that $T\psi=\phi$. We conclude that $\mathbf{H}$ is mapped isometrically onto itself by T, which contradicts the fact that Δ is closely connected. Hence Δ_1 is closely connected, which implies that $\theta_{\Delta_1} \in S_\kappa(\mathbf{F},\mathbf{G})$. Finally, we identify θ_{Δ_1} and θ_1. We amplify that PTP is a contraction in $\nu(G)$. This follows from the fact that Δ_1 is a unitary colligation. Next we note that $\sigma_p(PTP)\setminus\{0\}=\sigma_p(T)\setminus\{0\}$, so that if $z \in \mathbb{D}$ and $\frac{1}{z} \in \rho(T)$, we also have $\frac{1}{z} \in \rho(PTP)$. For such z the expression

$$I+zTP(I-zPTP)^{-1}P-(I-zT)^{-1}$$

is equal to 0 on $\mathbf{R}(G^*)$, but also on $\nu(G)$, because on $\nu(G)$ it is equal to

$$-(I-zT)^{-1}zT(I-P)(I+zTP)(I-zPTP)^{-1}=0,$$

as $TG^*=0$ and $I-P$ maps onto $R(G^*)$. We conclude that $\theta_{\Delta_1}(z)=GF+zGTP(I-zPTP)^{-1}PF=G(I-zT)^{-1}F=\frac{\theta(z)}{z} = \theta_1(z)$, which is what we wanted to show.

<u>**Corollary 10.7**</u>. <u>If</u> $\theta \in S_0(\mathbf{F},\mathbf{G})$ <u>and</u> $\theta(0)=0$, <u>then</u>

$$[\theta(z)\psi,\theta(z)\psi] \leqq |z|^2[\psi,\psi], \quad \psi \in \mathbf{F},\ z \in \mathbb{D},$$

<u>with equality if and only if</u> $TF\psi=0$, <u>in which case</u>

$$\theta(z)\psi=zGF\psi, \quad z \in \mathbb{D}.$$

Proof. The first statement follows from Proposition 10.6. We have equality if and only if $[\theta_1(z)\psi,\theta_1(z)\psi]=[\psi,\psi], z \in \mathbb{D}$, Using Proposition 8.2 and the closely connected unitary colligation Δ_1 of Proposition 10.6 for θ_1, we obtain equality if and only if $P_{\nu(G)}F\psi=0$. Next we show that $P_{\nu(G)}F\psi=0$ if and only if $TF\psi=0$. It is clear that $P_{\nu(G)}=I-G^*G$, and hence $P_{\nu(G)}F=(I-G^*G)F=T^*TF$. This shows that $TF\psi=0$ implies $P_{\nu(G)}F\psi=0$. Conversely, if $P_{\nu(G)}F\psi=0$, then $F\psi=G^*\chi$ for some $\chi \in \mathbf{G}$, but then $TF\psi=TG^*\chi=0$. Hence we have shown $\nu(P_{\nu(G)}F)=\nu(TF)$. The last statement follows by expanding the characteristic function θ in a Neumann series.

Now let us return to the situation described in Proposition 10.1 and let us assume, in addition, that $\hat{H} = \mathbf{H_L} \oplus \mathbf{L}$, where $\mathbf{L}$ is a Hilbert space. Then the connecting operator of the unitary colligation

$$(\hat{H},\mathbf{R}(V)^{\perp},\mathbf{D}(V)^{\perp};\ \tilde{V},I|_{\mathbf{R}(V)^{\perp}},P_{\mathbf{D}(V)^{\perp}},0)$$

has a matrix decomposition

$$\begin{pmatrix} P_{\mathbf{H_L}}VP_{\mathbf{D}(V)}|_{\mathbf{H_L}} & P_{\mathbf{H_L}}VP_{\mathbf{D}(V)}|_{\mathbf{L}} & P_{\mathbf{H_L}}I|_{\mathbf{R}(V)^{\perp}} \\ P_{\mathbf{L}}VP_{\mathbf{D}(V)}|_{\mathbf{H_L}} & P_{\mathbf{L}}VP_{\mathbf{D}(V)}|_{\mathbf{L}} & P_{\mathbf{L}}I|_{\mathbf{R}(V)^{\perp}} \\ P_{\mathbf{D}(V)^{\perp}}|_{\mathbf{H_L}} & P_{\mathbf{D}(V)^{\perp}}|_{\mathbf{L}} & 0 \end{pmatrix} : \begin{pmatrix} \mathbf{H_L} \\ \mathbf{L} \\ \mathbf{R}(V)^{\perp} \end{pmatrix} \to \begin{pmatrix} \mathbf{H_L} \\ \mathbf{L} \\ \mathbf{D}(V)^{\perp} \end{pmatrix}$$

In this way we obtain a new unitary colligation with inner space $\mathbf{H_L}$ and outer spaces $\begin{pmatrix} \mathbf{L} \\ \mathbf{R}(V)^{\perp} \end{pmatrix}$ and $\begin{pmatrix} \mathbf{L} \\ \mathbf{D}(V)^{\perp} \end{pmatrix}$, and with characteristic function

$$\Phi(z) = \begin{pmatrix} P_{\mathbf{L}}VP_{\mathbf{D}(V)}|_{\mathbf{L}} & P_{\mathbf{L}}I|_{\mathbf{R}(V)^{\perp}} \\ P_{\mathbf{D}(V)^{\perp}}|_{\mathbf{L}} & 0 \end{pmatrix} +$$

$$+z\begin{pmatrix} P_{\mathbf{L}}VP_{\mathbf{D}(V)}|_{\mathbf{H_L}} \\ P_{\mathbf{D}(V)^{\perp}}|_{\mathbf{H_L}} \end{pmatrix}(I - zP_{\mathbf{H_L}}VP_{\mathbf{D}(V)}|_{\mathbf{H_L}})^{-1}(P_{\mathbf{H_L}}VP_{\mathbf{D}(V)}|_{\mathbf{L}}\ \ P_{\mathbf{H_L}}I|_{\mathbf{R}(V)^{\perp}}).$$

If $P_{\mathbf{H_L}}VP_{\mathbf{D}(V)}|_{\mathbf{H_L}}$ does not map any nontrivial subspace of $\mathbf{H_L}$ isometrically onto itself, then

$$\Phi \in S_\kappa\left(\begin{pmatrix} L \\ \mathbf{R}(V)^\perp \end{pmatrix}, \begin{pmatrix} L \\ \mathbf{D}(V)^\perp \end{pmatrix}\right).$$

We have, in addition,

$$P_{\mathbf{D}(V)^\perp}\Phi(0)|_{\mathbf{R}(V)^\perp}=0.$$

Conversely, let $\Phi \in S_\kappa((\begin{smallmatrix}\mathbf{L}\\ \mathbf{M}\end{smallmatrix}),(\begin{smallmatrix}\mathbf{L}\\ \mathbf{N}\end{smallmatrix}))$, where $(\begin{smallmatrix}\mathbf{L}\\ \mathbf{M}\end{smallmatrix})$ and $(\begin{smallmatrix}\mathbf{L}\\ \mathbf{N}\end{smallmatrix})$ are (orthogonal sums of) Hilbert spaces. Then there exists a π_κ-space $\mathbf{H}_\mathbf{L}$ and a unitary operator U mapping

$$\begin{pmatrix} \mathbf{H}_\mathbf{L} \\ \mathbf{L} \\ \mathbf{M} \end{pmatrix} \text{ onto } \begin{pmatrix} \mathbf{H}_\mathbf{L} \\ \mathbf{L} \\ \mathbf{N} \end{pmatrix},$$

such that Φ is the characteristic function of the unitary colligation $(\mathbf{H}_\mathbf{L},(\begin{smallmatrix}\mathbf{L}\\ \mathbf{M}\end{smallmatrix}),(\begin{smallmatrix}\mathbf{L}\\ \mathbf{N}\end{smallmatrix});\ U)$. If, in addition, $P_\mathbf{N}\Phi(0)|_\mathbf{M}=0$, then it follows that $P_{\left(\begin{smallmatrix}\mathbf{H}_\mathbf{L}\\ \mathbf{L}\end{smallmatrix}\right)} U|_{\left(\begin{smallmatrix}\mathbf{H}_\mathbf{L}\\ \mathbf{L}\end{smallmatrix}\right)}$ is a partial isometry.

Remark. Let $\hat{\mathbf{H}}$ be a π_κ-space and let $S \subset \hat{\mathbf{H}}^2$ be a closed symmetric relation (subspace), such that $\nu(S^*-\bar{\mu})$ and $\nu(S^*-\mu)$ are Hilbert spaces for some $\mu \in \mathbb{C}_o$. Let $V=C_{\bar{\mu}}(S)$ be the Cayley transform of S, then V is an isometric operator from $\mathbf{D}(V)=\mathbf{R}(S-\bar{\mu})$ onto $\mathbf{R}(V)=\mathbf{R}(S-\mu)$. If we define $\tilde{V}$ to be the trivial extension of V to all of $\hat{\mathbf{H}}$, then $\tilde{V}=C_{\bar{\mu}}(S(\mu))$, where $S(\zeta)=S\dot{+}\{\{\phi,\zeta\phi\}|\phi \in \nu(S^*-\zeta)\},\zeta \in \mathbb{C}$. In this way we obtain a direct sum decomposition

$$\hat{\mathbf{H}}=\mathbf{R}(S-\ell)+\nu(S^*-\mu), \quad \ell \in \mathbb{C}_{\bar{\mu}}, \quad \frac{\ell-\mu}{\ell-\bar{\mu}} \in \rho(C_{\bar{\mu}}(S(\mu))),$$

and a corresponding parallel projection given by

$$P_{\nu(S^*-\mu)}(I-\frac{\ell-\bar{\mu}}{\ell-\mu}C_{\bar{\mu}}(S(\mu)))^{-1}.$$

If we denote the characteristic function of $\tilde{V}$ by θ, then

$$\theta(\frac{\ell-\bar{\mu}}{\ell-\mu})=\frac{\ell-\bar{\mu}}{\ell-\mu}P_{\nu(S^*-\mu)}(I-\frac{\ell-\bar{\mu}}{\ell-\mu}C_{\bar{\mu}}(S(\mu)))^{-1}|_{\nu(S^*-\bar{\mu})}$$

is called the characteristic function of the symmetric subspace S. Using

Proposition 10.5 we see that, for $\lambda \in \mathbb{R}\cup\{\infty\}$, $\mathbf{D}_0(\theta(\frac{\lambda-\bar{\mu}}{\lambda-\mu}))$ is the set of all

$\psi \in \nu(S^*-\bar{\mu})$ for which there exists a (unique) $\phi \in \nu(S^*-\mu)$ such that $\psi-\phi \in \mathbf{R}(S-\lambda)$ if $\lambda \in \mathbb{R}$, and such that $\psi-\phi \in \mathbf{D}(S)$ if $\lambda=\infty$, and then

$$\theta(\frac{\lambda-\bar{\mu}}{\lambda-\mu})\psi = \frac{\lambda-\bar{\mu}}{\lambda-\mu}\phi.$$

In the Hilbert space case it is not difficult to show that

$$\theta(\frac{\ell-\bar{\mu}}{\ell-\mu}) = C_\mu(S(\ell))|_{\nu(S^*-\bar{\mu})}, \qquad \ell \in \mathbb{C}_\mu^-,$$

and

$$\theta(\frac{\lambda-\bar{\mu}}{\lambda-\mu})\psi = C_\mu(S(\lambda))\psi, \qquad \psi \in \mathbf{D}_0(\theta(\frac{\lambda-\bar{\mu}}{\lambda-\mu})),\ \lambda \in \mathbb{R}\cup\{\infty\}.$$

Characteristic functions of partial isometries in Hilbert spaces were first studied by Livšic and Smuljan, see for instance the references in [1]. In the setting of π_κ-spaces this work was continued by Krein and Langer [13] and [15]. Proposition 10.6 allows a generalization, which is connected with results of Chamfy [7]. This will be treated elsewhere. The results mentioned after Corollary 10.7 in a Hilbert space setting are due to Arov and Grossman [1]. Characteristic functions of symmetric operators in Hilbert spaces were studied by Štraus [21], [24].

REFERENCES

[1] D.Z. Arov and L.Z. Grossman, Scattering matrices in the theory of dilations of isometric operators, Dokl. Akad. Nauk SSSR 270 (1983) 17-20 (English translation: Sov. Math. Dokl. 27 (1983) 518-522).

[2] T. Ya Azizov, Extensions of J-isometric and J-symmetric operators, Functional Anal. i Priložen 18 (1984) 57-58 (English translation: Functional Anal. Appl. 18 (1984) 46-48).

[3] J.A. Ball and J.W. Helton, A Beurling-Lax theorem for the Lie group U(m,n) which contains most classical interpolation theory, J. Operator Theory 9 (1983) 107-142.

[4] H. Bart, I. Gohberg and M.A. Kaashoek, Minimal factorization of matrix and operator functions, Birkhäuser, Basel-Boston-Stuttgart,1979.

[5] J. Bognar, Indefinite Inner Product Spaces, Springer-Verlag, Berlin-Heidelberg-New York, 1974.

[6] M.S. Brodskii, Unitary operator colligations and their characteristic functions, Uspekhi Mat. Nauk 33:4 (1978) 141-168 (English translation: Russian Math. Surveys 33:4 (1978) 159-191).

[7] C. Chamfy, Fonctions méromorphes dans le cercle-unité et leurs séries de Taylor, Annales de l'Institut Fourier 8 (1958) 211-261.

[8] A. Dijksma, H. Langer and H.S.V. de Snoo, Unitary colligations in π_κ-spaces, characteristic functions and Straus extensions, Pacific J. Math., to appear.

[9] O.P. Filimonov, The inverse problem of the theory of characteristic operator functions of metric colligations in the space Π_k, Dokl. Akad. Nauk. Ukrain. SSR Ser. A 1977, No. 9, 798-802.

[10] I.C. Gohberg and E.I. Sigal, An operator generalization of the logarithmic residue theorem and the theorem of Rouché, Mat. Sbornik 84 (1971) 607-630 (English translation: Math. USSR Sbornik 13 (1971) 603-625).

[11] I.S. Iohvidov, M.G. Krein and H. Langer, Introduction to the Spectral Theory of Operators in Spaces with an indefinite Metric, Akademie Verlag, Berlin, 1982.

[12] T. Kato, Perturbation Theory for Linear Operators, Springer Verlag, Berlin-Heidelberg-New York, 1966.

[13] M.G. Krein and H. Langer, Über die verallgemeinerten Resolventen und die charakteristische Funktion eines isometrischen Operators im Raume Π_k, Hilbert Space Operators and Operator Algebras (Proc. Int. Conf., Tihany, 1970) Colloquia Math. Soc. János Bolyai, no. 5, North-Holland, Amsterdam (1972) 353-399.

[14] M.G. Krein and H. Langer, Über die Q-Funktion eines π-hermiteschen Operators im Raume Π_k, Acta Sci. Math. 34 (1973) 191-230.

[15] M.G. Krein and H. Langer, Über einige Fortsetzungsprobleme, die eng mit der Theorie hermitescher Operatoren im Raume Π_k zusammen-hängen, Teil I: Einige Funktionenklassen und ihre Darstellungen, Math. Nachr., 77 (1977) 187-236.

[16] M.G. Krein and H. Langer, Some propositions on analytic matrix functions related to the theory of operators in the space Π_k, Acta Sci. Math. (Szeged) 43 (1981) 181-205.

[17] H. Langer, Spektraltheorie Linearer Operatoren in J-Räumen und einige Anwendungen auf die Schar $L(\lambda) = \lambda^2 I + \lambda B + C$, Habilitationsschrift, Technische Universität Dresden, 1965.

[18] H. Langer, Spectral functions of definitizable operators in Krein spaces, in "Functional Analysis, Proceedings Dubrovnik 1981", Lecture Notes in Mathematics 948, Springer-Verlag, Berlin-Heidelberg-New York, 1982.

[19] R. McKelvey, The spectra of minimal self-adjoint extensions of a symmetric operator, Pacific J. Math. 12 (1962) 1003-1022.

[20] P. Sorjonen, Pontryaginräume mit einem reproduzierenden Kern, Ann. Acad. Sci. Fennicae A.I. 594 (1975).

[21] A.V. Štraus, Characteristic functions of linear operators, Izv. Akad. Nauk SSSR Ser. Mat. 24 (1960) 43-74 (English translation: Amer. Math. Soc. Transl. (2) 40 (1964) 1-37).

[22] A.V. Štraus, On the extensions of symmetric operators depending on a parameter, Izv. Akad. Nauk SSSR Ser. Mat. 29 (1965) 1389-1416 (Englisch translation: Amer. Math. Soc. Transl. (2) 61 (1967) 113-141).

[23] A.V. Štraus, On one-parameter families of extensions of a symmetric operator, Izv. Akad. Nauk SSSR Ser. Mat. 30 (1966) 1325-1352 (English translation: Amer. Math. Soc. Transl. (2) 90 (1970) 135-164).

[24] A.V. Štraus, On the extensions and the characteristic function of a symmetric operator, Izv. Akad. Nauk SSSR Ser. Mat. 32 (1968) 186-207 (English translation: Math. USSR Izv. 2 (1968) 181-204).

[25] A.V. Štraus, Extensions and generalized resolvents of a symmetric operator which is not densely defined, Izv. Akad. Nauk SSSR Ser. Mat. 34 (1970) 175-202 (English translation: Math. USSR Izv. 4 (1970) 179-208.

[26] B. Sz.-Nagy and C. Foias, Harmonic Analysis of Operators on Hilbert Space, North-Holland Publishing Company, Amsterdam-London,1970.

A. Dijksma and H.S.V. de Snoo
Mathematisch Instituut
Rijksuniversiteit Groningen
Postbus 800
9700 AV GRONINGEN
Nederland

H. Langer
Technische Universität
Sektion Mathematik
DDR-8027 Dresden
Mommsenstrasse 13
D.D.R.

Operator Theory:
Advances and Applications, Vol. 19

THE MAXIMUM DISTANCE PROBLEM IN HILBERT SPACE

Robert L. Ellis, Israel Gohberg, and David C. Lay

In this paper we solve the following problem. Given an $(m+1)\times(m+1)$ positive definite Toeplitz matrix $R = (r_{j-k})$, we wish to find an infinite sequence of linearly independent vectors $x_1, x_2, \ldots$, in ℓ^2 such that $\langle x_k, x_j\rangle = r_{j-k}$ for $|j-k| \leq m$ and such that the distance from x_k to the linear span of $x_1, \ldots, x_{k-1}$ is maximal for $k \geq 2$. We construct a solution and show how to obtain all other solutions from it. This problem and its solution have close connections with maximum entropy in the mathematical theory of signal processing [3].

1. INTRODUCTION

We begin with some notation and terminology. For any vectors $x_1, \ldots, x_k$ in an inner product space (possibly infinite-dimensional), let $sp\langle x_1, \ldots, x_k\rangle$ denote the subspace spanned by $x_1, \ldots, x_k$ and let $dist(x_k, sp\langle x_1, \ldots, x_{k-1}\rangle)$ denote the distance from x_k to $sp\langle x_1, \ldots, x_{k-1}\rangle$. Given an integer $m \geq 0$ and a selfadjoint Toeplitz matrix $R = (r_{j-k})_{j,k=0}^{m}$, we shall say that an infinite sequence of vectors $x_1, x_2, \ldots$, in ℓ^2 is <u>admissible</u> if finite subsets of the sequence are linearly independent and if

$$\langle x_k, x_j\rangle = r_{j-k} \qquad (|j-k) \leq m) \tag{1.1}$$

We define

$$d_k = \sup dist(x_k, sp\langle x_1, \ldots, x_{k-1}\rangle) \qquad (k \geq 2) \tag{1.2}$$

where the supremum is taken over all admissible sequences of vectors in the space. The <u>maximum distance problem</u> (for R) is to describe all admissible sequences $\{x_k\}$ such that

$$dist(x_k, sp\langle x_1, \ldots, x_{k-1}\rangle) = d_k \qquad (k \geq 2)$$

Each such sequence is called a solution of the maximum distance problem for R. This maximum distance problem has close connections with maximum entropy in the mathematical theory of signal procesing [3,5,6].

We denote the Gram matrix of vectors $x_1,\dots,x_n$ by

$$\Gamma = \Gamma(x_1,\dots,x_n) = (\langle x_k,x_j\rangle)_{j,k=1}^{n}.$$

For any n, the constraints in (1.1) specify the entries in $\Gamma(x_1,\dots,x_n)$ that lie in a band centered along the main diagonal. Clearly, if the maximum distance problem with constraints (1.1) has a solution, then the Toeplitz matrix R is positive definite. We shall prove the converse and characterize all solutions of the problem.

In Section 2 we use some of the results of [2] to describe the maximum distance problem in a finite dimensional space and show how the vectors in the solution of the maximum distance problem change as the dimension increases. Then, in Section 3, we solve the maximum distance problem in ℓ^2. The results of Section 2 allow us to build the vectors in the solution by solving finite dimensional problems.

2. THE FINITE-DIMENSIONAL PROBLEM

We restrict our attention here to vectors that belong to $\mathbf{C}^n$, where $n > m$. Given an $(m+1)\times(m+1)$ Toeplitz matrix R as before, we say that a set $\{x_1,\dots,x_n\}$ in $\mathbf{C}^n$ is admissible if it is linearly independent and (1.1) holds. For $2 \leq k \leq n$, we define the numbers d_k as in (1.2), taking the supremum over all admissible sets in $\mathbf{C}^n$. The maximum distance problem in this case was posed and solved in [2], where the matrix R was any positive definite matrix. We shall recall the main result of that paper, here stated for the case in which R is a Toeplitz matrix. For any matrix $B = (b_{pq})_{p,q=1}^{n}$ and for $1 \leq j \leq k \leq n$, we denote by $B(j,\dots,k)$ the principal submatrix $(b_{pq})_{p,q=j}^{k}$.

THEOREM 1. *Let* m *be a positive integer and let* $R = (r_{j-k})_{j,k=0}^{m}$ *be a positive definite Toeplitz matrix. For any integer* $n > m$ *the system of equations*

$$(2.1)\qquad R\begin{bmatrix} w_{jj} \\ w_{j+1,j} \\ \vdots \\ w_{j+m,j} \end{bmatrix} = \begin{bmatrix} w_{jj}^{-1} \\ 0 \\ \vdots \\ 0 \end{bmatrix} \qquad (1 \leqslant j \leqslant n-m)$$

$$(2.2)\qquad R(1,\ldots,n+1-j)\begin{bmatrix} w_{jj} \\ w_{j+1,j} \\ \vdots \\ w_{nj} \end{bmatrix} = \begin{bmatrix} w_{jj}^{-1} \\ 0 \\ \vdots \\ 0 \end{bmatrix} \qquad (n-m < j \leqslant n)$$

defines an invertible lower-triangular band matrix W_n *with bandwidth* m *whose entries inside the lower halfband are defined by (2.1) and (2.2), with* $w_{jj} > 0$ *for* $1 \leq j \leq n$. *Let* $X_n = W_n^{-1}$, *and denote the columns of* X_n *by* $x_1,\ldots,x_n$. *Then* $\{x_k\}_{k=1}^n$ *is a solution of the maximum distance problem for* R. *Any other set of vectors is a solution of the maximum distance problem for* R *if and only if it is of the form* $Ux_1,\ldots,Ux_n$, *where* U *is an n×n unitary matrix. In particular, all solutions of the maximum distance problem for* R *have the same Gram matrix.*

If $\{x_1,\ldots,x_n\}$ is the solution of the maximum distance problem constructed in Theorem 1, then the Gram matrix $\Gamma = \Gamma(x_1,\ldots,x_n)$ has a Toeplitz structure within the band, since $\langle x_k, x_j\rangle = r_{j-k}$ for $|j-k| \leqslant m$. In fact, Γ is itself a Toeplitz matrix. This was demonstrated in [5] in a slightly different context. We can show this by obtaining recursion formulas for the entries in Γ outside the band. Observe that $\Gamma W = (X^*X)X^{-1} = X^* = W^{*-1}$. For $1 \leqslant k < j \leqslant n$, the product of row j of Γ and column k of W is zero because ΓW is upper triangular. Therefore, if $\Gamma = (g_{jk})$ and $k \leqslant n-m$, then

$$g_{jk}w_{kk} + g_{j,k+1}w_{k+1,k} + \cdots + g_{j,k+m}w_{k+m,k} = 0$$

so that

$$(2.3) \qquad g_{jk} = -(g_{j,k+1}w_{k+1,k} + \cdots + g_{j,k+m}w_{k+m,k})w^{-1}_{kk}$$

for $k<j$, $k\leq n-m$. Since $k \leq n-m$, the terms $w_{i,k}$ appearing in (2.3) depend only on k-i, because of (2.1). If $j-k = m+1$, then the terms g_{ji} on the right side of (2.3) equal r_{i-j}. It follows from (2.3) that g_{jk} depends only on k-j, when $j-k = m+1$. This implies that the terms on the right of (2.3) depend only on k-j when $j-k=m+2$. Hence g_{jk} depends only on k-j when $j-k=m+2$. Repeating this argument, we see that the entries in Γ are constant on the diagonals below the band. Since Γ is Hermitian, it follows that Γ is a Toeplitz matrix.

The next theorem describes the behavior of the solution of the maximum distance problem in $\mathbf{C}^n$ as the dimension n increases, while R remains fixed.

THEOREM 2. *Let* $R = (r_{j-k})^m_{j,k=0}$ *be a positive definite Toeplitz matrix. For* $n > m$, *let* Γ_n *be the Gram matrix of any solution of the maximum distance problem in* $\mathbf{C}^n$ *determined by* R. *Let* W_n *and* X_n *be the lower-triangular matrices, defined in Theorem 1, such that* $\Gamma_n = X^*_n X_n$ *and* $\Gamma^{-1}_n = W_n W^*_n$. *Then the leading principal submatrices* $W_n(1,\ldots,n-m)$ *and* $X_n(1,\ldots,n-m)$ *are lower-triangular Toeplitz matrices, and*

$$(2.4) \qquad W_q(1,\ldots,n-m) = W_n(1,\ldots,n-m) \qquad (q \geq n)$$

$$(2.5) \qquad X_q(1,\ldots,n-m) = X_n(1,\ldots,n-m) \qquad (q \geq n)$$

$$(2.6) \qquad \Gamma_q(1,\ldots,n) = \Gamma_n \qquad (q \geq n)$$

PROOF. It follows from (2.1) that the first m+1 entries in the first column of W_n are independent of n. (The rest of the entries in the first column are zeros.) Also, each of the first n-m columns of W_n except the first is a downward shift of the column to its left, with zeros in each column above the main diagonal. Hence $W_n(1,\ldots,n-m)$ is a lower-triangular Toeplitz matrix, and (2.4) is evident.

It follows from (2.2) that the last m columns of W_n are independent of n. Thus W_n has the form

$$W_n = \begin{bmatrix} T_n & 0 \\ U_n & V \end{bmatrix}$$

where $T_n = W_n(1,\ldots,n-m)$, U_n is an $m\times(n-m)$ matrix whose nonzero columns do not depend on n for $n > 2m$, and V is an $m\times m$ invertible lower-triangular matrix that does not depend on n. Therefore

$$X_n = W_n^{-1} = \begin{bmatrix} T_n^{-1} & 0 \\ -V^{-1}U_nT_n^{-1} & V^{-1} \end{bmatrix}.$$

Since T_n is a lower-triangular Toeplitz matrix, so is its inverse. Thus $X_n(1,\ldots,n-m)$ is a Toeplitz matrix.

To verify (2.5), let $p = n-m-1$ and let

$$T_n = (t_{j-k})_{j,k=0}^{p}, \quad \text{and} \quad T_n^{-1} = (\xi_{j-k})_{j,k=0}^{p}$$

where $t_k = 0$ for $m < k \leqslant p$ and $t_k = 0$ for $k < 0$. Since T_n and T_n^{-1} are lower-triangular and Toeplitz, the equation $T_n^{-1}T_n = I$ is equivalent to the equations

$$\begin{aligned} t_0\xi_0 &= 1 \\ t_0\xi_1 + t_1\xi_0 &= 0 \\ &\vdots \\ t_0\xi_p + \ldots + t_p\xi_0 &= 0. \end{aligned}$$

However, since $t_k = 0$ for $m < k \leqslant p$, these equations simplify to

$$\begin{aligned} t_0\xi_0 &= 1 \\ t_0\xi_1 + t_1\xi_0 &= 0 \\ &\vdots \\ t_0\xi_{m-1} + \ldots + t_{m-1}\xi_0 &= 0 \end{aligned} \tag{2.7}$$

and

$$t_0\xi_k + \dots + t_m\xi_{k-m} = 0 \qquad (m \leq k \leq p) \tag{2.8}$$

The entries $\xi_0,\dots,\xi_{m-1}$ are determined by the "initial conditions" in (2.7), and $\xi_m,\dots,\xi_p$ are uniquely determined by the mth order difference equation (2.8). Compare [1]. In fact, $t_0,\dots,t_m$ are independent of n, as we noted at the beginning of the proof. Hence there is an infinite sequence $\{\xi_k\}_{k=-\infty}^{\infty}$ such that $\xi_k = 0$ for $k < 0$ and for any $n > m$,

$$X_n(1,\dots,n-m) = T_n^{-1} = (\xi_{j-k})_{j,k=0}^{n-m-1}. \tag{2.9}$$

This implies (2.5).

To prove (2.6) we write $\Gamma_q^{-1} = YY^*$, where Y is upper-triangular and recall that Γ_q^{-1} is a band matrix, so that $Y = \Gamma_q^{-1}Y^{*-1}$ is an upper-triangular band matrix. Since Y^{*-1} is lower-triangular and Y^{-1} is upper triangular, it follows that

$$\begin{aligned}\Gamma_q(1,\dots,n) &= (Y^{*-1}Y^{-1})(1,\dots,n)\\ &= Y^{*-1}(1,\dots,n)Y^{-1}(1,\dots,n)\\ &= Y(1,\dots,n)^{*-1}Y(1,\dots,n)^{-1}.\end{aligned}$$

Consequently,

$$\Gamma_q(1,\dots,n)^{-1} = Y(1,\dots,n)Y(1,\dots,n)^*$$

Therefore $\Gamma_q(1,\dots,n)^{-1}$ is a band matrix. We may write $\Gamma_q(1,\dots,n)^{-1} = LL^*$, where L is a lower-triangular matrix with positive diagonal entries. Then L is banded, since $\Gamma_q(1,\dots,n)^{-1}$ is banded. Since Γ_q agrees with R in the band, it follows that the entries of L satisfy the same equations (2.1) and (2.2) as the entries of W_n. By the uniqueness of the solution of (2.1) and (2.2) with positive diagonal entries, we have $L = W_n$, so that $\Gamma_q(1,\dots,n)^{-1} = LL^* = \Gamma_n^{-1}$. Therefore $\Gamma_q(1,\dots,n) = \Gamma_n$.

3. THE INFINITE-DIMENSIONAL CASE

In this section we describe the maximum distance problem in ℓ^2. We fix a Toeplitz matrix $R = (r_{j-k})_{j,k=1}^{m}$ and we consider admissible sequences $\{x_k\}_{k=1}^{\infty}$ of vectors in ℓ^2 as defined in the introduction. Given such an admissible sequence, let X be the matrix whose columns are $x_1, x_2, \ldots$. For $k \geq 2$ let $d_k(X)$ be the ℓ^2-distance from x_k to the linear span of $x_1, \ldots, x_{k-1}$. Also, let

$$d_k = \sup\{d_k(X)\} \qquad (k \geq 2)$$

where the supremum is over all X whose columns form an admissible sequence. We wish to describe all admissible sequences $\{x_k\}$ of vectors in ℓ^2 such that

$$(3.1) \qquad d_k(X) = d_k \qquad (k \geq 2)$$

Theorem 2 suggests that one solution of the infinite-dimensional maximum distance problem should be formed by the columns of an infinite-dimensional lower-triangular Toeplitz matrix X. Our next theorem shows that this is indeed the case.

In solving the maximum distance problem we will use the Gram matrix $\Gamma = \Gamma(X)$ of an admissible sequence formed from the columns of the matrix X, that is,

$$\Gamma(X) = X^*X = (\langle x_k, x_j\rangle)_{j,k=1}^{\infty}.$$

It is known (cf. [7, page 14]) that the Gram determinant has the property that

$$(3.2) \qquad [d_k(X)]^2 = \frac{\det \Gamma(X)(1,\ldots,k)}{\det \Gamma(X)(1,\ldots,k-1)} \qquad (k \geq 2)$$

The distances in (3.2) are positive for all k, because the sequence $\{x_k\}$ is linearly independent; also, $\Gamma(X)(1,\ldots,1) = \langle x_1, x_1\rangle$, which is positive. It follows that every principal section of $\Gamma(X)$ is positive definite.

THEOREM 3. *Let* $R = (r_{k-j})_{j,k=0}^{m}$. *The maximum distance problem in* ℓ^2 *determined by* R *has a solution if and only if* R *is positive definite. In that case, let* $b = (b_0,\ldots,b_m)^T$ *be the unique solution of*

$$(3.3) \qquad Rb = \begin{bmatrix} b_0^{-1} \\ 0 \\ \vdots \\ 0 \end{bmatrix}$$

with $b_0 > 0$, *and define a polynomial* p *by*

$$(3.4) \qquad p(z) = \sum_{k=0}^{m} b_k z^k$$

Then $1/p(z)$ *has the form*

$$1/p(z) = \sum_{k=0}^{\infty} \xi_k z^k \qquad (|z| \leqslant 1)$$

for some ℓ^1*-sequence* $(\xi_k)_{k=0}^{\infty}$. *A sequence of vectors in* ℓ^2 *is a solution of the maximum distance problem for* R *if and only if it has the form* $Ux, USx, US^2x, \ldots,$ *where* $x = (\xi_0,\xi_1,\ldots)^T$, S *is the forward shift operator, and* U *is an isometry of* ℓ^2.

PROOF. If there is a solution $\{x_k\}$, then R is positive definite since it is a principal section of $\Gamma(X)$. Now assume that R is positive definite. For $n > m$, let X_n be the matrix in Theorem 2, and let $\Gamma_n = X_n^* X_n$. Then the columns of X_n form a solution of the maximum distance problem in $\mathbf{C}^n$ determined by R. We now construct a bounded linear operator $\tilde{X}_n$ on ℓ^2 by letting $\tilde{X}_n$ act as X_n on vectors of the form $(\eta_1,\eta_2,\ldots,\eta_n,0,0,\ldots)$ and as the zero operator on vectors of the form $(0,\ldots,0,\eta_{n+1},\eta_{n+2},\ldots)$. Operators $\tilde{X}_n^*$ and $\tilde{\Gamma}_n$ corresponding to X_n^*, and Γ_n, respectively, are constructed similarly. Then

(3.5) $$\tilde{\Gamma}_n = \tilde{X}_n^* \tilde{X}_n.$$

Let b and p be as in the statement of the theorem. By (2.1), $b_0,\ldots,b_m$ are the entries that lie in the first column of the band of X_n^{-1} for $n > m$. Let

(3.6) $$x = (\xi_0, \xi_1, \ldots)^T$$

be the solution of (2.7) and (2.8) with $t_k = b_k$ for $0 \leq k \leq m$, and define $\xi_k = 0$ for $k < 0$. Let X and X^* be the Toeplitz matrices defined by

$$X = (\xi_{j-k})_{j,k=0}^{\infty} \quad \text{and} \quad X^* = (\overline{\xi}_{k-j})_{j,k=0}^{\infty}.$$

If x is in ℓ^1 then it will follow immediately from a Theorem of Schur [9, page 226] that X and X^* represent bounded linear operators on ℓ^2, which we denote by $\tilde{X}$ and $\tilde{X}^*$, respectively. Then, from (2.9) it is clear that $\tilde{X}_n$ and $\tilde{X}_n^*$ converge pointwise on ℓ^2 to $\tilde{X}$ and $\tilde{X}^*$, respectively.

We shall prove next that x is in ℓ^1. It follows from a result due to Szegö that all zeros of the polynomial $p(z)$ lie outside the unit disk $|z| \leq 1$. (For recent proofs, see [8] and the proof of Theorem 6.1 in [4].) Hence $1/p(z)$ is analytic in $|z| \leq 1$ and therefore has the form

(3.7) $$1/p(z) = \sum_{k=0}^{\infty} c_k z^k \qquad (|z| \leq 1)$$

for some ℓ^1-sequence $\{c_k\}$. From (3.7) it is clear that $\{c_k\}$ is the unique solution of the difference equation in (2.7) and (2.8) with $t_k = b_k$ for $0 \leq k \leq m$, and hence $c_k = \xi_k$ for $k \geq 0$. Thus x is in ℓ^1.

Since $\tilde{X}_n$ and $\tilde{X}_n^*$ converge pointwise, the sequences of their norms are bounded, by the uniform boundedness principle.

From this and the identity

$$\tilde{X}_n^*\tilde{X}_n(z) - \tilde{X}^*\tilde{X}(z) = \tilde{X}_n^*(\tilde{X}_n(z) - \tilde{X}(z)) + (\tilde{X}_n^* - \tilde{X}^*)(\tilde{X}(z))$$

for any z in ℓ^2 it follows that $\tilde{X}_n^*\tilde{X}_n$ converges pointwise on ℓ^2 to $\tilde{X}^*\tilde{X}$. Let $x_1, x_2, \ldots$ be the columns of X. Then $\Gamma(X) = X^*X$. Hence from (3.5), $\tilde{\Gamma}_n$ converges pointwise to the operator corresponding to $\Gamma(X)$. Furthermore, from (2.6) it is clear that

$$\Gamma(X)(1,\ldots,n) = \Gamma_n \qquad (n > m) \tag{3.8}$$

In particular,

$$\Gamma(X)(1,\ldots,m+1) = \Gamma_{m+1} = R.$$

This implies that $\{x_k\}$ is an admissible sequence.

Let Y be a matrix whose columns $\{y_k\}$ form an admissible sequence in ℓ^2 and let $G = \Gamma(Y)$. For $n > m$, $G(1,\ldots,n)$ is the Gram matrix of some admissible set of vectors $z_1,\ldots,z_n$ for the maximum distance problem in $\mathbf{C}^n$ determined by R. Let d_n^o denote the nth maximum distance in $\mathbf{C}^n$, defined in (1.2). Then, by (3.2) and (3.8)

$$[d_n(Y)]^2 = \frac{\det G(1,\ldots,n)}{\det G(1,\ldots,n-1)} = [d_n(z_1,\ldots,z_n)]^2 \leq (d_n^o)^2$$

$$= \frac{\det \Gamma_n}{\det \Gamma_{n-1}} = \frac{\det \Gamma(X)(1,\ldots,n)}{\det \Gamma(X)(1,\ldots,n-1)} = [d_n(X)]^2.$$

Therefore X satisfies (3.1), so the columns of X form a solution of the maximum distance problem for R. Since X is represented by a Toeplitz matrix, the columns of X have the form $x, Sx, S^2x, \ldots,$ where x is given by (3.6) and S is the forward shift operator.

Clearly, the columns of UX also form a solution of the maximum distance problem for any isometry U. It remains to prove that every solution has this form. Let Y a matrix whose columns $\{y_k\}$ form a solution of the maximum distance problem. By (3.2)

$$\frac{\det \Gamma(Y)(1,\ldots,k)}{\det \Gamma(Y)(1,\ldots,k-1)} = \left(d_k^o\right)^2 = \frac{\det \Gamma(X)(1,\ldots,k)}{\det \Gamma(X)(1,\ldots,k-1)} \qquad (k \geq 2)$$

It follows from Theorem 1 that $\Gamma(Y) = \Gamma(X)$. Hence $\Gamma(Y)$ determines a bounded operator on ℓ^2. It is easy to see form the Cauchy-Schwarz inequality that Y determines a bounded linear operator. Also, $Y^*Y = X^*X$. Let $U = YX^{-1}$, so that $Y = UX$. Then U^*U is the identity matrix and so U represents an isometry on ℓ^2.

In general, the operator represented by U in Theorem 3 need not be unitary. For example, if U is the matrix for the right shift operator and $Y = UX$, then $\Gamma(Y) = \Gamma(X)$ and the columns of Y form a solution of the maximum distance problem.

REFERENCES

1. W. Barrett: Toeplitz matrices with banded inverses, Linear Algebra and Appl. 57 (1984), 131-145.

2. A. Ben-Artzi, R.L. Ellis, I. Gohberg, and D.Lay: The maximum distance problem and band sequences, Linear Algebra and Appl., to appear.

3. J. Burg: Maximum entropy spectral analysis, Ph.D. dissertation, Stanford University, Stanford, California, 1975.

4. H. Dym and I. Gohberg: Extensions of matrix valued functions with rational polynomial inverses, Integral Equations and Operator Theory, 2 (1979), 503-528.

5. H. Dym and I. Gohberg: Extensions of band matrices with band inverses, Linear Algebra and Appl. 36 (1981), 1-24.

6. R.L. Ellis, I. Gohberg, and D.C. Lay: Band extensions, maximum entropy and the permanence principle, in Maximum Entropy and Bayesian Methods in Applied Statistics, ed. by J. Justice, Cambridge Univ. Press, 1986.

7. I. Gohberg and S. Goldberg: Basic Operator Theory, Birkhauser, Boston, Massachusetts, 1981.

8. H. J. Landau: Maximum entropy and the moment problem, preprint.

9. A.E. Taylor and D.C. Lay: Introduction to Functional Analysis, 2ed., John Wiley & Sons, New York, 1980.

Robert L. Ellis and David C. Lay
Department of Mathematics
University of Maryland
College Park, MD 20740
U.S.A.

Israel Gohberg
School of Mathematical Sciences
Tel Aviv University
Ramat Aviv
69 978 Tel Aviv
ISRAEL

Operator Theory:
Advances and Applications, Vol. 19

BLOCK HANKEL MATRIX INVERSION - THE POLYNOMIAL APPROACH

Paul A. Fuhrmann

The paper provides an analysis of the problem of block Hankel matrix inversion. It places the problem in the context of partial realization theory. In particular the inverse of a block Hankel matrix is identified with a generalized Bezoutiant.

1. Introduction

The object of this paper is to extend the polynomial approach to the problem of inverting Hankel and Toeplitz matrices to the multivariable case. The advantage of this approach is that it is a conceptual one, makes contact with one of the fundamental problems of algebraic systems theory, namely the partial realization problem.

As can be expected from the study of the scalar problem, see in this connection Fuhrmann [1985,1986], generalized Bezoutiants which were originally introduced in Anderson and Jury [1976] play a significant role. The representation formulas for Bezoutiants obtained by Kravitsky [1980] and Ptak [1984] are extended to generalized Bezoutiants. The representation of the inverse of a block Hankel matrix as a generalized Bezoutiant is obtained. A more specialized result of this type has been obtained by Lerer [1985]. This extends the work of Lander [1974]. Finally a simple derivation of the result of Gohberg and Heinig [1974] is presented.

2. Generalized Bezoutiants as block Hankel inverses

Let us consider the block Hankel matrix

This research was partially supported by the NSF Grant No. ECS-8451519

$$H = \begin{bmatrix} G_1 & . & . & . & G_n \\ . & . & . & . & . \\ . & . & . & . & . \\ . & . & . & . & . \\ G_n & . & . & . & G_{2n-1} \end{bmatrix} \tag{2.1}$$

We assume that this matrix is nonsingular, In particular this implies the G_i are square, say $p \times p$, matrices.

A minimal rational extension of the sequence of matrices $G_1, \ldots, G_{2n-1}$ is an infinite sequence of matrices $G_1, \ldots, G_{2n-1}, G_{2n}, \ldots$ such that every integer $m > n$

$$\operatorname{rank} \begin{bmatrix} G_1 & . & . & . & G_m \\ . & . & . & . & . \\ . & . & . & . & . \\ . & . & . & . & . \\ G_m & . & . & . & G_{2m-1} \end{bmatrix} = np$$

This in turn implies that the matrix function

$$G(z) = \Sigma\, G_i z^{-1} \tag{2.2}$$

is rational of McMillan degree np.

A minimal (canonical) realization of a minimal rational extension of $G_1, \ldots, G_{2n-1}$ is called a minimal partial realization of the sequence. For the basic theory of partial realizations see Kalman [1979] and Gragg and Lindquist [1983].

It is obvious that a minimal rational extension of $G_1, \ldots, G_{2n-1}$, with the matrix H of (2.1) nonsingular, is completely determined by $X = G_{2n}$. We will denote the corresponding rational extension by $G_X(z)$. So

$$G_X(z) = G_1/z + \ldots + G_{2n-1}/z^{2n-1} + X/z^{2n} + \ldots \tag{2.3}$$

We will write $G(z)$ for the minimal rational extension corresponding to $X = G_{2n} = 0$.

Like any rational function G has a right coprime matrix fraction representation

$$G_X(z) = P_X(z) Q_X(z)^{-1} \tag{2.4}$$

The matrices P_X and Q_X are uniquely determined up to a common right unimodular factor. Without loss of generality we may assume Q_X to be monic of degree n, if we solve the equation.

$$\begin{pmatrix} G_1 & \cdot & \cdot & \cdot & G_n \\ \cdot & \cdot & \cdot & \cdot & \cdot \\ \cdot & \cdot & \cdot & \cdot & \cdot \\ \cdot & \cdot & \cdot & \cdot & \cdot \\ G_n & \cdot & \cdot & \cdot & G_{2n-1} \end{pmatrix} \begin{pmatrix} Q_0(X) \\ \cdot \\ \cdot \\ \cdot \\ Q_{n-1}(X) \end{pmatrix} = - \begin{pmatrix} G_{n+1} \\ \cdot \\ \cdot \\ G_{2n-1} \\ G_X \end{pmatrix} \qquad (2.5)$$

and define $Q_X(z)$ by

$$Q_X(z) = Iz^n + Q_{n-1}(X)z^{n-1} + \dots + Q_0(X) \qquad (2.6)$$

Then Q_X qualifies as a right denominator in a right coprime matrix fraction representation of G_X. With this choicc of Q_X, P_X is of course uniquely determined. We should return to its representation in the sequel.

With the right coprime matrix fraction representation (2.4) we have also a left comprime matrix fraction representation.

$$G_X(z) = \bar{Q}_X(z)^{-1}\bar{P}_X(z) \qquad (2.7)$$

with P_X, Q_X uniquely determined up to a common left unimodular factor.

Since the block Hankel matrix H of (2.1) is non-singular the reachability and observability indices of a canonical realization of any minimal rational extension is equal to n. This implies the existence of representation (2.7) with $\bar{Q}_X$ monic i.e.

$$\bar{Q}_X(z) = Iz^n + \bar{Q}_{n-1}(X)z^{n-1} + \dots + \bar{Q}_0(X) \qquad (2.8)$$

Thus $\bar{Q}_0(X), \dots, \bar{Q}_{n-1}(X)$ are solutions of the system

$$(\bar{Q}_0(X) \dots \bar{Q}_{n-1}(X)) \begin{pmatrix} G_1 & \cdot & \cdot & \cdot & G_n \\ \cdot & \cdot & \cdot & \cdot & \cdot \\ \cdot & \cdot & \cdot & \cdot & \cdot \\ \cdot & \cdot & \cdot & \cdot & \cdot \\ G_n & \cdot & \cdot & \cdot & G_{2n-1} \end{pmatrix} = - (G_{n+1} \dots G_{2n-1} \; X)$$

Clearly, in the scalar case, the systems (2.5) and (2.9) are identical. This is not the case when considering the matrix equations.

Thus we have the equality

$$P_X Q_X^{-1} = \bar{Q}_X^{-1} \bar{P}_X \qquad (2.10)$$

This is equivalent to the important relation

$$\bar{P}_X Q_X = \bar{Q}_X P_X \tag{2.11}$$

The last equality has an interpretation in terms of polynomial models. We refer to Fuhrmann [1976,1979,1981,1983] for background information. If X_{Q_X} and $X_{\bar{Q}_X}$ are the polynomial models associated with Q_X and $\bar{Q}_X$ respectively then the map $Z: X_{Q_X} \to X_{\bar{Q}_X}$ defined by

$$Zf = \pi_{\bar{Q}_X} \bar{P}_X f \quad \text{for } f \text{ in } X_{Q_X} \tag{2.12}$$

is an $F[z]$-module homomorphism, i.e. it intertwines the maps S_{Q_X} and $S_{\bar{Q}_X}$, that is

$$ZS_{Q_X} = S_{\bar{Q}_X} Z \tag{2.13}$$

holds. Under the assumption of right coprimeness of Q_X and P_X and the left coprimeness of $\bar{P}_X$ and $\bar{Q}_X$ the map Z is actually an isomorphism.

The Hankel map $H_{G_X}: F^p[z] \to z^{-1}F^p[[z^{-1}]]$ induced by G_X is defined by

$$H_{G_X} f = \pi_- G_X f \quad \text{for } f \text{ in } F^p[z] \tag{2.14}$$

Under our coprimeness assumption it is easy to check, see Fuhrmann [1976], that

$$\operatorname{Ker} H_{G_X} = Q_X F^p[z] \tag{2.15}$$

and

$$\operatorname{Im} H_{G_X} = X^{\bar{Q}_X} \tag{2.16}$$

Since $X_{Q_X}, Q_X F^p[z] = \{0\}$ the map H_{G_X} is injective on X_{Q_X} and hence its restriction $\bar{H}_{G_X}: X_{Q_X} \to X^{\bar{Q}_X}$ defined by

$$\bar{H}_{G_X} = H_{G_X} | X_{Q_X} \tag{2.17}$$

is invertible.

Moreover, if $\rho_{\bar{Q}_X^{-1}}: X_{\bar{Q}_X} \to X^{\bar{Q}_X}$ is the multiplication map defined by

$$\rho_{\bar{Q}_X^{-1}} f = \bar{Q}_X^{-1} f \tag{2.18}$$

then the following diagram

$$\begin{array}{ccc} X_{Q_X} & \longrightarrow & X_{\bar{Q}_X} \\ & {\scriptstyle \bar{H}_{G_X}} \searrow & \downarrow {\scriptstyle \rho_{\bar{Q}_X^{-1}}} \\ & & X^{\bar{Q}_X} \end{array} \tag{2.19}$$

is commutative.

Since P_X and Q_X are right coprime we can extend the matrix

$$\begin{pmatrix} P_X \\ Q_X \end{pmatrix}$$

to a unimodular matrix

$$\begin{pmatrix} P_X & -\bar{B}_1 \\ Q_X & \bar{A}_1 \end{pmatrix}$$

by a suitable choice of polynomial matrices $\bar{A}_1$ and $\bar{B}_1$. For this we refer to Rosenbrock [1970]. In the same way there eixst polynomial matrices A_1 and B_1 such that

$$\begin{pmatrix} A_1 & B_1 \\ -\bar{Q}_X & \bar{P}_X \end{pmatrix}$$

is also unimodular.

Multiplying the two matrices, and using (2.11), we obtain

$$\begin{pmatrix} A_1 & B_1 \\ -\bar{Q}_X & \bar{P}_X \end{pmatrix}\begin{pmatrix} P_X & -\bar{B}_1 \\ Q_X & A_1 \end{pmatrix} = \begin{pmatrix} U & W \\ 0 & V \end{pmatrix}$$

where U, V are necessarily unimodular. This in turn implies, letting $A_2 = U^{-1}A_1$, $B_2 = U^{-1}B_1$, $B_2 = B_1V^{-1}$ and $A_2 = A_1V^{-1}$, that

$$\begin{pmatrix} A_2 & B_2 \\ -\bar{Q}_X & \bar{P}_X \end{pmatrix}\begin{pmatrix} P_X & -\bar{B}_2 \\ Q_X & \bar{A}_2 \end{pmatrix} = \begin{pmatrix} I & W_1 \\ 0 & I \end{pmatrix}$$

with $W_1 = U^{-1}WV^{-1}$. Suitably redefining either the pair A_2, B_2 or $\bar{A}_2$, $\bar{B}_2$ one has the existence of polynomial matrices A, B, $\bar{A}$ and $\bar{B}$ for which

$$\begin{pmatrix} A & B \\ -\bar{Q}_X & \bar{P}_X \end{pmatrix}\begin{pmatrix} P_X & -\bar{B} \\ Q_X & \bar{A} \end{pmatrix} = \begin{pmatrix} I & 0 \\ 0 & I \end{pmatrix}$$

Apart from (2.11) equations (2.20) implies also the Bezout equations

$$AP_X + BQ_X = I \tag{2.21}$$

and

$$\bar{P}_X\bar{A} + \bar{Q}_X\bar{B} = I \tag{2.22}$$

also since clearly (2.20) implies

$$\begin{pmatrix} P_X & -\bar{B} \\ Q_X & \bar{A} \end{pmatrix}\begin{pmatrix} A & B \\ -\bar{Q}_X & \bar{P}_X \end{pmatrix} = \begin{pmatrix} I & 0 \\ 0 & I \end{pmatrix} \tag{2.23}$$

we have also

$$\bar{A}\bar{Q}_X = Q_XA . \tag{2.24}$$

Hence if $\bar{Z}: X_{\bar{Q}_X} \to X_{Q_X}$ is defined by

$$\bar{Z}g = \pi_{Q_X}\bar{A}g \quad \text{for } g \text{ in } X_{\bar{Q}_X}$$

then (2.20) and its implied Bezout equations

$$P_XA + \bar{B}\bar{Q}_X = I \tag{2.26}$$

and

$$Q_XB = \bar{A}\bar{P}_X = I \tag{2.27}$$

show that $\bar{A}$ and Q_X are left coprime and A and $\bar{Q}_X$ right coprime. This guarantees the invertibility of the map $\bar{Z}$.

Given any f in X_{Q_X} we compute

$$\bar{Z}Zf = \pi_{Q_X}\bar{A}\pi_{\bar{Q}_X}\bar{P}_Xf = \pi_{Q_X}\bar{A}\bar{P}_Xf = \pi_{Q_X}(I-Q_XB)f = f$$

and so we have shown that

$$\bar{Z} = Z^{-1} . \tag{2.28}$$

Now the commutativity of diagram (2.19) implies

$$HZ^{-1} = \rho_{(Q_X)^{-1}} = (\rho_{\bar{Q}_X})^{-1} \tag{2.9}$$

Let $\mathcal{B}_c$ denote the control direct sum from representations, see Fuhrmann [1983], and $\mathcal{B}^c$ its image under the homomorphism $\rho_{(Q_X)^{-1}}$.

Let $\mathcal{B}_0$ denote the standard direct sum representation. If one defines, following Anderson and Jury [1976] the generalized Bezoutiant corresponding to the polynomial matrices $\bar{A}$, $\bar{Q}_X$, Q_X, A satisfying (2.24), by

$$\Gamma(\bar{A},\bar{Q}_X,Q_X,A) = (B_{ij}) \tag{2.30}$$

where B_{ij} are the matrices defined by the equality

$$\frac{\bar{A}(z)\bar{Q}_X(w)-Q_X(z)A(w)}{z-w} = \Sigma\Sigma\ B_{ij}z^{i-1}w^{j-1}$$

then it has been shown in Fuhrmann [1983] that

$$[\bar{Z}]^{\mathcal{B}_0}_{\mathcal{B}_c} = \Gamma(\bar{A},\bar{Q}_X) , \tag{2.31}$$

Next we show that the matrix H of (2.1) has the identification

$$H = [\bar{H}] \tag{2.32}$$

Indeed if $[H] = (H_{ij})$ is the block matrix representation of $\bar{H}$ relative to the two direct sum representations, then from the definition it follows that for each constant vector ξ

$$Hz^{j-1}\xi = \pi_- G_X z^{j-1}\xi = \Sigma\ \bar{Q}^{-1}\bar{E}_j H_{ij}\xi$$

where

$$\bar{E}_j = \pi_+ z^{-j}\bar{Q}_X . \tag{2.33}$$

Hence

$$(H_{kj}\xi,\eta) = <\Sigma\ \bar{E}H_{ij}\xi, z^{k-1}\eta>$$
$$= [\Sigma\ \bar{Q}_X^{-1}\bar{E}H_{ij}\xi, z^{k-1}\eta]$$
$$= [\pi_- G_X z^{j-1}\xi z^{k-1}\eta] = [G_X z^{j-1}\xi, z^{k-1}\eta]$$
$$= (G_X z^{k+j-2}\xi,\eta) = (G_{k+j-1}\xi,\eta)$$

or

$$H_{ij} = G_{i+j-1} \tag{2.34}$$

and hence (2.32) follows. Taking (2.29), and (2.31) together with the obvious equality

$$\left[\rho_{(\bar{Q}_X)^{-1}}\right]_{\mathcal{B}_c}^{\mathcal{B}^c} = 1 \tag{2.35}$$

implies

$$H^{-1} = \Gamma(\bar{A},\bar{Q}_X,Q_X,A) \tag{2.36}$$

i.e. the inverse of the nonsingular block Hankel matrix H is a generalized Bezoutiant. This is a generalization of the theorem of Lander [1974] referred to before. We summarize it in the following form.

THEOREM 2.1. Let H be the block Hankel matrix (2.1) which is assumed to be nonsingular. Let $G_X = P_X Q_X^{-1}$ be a right coprime matrix fraction representation of the minimal rational extension of sequence $G_1,\ldots,G_{2n-1}$ determined by $G_{2n} = X$. Let A be the unique polynomial matrix which satisfies the Bezout equation.

$$AP_X + BQ_X = 1 \tag{2.37}$$

and for which $A\bar{Q}_X^{-1}$ is strictly proper.

Let $\bar{A}$, $\bar{Q}_X$ be determined through (2.23)

then

$$H^{-1} = \Gamma(\bar{A},\bar{Q}_X,Q_X,A). \tag{2.38}$$

3. Representation of Generalized Bezoutiants

In this section we present a simple derivation of representation formulas for generalized Bezoutiants. These generalize results of Ptak [1984]. The exposition follows Fuhrmann [1983] which dealt with the scalar case. We treat here only the case of interest to us namely that which is associated with a nonsingular Hankel matrix (2.1).

THEOREM 3.1. Let $G = PQ^{-1}$ be a right coprime matrix fraction representation with

$$Q(z) = Q_0 + Q_1 z + \dots + Q_{n-1} z^{n-1} + Iz^n \tag{3.1}$$

and G strictly proper.

Let $\bar{Q}^{-1}\bar{P}$ be the unique left coprime matrix fraction representation of G with

$$\bar{Q}(z) = \bar{Q}_0 + \bar{Q}_1 z + \dots + \bar{Q}_{n-1} z^{n-1} + Iz^n \tag{3.2}$$

Then the generalized Bezoutiant $\Gamma(\bar{P},Q,\bar{Q},P) = (B_{ij})$ has the following representation

$$\Gamma(\bar{P},Q,\bar{Q},P) =$$

$$\begin{pmatrix} \bar{P}_0 & & \\ \vdots & \ddots & \\ \bar{P}_{n-1} & \cdots & \bar{P}_0 \end{pmatrix} \begin{pmatrix} I & Q_{n-1} & \cdots & Q_1 \\ & \ddots & \ddots & \vdots \\ & & \ddots & Q_{n-1} \\ & & & I \end{pmatrix} - \begin{pmatrix} \bar{Q}_0 & & \\ \vdots & \ddots & \\ \bar{Q}_{n-1} & \cdots & \bar{Q}_0 \end{pmatrix} \begin{pmatrix} 0 & P_{n-1} & \cdots & P_1 \\ & \ddots & \ddots & \vdots \\ & & \ddots & P_{n-1} \\ & & & 0 \end{pmatrix}$$

$$= \begin{pmatrix} \bar{P}_0 & & \\ \vdots & \ddots & \\ \bar{P}_{n-1} & \cdots & \bar{P}_0 \end{pmatrix} \begin{pmatrix} Q_1 & \cdots & Q_{n-1} & I \\ \vdots & \iddots & \iddots & \\ Q_{n-1} & \iddots & & \\ I & & & \end{pmatrix} - \begin{pmatrix} \bar{Q}_0 & & \\ \vdots & \ddots & \\ \bar{Q}_{n-1} & \cdots & \bar{Q}_0 \end{pmatrix} \begin{pmatrix} P_1 & \cdots & P_{n-1} & 0 \\ \vdots & \iddots & \iddots & \\ A_{n-1} & \iddots & & \\ 0 & & & \end{pmatrix}$$

<u>Remark</u>: A left coprime matrix fraction representation as described exists as the nonsingularity of H implies all observability, as well as all reachability indices, are equal to n.

PROOF. The generalized Bezoutiant $\Gamma(\bar{P},Q,\bar{Q},P) = (B_{ij})$ corresponding to the quadruple of polynomial matrices $\bar{P}$, Q, $\bar{Q}$ and P that satisfy

$$\bar{P}Q = \bar{Q}P \tag{3.3}$$

is defined by

$$\frac{\bar{P}(z)Q(w) - \bar{Q}(z)P(w)}{z-w} = \Sigma\Sigma\, B_{ij} z^{i-1} w^{j-1} \tag{3.4}$$

Adapting an argument of Kravitsky [1980], equation (3.4) can be rewritten as

$$\bar{P}(z)Q(w) - \bar{Q}(z)P(w) = \Sigma\Sigma\, B_{ij} z^{i-1}(z-w)w^{j-1} \tag{3.5}$$

Equality (3.5) holds for any pair of noncommuting variables as long as they commute with multiplication by $p \times p$ matrices.

Let us denote now by S the $n \times n$ block matrix, of $p \times p$ blocks,

$$\begin{pmatrix} 0 & I & . & . & . & 0 \\ 0 & . & I & . & . & 0 \\ 0 & . & . & . & . & 0 \\ 0 & . & . & . & I & 0 \\ 0 & . & . & . & . & I \\ 0 & . & . & . & . & 0 \end{pmatrix} \tag{3.6}$$

and

$$J = \begin{pmatrix} 0 & . & . & . & I \\ 0 & . & . & I & 0 \\ 0 & . & . & . & 0 \\ 0 & I & . & . & 0 \\ I & . & . & . & 0 \end{pmatrix} \tag{3.7}$$

Then for every constant $p \times p$ matrix A we define

$$AS = SA = \begin{pmatrix} 0 & A & . & . & . & 0 \\ 0 & . & A & . & . & 0 \\ 0 & . & . & . & . & 0 \\ 0 & . & . & . & A & 0 \\ 0 & . & . & . & . & A \\ 0 & . & . & . & . & 0 \end{pmatrix}$$

Also for any polynomial matrix $A(z)$ of degree n we define $A^{\#}(z)$ by

$$A^{\#}(z) = A(z^{-1})z^{n} . \tag{3.8}$$

Also we will use the notation $A(S)$ for

$$A(S) = \begin{pmatrix} A_0 & . & . & . & A_{n-1} \\ & . & . & . & . \\ & & . & . & . \\ & & & . & . \\ & & & & A_0 \end{pmatrix} \tag{3.9}$$

From (3.5) we have directly

$$\begin{aligned} &\bar{P}(z)Q^{\#}(w) - \bar{Q}(z)P^{\#}(w) \\ &\quad = [\bar{P}(z)Q(w^{-1}) - \bar{Q}(z)P(w^{-1})]w^{n} \\ &\quad = \Sigma\Sigma\ B_{ij}z^{i-1}(z-w^{-1})w^{-j+1}w^{n} \\ &\quad = \Sigma\Sigma\ B_{ij}z^{i-1}(zw-1)w^{n-1} \end{aligned}$$

In this identity we substitute $z = \tilde{S}$ and $w = S$. Since

$$(\tilde{S}S-I) = -\begin{pmatrix} I & 0 & . & . & . & 0 \\ 0 & 0 & . & . & . & 0 \\ 0 & 0 & . & . & . & 0 \\ 0 & 0 & . & . & . & 0 \\ 0 & 0 & . & . & . & 0 \end{pmatrix}$$

and $\tilde{S}^{i-1}(\tilde{S}S-I)S^{n-j}$ is the $n\times n$ block matrix with $-I$ in the $i, n-j$ position the following identity is obtained.

$$\Gamma(\bar{P},Q,\bar{Q},P) = (B_{ij}) = -\ [\bar{P}(\tilde{S})Q^{\#}(S) - \bar{Q}(\tilde{S})P^{\#}(S)]J \tag{3.11}$$

This is clearly equivalent to the statement of the theorem.

Using the previously representation for Bezoutiants and Theorem 2.1 on the inverse of block Hankel matrices we can state now the following.

THEOREM 3.2. Let H be the block Hankel matrix (2.1) which is assumed to be nonsingular. Let $G_X = P_X Q_X^{-1}$ be the minimal rational extention of the sequence $G_1,\ldots,G_{2n-1}$ corresponding to $G_{2n} = X$. Let A be the unique polynomial matrix such that $A\bar{Q}_X^{-1}$ is strictly proper and which satisfies the Bezout equation

$$AP_X + BQ_X = 1 \tag{3.12}$$

Then

$$H^{-1} = \begin{pmatrix} \bar A_0 & & & \\ \cdot & \cdot & & \\ \cdot & \cdot & \cdot & \\ \cdot & \cdot & \cdot & \cdot \\ \bar A_{n-1} & \cdot & \cdot & \bar A_0 \end{pmatrix} \begin{pmatrix} Q_1 & \cdot & \cdot & \cdot & \bar Q_{n-1} & I \\ \cdot & \cdot & \cdot & \cdot & \cdot & \\ \cdot & \cdot & \cdot & \cdot & & \\ \bar Q_{n-1} & \cdot & & & & \\ I & & & & & \end{pmatrix} -$$

$$\begin{pmatrix} Q_0 & & & \\ \cdot & \cdot & & \\ \cdot & \cdot & \cdot & \\ \cdot & \cdot & \cdot & \cdot \\ Q_{n-1} & \cdot & \cdot & Q_0 \end{pmatrix} \begin{pmatrix} A_1 & \cdot & \cdot & \cdot & A_{n-1} & 0 \\ \cdot & \cdot & & & & \\ \cdot & \cdot & \cdot & & & \\ A_{n-1} & & & & & \\ 0 & & & & & \end{pmatrix} \tag{3.13}$$

Similar formulas can be obtained as in Ptak [1984].

4. Parametrization of Minimal Rational Extension

In this section we proceed to give a parametrization of the set of all minimal rational extensions to the sequence $G_1, \dots, G_{2n-1}$ under the assumption that the block Hankel (2.1) is nonsingular.

We will use G to denote the minimal rational extensions corresponding to the choice $G_{2n} = X$ and G to the choice $G_{2n} = X = 0$. We shall write

$$G_X = P_X Q_X^{-1} = \bar Q_X^{-1} \bar P_X \quad \text{and} \quad G = PQ^{-1} = \bar Q^{-1} \bar P$$

with the assumption that Q_X, $\bar Q_X$ are always taken to be monic. This determines them uniquely.

We saw already that with

$$Q_X(z) = Iz^n + Q_{n-1}(X) z^{n-1} + \dots + Q_0(X)$$

and

$$\bar Q_X(z) = Iz^n + Q_{n-1}(X) z^{n-1} + \dots + \bar Q_0(X)$$

The coefficients $Q_i(X)$ and $\bar Q_i(X)$ are the solutions of the systems (2.5) and (2.9) respectively.

Let A, B and $\bar A$, $\bar B$ be those polynomial matrices appearing in the Bezout equations (2.21) and (2.22). Clearly since $A\bar Q_X^{-1}$ and $Q_X^{-1}\bar A$ are strictly proper both A and $\bar A$ are polynomial matrices of degree $< n-1$.

Starting from the Bezout equation

$$AP_X + BQ_X = I \tag{4.1}$$

rewritten as

$$AP_XQ_X^{-1} + B = AG_X + B = Q_X^{-1} = I/z^n + \Omega_1/z^{n+1} \ \ldots \tag{4.2}$$

it is clear that $B(z)$ has degree less than or equal to $n-2$.

Equating coefficients in

$$(A_0+A_1z+\ldots+A_{n-1}z^{n-1})(G_1/z+\ldots+G_{2n-1}/z^{2n-1}+X/z^{2n}+\ldots)$$

$$+ (B_0+B_1z+\ldots+B_{n-2}z^{n-2}) = I/z^n + \Omega_1/z^{n+1} + \ldots \tag{4.3}$$

yields, for nonnegative coefficients, the equations

$$\begin{aligned} A_{n-1}G_1 &= -B_{n-2} \\ A_{n-1}G_2 + A_{n-2}G_1 &= -B_{n-3} \\ &\vdots \\ A_{n-1}G_{n-1} + \ldots + A_1G_1 &= -B_0 \end{aligned}$$

and for the first n negative coefficients

$$\begin{aligned} A_0G_1 + \ldots + A_{n-1}G_n &= 0 \\ A_0G_2 + \ldots + A_{n-1}G_{n+1} &= 0 \\ &\vdots \\ A_0G_{n-1} + \ldots + A_{n-1}G_{2n-2} &= 0 \\ A_0G_n + \ldots + A_{n-1}G_{2n-1} &= I \end{aligned}$$

These can be rewritten in matrix form as

$$\begin{pmatrix} A_{n-1} & & & \\ \vdots & \ddots & & \\ \vdots & & \ddots & \\ A_1 & \cdots & \cdots & A_{n-1} \end{pmatrix} \begin{pmatrix} G_0 \\ \vdots \\ \vdots \\ G_{n-1} \end{pmatrix} = - \begin{pmatrix} B_{n-2} \\ \vdots \\ \vdots \\ B_0 \end{pmatrix} \tag{4.4}$$

and

$$(A_0 \ \ldots \ A_{n-1}) \begin{pmatrix} G_1 & \cdot & \cdot & \cdot & G_n \\ \vdots & \cdot & \cdot & \cdot & \\ \vdots & \cdot & \cdot & \cdot & \\ G_n & \cdot & \cdot & \cdot & G_{2n-1} \end{pmatrix} = (0 \ \ldots \ 0 \quad I) \tag{4.5}$$

respectively.

Thus (4.5) can be solved for the A_i and those be used in (4.4) to obtain the B_i.

In particular if the polynomials A^i are defined by

$$A^{(i)}(z) = \pi_+ z^{-1} A \tag{4.6}$$

Then (4.4) can be rewritten as

$$\begin{aligned} B(z) &= B_0 + B_1 z + \dots + B_{n-2} z^{n-2} \\ &= - \Sigma A^i(z) G_i \end{aligned} \tag{4.7}$$

Incidently, as in the scalar case, (4.5) and (4.7) prove that A and B in the Bezout equation (2.21) and $\bar{A}$ and $\bar{B}$ in (2.22) are independent of the choice of $G_{2n} = X$.

In a completely analogous manner we can start from the Bezout equation.

$$\bar{P}_X \bar{A} + \bar{Q}_X \bar{B} = I \tag{4.8}$$

and the resulting equation

$$G_X \bar{A} + \bar{B} = \bar{Q}_X^{-1} \tag{4.9}$$

Equating coefficients in the last equation yields the system

$$\begin{pmatrix} G_1 & \cdot & \cdot & \cdot & G_n \\ \cdot & \cdot & \cdot & \cdot & \cdot \\ \cdot & \cdot & \cdot & \cdot & \cdot \\ \cdot & \cdot & \cdot & \cdot & \cdot \\ G_n & \cdot & \cdot & \cdot & G_{2n-1} \end{pmatrix} \begin{pmatrix} \bar{A}_0 \\ \cdot \\ \cdot \\ \cdot \\ \bar{A}_{n-1} \end{pmatrix} = \begin{pmatrix} 0 \\ \cdot \\ \cdot \\ 0 \\ I \end{pmatrix} \tag{4.10}$$

and $\bar{B}(z) = \bar{B}_0 + \bar{B}_1 z + \dots + \bar{B}_{n-2} z^{n-2}$ is determined through

$$(G_1 \dots G_{n-1}) \begin{pmatrix} \bar{A}_{n-1} & \cdot & \cdot & \cdot & A_1 \\ & \cdot & \cdot & \cdot & \\ \bigcirc & & \cdot & \cdot & \\ & & & \cdot & \bar{A}_{n-1} \end{pmatrix} = - (\bar{B}_{n-2} \dots \bar{B}_0) \tag{4.11}$$

Next we pass to the computations of P_X, Q_X and $\bar{P}_X$, $\bar{Q}_X$. Since

$$G_X = P_X Q_X^{-1} = \bar{Q}_X^{-1} \bar{P}_X \tag{4.12}$$

we obtain

$$P_X = G_X Q_X \tag{4.13}$$

and

$$\bar{P}_X = \bar{Q}_X G_X \tag{4.14}$$

By equating coefficients we obtain

$$(P_0(X) \; \dots \; P_{n-1}(X)) = (G_1 \; \dots \; G_n) \begin{pmatrix} Q_1(X) & \cdot & \cdot & \cdot & Q_{n-1}(X) & I \\ \cdot & & \cdot & \cdot & \cdot & \cdot \\ \cdot & & \cdot & \cdot & \cdot & \\ Q_{n-1}(X) & \cdot & & & 0 & \\ I & & & & & \end{pmatrix} \tag{4.15}$$

and

$$\begin{pmatrix} \bar{P}_0(X) \\ \cdot \\ \cdot \\ \cdot \\ \bar{P}_{n-1}(X) \end{pmatrix} \equiv \begin{pmatrix} \bar{Q}_1(X) & \cdot & \cdot & \cdot & I \\ \cdot & \cdot & \cdot & \cdot & \\ \cdot & \cdot & \cdot & & \\ \cdot & \cdot & & & \\ \bar{Q}_{n-1}(X) & \cdot & & & \end{pmatrix} \begin{pmatrix} G_1 \\ \cdot \\ \cdot \\ \cdot \\ G_n \end{pmatrix}$$

respectively. This implies the formulas

$$P_X(z) = \Sigma \; G_i E_i(X,z) \tag{4.17}$$

$$\bar{P}_X(z) = \Sigma \; \bar{E}_i(X,z) G_i \tag{4.18}$$

where

$$E_i(X,z) - \pi_+ z^{-1} Q_X \tag{4.19}$$

and

$$\bar{E}_i(X,z) = \pi_+ z^{-i} \bar{Q}_X \; .$$

Thus (4.17) and (4.18) can be rewritten as

$$P_X(z) = \Sigma \; G_i E_i(X,z) = \Sigma \; G_i \pi_+ z^{-i} Q_X(z) =$$

$$= \Sigma \; \pi_+ z^{-i} G_i Q_X(z) = \pi_+ G Q_X \tag{4.21}$$

and

$$\bar{P}_X(z) = \Sigma \; \bar{E}_i(X,z) G_i = \Sigma \; \pi_+ z^{-i} \bar{Q}_X(z) G_i = \pi_+ \bar{Q}_X G \; , \tag{4.22}$$

In accordance with (4.13) and (4.14).

Since $Q_i(X)$ and Q_i solve

$$\begin{pmatrix} G_1 & \cdot & \cdot & \cdot & G_n \\ \cdot & \cdot & \cdot & \cdot & \cdot \\ \cdot & \cdot & \cdot & \cdot & \cdot \\ \cdot & \cdot & \cdot & \cdot & \cdot \\ G_n & \cdot & \cdot & \cdot & G_{2n-1} \end{pmatrix} \begin{pmatrix} Q_0(X) \\ \cdot \\ \cdot \\ \cdot \\ Q_{n-1}(X) \end{pmatrix} = - \begin{pmatrix} G_{n+1} \\ \cdot \\ \cdot \\ \cdot \\ G_{2n-1} \\ X \end{pmatrix}$$

and

$$\begin{pmatrix} G_1 & \cdot & \cdot & \cdot & G_n \\ \cdot & \cdot & \cdot & \cdot & \cdot \\ \cdot & \cdot & \cdot & \cdot & \cdot \\ \cdot & \cdot & \cdot & \cdot & \cdot \\ G_n & \cdot & \cdot & \cdot & G_{2n-1} \end{pmatrix} \begin{pmatrix} Q_0(X) & -Q_0 \\ \cdot & \\ \cdot & \\ \cdot & \\ Q_{n-1}(X) & -Q_{n-1} \end{pmatrix} = - \begin{pmatrix} 0 \\ \vdots \\ \cdot \\ 0 \\ X \end{pmatrix}$$

Since $(\bar{A}_0, \ldots, \bar{A}_{n-1})^-$ solves (4.10) we -mmediately obtain the equalities

$$Q_i(X) = Q_i - \bar{A}_i X \ . \tag{4.23}$$

or in polynomial form

$$Q_X(z) = Q(z) - \bar{A}(z)X \ . \tag{4.24}$$

A similar computation yields

$$\bar{Q}_X(z) = \bar{Q}(z) - XA(z) \ . \tag{4.25}$$

We can use (4.24) in (4.21) to compute

$$\begin{aligned} P_X(z) &= \pi_+ GQ_X = \pi_+ G(Q - \bar{A}X) \\ &= P - \pi_+ \bar{Q}^{-1}\bar{P}\bar{A}X \\ &= P - \pi_+ \bar{Q}^{-1}(I - \bar{G}\bar{B}X) \\ &= P + \bar{B}X \ , \end{aligned}$$

i.e.

$$P_X(z) = P(z) + \bar{B}(z)X \tag{4.26}$$

and again its analogue

$$\bar{P}_X(z) = \bar{P}(z) + XB(z) \ . \tag{4.27}$$

We can summarize the proceeding analysis in the following.

THEOREM 4.1. Given the nonsingular Hankel matrix.

$$\begin{pmatrix} G_1 & \cdot & \cdot & \cdot & G_n \\ \cdot & \cdot & \cdot & \cdot & \cdot \\ \cdot & \cdot & \cdot & \cdot & \cdot \\ \cdot & \cdot & \cdot & \cdot & \cdot \\ G_n & \cdot & \cdot & \cdot & G_{2n-1} \end{pmatrix}$$

Let $G = PQ^{-1} = \bar{Q}^{-1}\bar{P}$ and $G_X = P_X Q_X^{-1} = Q_X^{-1}\bar{P}_X$ be the minimal rational extensions of $G_1, \ldots, G_{2n-1}$, corresponding to the choice $G_{2n} = 0$ and $G_{2n} = X$ respectively and all matrix fraction taken coprime with monic denominators. Then Λ, B and $\bar{\Lambda}$, $\bar{B}$ that appear in the Bezout equations (2.21) and (2.22) are determined through equations (4.5), (4.4), (4.10) and (4.11). Q, $\bar{Q}$, Q_X and $\bar{Q}_X$ by (2.5) and (2.9) and $\bar{P}$, P_X, $\bar{P}_X$ by (4.15) and (4.16). In particular all minimal rational extensions are parametrized by,

$$G_X = (P+\bar{B}X)(Q-\bar{A}X)^{-1} = (\bar{Q}-XA)^{-1}(\bar{P}+XB) \ . \tag{4.29}$$

Remark: It should be remarked that a much more general parametrization of rational extensions, or equivalently of partial realizations which does not use the very restrictive assumption that the block Hankel matrix in (2.1) is nonsingular has been recently derived by Antoulas [1985].

Remark: An interesting quality relation follows from equation (4.29). Under our assumption of the nonsingularity of the Hankel matrix (2.1) the sequence $G_1, \ldots, G_{2n-1}$ has a stable partial realization if and only if the transfer function $Q^{-1}\bar{A}$ is stabilizable by constant output feedback. Similar ideas appear already in Byrnes and Lindquist [1982].

Since G_X and G are both minimal rational extensions of $G_1, \ldots, G_{2n-1}$ we have an expansion of the form

$$G_X(z) - G(z) = X/z^{2n} + \Gamma_1/z^{2n+1} + \ldots \tag{4.30}$$

For the difference. It is of interest to derive this from the parametrization (4.29).

Indeed

$$\begin{aligned} G_X - G &= P_X Q_X^{-1} - PQ^{-1} \\ &= \bar{Q}_X^{-1}\bar{P}_X - PQ^{-1} \\ &= \bar{Q}_X^{-1}(\bar{P}_X Q - \bar{Q}_X P)Q^{-1} \\ &= \bar{Q}_X^{-1}[(\bar{P}+XB)Q - (\bar{Q}-XA)P]Q^{-1} \\ &= \bar{Q}_X^{-1}[\bar{P}Q-\bar{Q}P) + X(AP+BQ)]Q^{-1} = \bar{Q}_X^{-1}XQ^{-1} \end{aligned}$$

by (2.11) and the Bezout equation (2.21). Since both Q and $\bar{Q}_X$ are monic of degree n the expansion (4.30) is verified.

It is clear that the choice of a particular minimal rational extension of the sequence $G_1,\ldots,G_{2n-1}$ does not influence the inverse of the block Hankel matrix (2.1). The inversion formula (2.38) in conjunction with equations (4.24) and (4.25) explain the mechanism. Indeed

$$\frac{\bar{A}(z)\bar{Q}_X(w)-Q_X(z)A(w)}{z-w} = \frac{\bar{A}(z)[(\bar{Q}(w)-XA(w))-(Q(z)-\bar{A}(z)X)]A(w)}{z-w}$$

$$= \frac{\bar{A}(z)\bar{Q}(w)-Q(z)A(w)}{z-w}$$

or

$$\Gamma(\bar{A},\bar{Q}_X,Q_XA) = \Gamma(\bar{A},\bar{Q},Q,A) \quad . \tag{4.31}$$

5. Inversion of Block Hankel Matrices

Generally given an $n\times n$ block matrix A we need to solve n systems of equations with $n\times 1$ (or $1\times n$) right handed side to obtain its inverse. In the case of block Hankel, or block Toeplitz, the special structure allows us to get away with solving four such systems. A further reduction to two holds for the symmetric cose, namely when $\bar{G}_i = G_i$ for $i = 1,\ldots,2n-1$. These results go back to the work of Gohberg and Heinig [1974]. In this section we redrive some of the results of this type using polynomial methods. The method used seems to be particularly effective and illuminating.

That one expects results of this type is clear from the identification

$$H^{-1} = \Gamma(\bar{A},\bar{Q}_X,Q_X,A)$$

and the fact that the coefficients of the polynomial matrices involved can be obtained by solving equations (4.10), (2.5), (2.9) and (4.5).

It is convenient at this stage to get functional representations for the solution of this type of equation.

Consider

$$\begin{pmatrix} G_1 & \cdot & \cdot & \cdot & G_n \\ \cdot & \cdot & \cdot & \cdot & \cdot \\ \cdot & \cdot & \cdot & \cdot & \cdot \\ \cdot & \cdot & \cdot & \cdot & \cdot \\ G_n & \cdot & \cdot & \cdot & G_{2n-1} \end{pmatrix} \begin{pmatrix} Y_0 \\ \cdot \\ \cdot \\ \cdot \\ Y_{n-1} \end{pmatrix} = \begin{pmatrix} R_1 \\ \cdot \\ \cdot \\ \cdot \\ R_n \end{pmatrix} \tag{5.1}$$

For a solution vector $(Y_0 \ldots Y_{n-1})$ we define the polynomial matrix $Y(z)$ by

$$Y(z) = \Sigma Y_i z^i . \tag{5.2}$$

Clearly, by the identification of the Bezoutiant as a block matrix representation, as in (2.31), taken together with (2.28) it follows that for each constant vector ξ we have

$$Y(z)\xi = \bar{Z}\, \Sigma\, \bar{E}_i(z) R_i \xi \tag{5.3}$$

Here $\bar{E}_i$ are defined by (4.20) and we choose, for simplicity, the minimal rational extension corresponding to $X=0$.

We recall that a functional representation for $\bar{Z}$ is given by (2.25) and hence

$$Y(z)\xi = \pi_{Q_X} \bar{A}\, \Sigma\, \bar{E}_i R_i \xi . \tag{5.4}$$

For the special case that

$$R_i = \begin{cases} 0 & i = 1,\ldots,n-1 \\ I & i = n \end{cases} \tag{5.5}$$

that is the case corresponding to the solution of (4.10), we have, noting that $E_n(z) = I$,

$$Y(z)\xi = \pi_{Q_X} \bar{A}\bar{E}_n \xi = \bar{A}(z)\xi$$

or

$$Y(z) = \bar{A}(Z) \tag{5.6}$$

in agreement with (4.10). Here we used the fact that $Q^{-1}\bar{A}$ is strictly proper. Incidentally this shows once again that $\bar{A}$ is independent of the choice of X so long as we require $Q_X^{-1}\bar{A}$ to be strictly proper.

Next instead of (5.5) we consider

$$R_i = \begin{cases} I & i = 1 \\ 0 & i = 2,\ldots,n\ . \end{cases} \tag{5.7}$$

To solve the system we compute

$$W(z)\xi = Z^{-1}\,\Sigma\,\bar{E}_i R_i \xi = \bar{Z}\bar{E}_1\xi = \bar{Z}\bar{E}_1\xi = \pi_Q \bar{A}\bar{E}\ \xi$$

Now

$$\pi_Q \bar{A}\bar{E}_1\xi = \bar{A}\bar{E}_1\xi - QS\xi$$

For some polynomial matrix S. As

$$\bar{E}_1(z) = \frac{\bar{Q}(z)-Q(0)}{z}$$

we have

$$W(z) = A(z)\,\frac{\bar{Q}(z)-\bar{Q}(0)}{z} - Q(z)S(z)\ .$$

Using equality (2.24) we can rewrite this as

$$W(z) = Q(z)\left[\frac{A(z)-A(0)}{z} - S(z)\right] + \frac{Q(z)A(0)-\bar{A}(z)Q(0)}{z}$$

Necessarily, as $Q^{-1}W$ is strictly proper.

We must have $S = (A(z)-A(0))/z$ and hence

$$W(z) = \frac{Q(z)A(0)-\bar{A}(z)\bar{Q}(0)}{z} \tag{5.8}$$

If $A(0)$ is nonsingular, which generically is the case, we can define a polynomial matrix Y by

$$Y(z) = zW(z)A(0)^{-1} \tag{5.9}$$

i.e.

$$Y(z) = Q(z) - \bar{A}(z)\Gamma \tag{5.10}$$

with

$$\Gamma = \bar{Q}(0)A(0)^{-1}\ . \tag{5.11}$$

We proceed now to analyze the dual set of equations namely those of the form

$$(\bar{V}_0 \;\ldots\; \bar{V}_{n-1}) \begin{pmatrix} G_1 & \cdot & \cdot & \cdot & G_n \\ \cdot & \cdot & \cdot & \cdot & \cdot \\ \cdot & \cdot & \cdot & \cdot & \cdot \\ \cdot & \cdot & \cdot & \cdot & \cdot \\ G_n & \cdot & \cdot & \cdot & G_{2n-1} \end{pmatrix} = (\bar{R}_1 \;\ldots\; \bar{R}_n) \tag{5.12}$$

by transposing

$$\begin{pmatrix} \tilde{G}_1 & \cdot & \cdot & \cdot & \tilde{G}_n \\ \cdot & \cdot & \cdot & \cdot & \cdot \\ \cdot & \cdot & \cdot & \cdot & \cdot \\ \cdot & \cdot & \cdot & \cdot & \cdot \\ \tilde{G}_n & \cdot & \cdot & \cdot & \tilde{G}_{2n-1} \end{pmatrix} \begin{pmatrix} \tilde{\bar{V}}_0 \\ \\ \\ \\ \tilde{\bar{V}}_{n-1} \end{pmatrix} = \begin{pmatrix} \tilde{\bar{R}}_1 \\ \cdot \\ \cdot \\ \cdot \\ \tilde{\bar{R}}_n \end{pmatrix} \tag{5.13}$$

Note that for a choice of $G_m = X$ the minimal rational extensions of $\tilde{G}_1, \ldots, \tilde{G}_{2n-1}$ are given by

$$\tilde{G}_X(z) = \tilde{Q}_X^{-1}\tilde{P}_X = \tilde{\bar{P}}_X\tilde{\bar{Q}}_X^{-1} \tag{5.14}$$

From (5.14) we obtain

$$\tilde{P}_X\tilde{\bar{Q}}_X = \tilde{Q}_X\tilde{\bar{P}}_X \tag{5.15}$$

and the isomorphism $(Z^*)^{-1}: X_{\tilde{Q}_X} \to X_{\tilde{\bar{Q}}_X}$ is given by

$$(Z^*)^{-1}f = \pi_{\tilde{\bar{Q}}_X}\tilde{\bar{A}}f \tag{5.16}$$

From (5.13) taking $X = 0$, it is clear that

$$\tilde{\bar{V}}(z)\xi = \pi_{\tilde{\bar{Q}}}\tilde{\bar{A}}\, \Sigma\, \tilde{\bar{E}}_i\tilde{\bar{Q}}_i\xi$$

Again we consider two special cases. In the first one we take

$$\bar{Q}_i = \begin{cases} 0 & i = 1,\ldots,n-1 \\ I & i = n \end{cases} \tag{5.17}$$

to obtain

$$\tilde{\bar{V}}(z) = \pi_{\tilde{\bar{Q}}}\tilde{\bar{A}}\tilde{\bar{E}}_n = \rho_{\tilde{\bar{Q}}}\tilde{\bar{A}} = \tilde{A}$$

or

$$\bar{V}(z) = A(z) \tag{5.18}$$

The second case is obtained by choosing

$$\bar{R}_i = \begin{cases} I & i = 1 \\ 0 & i = 2,\dots,n \,. \end{cases} \tag{5.19}$$

and leads to

$$\tilde{\bar{V}}(z) = \frac{\tilde{\bar{Q}}(z)\tilde{\bar{A}}(0) - \tilde{\bar{A}}(z)\tilde{\bar{Q}}(0)}{z}$$

or

$$\bar{V}(z) = \frac{\bar{A}(0)\bar{Q}(z) - Q(0)A(z)}{z} \tag{5.20}$$

Define now a polynomial matrix Y by , assuming $A(0)$ is nonsingular.

$$\bar{Y}(z) = z\bar{A}(0)^{-1}\bar{V}(z) \tag{5.21}$$

i.e.

$$\bar{Y}(z) = \bar{Q}(z) - \bar{\Gamma}A(z) \tag{5.22}$$

where

$$\bar{\Gamma} = \bar{A}(0)^{-1}Q(0) \,. \tag{5.23}$$

Note however that equality (2.24) implies the equality

$$\bar{A}(0)^{-1}Q(0) = \bar{Q}(0)A(0)^{-1} \tag{5.24}$$

or

$$\bar{\Gamma} = \Gamma \,. \tag{5.25}$$

A computation identical to one performed in section 4 leads to

$$\Gamma(\bar{A},\bar{Y},Y,A) = \Gamma(\bar{A},\bar{Q},Q,A) = H^{-1}$$

We can summarize this in the following theorem which but for changes in notation as due to Gohberg and Heinig [1974].

THEOREM 5.1. Given the block Hankel matrix H of (2.1) let the polynomial matrices A, $\bar{A}$, V, $\bar{V}$ be given in terms of the solutions of the systems (4.5), (4.10), (5.12) and (5.13) respectively. Then if A_0 is nonsingular so is H and its inverse is given by

$$H^{-1} = \Gamma(\bar{A},\bar{Y},Y,A) \tag{5.26}$$

with Y, $\bar{Y}$ defined by (5.10) and (5.22) respectively.

REFERENCES

[1985] A.C. Antoulas, "On recursiveness and related topics in linear systems", Technical Report Number TR 8507, Rice University.

[1976] B.D.O. Anderson and E.I. Jury, "Generalized Bezoutian and Sylvester matrices in multivatiable linear controls", IEEE Trans. Aut. Contr. AC-21, 551-556.

[1982] C.I. Byrnes and A. Lindquist, "The stability and instability of partial realizations", Sys. and Contr. Let. 2,99-105.

[1984] M. Fiedler, "Hankel and Loewner matrices: Linear Algebra and Appl., 58, 75-95.

[1976] P.A. Fuhrmann, "Algebraic system theory: an analyst's point of view", J. Franklin Inst., 301, 521-540.

[1979] P.A. Fuhrmann, "Linear feedback via polynomial models", Int. J. Contr. 30, 363-377.

[1981] P.A. Fuhrmann, "Duality in polynomial models with some applications to geometric control theory", IEEE Trans. Aut. Control, AC-26, 284-295.

[1983] P.A. Fuhrmann, "On symmetric rational transfer functions", Linear Algebra and Appl., 50, 167-250.

[1985] P.A. Fuhrmann, "On the partial realization problem and the recursive inversion of Hankel and Toeplitz matrices", Contemporary Mathematics; "Linear Algebra and its Role in System Theory", Amer. Math. Soc.,

[1986] P.A. Fuhrmann, "Remarks on the Inversion of Hankel matrices", Lin. Alg. Appl., to appear.

[1974] I.C. Gohberg and G. Heinig, "Inversion of finite section matrices consisting of elements of a non-commutative algebra", (Russian), Rev. Roum. Math. Pures et Appl., 19, 623-663.

[1983] W.B. Gragg and A. Lindquist, "On the partial realization problem", Linear Algebra and Appl., 50, 277-319.

[1979] R.E. Kalman, "On partial realizations, transfer functions and canonical forms",Acta Polyt. Scand., 31, 9-32.

[1980] N. Kravitsky, "On the discriminant function of two non-commuting nonselfadjoint operators", Integral Eq. and Operator Theory, 3, 97-124.

[1974] F.I. Lander, "The Bezoutiant and inversion of Hankel and Toeplitz matrices" (Russian), Mat. Issled. IX(32), 69-87.

[1985] L. Lerer and M. Tismenetsky, Private communication.

[1984] V. Ptak, "Explicit expression for Bezoutians", Linear Algebra and Appl., 59, 43-54.

[1970] H.H. Rosenbrock, State Space and Multivariable Theory, J. Wiley, New York.

P.A. Fuhrmann
Department of Mathematics
Ben-Gurion University of the Negev
Beer Sheva, Israel

Operator Theory:
Advances and Applications, Vol. 19

HANKEL MATRICES GENERATED BY MARKOV PARAMETERS, HANKEL MATRIX EXTENSION, PARTIAL REALIZATION, AND PADE-APPROXIMATION

Georg Heinig and Uwe Jungnickel

In a former paper of the authors [8] Hankel matrices $[s_{i+j}]$ generated by the Markov parameters of rational functions $f(\lambda) = s_0\lambda^{-1} + s_1\lambda^{-2} + \ldots$ are considered. The present article deals with the inverse problem of finding $f(\lambda)$ for a given sequence $(s_0,\ldots,s_{N-1})$. A parametrization of all solution is given. Further, it is shown that this problem is closely related to other ones, such as Hankel matrix extension problems, partial realization and Padé-approximation.

INTRODUCTION

Let $\mathbb{C}_n(\lambda)$ denote the class of polynomials with degree n at most and $\mathbb{C}_n^o(\lambda)$ the subclass consisting of all monic polynomials of degree n. The <u>Markov parameters</u> of a proper rational function $f(\lambda) = p(\lambda)/a(\lambda)$ $(p(\lambda) \in \mathbb{C}_{n-1}(\lambda),\ a(\lambda) \in \mathbb{C}_n^o(\lambda))$ are, by definition, the coefficients s_k of the Laurent expansion at infinity

$$f(\lambda) = p(\lambda)/a(\lambda) = s_0\lambda^{-1} + s_1\lambda^{-2} + \ldots \quad . \tag{0.1}$$

Hankel matrices generated by $f(\lambda)$ are, by definition, the k×l matrices

$$H_{kl}(f) := [s_{i+j}]_{\substack{i=0,\ldots,k-1\\ j=0,\ldots,l-1}} \qquad (k,l=1,2,\ldots,\infty).$$

In the paper of the present authors [8] the problem of inversion of a regular matrix $H_{nn}(f)$ in terms of the polynomials $p(\lambda)$ and $a(\lambda)$ without computing the entrics s_{i+j} is studied. Let us quote the main results, because we make use of them in the sequel.

THEOREM 0.1. *The matrix $H_{nn}(f)$ is regular iff $a(\lambda)$ and $p(\lambda)$ are coprime. The inverse is given by*

$$H_{nn}(f)^{-1} = B(a,b) := \left[b_{ij}\right] ;^{1)} \tag{0.2}$$
$$\sum_{i,j=0}^{n-1} b_{ij}\lambda^i\mu^j = \frac{a(\lambda)b(\mu) - a(\mu)b(\lambda)}{\lambda - \mu},$$

<u>where</u> $b(\lambda)$ <u>is the solution of the congruence</u>

$$p(\lambda)b(\lambda) = 1 \mod a(\lambda) . \tag{0.3}$$

The second main result of [8] concerns the description of the kernel of $H_{kl}(f)$. In order to simplify the formulation we introduce the designations

$$x(\lambda) := x_0 + x_1\lambda + \ldots + x_n\lambda^n$$

for $x = (x_k)_0^n \in \mathbb{C}^{n+1}$ and

$$M(\lambda) := \{x(\lambda) : x \in M\}$$

for $M \subseteq \mathbb{C}^{n+1}$. In particular, we have $\mathbb{C}_n(\lambda) = \mathbb{C}^{n+1}(\lambda)$. Furthermore we introduce the notation

$$\mathcal{A}_{kl}(f) := (\ker H_{kl}(f))(\lambda) \ (\subseteq \mathbb{C}_{l-1}(\lambda)).$$

THEOREM 0.2. <u>Let</u> $f(\lambda)$ <u>be given by</u> (0.1), <u>where</u> $p(\lambda)$ <u>and</u> $a(\lambda)$ <u>are coprime, and let</u> $b(\lambda) \in \mathbb{C}_{n-1}(\lambda)$ <u>be the solution of</u> (0.3).

a) <u>If</u> $k,l \geq n$, <u>then</u>

$$\mathcal{A}_{kl} = \{a(\lambda)u(\lambda) : u(\lambda) \in \mathbb{C}_{l-n-1}(\lambda)\} .$$

b) <u>If</u> $n \geq k, l \geq m$, <u>where</u> m <u>is the degree of</u> $b(\lambda)$, <u>then</u>

$$\mathcal{A}_{kl} = \{b(\lambda)v(\lambda) : v(\lambda) \in \mathbb{C}_{r-1}(\lambda)\} ,$$

<u>where</u> $r = \min(l-m, n-k)$.

c) <u>If</u> $l \geq n \geq k$ <u>and</u> $k+l \geq m+n$, <u>then</u>

$$\mathcal{A}_{kl} = \{a(\lambda)u(\lambda) + b(\lambda)v(\lambda) : u(\lambda) \in \mathbb{C}_{l-n-1}(\lambda),\ v(\lambda) \in \mathbb{C}_{n-k-1}(\lambda)\} .$$

d) <u>If</u> $k \geq n \geq l$ <u>and</u> $k+l \geq m+n$, <u>then</u>

$$\mathcal{A}_{kl} = \{0\}.$$

In particular, we shall need the following

COROLLARY 0.1.

$$a(\lambda) \in \mathcal{A}_{k,n+1} \qquad (k=1,2,\ldots) .$$

The present note deals with the inverse problem of finding a rational function $f(\lambda)$ such that $H_{kl}(f) = H$ for a given

---

1) The matrix $B(a,b)$ is usually called the Bezoutian of the polynomials $a(\lambda)$ and $b(\lambda)$.

$k \times l$ Hankel matrix H. It is easily seen that this problem has always infinitely many solutions. In order to get a classification of all these solutions we formulate the problem more precisely.

PROBLEM 1. Given a $k \times l$ Hankel matrix H and a natural number n; describe all proper rational functions $f(\lambda) = p(\lambda)/a(\lambda)$ with $a(\lambda) \in \mathbb{C}_n^o(\lambda)$, $p(\lambda)$ and $a(\lambda)$ coprime, and $H_{kl}(f) = H$.

Functions solving Problem 1 will be called <u>generating for H</u>. The number n is usually called the <u>McMillan degree</u> of $f(\lambda)$.

Problem 1 was solved for the case of a regular Hankel matrix in our paper [8] and independently in [3]. Below we shall give a complete parametrization of the solution of Problem 1 in the general case.

Clearly, Problem 1 can be formulated in the following form, too.

PROBLEM 1'. Given complex numbers $s_0, s_1, \ldots, s_{N-1}$ and a natural number n; describe all rational functions $f(\lambda) = p(\lambda)/a(\lambda)$ with $a(\lambda) \in \mathbb{C}_n^o(\lambda)$ and

$$f(\lambda) = s_0\lambda^{-1} + s_1\lambda^{-2} + \ldots + s_{N-1}\lambda^{-N} + o(\lambda^{-N})$$

in a neighbourhood of infinity.

Obviously, Problem 1 is equivalent to Problem 1' for $N=l+k-1$.

It is well-known to specialists that there exist close relations between Problem 1' and some other important problems, such as

- the problem of extending a Hankel matrix to an infinite Hankel matrix with given rank n (Problem 2),
- the partial realization problem of linear system theory (Problem 3), and
- the Padé-approximation problem (Problem 4).

In Section 2 we shall give an insight into the connections between these problems. Moreover, with the help of our main result, we shall present a parametrization of the class of all solutions of these problems. Despite of a vast literature dedicated to partial realization and Padé-approximation problems we could not find such a representation.

It was already known in the 19th century that there is a close connection between positive definite Hankel matrices and the algebraic moment problem. A (in some sense) more general problem is the problem of „canonical representation" of Hankel matrices as a linear combination of simplest Hankel matrices (see [9], [7]). The approach presented below enables us (in Subsection 2.4) to solve a variant of the canonical representation problem (slightly different from that one considered in [9]) in a quite elementary manner, using the partial fraction decomposition of rational functions.

It is well-known that there are a lot of analogies between Hankel matrices and Toeplitz matrices, i.e. matrices $[a_{i-j}]$, reflecting, for example, the correspondence between the algebraic and trigonometric moment problems. Therefore, the natural question of transmitting all concepts, problems and results of this paper and the paper [8] as well to Toeplitz matrices arises. This has been done in the recent thesis [11] and will be published elsewhere.

Another question of great interest consists in the generalization of all results to the block case. It turns out that the block case is much more involved than the scalar one. However, some of the main generalization steps are already done and are described in the thesis [11], too.

1. MAIN RESULTS

1.1 Corresponding polynomials

Suppose we are given a sequence of N complex numbers $s=(s_0,s_1,\ldots,s_{N-1})$ considered as a column vector. We introduce the k×l Hankel matrices, $k+l=N+1$,

$$H_{kl}[s] := \left[s_{i+j}\right]_{\substack{i=0,\ldots,k-1\\ j=0,\ldots,l-1}} .$$

DEFINITION. The polynomial $a(\lambda)\in \mathbb{C}_n^o(\lambda)$ is said to be <u>corresponding to $H_{kl}[s]$</u> iff there exists a $p(\lambda)\in\mathbb{C}_{n-1}(\lambda)$ such that

$$H_{kl}[s] = H_{kl}(p(\lambda)/a(\lambda)) . \tag{1.1}$$

Clearly, (1.1) holds iff

$$s = H_{N1}(p(\lambda)/a(\lambda))$$

or, what is the same,

$$p(\lambda)/a(\lambda) = s_0\lambda^{-1} + \ldots + s_{N-1}\lambda^{-N} + o(\lambda^{-N}). \qquad (1.2)$$

For this reason $a(\lambda)$ is also called corresponding to s. In other words, if $a(\lambda)$ corresponds to s, then $a(\lambda)$ corresponds to all matrices $H_{kl}[s]$ with $k+l=N+1$.

Our first observation is the fact that for any corresponding polynomial there is exactly one generating function.

PROPOSITION 1.1. <u>If</u> $a(\lambda) = \sum_{k=0}^{n} a_k\lambda^k \in \mathbb{C}_n^o(\lambda)$ <u>is a corresponding to</u> s <u>polynomial and</u> $f(\lambda) = p(\lambda)/a(\lambda)$ <u>is a generating function, then the column vector</u> p <u>of coefficients of</u> $p(\lambda)$ <u>is given by</u>

$$p = \begin{bmatrix} s_{n-1} & \cdots & s_0 \\ \vdots & \iddots & \\ s_0 & & 0 \end{bmatrix} \begin{bmatrix} a_n \\ \vdots \\ a_1 \end{bmatrix} \qquad (1.3a)$$

<u>or</u>

$$p = \begin{bmatrix} a_1 & \cdots & a_n \\ \vdots & \iddots & \\ a_n & & 0 \end{bmatrix} \begin{bmatrix} s_0 \\ \vdots \\ s_{n-1} \end{bmatrix}. \qquad (1.3b)$$

PROOF. Since $f(\lambda)$ is a generating function, (1.2) holds. The relation (1.2) can be written in the form

$$p(\lambda) = a(\lambda)(s_0\lambda^{-1} + \ldots + s_{N-1}\lambda^{-N}) + o(\lambda^{n-N}).$$

Comparing coefficients one immediately obtains (1.3a) or, what is the same, (1.3b). □

In view of Proposition 1.1 it suffices to describe the class of all corresponding to s polynomials in order to solve Problem 1. To this end we proceed with another elementary observation.

PROPOSITION 1.2. <u>Suppose</u> $s \in \mathbb{C}^N$, <u>then any monic polynomial with degree</u> $n \geq N$ <u>is corresponding to</u> s.

PROOF. Obviously, the column vectors

$$h_i = H_{N1}(\lambda^i/a(\lambda)) \qquad (i=n-N,\ldots,n-1)$$

form a basis of $\mathbb{C}^N$. In particular, s is a linear combination of the h_i, $s = \sum_{i=n-N}^{n-1} p_i h_i$. This implies

$$s = H_{N1}(p(\lambda)/a(\lambda)) \ ,$$

$p(\lambda)$ being $\sum p_i \lambda^i$. Thus $a(\lambda)$ is corresponding to s. □

Now we formulate the main result of this section.

THEOREM 1.1. *The monic polynomial* $a(\lambda) = \sum_{k=0}^{n} a_k \lambda^k$, $(n<N)$, *is corresponding to* $s \in \mathbb{C}^N$ *iff*

$$H_{m-1,n+1}[s]\, a = 0 \ , \tag{1.4}$$

where $a = (a_i)_0^n$ *and* $m+n=N+1$.

For the proof we need a kernel description of the rectangular matrices

$$D_m(a) := \left.\begin{bmatrix} a_0 & a_1 \cdots a_n & & & 0 \\ & a_0 & a_1 \cdots a_n & & \\ & & \ddots & \ddots & \\ 0 & & & a_0 & a_1 \cdots a_n \end{bmatrix}\right\} m-1 \quad .$$

LEMMA 1.1. *The column vectors*

$$h_i := H_{N1}(\lambda^i/a(\lambda)) \qquad (i=0,1,\ldots,n-1)$$

form a basis of $\ker D_m(a)$.

PROOF. First we note the equality

$$D_m(a)\, x = X\, a \tag{1.5}$$

for vectors $x = (x_i)_0^{N-1}$, where X denotes the $(m-1)\times(n+1)$ Hankel matrix $[x_{i+j}]$. In particular, we have

$$D_m(a)\, h_i = H_{m-1,n+1}(\lambda^i/a(\lambda))\, a \qquad (i=0,\ldots,n-1) \ .$$

Owing to Corollary 0.1, the right-hand side of the latter equality vanishes; hence $h_i \in \ker D_m(a)$. On the other hand, the vectors h_i are, obviously, linearly independent and $\dim \ker D_m(a) = n$. Thus, the h_i form a basis of $\ker D_m(a)$. □

PROOF OF THEOREM 1.1. First we assume that $a(\lambda) \in \mathbb{C}_n^o(\lambda)$ is corresponding to s. Then, by Corollary 0.1, the vector of coefficients a is a solution of (1.4).

Now let $a = (a_i)_0^n$, $a_n=1$, fulfil (1.4). We have to show that $a(\lambda) = \sum a_i \lambda^i$ is corresponding to s. In view of (1.5) we have

$$D_m(a)\, s = 0.$$

Owing to Lemma 1.1, s is a linear combination of the vectors h_i, $s=\sum_{i=0}^{n-1} p_i h_i$. Therefore,

$$s = H_{N1}(p(\lambda)/a(\lambda)) ,$$

where $p(\lambda) := \sum_{i=0}^{n-1} p_i \lambda^i \in \mathbb{C}_{n-1}(\lambda)$. The latter means that $a(\lambda)$ is corresponding to s, and the theorem is proved. □

1.2 Parametrization of the class of corresponding polynomials

For a more profounder analysis of the class of corresponding polynomials and an explicit parametrization we introduce some concepts and note some results from the monograph [9]. Let $s \in \mathbb{C}^N$, $s \neq 0$. We are interested in the subspaces of $\mathbb{C}_{l-1}(\lambda)$

$$\mathcal{A}_l := (\ker H_{kl}[s])(\lambda) \qquad (k+l=N+1)$$

and

$$\mathcal{B}_l := (\ker H_{kl}[s]^T)(\lambda) ,$$

where „T" denotes transpositon. Clearly, there exists a largest integer n(s) such that $\mathcal{A}_l = \{0\}$ for all $l \leq n(s)$ and a smallest integer m(s), $m(s) \geq n(s)$, such that $\mathcal{B}_l = \{0\}$ for all $l \geq m(s)$. We have, in particular, m(s)=n(s) iff the Hankel matrix $H_{ll}[s]$, $l=\frac{1}{2}(N+1)$, is regular. The integers n(s) and m(s) are referred to as <u>characteristic degrees of s</u>. In view of

$$H_{kl}[s]^T = H_{lk}[s]$$

we have $\mathcal{A}_l = \mathcal{B}_k$ with k+l=N+1. This implies the relation

$$n(s) + m(s) = N + 1 .$$

Therefore, the second characteristic degree is uniquely defined by the first one.

Now we introduce the concept of characteristic polynomials. In the regular case n(s)=m(s) we have $\mathcal{A}_{n(s)} = \{0\}$ and $\dim \mathcal{A}_{n(s)+1} = 2$. A pair of polynomials $(\alpha(\lambda),\beta(\lambda))$, $\alpha(\lambda)$, $\beta(\lambda) \in \mathbb{C}_{n(s)}(\lambda)$, forming a basis of $\mathcal{A}_{n(s)+1}$ will be called <u>pair of characteristic polynomials</u> of the vector s or of any of the Hankel matrices $H_{kl}[s]$, k+l=N+1. In case n(s)<m(s) there is a polynomial $\alpha(\lambda) \in \mathbb{C}_{n(s)}(\lambda)$ such that

$$\mathcal{A}_{n(s)+1} = \text{lin}\,\{\alpha(\lambda)\} \qquad (1.6)$$

and, furthermore, a polynomial $\beta(\lambda) \in \mathbb{C}_{m(s)}(\lambda)$ such that

$$\mathcal{A}_{m(s)+1} = \mathrm{lin}\{\alpha(\lambda),\dots,\lambda^{m(s)-n(s)}\alpha(\lambda),\beta(\lambda)\} \qquad (1.7)$$

(see [9],I,5). A pair of polynomials $(\alpha(\lambda),\beta(\lambda))$ satisfying (1.6), (1.7) will be called <u>pair of characteristic polynomials</u> in the case $n(s)<m(s)$.

Let us note a property of characteristic polynomials, proved in [9], (Prop. 5.5).

LEMMA 1.2. *Suppose $(\alpha(\lambda), \beta(\lambda))$ is a pair of characteristic polynomials. Then $\alpha(\lambda)$ and $\beta(\lambda)$ are coprime and the highest order coefficient of one of the polynomials is different from 0 (i.e., in the sense of [9], ∞ is no common root of $\alpha(\lambda)$ and $\beta(\lambda)$).*

Next we quote a kernel structure theorem, wich is the convenient for us formulation of Theorem 5.1 in [9].

LEMMA 1.3. *Suppose $n(s)$, $m(s)$ to be the characteristic degrees of s and $(\alpha(\lambda), \beta(\lambda))$ a pair of characteristic polynomials. Then*

$$\mathcal{A}_{n+1} = \begin{cases} \{0\} & : n<n(s) \\ \{u(\lambda)\alpha(\lambda) \;:\; u(\lambda) \in \mathbb{C}_{n-n(s)}(\lambda)\} & : n(s)\le n<m(s) \\ \{u(\lambda)\alpha(\lambda) + v(\lambda)\beta(\lambda) : u(\lambda) \in \mathbb{C}_{n-n(s)}(\lambda),\ v(\lambda) \in \mathbb{C}_{n-m(s)}(\lambda)\} & : m(s)\le n \;. \end{cases}$$

In order to describe the class of corresponding polynomials in terms of $\alpha(\lambda)$ and $\beta(\lambda)$ one has to look for monic polynomials belonging to $\mathcal{A}_{n+1}$. For simplifying the formulation of the corresponding result it is convenient to choose pairs of characteristic polynomials with some additional properties.

The pair $(\alpha(\lambda), \beta(\lambda))$ of characteristic polynomials will be called <u>canonoical</u> iff either

i) $\alpha(\lambda) \in \mathbb{C}^{o}_{n(s)}(\lambda)$ and $\deg \beta(\lambda) < m(s)$

or

ii) $\deg \alpha(\lambda) < n(s)$ and $\beta(\lambda) \in \mathbb{C}^{o}_{m(s)}(\lambda)$.

Let us explain that canonical pairs of characteristic polynomials do always exist. In fact, according to the second assertion of Lemma 1.2 we have $\deg \alpha(\lambda) = n(s)$ or $\deg \beta(\lambda) = m(s)$. In the first case, after multiplying $\alpha(\lambda)$ by a constant, we obtain a monic characteristic polynomial $\widetilde{\alpha}(\lambda)$. Furthermore, since

$\tilde{\beta}(\lambda) = \beta(\lambda) - \lambda^{m(s)-n(s)}\alpha(\lambda)$ is characteristic together with $\beta(\lambda)$, too, and $\deg \tilde{\beta}(\lambda) < m(s)$ we get a pair satisfying i). In the case $\deg \alpha(\lambda) < n(s)$, $\deg \beta(\lambda) = m(s)$ we obtain, after multiplying $\beta(\lambda)$ by a constant, a pair satisfying ii).

In the sequel we have to distinguish between the cases i) and ii). For this it is convenient to introduce the number

$$\delta(s) := n(s) - \deg \alpha(\lambda) .$$

We have case i) if $\delta(s) = 0$ and case ii) if $\delta(s) > 0$.

Besides, let us remark that the pair $(n(s), \delta(s))$ is just the characteristic of the Hankel matrices $H_{kl}[s]$ introduced in another way by I.S.IOHVIDOV (see [10], [9]).

THEOREM 1.2. _Let_ $n(s)$, $m(s)$ _be the characteristic degrees of_ $s \in \mathbb{C}^N$, _and_ $(\alpha(\lambda), \beta(\lambda))$ _a canonical pair of characteristic polynomials. Then the class_ $\mathcal{C}_n$ _of corresponding to_ s _polynomials with degree_ n _can be described in the following way:_

a) _If_ $n(s) = m(s)$, _then_

$$\mathcal{C}_n = \begin{cases} \emptyset & : n < n(s) \\ \{u(\lambda)\alpha(\lambda) + v(\lambda)\beta(\lambda) :\ u(\lambda) \in \mathbb{C}^0_{n-n(s)}(\lambda),\ v(\lambda) \in \mathbb{C}_{n-n(s)}(\lambda)\} & : n \geq n(s) . \end{cases}$$

b) _If_ $n(s) < m(s)$ _and_ $\delta(s) = 0$, _then_

$$\mathcal{C}_n = \begin{cases} \emptyset & : n < n(s) \\ \{u(\lambda)\alpha(\lambda) :\ u(\lambda) \in \mathbb{C}^0_{n-n(s)}(\lambda)\} & : n(s) \leq n < m(s) \\ \{u(\lambda)\alpha(\lambda) + v(\lambda)\beta(\lambda) :\ u(\lambda) \in \mathbb{C}^0_{n-n(s)}(\lambda),\ v(\lambda) \in \mathbb{C}_{n-m(s)}(\lambda)\} & : n \geq m(s) . \end{cases}$$

c) _If_ $n(s) < m(s)$ _and_ $\delta(s) > 0$, _then_

$$\mathcal{C}_n = \begin{cases} \emptyset & : n < m(s) \\ \{u(\lambda)\alpha(\lambda) + v(\lambda)\beta(\lambda) :\ u(\lambda) \in \mathbb{C}_{n-n(s)}(\lambda),\ v(\lambda) \in \mathbb{C}^0_{n-m(s)}(\lambda)\} & : n \geq m(s) \end{cases}$$

PROOF. According to Theorem 1.1 we have

$$\mathcal{C}_n = \mathcal{A}_{n+1} \cap \mathbb{C}^0_n(\lambda) .$$

From this relation and Lemma 1.3 we immediately obtain all statements of the theorem, respecting the definition of the canonical pair of characteristic polynomials. □

COROLLARY 1.1. _Let_ min deg s _denote the minimal degree of a polynomial corresponding to_ s. _Then_

$$\min \deg s = \begin{cases} n(s) & \text{if } \delta(s) = 0 \\ m(s) & \text{if } \delta(s) > 0 \,. \end{cases}$$

The classes of corresponding polynomials with minimal degree $\mathcal{C}_{\min}$ are given by

a) $n(s) = m(s)$: $\quad \alpha(\lambda) + c\beta(\lambda) \quad (c \in \mathbb{C})$

b) $n(s) < m(s)$, $\delta(s) = 0$: $\quad \alpha(\lambda)$

c) $n(s) < m(s)$, $\delta(s) > 0$: $\quad \beta(\lambda) + c(\lambda)\alpha(\lambda)$
$(c(\lambda) \in \mathbb{C}_{m(s)-n(s)}(\lambda))$.

COROLLARY 1.2. Let γ_n denote the dimension of the linear hull of $\mathcal{C}_n$.

a) If $n(s) = m(s)$, then

$$\gamma_n = \begin{cases} 0 & : n < n(s) \\ 2(n-n(s)+1) & : n(s) \le n < N \\ n+1 & : N \le n \,. \end{cases}$$

b) If $n(s) < m(s)$ and $\delta(s) = 0$, then

$$\gamma_n = \begin{cases} 0 & : n < n(s) \\ n-n(s)+1 & : n(s) \le n < m(s) \\ 2(n+1)-n(s)-m(s) & : m(s) \le n < N \\ n+1 & : N \le n \,. \end{cases}$$

c) If $n(s) < m(s)$ and $\delta(s) > 0$, then

$$\gamma_n = \begin{cases} 0 & : n < m(s) \\ 2(n+1)-n(s)-m(s) & : m(s) \le n < N \\ n+1 & : N \le n \,. \end{cases}$$

1.3. Generating functions

With the help of Theorem 1.2 and Proposition 1.1 all solutions of Problem 1, i.e. all generating functions of s with McMillan degree n can be described. To obtain a more explicit representation of the numerator we prove another auxiliary statement.

LEMMA 1.4. Suppose $(\alpha(\lambda), \beta(\lambda))$ is a pair of characteristic polynomials of $s \in \mathbb{C}^N$ with degrees n and m, respectively. Let us define the vectors

$$\gamma := \begin{bmatrix} \alpha_1 & \cdots & \alpha_n \\ \vdots & ⋰ & \\ \alpha_n & & 0 \end{bmatrix} \begin{bmatrix} s_0 \\ \vdots \\ s_{n-1} \end{bmatrix}, \quad \delta := \begin{bmatrix} \beta_1 & \cdots & \beta_m \\ \vdots & ⋰ & \\ \beta_m & & 0 \end{bmatrix} \begin{bmatrix} s_0 \\ \vdots \\ s_{m-1} \end{bmatrix} . \qquad (1.8)$$

Then $\sigma := \beta_0\gamma_0 - \alpha_0\delta_0 \neq 0$, and the polynomials $\gamma(\lambda)$, $\delta(\lambda)$ represent a solution of the Diophantine equation

$$\beta(\lambda)\gamma(\lambda) - \alpha(\lambda)\delta(\lambda) = \sigma . \qquad (1.9)$$

PROOF. From the definition of the characteristic polynomials we obtain

$$\alpha \in \ker H_{m-1,n+1}[s'] \quad , \beta \in \ker H_{n-1,m+1}[s'] \ ,$$

where $s' := (s_i)_0^{n+m-2} \in \mathbb{C}^{n+m-1}$. Therefore, by Theorem 1.1, $\alpha(\lambda)$ and $\beta(\lambda)$ are polynomials corresponding to s'. In view of (1.8) and Proposition 1.1 both the functions $\gamma(\lambda)/\alpha(\lambda)$ and $\delta(\lambda)/\beta(\lambda)$ are generating for s', wich implies

$$\gamma(\lambda)/\alpha(\lambda) - \delta(\lambda)/\beta(\lambda) = o(\lambda^{1-n-m}) \ .$$

Hence

$$\beta(\lambda)\gamma(\lambda) - \alpha(\lambda)\delta(\lambda) = o(\lambda) \ ,$$

and (1.9) holds. According to Lemma 1.2, the constant σ does not vanish. □

Now we shall give the complete parametrization of the class $\mathcal{F}_n$ of generating functions with McMillan degree n.

THEOREM 1.3. Let $n(s)$, $m(s)$ be the characteristic degrees of $s \in \mathbb{C}^N$, and $(\alpha(\lambda), \beta(\lambda))$ a canonical pair of characteristic polynomials. Further, define $\gamma(\lambda)$, $\delta(\lambda)$ via (1.8).

a) If $n(s) = m(s)$, then

$$\mathcal{F}_n = \begin{cases} \emptyset & : n < n(s) \\ \left\{ \dfrac{u(\lambda)\gamma(\lambda) + v(\lambda)\delta(\lambda)}{u(\lambda)\alpha(\lambda) + v(\lambda)\beta(\lambda)} \text{, where } u(\lambda) \in \mathbb{C}^0_{n-n(s)}(\lambda) \right. & \\ \quad \left. \text{and } v(\lambda) \in \mathbb{C}_{n-n(s)}(\lambda) \text{ are coprime} \right\} & : n \geq n(s) . \end{cases}$$

b) If $n(s) < m(s)$ and $\delta(s) = 0$, then

$$\mathcal{F}_n = \begin{cases} \emptyset & : n < m(s),\ n \neq n(s) \\ \{\gamma(\lambda)/\alpha(\lambda)\} & : n = n(s) \\ \left\{ \dfrac{u(\lambda)\gamma(\lambda) + v(\lambda)\delta(\lambda)}{u(\lambda)\alpha(\lambda) + v(\lambda)\beta(\lambda)} \text{, where } u(\lambda) \in \mathbb{C}^0_{n-n(s)}(\lambda) \right. & \\ \quad \left. \text{and } v(\lambda) \in \mathbb{C}_{n-m(s)}(\lambda) \text{ are coprime} \right\} & : n \geq m(s) . \end{cases}$$

c) If $n(s) < m(s)$ and $\delta(s) > 0$, then

$$\mathcal{F}_n = \begin{cases} \emptyset & : n < m(s) \\ \left\{ \dfrac{u(\lambda)\gamma(\lambda) + v(\lambda)\delta(\lambda)}{u(\lambda)\alpha(\lambda) + v(\lambda)\beta(\lambda)} \text{, where } u(\lambda) \in \mathbb{C}_{n-n(s)}(\lambda) \right. & \\ \quad \left. \text{and } v(\lambda) \in \mathbb{C}^0_{n-m(s)}(\lambda) \text{ are coprime} \right\} & : n \geq m(s) . \end{cases}$$

PROOF. If $\delta(s)=0$, then, by Corollary 1.1 and Proposition 1.1, $\gamma(\lambda)/\alpha(\lambda)$ is a generating function for s of McMillan degree n(s). Therefore, a rational function $p(\lambda)/a(\lambda)$, $p(\lambda)$ and $a(\lambda)$ coprime, is generating s iff

$$p(\lambda)/a(\lambda) - \gamma(\lambda)/\alpha(\lambda) = o(\lambda^{-N}) \ . \tag{1.10a}$$

Analogously, if $\delta(s)>0$, then $p(\lambda)/a(\lambda)$ is generating s iff

$$p(\lambda)/a(\lambda) - \delta(\lambda)/\beta(\lambda) = o(\lambda^{-N}) \ . \tag{1.10b}$$

Now let $n \geq m(s)$ and

$$\begin{aligned} p(\lambda) &= u(\lambda)\gamma(\lambda) + v(\lambda)\delta(\lambda) \ , \\ a(\lambda) &= u(\lambda)\alpha(\lambda) + v(\lambda)\beta(\lambda) \in \mathbb{C}_n^o(\lambda) \ , \end{aligned} \tag{1.11}$$

where $u(\lambda) \in \mathbb{C}_{n-n(s)}(\lambda)$ and $v(\lambda) \in \mathbb{C}_{n-m(s)}(\lambda)$ are coprime. We have to show that $p(\lambda)/a(\lambda) \in \mathcal{F}_n$. Because of (1.9), the equations (1.11) are equivalent to

$$\begin{aligned} u(\lambda) &= \sigma^{-1}(p(\lambda)\beta(\lambda) - a(\lambda)\delta(\lambda)) \ , \\ v(\lambda) &= \sigma^{-1}(a(\lambda)\gamma(\lambda) - p(\lambda)\alpha(\lambda)) \ . \end{aligned} \tag{1.12}$$

Moreover, by (1.9) $p(\lambda)$ and $a(\lambda)$ are coprime iff $u(\lambda)$ and $v(\lambda)$ are so. Hence $p(\lambda)/a(\lambda)$ is a generating function with McMillan degree n, since (1.12) implies the relations (1.10a) or (1.10b), dependent on $\delta(s)$.

On the other hand, let $p(\lambda)/a(\lambda) \in \mathcal{F}_n$ and define $u(\lambda)$, $v(\lambda)$ with the help of (1.12). Then (1.11) holds, and it remains to check the properties required for $u(\lambda)$ and $v(\lambda)$. In the case $\delta(s)=0$, from (1.10a) it follows that $v(\lambda) \in \mathbb{C}_{n-m(s)}(\lambda)$, $n \geq m(s)$ or $v(\lambda) \equiv 0$, $n<m(s)$. Hence, from (1.11) and Theorem 1.2 we obtain $u(\lambda) \in \mathbb{C}^o_{n-n(s)}(\lambda)$. Coprimeness of $p(\lambda)$ and $a(\lambda)$ implies that of $u(\lambda)$ and $v(\lambda)$ and, in particular, if $n<m(s)$ we must have $u(\lambda) \equiv 1$ and $n=n(s)$. In the case $\delta(s)>0$ we have, according to (1.10b), $u(\lambda) \in \mathbb{C}_{n-n(s)}(\lambda)$ and therefore, again by (1.11) and Theorem 1.2, $v(\lambda) \in \mathbb{C}^o_{n-m(s)}(\lambda)$. □

2. APPLICATIONS

2.1 Hankel matrix extension

We are going to apply the results of Section 1 to the following problem.

PROBLEM 2. Given an k×l Hankel matrix $H=\left[s_{i+j}\right]$ and a natural number n; describe all extensions of H to an infinite

Hankel matrix $H_\infty = [s_{i+j}]_{i,j=0}^{\infty}$ with rank n.

The connection between Problem 2 and Problem 1 is realized by the following well-known

THEOREM OF KRONECKER. The infinite Hankel matrix H_∞ has rank n iff there is a proper rational function $f(\lambda) = p(\lambda)/a(\lambda)$, $a(\lambda) \in \mathbb{C}_n^0(\lambda)$ and $p(\lambda)$ coprime, such that

$$H_\infty = H_\infty(f) \ .$$

The function $f(\lambda)$ is uniquely determined. (see [4]).

Let us recall that $H_\infty(f)$ is defined as $H_\infty(f) := [s_{i+j}]_0^\infty$, where $f(\lambda) = s_0\lambda^{-1} + s_1\lambda^{-2} + \ldots$ in a neighbourhood of ∞.

Kronecker's theorem offers a one-to-one correspondence between generating functions of H with McMillan degree n and infinite extensions of H with rank n. Therefore, Theorem 1.3 presents the full solution of Problem 2. Let us describe all infinite extensions of H, if there exists one, with the property rank H_∞ = rank H. Such an extension will be refered to as a singular extension of H.

COROLLARY 2.1. Let $(\alpha(\lambda), \beta(\lambda))$ be a canonical pair of characteristic polynomials of $s \in \mathbb{C}^N$, and let $(\gamma(\lambda), \delta(\lambda))$ be the polynomials defined by (1.8). There exists a singular extension of $H = H_{kl}[s]$ iff $\delta(s) = 0$ and $n(s) \le k, l \le m(s)$. In this case the set of singular extensions is given by

$$\text{a)} \quad H_\infty\left(\frac{\gamma(\lambda) + c\delta(\lambda)}{\alpha(\lambda) + c\beta(\lambda)}\right), \quad (c \in \mathbb{C}) \qquad : n(s)=m(s)$$

$$\text{b)} \quad H_\infty\left(\frac{\gamma(\lambda)}{\alpha(\lambda)}\right) \qquad : n(s)<m(s) \ .$$

PROOF. It was shown in [9] that rank $H_{mn}[s] \le n(s)$ for all $m+n=N+1$ and, moreover, rank $H_{mn}[s] = n(s)$ iff $n(s) \le m, n \le m(s)$. By Theorem 1.3, rank $H_\infty \ge n(s)$ for all infinite extensions of H and there exists an extension with rank n(s) iff $\delta(s) = 0$. This proves the first assertion. The second one follows from Theorem 1.3. □

The matrices $H_\infty(f)$, $f(\lambda) = p(\lambda)/a(\lambda)$, generate linear operators in the space l_2 (and other Banach spaces of sequences), which are not necessarily bounded. For example, $H_\infty((\lambda - \lambda_0)^{-1}) = [\lambda_0^{i+j}]$ is bounded iff $|\lambda_0| < 1$. More general, as it is easily proved, $H_\infty(f)$ is bounded iff all roots of the polynomial $a(\lambda)$, i.e. all poles of $f(\lambda)$ lie inside the unit disk. Therefore, for a finite Hankel matrix $H_{kl}[s]$ there exists an infinite extension H_∞ with rank n generating a bounded operator in l_2 iff there is a corresponding to s polynomial with degree n, the roots of which are lying inside the unit disk. It is an open problem (at least for us) how one can decide effectively whether there is corresponding polynomial with such a property. In this connection it seems to be of interest to ask about the minimal rank of an infinite extension of a given Hankel matrix generating a bounded operator. Especially we ask whether for a given regular $n\times n$ Hankel matrix H there exists a rank n extension H_∞ generating a bounded operator. In some cases we can state a necessary condition.

PROPOSITION 2.1. Suppose H is a (real) positive definite Hankel matrix, $(\alpha(\lambda), \beta(\lambda))$ a canonical pair of characteristic polynomials of H. If there is a real infinite extension of H with rank n, then the zeros of $\beta(\lambda)$ are lying in the interval $(-1,1)$.

PROOF. The inverse of the Hankel matrix H is up to a multiplicative constant the Bezoutian B of $\alpha(\lambda)$ and $\beta(\lambda)$ (see [9]), which is also positive definite. The signature of B coincides with the Cauchy index of $\beta(\lambda)/(\alpha(\lambda)+c\beta(\lambda))$ (c real). Therefore, the roots of $\beta(\lambda)$ are simple and lie between the roots of $\alpha(\lambda)+c\beta(\lambda)$, which are real and simple too. Provided that at least one root of $\beta(\lambda)$ lies off $(-1,1)$ we would have at least one root of $\alpha(\lambda)+c\beta(\lambda)$ outside the unit disk. Hence H could not have a real extension generating a bounded operator. This leads to the assertion of the proposition. □

2.2 Partial realization

Suppose we are given a sequence $s=(s_0,\dots,s_{N-1})$ of complex (or real) numbers. The triple (A,b,c^T) consisting of an $n\times n$ matrix A, a column vector b and a row vector c^T $(b,c\in\mathbb{C}^n)$ is said to be a partial realization with dimension n of s iff

$$s_i = c^T A^i b \qquad (i=0,1,\dots,N-1) .$$

The partial realization problem is stated in the following way.

PROBLEM 3. Given a sequence $s \in \mathbb{C}^N$ and a natural number n; describe all partial realizations of s with dimension n.

For some background information concerning this problem we refer to [13], [12], [5] and references therein. Let us remark, furthermore, that in [5] an analysis of algorithms for solving Problem 3 is presented. The aim of our note is to represent the general solution of this problem in terms of characteristic polynomials of s. To this end we show some connection between the partial realization problem and Problem 1'.

Suppose (A,b,c^T) is a partial realization of s with dimension n. Then

$$f(\lambda) = c^T(\lambda I - A)^{-1}b$$

is a proper rational function, such that $s_0,\dots,s_{N-1}$ are just the first N Markov parameters of $f(\lambda)$. Therefore, $f(\lambda)$ is a generating function of s with McMillan degree $\leq$ n. The realization (A,b,c^T) is called <u>minimal</u> iff $f(\lambda)$ has McMillan degree n. That means any partial realization which is minimal corresponds with a solution of Problem 1'.

We show that the converse is, in a sense, also true. Suppose $f(\lambda) = p(\lambda)/a(\lambda)$ ($a(\lambda) \in \mathbb{C}_n^o(\lambda)$, $p(\lambda) \in \mathbb{C}_{n-1}(\lambda)$) is a generating function of s, $a(\lambda) = a_0 + \dots + a_n\lambda^n$. We consider the companion matrix

$$C_a := \begin{bmatrix} 0 & 1 & & \\ & & \ddots & \\ & & & 1 \\ -a_0 & \dots & & -a_{n-1} \end{bmatrix} .$$

Then, as it is easily verified, we have

$$f(\lambda) = p^T(\lambda I - C_a)^{-1}e_{n-1} ,$$

where $e_{n-1} = (0,\dots,0,1)$ and p is the vector of coefficients of $p(\lambda)$. Hence (C_a, e_{n-1}, p^T) is a minimal partial realization of s. Putting

$$A = U C_a U^{-1} , \quad b = Ue_{n-1} , \quad c^T = p^T U^{-1} , \qquad (2.1)$$

where U is a regular n×n matrix, we obtain further partial realizations of s with dimension n. It remains to remark that it is a well-known fact that two minimal realizations of $f(\lambda)$ are connected by a similarity relation, which implies that all solutions of Problem 3 are given by (2.1).

2.3 Padé - approximation

It is well-known that there are close relations between Hankel (or Toeplitz) matrices and the Padé-approximation problem. There are a lot of papers concerning the connections between algorithms of Hankel and Toeplitz matrix inversion and Padé-approximation (see for instance [1], [2], [6]). However, we think that the connection described below between Padé-approximation and the Problem 1 offers some new view-point and makes the relations more transparent.

Let us formulate the Padé-approximation problem. Suppose we are given a function $F(\lambda)$ analytic at $\lambda = 0$,

$$F(\lambda) = s_0 + s_1\lambda + s_2\lambda^2 + \dots \tag{2.2}$$

(or, more general, a formal power series (2.2)).

DEFINITION. A rational function

$$F^{m/n}(\lambda) = \frac{P^{m/n}(\lambda)}{A^{m/n}(\lambda)}$$

is said to be an m/n - Padé - approximant [1] of $F(\lambda)$ iff

(i) $A^{m/n}(\lambda) \in \mathbb{C}_n^{co}(\lambda)$, $P^{m/n}(\lambda) \in \mathbb{C}_m(\lambda)$, $\gcd(A^{m/n}, P^{m/n}) = 1$

(ii) $F(\lambda) = F^{m/n}(\lambda) + o(\lambda^{m+n})$.

Here $\mathbb{C}_n^{co}(\lambda)$ denotes the class of comonic polynomials of degree n at most (i.e. $A^{m/n}(0) = 1$).

PROBLEM 4 (PADE - APPROXIMATION PROBLEM). Given complex numbers $s_0, \dots, s_{m+n}$, ask for a rational function $F^{m/n}(\lambda)$ satisfying (i) and (ii).

Let us formulate our main result concerning the Padé-approximation problem.

THEOREM 2.1. <u>Let</u> s <u>be a given sequence of complex numbers</u>, $s = (s_0, \dots, s_{m+n})$, <u>and let</u> $s^{m/n} \in \mathbb{C}^{2n}$ <u>denote the sequence</u>

---

1) For brevity we shall write „m/n - PA".

$$s^{m/n} := \begin{cases} (s_{m-n+1}, \ldots\ldots, s_{m+n}) & : n \leq m \\ (0, \ldots, 0, s_0, \ldots, s_{m+n}) & : n > m \, . \end{cases} \qquad (2.3)$$

Further, let $r = n(s^{m/n})$ and $m(s^{m/n})$ be the characteristic degrees of $s^{m/n}$, $\alpha(\lambda) \in \mathbb{C}_r^o(\lambda)$ its first characteristic polynomial and $\gamma(\lambda) \in \mathbb{C}_{r-1}(\lambda)$ the polynomial whose column vector of coefficients is

$$\gamma := \begin{bmatrix} \alpha_1 & \cdots & \alpha_r \\ \vdots & \cdot^{\cdot^{\cdot}} & \\ \alpha_r & & 0 \end{bmatrix} \begin{bmatrix} s_{m-n+1} \\ \vdots \\ s_{m-n+r} \end{bmatrix} \text{ [1)]} .$$

There exists an m/n - PA of s iff $\delta(s^{m/n}) = 0$. If the latter condition is fulfilled, then the m/n - PA

$$F^{m/n}(\lambda) = P^{m/n}(\lambda)/A^{m/n}(\lambda)$$

is uniquely determined and given by

a) $n > m$:

$$A^{m/n}(\lambda) = \lambda^r \alpha(\lambda^{-1}) \, , \quad P^{m/n}(\lambda) = \lambda^{m-n+r}\gamma(\lambda^{-1}) \qquad (2.4a)$$

b) $n \leq m$:

$$A^{m/n}(\lambda) = \lambda^r \alpha(\lambda^{-1}) \, , \qquad (2.4b)$$
$$P^{m/n}(\lambda) = \lambda^{m-n+r}\gamma(\lambda^{-1}) + (s_0 + \ldots + s_{m-n}\lambda^{m-n})\lambda^r\alpha(\lambda^{-1}) .$$

PROOF. Suppose that there exists an m/n - PA $F^{m/n}(\lambda)$ of s. We set $P(\lambda) := P^{m/n}(\lambda)$, $A(\lambda) := A^{m/n}(\lambda)$ and observe that (ii) is equivalent to

a) $n > m$ b) $n \leq m$

$$\begin{bmatrix} P_0 \\ \vdots \\ P_m \\ 0 \\ \vdots \\ 0 \end{bmatrix} = \begin{bmatrix} 0 & & s_0 \\ & \cdot^{\cdot^{\cdot}} & \vdots \\ \hline 0 & s_0 \cdots & s_{m+1} \\ \cdot^{\cdot^{\cdot}} & & \vdots \\ s_0 & \cdots\cdots & s_n \\ \vdots & & \vdots \\ s_m & \cdots\cdots & s_{m+n} \end{bmatrix} \begin{bmatrix} A_n \\ \vdots \\ A_0 \end{bmatrix} , \qquad \begin{bmatrix} P_0 \\ \vdots \\ P_m \\ 0 \\ \vdots \\ 0 \end{bmatrix} = \begin{bmatrix} 0 & & s_0 \\ & \cdot^{\cdot^{\cdot}} & \vdots \\ s_0 & \cdots\cdots & s_n \\ \vdots & & \vdots \\ \hline s_{m-n+1} & \cdots & s_{m+1} \\ \vdots & & \vdots \\ s_m & \cdots\cdots & s_{m+n} \end{bmatrix} \begin{bmatrix} A_n \\ \vdots \\ A_0 \end{bmatrix} .$$

In particular, $(A_{n-i})_0^n \in \ker H_{n,n+1}[s^{m/n}]$. By Theorem 1.1, the polynomial $\lambda^n A(\lambda^{-1})$ is corresponding to $s^{m/n}$, which implies that $\min \deg s^{m/n} = n < 2n-r+1 = m(s^{m/n})$. Due to Corollary 1.1, it follows that $\min \deg s^{m/n} = r$ and $\delta(s^{m/n}) = 0$.

---

1) $s_k = 0$, $k<0$.

Now let $\delta(s^{m/n}) = 0$. A rational function $G(\lambda)$ has the property $F(\lambda) = G(\lambda) + o(\lambda^{m+n})$ iff

$$\lambda^{m-n}G(\lambda^{-1}) = s_0\lambda^{m-n} + \dots + s_{m+n}\lambda^{-2n} + o(\lambda^{-2n})$$

in a neighbourhood of infinity. That means, any m/n - PA $F^{m/n}(\lambda)$ of s corresponds to a generating function

$$\begin{array}{ll} \lambda^{m-n}F^{m/n}(\lambda^{-1}) & : n>m \\ \lambda^{m-n}F^{m/n}(\lambda^{-1}) - s_0\lambda^{m-n} - \dots - s_{m-n} & : n\le m \end{array}$$

of $s^{m/n}$. Therefore, let us consider the classes $\mathcal{F}_k$ of generating functions of $s^{m/n}$ with McMillan degree $k < m(s^{m/n})$. By Theorem 1.3 there is an unique generating function $f(\lambda) = \gamma(\lambda)/\alpha(\lambda) \in \mathcal{F}_r$ and $\mathcal{F}_k = \emptyset$ if $k \neq r$. Consequently, the functions

$$F^{m/n}(\lambda) = \lambda^{m-n}f(\lambda^{-1}) = \frac{\lambda^{m-n+r}\gamma(\lambda^{-1})}{\lambda^r\alpha(\lambda^{-1})} \qquad : n>m,$$

$$F^{m/n}(\lambda) = \lambda^{m-n}f(\lambda^{-1}) + s_0 + \dots + s_{m-n}\lambda^{m-n}$$

$$= \frac{\lambda^{m-n+r}\gamma(\lambda^{-1}) + (s_0 + \dots + s_{m-n}\lambda^{m-n})\lambda^r\alpha(\lambda^{-1})}{\lambda^r\alpha(\lambda^{-1})} \qquad : n\le m$$

fulfill (ii). In virtue of $\alpha(\lambda) \in \mathbb{C}_r^o(\lambda)$, $\gamma(\lambda) \in \mathbb{C}_{r-1}(\lambda)$, $r \le n$, and $\gcd(\alpha(\lambda), \gamma(\lambda)) = 1$, the numerators and denominators fulfill the conditions (i), too, so that the m/n - PA exists and is given by (2.4a) and (2.4b), respectively. To show uniqueness consider another generating function $g(\lambda) = p(\lambda)/a(\lambda) \in \mathcal{F}_k$, $k \ge m(s^{m/n}) > n$. In this case we would have

$$\lambda^{m-n}g(\lambda^{-1}) = \frac{\lambda^{m-n+k}p(\lambda^{-1})}{\lambda^k a(\lambda^{-1})} ,$$

and the conditions $\lambda^k a(\lambda^{-1}) \in \mathbb{C}_n^{co}(\lambda)$, $\lambda^{m-n+k}p(\lambda^{-1}) \in \mathbb{C}_m(\lambda)$ would imply $a(0) = p(0) = 0$, which is a contradiction. Hence, the m/n - PA is uniquely determined, and the theorem is proved. □

The Padé-approximants of a given formal power series (2.2) are usually arranged in a matrix $\mathcal{P}(F) = \left[F^{m/n}(\lambda)\right]_{m,n=0}^{\infty}$, which is called the <u>Padé table</u> of $F(\lambda)$. The structure properties of $\mathcal{P}(F)$ are determined by the elements $F^{m/n}(\lambda)$ for which the Hankel matrices $H_{nn}[s^{m/n}]$ are regular. Let us give a characterization of those elements.

THEOREM 2.2. Let $s = (s_0,\ldots,s_{m+n})$ be a given sequence of complex numbers and let r, $m(s^{m/n})$ be the characteristic degrees of $s^{m/n}$, the vector defined by (2.3). The matrix $H_{nn}[s^{m/n}]$ is regular iff $\delta(s^{m/n}) = 0$ and $r = n$. If $(\alpha(\lambda), \beta(\lambda))$ denotes a pair of characteristic polynomials of $s^{m/n}$ with $\alpha(\lambda) \in \mathbb{C}_n^o(\lambda)$, $\beta(\lambda) \in \mathbb{C}_n(\lambda)$, then

$$H_{nn}[s^{m/n}]^{-1} = \sigma^{-1}B(\alpha,\beta) \ , \tag{2.5}$$

where $\sigma := \sum_{i=1}^{n} (\beta_0\alpha_i - \alpha_0\beta_i)s_{m-n+i}$.

PROOF. It is easily seen that the conditions $r = n$ and $\delta(s^{m/n}) = 0$ are necessary. We prove sufficiency. If $\delta(s^{m/n})=0$ and $r = n$, then, according to Theorem 1.3, $\gamma(\lambda)/\alpha(\lambda)$ is a generating function of $s^{m/n}$, where $\gamma(\lambda)$ is defined by (1.8). Lemma 1.4 states that

$$\beta(\lambda)\gamma(\lambda) \equiv \sigma \mod \alpha(\lambda) \ .$$

In view of Theorem 0.1 $H_{nn}[s^{m/n}] = H_{nn}(\frac{\gamma}{\alpha})$ is regular and its inverse is given by (2.5). □

REMARK 2.1. The conditions $\delta(s^{m/n}) = 0$ and $r = n$ are equivalent to the existence of the m/n - PA $F^{m/n}(\lambda)$ and the additional property

$$(m - \deg P^{m/n}(\lambda))(n - \deg A^{m/n}(\lambda)) = 0$$

(see (2.4)).

The elements of $\mathcal{P}(F)$ for which $m = \deg P^{m/n}(\lambda)$ and $n = \deg A^{m/n}(\lambda)$ are called the top elements of the table. From Theorems 2.1 and 2.2 the following structure of $\mathcal{P}(F)$ can be deduced.

1° Any m/n - PA is also an m_0/n_0 - PA, where $m_0 := \deg P^{m/n}(\lambda)$, $n_0 := \deg A^{m/n}(\lambda)$.

2° Let F^{m_0/n_0} be a top element and $z \geq 0$ the largest integer such that

$$F(\lambda) = F^{m_0/n_0}(\lambda) + o(\lambda^{m_0+n_0+z}) \ .$$

Then the m/n - PA with

$$m \geq m_0, \quad n \geq n_0, \quad m + n \leq m_0 + n_0 + z$$

exist and are equal to $F^{m_0/n_0}(\lambda)$.

3°. If

$$m \le m_0 + z, \quad n \le n_0 + z, \quad m + n > m_0 + n_0 + z,$$

then the m/n - PA does not exist.

4°. The matrix $H_{nn}[s^{m/n}]$ is regular iff the pair m/n lies on the periphery of the square

$$m_0 \le m \le m_0 + z + 1 \, , \quad n_0 \le n \le n_0 + z + 1 \, .$$

Let us illustrate these facts by a figure.

FIGURE 2.1.

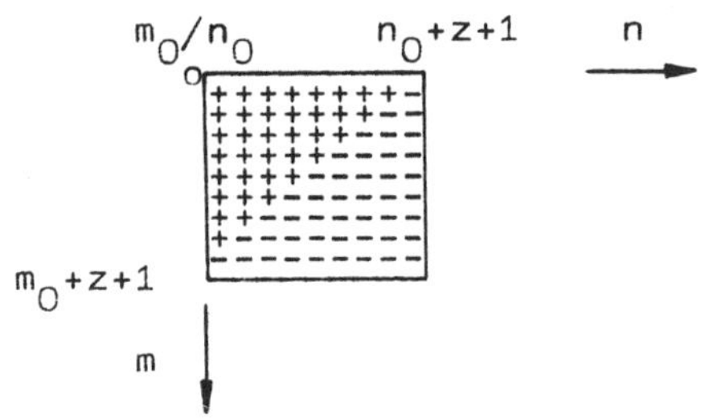

- o top element
- — $H_{nn}[s^{m/n}]$ regular
- ++ $F^{m/n}(\lambda) = F^{m_0/n_0}(\lambda)$
- -- m/n - PA does not exist

2.4 Canonical decomposition

As a further application of the results of Section 1 we consider the problem of canonical decomposition of Hankel matrices, which is a representation of Hankel matrices as a linear combination of simplest Hankel matrices. We adopt the following designation

$$1_j^N(\lambda) := \left(j!\binom{i}{j}\lambda^{i-j}\right)_{i=0}^{N-1} \quad (\lambda \ne 0) \; ; \qquad (j=0,1,\dots,N-1)$$
$$1_j^N(0) := e_j \; ,$$

where e_j denotes the unit vector with 1 at its jth component. The matrices $\mathcal{H}_j(\lambda) = H_{kl}[1_j^N(\lambda)]$ will be called <u>elementary</u> Hankel matrices. A representation of a given Hankel matrix H in the form

$$H = \sum_{i=1}^{r} \sum_{j=0}^{\nu_i - 1} \alpha_{ij} \mathcal{H}_j(\lambda_i)$$

is called <u>canonical decomposition of H</u> and the number $\sum \nu_i = \nu$ is said to be the rank of this representation. The decomposition is called <u>simple</u> iff all ν_i are equal to 1.

PROBLEM 5. Given a kxl Hankel matrix and a natural number n; describe all canonical decomposition (simple canonical decomposition) of rank n.

Problem 5 is equivalent to the following generalized moment problem.

PROBLEM 5'. Given a sequence $(s_0,\dots,s_{N-1})$ and a natural number n; describe all distribution on $\mathbb{C}$

$$\omega = \sum \alpha_{ij}\delta^{(j)}(\lambda - \lambda_i)$$

the support of which consists of exactly n points such that

$$s_i = \int_{\mathbb{C}} \lambda^i \omega d\lambda \qquad (i=0,1,\dots,N-1) .$$

Here $\delta^{(j)}$ denotes the jth derivative of the Dirac distribution.

Looking only for simple canonical decompositions Problem 5 is equivalent to

PROBLEM 5''. Given a sequence $(s_0,\dots,s_{N-1})$ and a natural number n; describe all measures μ on C, the support of which consists of exactly n points such that

$$s_i = \int_{\mathbb{C}} \lambda^i \, d\mu = \sum_{j=1}^{n} \alpha_j \lambda_j^i .$$

The problem of canonical decomposition was stated in a slightly different form for Toeplitz matrices in [7] and for Hankel matrices in [9]. In these papers one can find other equivalent formulations of the problem, e.g. in the language of quadratic forms.

The crucial point in order to apply the results of Section 1 is the observation

$$\mathcal{H}_j(\lambda_0) = H_{kl}((\lambda - \lambda_0)^{-j-1}) .$$

Therefore, if

$$f(\lambda) = \sum_{i=1}^{r} \sum_{j=1}^{\nu_i} \alpha_{ij}(\lambda - \lambda_i)^{-j} , \tag{2.6}$$

then

$$H_{kl}(f) = \sum_{i=1}^{r} \sum_{j=0}^{\nu_i - 1} \alpha_{ij}\mathcal{H}_j(\lambda_i) .$$

PROPOSITION 2.1. *Suppose* H *is a given* $k\times l$ *Hankel matrix*, $f(\lambda)$ *is a generating function of* H *with McMillan degree* n, *and* (2.6) *is the partial fraction decomposition of* $f(\lambda)$ $(\sum \nu_i = n)$. *Then*

$$H = \sum_{i=1}^{r} \sum_{j=0}^{\nu_i - 1} \alpha_{ij}\mathcal{H}_j(\lambda_i) \tag{2.7}$$

is a canonical decomposition of H *with rank* n. *On the other hand, if* (2.7) *is a canonical decomposition of* H, $n = \sum \nu_i$, *then* $f(\lambda)$ *defined by* (2.6) *is a generating function of* H *with McMillan degree* n.

REFERENCES

1. Baker, G.A., Jr.: Essentials of Padé approximants. Academic Press, New York 1975.

2. Bultheel, A.: Fast factorization algorithms and Padé approximation: a survey. K.U. Leuven, Report TW 42 (1978).

3. Fuhrmann, P.A.: Remarks on the inversion of Hankel matrices. (to appear).

4. Gantmacher, F.R.: The theory of matrices, vol. 2, Chelsea, New York 1960.

5. Gragg, W.B. and Lindquist, A.: On the partial realization problem. Lin. Alg. and its Appl. 50 (1983), 277 - 319.

6. Gilewicz, J.: Approximants de Padé. Lecture notes in math. 667, Springer Verlag, Berlin 1978.

7. Heinig, G.: Endliche Toeplitzmatrizen und zweidimensionale Wiener-Hopf-Operatoren mit homogenem Symbol I, II. Math. Nachr. 82 (1978), 29 - 68.

8. Heinig, G. and Jungnickel, U.: Hankel matrices generated by the Markov parameters of rational functions. (to appear in Lin. Alg. and its Appl.).

9. Heinig, G. and Rost, K.: Algebraic Methods for Toeplitz - like Matrices and Operators. Birkhäuser Verlag, Basel - Boston - Stuttgart 1984.

10. Iohvidov, I.S.: Hankel and Toeplitz matrices and forms (Russian). Nauka, Moscow 1974.

11. Jungnickel, U.: Zur algebraischen Theorie der endlichen Hankel - und Toeplitzmatrizen und ihren Anwendungen. Diss. A, TH Karl - Marx - Stadt 1985.

12. Kalman, R.E.: On partial realizations, transfer functions and canonical forms. Acta Polyt. Scand. 31 (1979), 9 - 32.

13. Kalman, R.E., Falb, P. and Arbib, M.: Topics in Mathematical System Theory. McGraw - Hill, New York 1969.

G. Heinig, U. Jungnickel
Technische Hochschule Karl-Marx-Stadt
Sektion Mathematik
9010 Karl-Marx-Stadt
German Democratic Republic

Operator Theory:
Advences and Applications, Vol. 19

COMMUTING NONSELFADJOINT OPERATORS AND MAPPINGS OF VECTOR BUNDLES ON ALGEBRAIC CURVES

M.S. Livšic

The definition of a characteristic function of two commuting operators which had been suggested by the theory of operator colligations was proved to be a fruitful definition. It leads to a family of operator valued functions of a complex variable. All functions of this family are connected by a *joint characteristic function* (JCF) which is a function of a point of a real algebraic curve. The JCF considered as an operator is defined on a vector bundle. In the case when one of the operators is dissipative we obtain a complete description of the class of characteristic functions in terms of mappings of vector bundles on the above mentioned algebraic curve.

1. Basic definitions and properties

Let us recall [2,4,5] the definition of a colligation

DEFINITION. Let H, E be Hilbert spaces, A_k $(k=1,2,\dots,m)$ be bounded linear operators in H, σ_k $(k=1,2,\dots,m)$ be bounded selfadjoint operators in E, Φ be a bounded linear mapping of H into E. A set
$X = (A_1,\dots,A_m; H,\Phi,E; \sigma_1,\dots,\sigma_m)$
is said to be a *colligation* if

$$\frac{1}{i}(A_k - A_k^*) = \Phi^*\sigma_k\Phi, \qquad (k = 1,2,\dots,m) \tag{1}$$

In the following we assume that $\dim E = n < \infty$ and $\cap \operatorname{Ker} \sigma_k = 0$. If $\operatorname{range} \Phi = E$ the colligation is called a *strict* colligation.

Let $(A_1,\dots,A_m)$ be a set of bounded operators with finite dimensional imaginary parts and $G_k = \operatorname{range}(A_k - A_k^*)$, $G = G_1 + \dots + G_m$.
Then

$$\frac{1}{i}(A_k - A_k^*) = P_G\sigma_k P_G ,$$

where $\sigma_k = \frac{1}{i}(A_k - A_k^*)|_G$ are selfadjoint operators in G and

P_G is the orthogonal projection onto G. Hence the set $(A_1,\ldots,A_m)$ can always be embedded in a strict colligation with $E = G$, $\Phi = P_G$.

Remark. If X is strict then $\Phi^*E = G$, $\Phi G = E$ and $\dim G = \dim E$. Indeed, relations (1) imply that range $(A_k - A_k^*) =$ range $\Phi^*\sigma_k$ and $\Phi^*(\Sigma \text{ range } \sigma_k) = G$. From Σ range $\sigma_k = E$ it follows that $\Phi^*E = G$. Assume that $(\Phi G, u_0) = 0$ then $(G,\Phi^*u_0) = 0$ and $\Phi^*u_0 = 0$. Hence $(H,\Phi^*u_0) = (\Phi H,u_0) = 0$ and $u_0 = 0$.

DEFINITION. The operator-function in E

$$S(x,z) = I - i\Phi(x_1A_1 + \ldots + x_mA_m - zI)^{-1}\Phi^*(x_1\sigma_1 + \ldots + x_m\sigma_m) , \tag{2}$$

where $x = (x_1,\ldots,x_m) \in R_m$, $z \in \mathbb{C}$ is said to be the *complete characteristic function* (CCF) of the colligation X.
This function is a regular analytic function of z and of parameters $x_1,\ldots,x_m$ at all points for which there exists the resolvent $((xA)-z])^{-1}$, where (xA) denotes $x_1A_1 + \ldots + x_mA_m$. It is obvious that $S(tx,tz) = S(x,z)$. In the case of a single-operator colligation $(A; H,\Phi,E; \sigma)$ the CCF is

$$S(x_1,z) = S(1,zx_1^{-1})$$

and it is enough to consider only the case $x_1 = 1$:

$$S(z) = I - i\Phi(A-zI)^{-1}\Phi^*\sigma \tag{3}$$

The class $\Omega(\sigma)$. An operator-function $W(z)\colon E \to E$ is said to be a function of the class $\Omega(\sigma)$ if it has the following properties:

1) $W(z)$ is a meromorphic function in the open upper half plane $\operatorname{Im} z > 0$.

2) $W(z)$ is holomorphic in a neighbourhood $|z| > a$ of $z = \infty$ and $W(\infty) = I$.

3) $W^*(z)\sigma W(z) > \sigma \quad (\operatorname{Im} z > 0)$

4) $W^*(z)\sigma W(z) = \sigma \quad (\operatorname{Im} z = 0,\ |z| > a$

The following theorems hold [1]:

THEOREM 1. The characteristic function $S(z)$ of a single-operator colligation belongs to the class $\Omega(\sigma)$.

THEOREM 2. Let $\sigma=\sigma^*$ be a given invertible operator in E. If a given function $W(z)$ belongs to the class $\Omega(\sigma)$ then $W(z)$ is the characteristic function of a colligation $(A; H,\Phi,E; \sigma)$. It is worth noting that the CCF $S(x,z)$ of a colligation $(A_1,\ldots,A_m; H,\Phi,E; \sigma_1,\ldots,\sigma_m)$ coincides with the CF of the single-operator colligation

$$((xA); H,\Phi,E; (x\sigma)) \tag{4}$$

COROLLARY. The restriction of the CCF to the fixed point x $(x \in R_m)$ belongs to the class $\Omega[(x\sigma)]$. A colligation X is said to be *commutative* if $A_kA_j = A_jA_k$ $(k,j=1,2,\ldots,m)$. It is known [4] that in this case the subspace

$$H = \text{span}\{A_1^{k_1}\ldots A_m^{k_m}\Phi^*(E)\} = \text{span}\{A_1^{*k_1}\ldots A_m^{*k_m}\Phi^*(E)\}\ ,$$
$$(0 \leqslant k_1,\ldots,k_m < \infty) \tag{5}$$

reduces $A_1,\ldots,A_m$ and that the restrictions of $A_1,\ldots,A_m$ to the orthogonal component $H_0 = H \ominus \tilde{H}$ are selfadjoint. The subspace $\tilde{H}$ is said to be the *principal* subspace of the commutative colligation.

THEOREM 3 [4]. Let a space E and selfadjoint operators σ_k $(k=1,2,\ldots,m)$ be given. Assume that $\det(x\sigma) \not\equiv 0$. Then the CCF $S(x,z)$ determines the corresponding commutative colligation up to an unitary transformation of its principal subspace.

Now the question arises: under what conditions is our operator-valued function $W(x,z)\colon E \to E$ the CCF of several commuting operators? Theorem 1 together with the Corollary gives some necessary conditions but they do not reflect the commutativity of operators $A_1,\ldots,A_m$. In the following we will investigate the case $m=2$ of two commuting operators. It turns out that the CCF must satisfy additional conditions on an algebraic curve, which play an important role in the whole theory.

2. Commutative pairs of operators

Let $X = (A_1,A_2; H,\Phi,E; \sigma_1,\sigma_2)$ be a commutative colligation and $\dim E = n < \infty$. Then the ranges of operators

$$A_1A_2^*-A_2A_1^* = (A_1-A_1^*)A_2^*-(A_2-A_2^*)A_1^*$$
$$A_2^*A_1-A_1^*A_2 = (A_1-A_1^*)A_2-(A_2-A_2^*)A_1 \tag{6}$$

belong to the "nonhermitian" subspace

$$G = \text{rang}(A_1-A_1^*) + \text{range}(A_2-A_2^*) \tag{7}$$

For strict colligations this implies that operators

$\gamma^{in} = (\Phi_0^{-1})^*[\frac{1}{i}(A_1A_2^*-A_2A_1^*)]\Phi_0^{-1}$, $\gamma^{out} = (\Phi_0^{-1})^*[\frac{1}{i}(A_2^*A_1-A_1^*A_2)]\Phi_0^{-1}$,

where $\Phi_0 = \Phi|_G$ are selfadjoint and

$$\frac{1}{i}(A_1A_2^*-A_2A_1^*) = \Phi^*\gamma^{in}\Phi ,$$
$$\frac{1}{i}(A_2^*A_1-A_1^*A_2) = \Phi^*\gamma^{out}\Phi . \tag{8}$$

Operators γ^{in}, γ^{out} play an important role but the condition $\Phi(H) = E$ is too restrictive, the projection of a strict colligation on a subspace of H is not always strict. Moreover, in the simplest case when $\dim H = 1$, colligations are not strict, unless $\dim E = 1$. To overcome this difficulty we use the notion of a regular colligation, introduced by N. Kravitsky [2].

DEFINITION. A commutative colligation is called a *regular* colligation if there exists a selfadjoint operator γ^{in} in E such that

$$A_2\Phi^*\sigma_1 - A_1\Phi^*\sigma_2 = \Phi^*\gamma^{in} \tag{9}$$

Define

$$\gamma^{out} = \gamma^{in} + i(\sigma_1\Phi\Phi^*\sigma_2-\sigma_2\Phi\Phi^*\sigma_1) \tag{10}$$

Then, using (1), we obtain

$$A_2^*\Phi^*\sigma_1 - A_1^*\Phi^*\sigma_2 = \Phi^*\gamma^{out} \tag{11}$$

Remarks. 1) Strict colligations are regular
2) The operator γ^{in} is defined uniquely for strict colligations but for regular colligation it is not so. We will include operators γ^{in} and γ^{out} in the notation of a regular colligation:

$$X = (A_1,A_2;\ H,\Phi,E;\ \sigma_1,\sigma_2,\gamma^{in},\gamma^{out}) \tag{12}$$

3) <u>α-transformations</u>. Let $\alpha = \begin{pmatrix} \alpha_{11} & \alpha_{12} \\ \alpha_{21} & \alpha_{22} \end{pmatrix}$ be a real matrix with $\det(\alpha) = 1$. Then the set

$$\alpha(X) = (A_1', A_2'; H, \Phi, E; \sigma_1', \sigma_2', \gamma^{in}, \gamma^{out}) \tag{13}$$

where

$$\begin{aligned} A_1' &= \alpha_{11}A_1 + \alpha_{12}A_2, \quad & \sigma_1' &= \alpha_{11}\sigma_1 + \alpha_{12}\sigma_2 \\ A_2' &= \alpha_{21}A_1 + \alpha_{22}A_2, \quad & \sigma_2' &= \alpha_{21}\sigma_1 + \alpha_{22}\sigma_2 \end{aligned} \tag{14}$$

is also a regular colligation.

THEOREM 4. *For regular colligations the following relation holds:*

$$(\lambda_1\sigma_2 - \lambda_2\sigma_1 + \gamma^{out})S(x_1, x_2, x_1\lambda_1 + x_2\lambda_2) =$$

$$\widetilde{S}(x_1, x_2, x_1\lambda_1 + x_2\lambda_2)(\lambda_1\sigma_2 - \lambda_2\sigma_1 + \gamma^{in}), \tag{15}$$

where $S(x,z)$ is the CCF,

$$\widetilde{S}(x,z) = I - i(x\sigma)\Phi((xA) - z)^{-1}\Phi^*,$$

$$(x = (x_1, x_2) \in R_2, \ z \in \mathbb{C}) \tag{16}$$

and $(x\lambda) = x_1\lambda_1 + x_2\lambda_2 \notin \text{spectrum}(x_1A_1 + x_2A_2)$, $\lambda = (\lambda_1, \lambda_2) \in C_2$.

PROOF. Using regularity conditions (9), (11) we obtain

$$(A_2 - \lambda_2 I)\Phi^*\sigma_1 - (A_1 - \lambda_1 I)\Phi^*\sigma_2 = \Phi\,(\gamma^{in} + \lambda_1\sigma_2 - \lambda_2\sigma_1), \tag{17}$$

$$\sigma_1\Phi(A_2 - \lambda_2 I) - \sigma_2\Phi(A_1 - \lambda_1 I) = (\gamma^{out} + \lambda_1\sigma_2 - \lambda_2\sigma_1)\Phi \tag{18}$$

Multiplying (17) by $\sigma_1\Phi(A_1 - \lambda_1 I)^{-1}$ from the left, (18) by $(A_1 - \lambda_1 I)^{-1}\Phi^*\sigma_1$ from the right and substructing we obtain

$$\begin{aligned} &-\sigma_1\Phi\Phi^*\sigma_2 + \sigma_2\Phi\Phi^*\sigma_1 = \\ &= \sigma_1\Phi(A_1 - \lambda_1 I)^{-1}\Phi^*(\gamma^{in} + \lambda_1\sigma_2 - \lambda_2\sigma_1) - \\ &- (\gamma^{out} + \lambda_1\sigma_2 - \lambda_2\sigma_1)\Phi(A_1 - \lambda_1 I)^{-1}\Phi^* \end{aligned}$$

Using (10), we obtain

$$\begin{aligned} &\gamma^{in} - i\sigma_1\Phi(A_1 - \lambda_1 I)^{-1}\Phi^*(\gamma^{in} + \lambda_1\sigma_2 - \lambda_2\sigma_1) = \\ &= \gamma^{out} - i(\gamma^{out} + \lambda_1\sigma_2 - \lambda_2\sigma_1)\Phi(A_1 - \lambda_1 I)^{-1}\Phi^*\sigma_1, \end{aligned}$$

$$(\gamma^{out}+\lambda_1\sigma_2-\lambda_2\sigma_1)S(1,0,\lambda_1) =$$
$$\tilde{S}(1,0,\lambda_1)(\gamma^{in}+\lambda_1\sigma_2-\lambda_2\sigma_1)$$

Performing the α-transformation with $\alpha = \begin{pmatrix} x_1, & x_2 \\ -x_2, & x_1 \end{pmatrix}\left(x_1^2+x_2^2\right)^{-\frac{1}{2}}$, $\lambda_1' = \alpha_{11}\lambda_1+\alpha_{12}\lambda_2$, $\lambda_2' = \alpha_{21}\lambda_1+\alpha_{22}\lambda_2$ we obtain (15).

COROLLARY. For regular colligations the following relation holds:

$$\det(\lambda_1\sigma_2-\lambda_2\sigma_1+\gamma^{in}) = \det(\lambda_1\sigma_2-\lambda_2\sigma_1+\gamma^{out}) ,$$
$$\lambda = (\lambda_1,\lambda_2) \in \mathbb{C}_2 . \quad (19)$$

The polynomial $D(\lambda_1,\lambda_2) = \det(\lambda_1\sigma_2-\lambda_2\sigma_1+\gamma^{in})$ is said to be the *discriminant* of the colligation and

$$\Gamma = \{(\lambda_1,\lambda_2) \in C_2 \mid D(\lambda_1,\lambda_2) = 0\}$$

is said to be the *discriminant manifold* of the colligation. There are two cases:

1) $D(\lambda_1,\lambda_2) \not\equiv 0$, Γ is an algebraic curve
2) $D(\lambda_1,\lambda_2) \equiv 0$, $\Gamma = \mathbb{C}_2$

Let us consider two vector bundles

$$E^{in}(\lambda) = \mathrm{Ker}(\lambda_1\sigma_2-\lambda_2\sigma_1+\gamma^{in}) ,$$
$$E^{out}(\lambda) = \mathrm{Ker}(\lambda_1\sigma_2-\lambda_2\sigma_1+\gamma^{out}) \quad (\lambda \in \Gamma)$$

Let N be the set of all points $(x,\lambda) \in R_2\times\Gamma$ such that $(xA)-(x\lambda)I = x_1(A_1-\lambda_1 I) + x_2(A_2-\lambda_2 I)$ is invertible, $\pi(N)$ be the projection of N into Γ ($\lambda \in \pi(N)$ iff there exists $(x,\lambda) \in N$). Let N^* be the set of all points $(x,\lambda) \in R_2\times\Gamma$ such that $(x\lambda) \notin \mathrm{spectrum}(xA^*)$.

THEOREM 5. *If* $(A_1, A_2; H,\Phi,E; \sigma_1,\sigma_2,\gamma^{in},\gamma^{out})$ *is a regular colligation then*

1) $S(x,(x\lambda))$ *maps* $E^{in}(\lambda)$ *into* $E^{out}(\lambda)$, $(x,\lambda \in N)$
2) $S(x,(x\lambda))$ *maps* $E^{in}(\lambda)$ *onto* $E^{out}(\lambda)$, $(x,\lambda \in N \cap N^*)$
3) *The restriction* $S(x,(x\lambda))\big|_{E^{in}(\lambda)}$ *to the subspace* $E^{in}(\lambda)$ *does not depend on* x $(x,\lambda \in N)$.

PROOF. The assertion 1) follows immediately from (15).

To prove 2) we will use the relation

$$[I+i\Phi(A^*-zI)^{-1}\Phi^*\sigma][I-i\Phi(A-zI)^{-1}\Phi^*\sigma] = I , \qquad (20)$$

which holds for a single-operator colligation $(A; H,\Phi,E; \sigma)$, and which can be easily checked. The relation (20) implies that

$$[\tilde{S}(x,z^*)]^* = S^{-1}(x,z) \qquad (21)$$

From (16) it follows that

$$(\gamma^{in}+\lambda_1\sigma_2-\lambda_2\sigma_1)[\tilde{S}(x,(x\lambda^*)]^* =$$
$$[S(x,(x\lambda^*)]^*(\gamma^{out}+\lambda_1\sigma_2-\lambda_2\sigma_1) \qquad (22)$$

Relations (22), (21) imply that $S^{-1}(x,x\lambda))$ maps $E^{out}(\lambda)$ into $E^{in}(\lambda)$, hence $S(x,(x\lambda))E^{in}(\lambda) = E^{out}(\lambda)$, $(\lambda \in N \cap N^*)$.

To prove the assertion 3) let us assume that $(\xi_1,\xi_2,\xi,\lambda_1+\xi_2\lambda_2) \in N$, $(\eta_1,\eta_2,\eta_1\lambda_1+\eta_2\lambda_2) \in N$ and $\xi_1\eta_2-\xi_2\eta_1 > 0$. Performing the α-transformation

$$\begin{aligned} \rho\lambda_1' &= \xi_1\lambda_1+\xi_2\lambda_2 , \quad \rho A_1' = \xi_1 A_1+\xi_2 , \quad \rho\sigma_1' = \xi_1\sigma_1+\xi_2\sigma_2 \\ \rho\lambda_2' &= \eta_1\lambda_1+\eta_2\lambda_2 , \quad \rho A_2' = \eta_1 A_1+\eta_2 A_2, \quad \rho\sigma_2' = \eta_1\sigma_1+\eta_2\sigma_2 , \end{aligned} \qquad (23)$$

where $\rho = (\xi_1\eta_2-\xi_2\eta_1)^{\frac{1}{2}}$, we obtain $E^{in}(\lambda_1,\lambda_2) = E'^{in}(\lambda_1',\lambda_2')$ and

$$(\lambda_1'\sigma_2'-\lambda_2'\sigma_1'+\gamma^{in})u = 0 , \quad (u \in E^{in}(\lambda)) , \qquad (24)$$

This relation implies

$$(\lambda_2'\Phi^*\sigma_1'-\lambda_1'\Phi^*\sigma_2')u = \Phi^*\gamma^{in}u$$

The regularity condition (9) implies

$$(A_2'\Phi^*\sigma_1'-A_1'\Phi^*\sigma_2')u = \Phi^*\gamma^{in}u$$

Hence

$$(A_2'-\lambda_2'I)\Phi^*\sigma_1' = (A_1'-\lambda_1'I)\Phi^*\sigma_2' ,$$
$$(A_1'-\lambda_1 I)^{-1}\Phi^*\sigma_1'u = (A_2'-\lambda_2'I)^{-1}\Phi^*\sigma_2'u$$

and $S(\xi_1,\xi_2,\xi_1\lambda_1+\xi_2\lambda_2)u = S(\eta_1,\eta_2,\eta_1\lambda_1+\eta_2\lambda_2)u$

We can now summarize properties of the CCF in the following.

THEOREM 6. *The CCF of a regular colligation possesses the following properties:*

I) a) $S(x_1,x_2,z)$ is a holomorphic function of x_1,x_2,z in the region K_a: $|z| > a(|x_1|^2+|x_2|^2)^{\frac{1}{2}}$ ($a = \operatorname{Sup}||x_1A_1+x_2A_2||$, $|x_1|^2+|x_2|^2 = 1$)

b) $S(tx_1,tx_2,tz) = S(x_1,x_2,z)$

c) $S(x_1,x_2,\infty) = I$

II) $S(x_1,x_2,z)$ is a meromorphic function of z in the region $\operatorname{Im} z > 0$, $(x_1,x_2) \in R_2$, belonging to the class $\Omega[(x\sigma)]$

III) a) The restriction of $S(x_1,x_2,x_1\lambda_1+x_2\lambda_2)$ to $E^{in}(\lambda)$ does not depend on $x = (x_1,x_2)$

b) $S(x_1,x_2,x_1\lambda_1+x_2\lambda_2)$ maps $E^{in}(\lambda)$ into $E^{out}(\lambda)$, $(x, \lambda \in N)$

c) $S(x_1,x_2,x_1\lambda_1+x_2\lambda_2)$ maps $E^{in}(\lambda)$ onto $E^{out}(\lambda)$, $(x, \lambda \in N \cap N^*)$.

DEFINITION. The mapping $\hat{S}(\lambda) = S(x_1,x_2,x_1\lambda_1+x_2\lambda_2)\big|_{E^{in}(\lambda)}$ is said to be the *joint characteristic function* (JCF) of a regular colligation.

As an operator $\hat{S}(\lambda)$ is defined on the subspace $E^{in}(\lambda) = \ker(\lambda_1\sigma_2-\lambda_2\sigma_1+\gamma^{in})$ and maps this subspace into $E^{out}(\lambda) = \ker(\lambda_1\sigma_2)-\lambda_2\sigma_1+\gamma^{out})$. As a function of point λ of the discriminant manifold Γ it is defined in the region $\pi(N)$.

Remarks: 1) If $\dim H < \infty$ then $S(x_1,x_2,z)$ is a rational function of x_1,x_2,z

2) If A_1,A_2 are compact operators then $S(x_1,x_2,z)$ is a meromorphic function of z in the region $\mathbb{C} \setminus \{0\}$.

The consonance *between operator-functions and vector bundles.*

Let $f(\lambda_1,\lambda_2) = 0$ be an algebraic curve Γ which possesses a determinantal representation

$$f(\lambda_1,\lambda_2) = \det(\lambda_1\sigma_2-\lambda_2\sigma_1+\gamma) , \tag{25}$$

where $\sigma_1 = \sigma_1^*$, $\sigma_2 = \sigma_2^*$, $\gamma = \gamma^*$ and let

$$F(\lambda) = \ker(\lambda_1\sigma_2-\lambda_2\sigma_1+\gamma) \qquad (\lambda \in \Gamma) \tag{26}$$

be the corresponding vector bundle.

Let $W(x_1,x_2,z) : E \to E$ be an operator-functions, holomorphic in a region $K_a = \{(x,z) \in C_3 \mid |z| > a(|x_1|^2+|x_2|^2)^{\frac{1}{2}}$

and homogeneous with respect to variable x, z

$$W(tx_1, tx_2, tz) = W(x_1, x_2, z)$$

DEFINITION. The operator function $W(x,z)$ and the vector bundle $F(\lambda) = \ker(\sigma_1\lambda_2 - \lambda_2\sigma_1 + \gamma)$ are said to be *consonant* if the restriction

$$\hat{W}(\lambda) = W(x_1, x_2, x_1\lambda_1 + x_2\lambda_2) \mid \ker(\lambda_1\sigma_2 - \lambda_2\sigma_1 + \gamma) \tag{28}$$

to the vector bundle $F(\lambda)$ does not depend on $x = (x_1, x_2)$, $(x,\lambda) \in K_a$. The operator-function $\hat{W}(\lambda)$ will be called the *trunk* of $W(x,z)$ with respect to $F(\lambda)$.

THEOREM 7. *If* $W(x,z)$ *is consonant with the bundle* $F(\lambda) = \ker(\lambda_1\sigma_2 - \lambda_2\sigma_1 + \gamma)$ *and* $\sigma > 0$ *then* $W(x,z)$ *is defined uniquely by its trunk* $\hat{W}(\lambda)$ *with respect to* $F(\lambda)$.

PROOF. Let t be a real number such that $|t| < \varepsilon$ where $\varepsilon = ||\sigma_2|| \cdot ||\sigma_1||^{-1}$. Let us consider the curve $f_t(\lambda_1, z) = 0$ in the (λ_1, z) plane C_2, where

$$f_t(\lambda_1, z) = \det[\lambda_1(t\sigma_1 + \sigma_2) - z\sigma_1 + \gamma] \tag{29}$$

It is evident that $f_t(\lambda_1, z) = f(\lambda_1, z - t\lambda_1)$ and that $F(\lambda_1, z - t\lambda_1) = F_t(\lambda_1, z)$, where

$$F_t(\lambda_1, z) = \ker[\lambda_1(t\sigma_1 + \sigma_2) - z\sigma_1 + \gamma] \tag{30}$$

If z and t are real and $|t| < \varepsilon$ then $t\sigma_1 + \sigma_2 > 0$. Hence all different roots $\lambda_1^{(k)} = \lambda_1^{(k)}(t,z)$ of the equation $f_t(\lambda_1, z) = 0$ are real and the corresponding subspaces $f_t(\lambda^{(k)}, z)$ are orthogonal with respect to positive matrix $\sigma_2 + t\sigma_1$:

$$E = \Sigma \oplus F_t(\lambda_1^{(k)}, z) = \Sigma \oplus F(\lambda_1^{(k)}, \lambda_2^{(k)}) \tag{31}$$

where $F(\lambda_1^{(k)}, \lambda_2^{(k)}) = 0$ and $t\lambda_1^{(k)} + \lambda^{(k)} = z$.

Hence

$$W(t,1,z) = \Sigma W(t,1,z) P^t \lambda^{(k)}, z) ,$$

where $P^t(\lambda_1^{(k)}, z)$ is the orthoprojector onto $F_t(\lambda^{(k)}, z)$ with respect to $\sigma_2 + t\sigma > 0$. Thus, for $W(t,1,z)$ we obtain the following *restoration* formula:

$$W(t,1,z) = \sum_{t\lambda_1^{(k)} + \lambda_2^{(k)} = z} \hat{W}(\lambda_1^{(k)}, \lambda^{(k)}) P^t(\lambda^{(k)}, z) \tag{32}$$

COROLLARY 1. If $\ker(\lambda_1\sigma_2-\lambda_2\sigma_1+\gamma') \equiv \ker(\lambda_1\sigma_2-\lambda_2\sigma_1+\gamma)$ and $\sigma_2 > 0$ then $\gamma'=\gamma$. Indeed, let $W=\gamma'$. Then

$$\gamma' \mid \ker(\lambda_1\sigma_2-\lambda_2\sigma_1+\gamma') = (\lambda_2\sigma_1-\lambda_2\sigma_2) \mid \ker(\lambda_1\sigma_2-\lambda_2\sigma_1+\gamma')$$

$$\gamma \mid \ker(\lambda_1\sigma_2-\lambda_2\sigma_1+\gamma) = (\lambda_2\sigma_1-\lambda_1\sigma_2) \mid \ker(\lambda_1\sigma_2-\lambda_2\sigma_1+\gamma)$$

Hence,

$$\gamma'|\ker(\lambda_1\sigma_2-\lambda_2\sigma_1+\gamma) = \gamma|\ker(\lambda_1\sigma_2-\lambda_2\sigma_1+\gamma)$$

and $\gamma'=\gamma$.

COROLLARY 2. The CCF of a regular colligation with $\sigma_2 > 0$ is uniquely determined by the JCF of this colligation.

COROLLARY 3. A regular colligation with $\sigma_2 > 0$ is determined up to a unitary transformation of its principal subspace by the JCF $\hat{W}(\lambda)$: if

$$X' = (A_1',A_2';\ H',\Phi'.E;\ \sigma_1,\sigma_2,\gamma^{in},\gamma^{out})$$

and

$$X'' = (A_2'',A_2'';\ H'',\Phi'',E;\ \sigma_1,\sigma_2,\gamma^{in},\gamma^{out})$$

are two regular commutative colligations with the same JCF $W(\lambda)$ then there exists an isometric transformation T of the principal subspace $\tilde{H}'$ onto the subspace $\tilde{H}''$ such that $A_k''T = TA_k'$, $\Phi' = \Phi''T$.

COROLLARY 4. A CCF $S(x,z)$ is determined uniquely by σ_1, $\sigma_2 > 0$, γ^{in} and its values $S(x_0,z)$ at an arbitrary fixed point x_0.

COROLLARY 5. If the trunk $\hat{W}(\lambda)$ of $W(x,z)$ is isometric with respect to σ_1, σ_2 for real $\lambda = (\lambda_1,\lambda_2)$ and $\sigma_2 > 0$ then $W(x,z)$ is isometric with respect to $x_1\sigma_1+x_2\sigma_2$ for real x, z in the region K_a.

THEOREM 8. *If* $W(x,z)$ *is consonant with* $F(\lambda) = \ker(\lambda_1\sigma_2-\lambda_2\sigma_1+\gamma)$ *and* $\sigma_2 > 0$ *then* $W(x,z)$ *satisfies the following* PDE:

$$\frac{\partial W}{\partial x_1} + \frac{\partial W}{\partial z}(x_1\sigma_1+x_2\sigma_2)^{-1}(z\sigma_1-x_2\gamma) = 0 \ ,$$

$$(|x_1| < \varepsilon\ |x_2|) \ . \tag{33}$$

PROOF. The restriction $W(t,1,t\lambda_1+\lambda_2)|F(\lambda)$ does not depend on t. Hence $\frac{d}{dt} W(t,1,t\lambda_1+\lambda_2)|F(\lambda) = 0$ and

$$\left(\frac{\partial W}{\partial t} + \frac{\partial W}{\partial z}\lambda_1\right) \mid F(\lambda) = 0 \ , \quad z = t\lambda_1 + \lambda_2 \ .$$

It is clear that $(\lambda_1\sigma_2 - (z - t\lambda_1)\sigma_1 + \gamma) \mid F(\lambda) = 0$. Hence,

$$\lambda_1 \mid F(\lambda) = (t\sigma_1 + \sigma_2)^{-1}(z\sigma_1 - \gamma) \mid F(\lambda) \ ,$$

$$\lambda_2 = z - t\lambda_1$$

and

$$\left|\frac{\partial W}{\partial t} + \frac{\partial W}{\partial z}(t\sigma_1 + \sigma_2)^{-1}(z\sigma_1 - \gamma)\right] F(\lambda) = 0 \ ,$$

$$z = t\lambda_1 + \lambda_2$$

Using the restoration formula we obtain

$$\frac{\partial W}{\partial t} + \frac{\partial W}{\partial z}(t\sigma_1 + \sigma_2)^{-1}(z\sigma_1 - \gamma) = 0 \qquad (34)$$

3. Description of a class of characteristic functions

Let us recall [1,4] that a single-operator colligation $(A; H,\Phi,E; \sigma)$ is said to be *dissipative* if $\sigma > 0$. In this case $\frac{1}{i}(A - A^*) \geqslant 0$ and $||e^{itA}||$ is a decreasing function of t. We will say that colligation is *completely dissipative* if $\sigma > 0$ and $\lim_{t\to+\infty} e^{itA} h = 0 \quad (h \in H)$

THEOREM 9 [8,9]. A dissipative colligation is completely dissipative iff its characteristic function $S(z)$ is an inner function in $\operatorname{Im} z \geqslant 0$, which means that $S(\lambda+i0)$ is an isometry with respect to σ for almost all real λ.

Let $f(\lambda_1,\lambda_2) = 0$ be an algebraic curve and $f(\lambda_1,\lambda_2) = \det(\lambda_1\sigma_2-\lambda_2\sigma_1+\gamma)$ be its determinantal representation $(\sigma_1=\sigma_1^*$, $\sigma_2 > 0$, $\gamma=\gamma^*)$. The set of all possible γ in the above mentioned representation with fixed σ_1, σ_2 we will denote by $\Delta_f(\sigma_1,\sigma_2)$.

The class $\Omega_f(\sigma_1,\sigma_2)$. An operator-function $W(x,z)\colon E \to E$ is said to be a function of the class $\Omega_f(\sigma_1,\sigma_2)$ if $W(x,z)$ satisfies the following conditions:

1) $W(x,z)$ has the form

$$W(x,z) = I-iR(x,z)(x\sigma) \ , \qquad (35)$$

where $R(x,z)$ is holomorphic in a region

$K_a = \{(x,z) \in \mathbb{C}_3 \mid |z| > (|x_1|^2+|x_2|^2)^{\frac{1}{2}}$ and $R(tx,tz) = t^{-1}R(x,z)$.

2) $W(x,z)$ is consonant with a bundle $F(\lambda) = \ker(\lambda_1\sigma_2-\sigma_2\sigma_1+\gamma)$, where $\gamma \in \Delta_f(\sigma_1,\sigma_2)$, and the trunk $\hat{W}(\lambda) = W(x,(x\lambda) \mid F(\lambda)$ is an isometric mapping of $F(\lambda)$ onto a bundle of the form $\tilde{F}(\lambda) = \det(\lambda_1\sigma_2-\lambda_2\sigma_1+\tilde{\gamma})$ $(\tilde{\gamma} \in \Delta_f(\sigma_1,\sigma_2))$ with respect to σ_1 and σ_2 $(x,(x\lambda) \in Ka)$

3) $W(0,1,z)$ is an inner function of the class $\Omega(\sigma)$.

THEOREM 10. *A given function* $W(x,z)$ *is the* CCF *of a commutative, regular colligation* $(A_1,A_2;\ H,\Phi,E;\ \sigma_1,\sigma_2,\gamma,\tilde{\gamma})$ *such that* $(A_2;H,\Phi,E;\sigma_2)$ *is completely dissipative iff* $W(x,z)\in\Omega_f(\sigma_1,\sigma_2)$.

To prove this theorem we need some lemmas.

LEMMA 1. The function $R(x,\ )$, defined by (35), satisfies the following PDE:

$$\frac{\partial R}{\partial x_1}\sigma_2 - \frac{\partial R}{\partial x_2}\sigma_1 - \frac{\partial R}{\partial z}\gamma = 0 , \tag{36}$$

$$\sigma_2\frac{\partial R}{\partial x_1} - \sigma_1\frac{\partial R}{\partial x_2} - \tilde{\gamma}\frac{\partial R}{\partial z} = 0 \tag{37}$$

PROOF. Equation (33) implies

$$\frac{\partial}{\partial x_1}[R\cdot(x_1\sigma_1+x_2\sigma_2] + \frac{\partial R}{\partial z}(z\sigma_1-x_2\gamma) = 0 . \tag{38}$$

The condition $R(tx,tz) = t^{-1}R(x,z)$ implies

$$x_1\frac{\partial R}{\partial x_1} + x_2\frac{\partial R}{\partial x_2} + z\frac{\partial R}{\partial z} + R = 0 \tag{39}$$

Using (38) and (39) we obtain (36). Condition 2) together with Corollary 5 leads to the following relation:

$$W^*(x,z)(x\sigma)W(x,z) = (x\sigma) , \tag{40}$$

where x and z are real, and

$$[W(x,z)]^{-1} = I+i[R(x,z)]^*(x\sigma) \tag{41}$$

Condition 2) implies that $[W(x,z)]^{-1}$ is consonant with $\tilde{F}(\lambda)$ which leads to the PDE

$$\frac{\partial(W^{-1})}{\partial x_1} + \frac{\partial(W^{-1})}{\partial}(x_1\sigma_1+x_2\sigma_2)^{-1}(z\sigma_1-x_2\tilde{\gamma}) = 0 \tag{42}$$

Using (39), (41), (42) we obtain

$$\frac{\partial R^*}{\partial x_1}\sigma_2 - \frac{\partial R^*}{\partial x_2}\sigma_1 - \frac{\partial R^*}{\partial z}\tilde{\gamma} = 0$$

which is equivalent to (37).

We define the following function

$$V(x) = -\frac{1}{2\pi i}\int_{|z|=r} e^{iz}R(x,z)dz \ , \quad (r > a(|x_1|^2+|x_2|^2)^{\frac{1}{2}}$$

As $R(x,z)$ is holomorphic in K_a, $V(x)$ is an entire function.

LEMMA 2. $V(x)$ satisfies the following PDE:

$$\frac{\partial V}{\partial x_i}\sigma_2 - \frac{\partial V}{\partial x_2}\sigma_1 + iV\gamma = 0 \ , \tag{43}$$

$$\sigma_2\frac{\partial V}{\partial x_1} - \sigma_1\frac{\partial V}{\partial x_2} + i\tilde{\gamma}V = 0 \tag{44}$$

PROOF. These equations follow from PDE (36), (37).

We will need the following.

THEOREM 11 [7]. *Let* H *be a set of solutions of the* PDE

$$\sigma_2\frac{\partial v}{\partial t} = \sigma_1\frac{\partial v}{\partial \xi} - i\beta v \ , \tag{45}$$

where $\sigma_1=\sigma_1^*$, $\sigma_2 > 0$, $\beta=\beta^*$ *are given operators in a finite dimensional space* E. *Assume that* H *satisfies the following conditions:*

1) H *is a Hilbert space with respect to the scalar product*

$$[v_1,v_2] = \int_0^\infty (\sigma_2 v_1(0,\xi),\ v_2(0,\xi))d\xi \tag{46}$$

2) *If* v *belongs to* H *then* $\frac{\partial v}{\partial \xi}$ *and* $\frac{\partial v}{\partial \tau}$ *belong to* H.

3) *Operator* $B = -i\frac{\partial}{\partial \xi}$ *is bounded in* H.

Then the set

$$(-i\frac{\partial}{\partial \tau}\ ,\ -i\frac{\partial}{\partial \xi}\ ;\ H,\Phi v(\tau,\xi) = v(0,0),E;\ \sigma_1,\sigma_2) \tag{47}$$

forms a regular colligation such that

$$\lim_{t\to+\infty} e^{itB}v = 0 \qquad (v \in H) \tag{48}$$

PROOF OF THEOREM 10. Let $W(x,z)$ be a function of class $\Omega_f(\sigma_1,\sigma_2)$. Condition 3) implies that $W(0,1,z)$ is the characteristic function of a colligation $(A_2'; H',\Phi',E;\sigma_2)$ such that

$$\lim_{t\to+\infty} e^{itA_2'}h' = 0 \tag{49}$$

Hence,

$$W(0,1,z) = I - i\Phi'(A_2' - zI)^{-1}\Phi'^*\sigma_2 \tag{50}$$

$$R(0,1,z) = \Phi'(A_2' - zI)^{-1}\Phi'^*$$

$$V(0,x_2) = -\frac{1}{2\pi i}\,\Phi' \int_{|z|=r} e^{iz}(x_2A_2' - zI)^{-1}dz\;\Phi'^* = \Phi' e^{ix_2A_2'}\Phi'^* \tag{51}$$

Let us consider the set H of solutions of PDE (37) satisfying the initial conditions of the form

$$v_{h'}(x_1,x_2)\big|_{x_1=0} = \Phi' e^{ix_2A_2'}h' \qquad (h' \in H') \tag{52}$$

We introduce in H the following norm

$$||v(x_1,x_2)||^2 = \int_0^\infty (\sigma_2 v(0,x_2),\ v(0,x_2))dx_2 \tag{53}$$

The scalar product

$$(\sigma_2 v_{h'}(0,x_2),\ v_{h'}(0,x_2)) = (\sigma_2\Phi' e^{ix_2A_2'}h',\ \Phi' e^{ix_2A_2'}h') =$$

$$(\Phi'^*\sigma_2\Phi' e^{ix_2A_2'}h',\ e^{ix_2A_2'}h') =$$

$$\frac{1}{i}\,((A_2' - A_2'^*)e^{ix_2A_2'}h',\ e^{ix_2A_2'}h') =$$

$$-\frac{d}{dx_2}\,(e^{ix_2A_2'}h',\ e^{ix_2A_2'}h') .$$

Hence,

$$||v_{h'}(x_1,x_2)||^2 = -\int_0^\infty \frac{d}{dx_2}\,(e^{ix_2A_2'}h',\ e^{ix_2A_2'}h')dx_2 = (h',h') \tag{54}$$

This equality means that the mapping T of H' onto H given by (52) is an isometry. Operator A_2' in H' is transformed by T into the operator $-i\frac{\partial}{\partial x_2}$ in H.

As $H' = \text{span}\ e^{itA_2'}\Phi'^*(E)$ $(-\infty < t < \infty)$ so

$$H = \operatorname*{span}_{-\infty<t<\infty} V(x_1,x_2+t)(E) \tag{55}$$

Consider now $A_1 = -i\frac{\partial}{\partial x_1}$. From PDE (43) and from (55) it follows that H is invariant under A_1 and that A_1 is bounded. Then, using Theorem 11 we can construct a regular commutative colligation

$$X = (-i \frac{\partial}{\partial x_1}, -i \frac{\partial}{\partial x_2}; H,\Phi,E; \sigma_1,\sigma_2; \gamma,\tilde{\gamma}) \tag{56}$$

Let $S(x_1,x_2,z)$ be the CCF of X. Then $S(0,1,z) = W(0,1,z)$. Using Corollary 4 we obtain $S(x_1,x_2,z) = W(x_1,x_2 z)$.

Remark. For a colligation X, which satisfies the conditions of Theorem 10 the relation

$$H = \text{span}\{A_2^k\Phi^*(E)\} , \quad (k = 0,1,2,...) , \tag{57}$$

holds. Indeed, let $H = \tilde{H} \oplus H_0$, where $H = \text{span}\{A_2^k\Phi^*(E)\}$ and $A_2|H_0$ is selfadjoint. Then $||e^{itA_2}h_0|| = ||h_0||$ $(h_0 \in H_0)$. The condition $\lim_{t\to+\infty} e^{itA_2}h_0 = 0$ implies $h_0 = 0$.

COROLLARY. Let $W(x,z) \in \Omega_f(\sigma_1,\sigma_2)$ and

$$X = (A_1,A_2; H,\Phi,E; \sigma_1,\sigma_2; \gamma^{in},\gamma^{out})$$

be a corresponding commutative regular colligation such that (A_2,H,Φ,E,σ_2) is completely dissipative and

$$W(x,z) = I-i\Phi(x_1A_1+x_2A_2-zI)^{-1}\Phi^*(x_1\sigma_1+x_2\sigma_2)$$

Then H is finite dimensional iff $W(0,1,z)$ is a rational function of z [1].

COROLLARY. If $W(x,z) \in \Omega_f(\sigma_1,\sigma_2)$ and $W(0,1,z)$ is a rational function then $W(x_1,x_2,z)$ is a rational function of x_1, x_2, z. In this case $W(0,1,z)$ is holomorphic on the real axis and

$$W^*(0,1,z)\sigma_2W(0,1,z) = \sigma_2 , \quad (\text{Im } z = 0)$$

4. Elementary factors and a factorization theorem

The simplest functions belonging to $\Omega_f(\sigma_1,\sigma_2)$ can be obtained as CCF of one-point regular colligations

$$(\zeta_1,\zeta_2; H,\Phi,E; \sigma_1,\sigma_2,\gamma^{in},\gamma^{out}) \tag{58}$$

where dim H = 1, ζ_1, ζ_2 are complex numbers and $A_jh = \zeta_jh$ $(h \in H, j=1,2)$. Denoting these elementary factors by $E(\zeta,x,z)$

we obtain

$$E(\zeta,x,z) = I-i(x_1\zeta_1+x_2\zeta_2-z)^{-1}\Phi\Phi^*(x_1\sigma_1+x_2\sigma_2) \tag{59}$$

the mapping Φ and the point ζ can not be chosen arbitrarily. Using conditions (1) and (9) we obtain:

$$\frac{1}{i}(\zeta_j-\bar{\zeta}_j) = \Phi^*\sigma_j\Phi\ , \quad j = 1,2 \tag{60}$$

$$(\bar{\zeta}_1\sigma_2-\bar{\zeta}_2\sigma_1+\gamma^{in})\Phi = 0\ , \tag{61}$$

$$\gamma^{out} = \gamma^{in} + i(\sigma_1\Phi\Phi^*\sigma_2-\sigma_2\Phi\Phi^*\sigma_1) \tag{62}$$

Hence the point ζ is on the curve $f(\lambda_1,\lambda_2) = 0$, where

$$f(\lambda_1,\lambda_2) \equiv \det(\lambda_1\sigma_2-\lambda_2\sigma_1+\gamma^{in}) \equiv \det(\lambda_1\sigma_2-\lambda_2\sigma_1+\gamma^{out}) \tag{63}$$

is a polynomial with real coeffcients. From (61) we conclude that

$$(\text{Im}\ \zeta_1)\Phi^*\sigma_2\Phi = (\text{Im}\ \zeta_2)\Phi^*\sigma_1\Phi \tag{64}$$

As $\sigma_2 > 0$ then $\text{Im}\ \zeta_2 > 0$. Multiplying the mapping Φ by a scalar we will satisfy both normalization conditions (60). The elementary factor $E(\zeta,x,z)$ is consonant with $\ker(\lambda_1\sigma_2-\lambda_2\sigma_1+\gamma^{in})$ and

$$E(\zeta,x,x_1\lambda_1+x_2\lambda_2)|\ker(\lambda_1\sigma_2-\lambda_2\sigma_1+\gamma^{in}) = \ker(\lambda_1\sigma_2-\lambda_2\sigma_1+\gamma^{out}) \tag{65}$$

where $\lambda \neq \zeta$, $\lambda \neq \bar{\zeta}$.

The mapping Φ has the form $\Phi h = (h,h_0)u_0$, where $h_0 \in H$, $|h_0| = 1$ and $u_0 \in E$. Hence in matrix form

$$\Phi\Phi^* = \begin{pmatrix}\xi_1\\ \vdots\\ \xi_n\end{pmatrix}(\bar{\xi}_1,\dots,\bar{\xi}_n) = \begin{matrix}\xi_1\bar{\xi}_1,\dots,\xi_1\bar{\xi}_n\\ \cdot\ \cdot\ \cdot\ \cdot\ \cdot\ \cdot\ \cdot\\ \xi_n\bar{\xi}_1,\dots,\xi_n\bar{\xi}_n\end{matrix}$$

Remark. If W_1, W_2 belong to $\Omega_f(\sigma_1,\sigma_2)$ and

$$\ker(\lambda_1\sigma_2-\lambda_2\sigma_1+\gamma^{(1)}) \xrightarrow{\hat{W}_1(\lambda)} \ker(\lambda_1\sigma_2-\lambda_2\sigma_1+\gamma^{(2)})$$

$$\xrightarrow{\hat{W}_2(\lambda)} \ker(\lambda_1\sigma_2-\lambda_2\sigma_1+\gamma^{(3)})$$

then the product $W = W_2W_1$ belongs to $\Omega_f(\sigma_1,\sigma_2)$. Using this remark we can construct products of elementary factors:

$$E(\zeta^{(N)},x,z)\ \dots\ E(\zeta^{(2)},x,z)E(\zeta^{(1)},x,z)$$

belonging to $\Omega_f(\sigma_1,\sigma_2)$, where mappings $\Phi^{(k)}$ and $\gamma^{(k)}$ are connected by recursive formulas

$$\gamma^{(1)} = \gamma \qquad (\gamma \in \Delta_f(\sigma_1,\sigma_2))$$

$$(\bar{\zeta}_1^{(k)}\sigma_2-\bar{\zeta}_2^{(k)}\sigma_1+\gamma^{(k)})\Phi^{(k)} = 0, \quad \mathrm{Im}\ \zeta_2^{(k)} = \Phi^{(k)*}\sigma_2\Phi^{(k)} \tag{66}$$

$$\gamma^{(k+1)} = \gamma^{(k)}+i(\sigma_1\Phi^{(k)}\Phi^{(k)*}\sigma_2-\sigma_2\Phi^{(k)}\Phi^{(k)*}\sigma_1), \quad k=1,2,\dots,N$$

$$\gamma^{(N=1)} = \tilde{\gamma}$$

These equations are always solvable step by step and

$$\det(\lambda_1\sigma_2-\lambda_2\sigma_1+\gamma^{(k)}) \equiv \det(\lambda_1\sigma_2-\lambda_2\sigma_1+\gamma^{(k+1)} \equiv f(\lambda_1,\lambda_2)$$

does not depend on the number k $(k=1,2,\dots,N+1)$.
Let $W(x,z)$ be a rational function which belongs to $\Omega_f(\sigma_1,\sigma_2)$. Then by Theorem 10 there exists a colligation $X = (A_1,A_2;\ H,\Phi,E;\ \sigma_1,\sigma_2,\gamma,\tilde{\gamma})$, $\dim H < \infty$, such that its CCF coincides with $W(x,z)$. Let $H = H_0 \supset H_1 \supset H_2 \supset \dots \supset H_{N-1} \supset H_N = 0$ be a chain of joint invariant subspaces of operators A_1, A_2 such that $\dim(H_{k-1} \ominus H_k) = 1$ $(k=1,2,\dots,N)$. Then X is a coupling $X = X_1 \vee X_2 \vee \dots \vee X_N$ of one-dimensional colligations, X_k is the projection of X onto $H_{k-1} \ominus H_k$ [4,2,6]. Hence the CCF $W(x,z)$ is the product of CCF of X_k

$$W(x,z) = E(\zeta^{(N)},x,z)\ \dots\ E(\zeta^{(2)},x,z)E(\zeta^{(1)},x,z)\ ,$$

which satisfies the conditions (66). Thus we have proved the following factorization theorem.

THEOREM 12. *A rational function* $W(x,z)$ *belongs to* $\Omega_f(\sigma_1,\sigma_2)$ *iff it is a product of elementary factors, satisfying the recurrent equations* (66).

5. Transcendental singularities in the real region

An elementary factor $E(\zeta,x,z)$ with fixed x has single pole $z = x_1\zeta_1+x_2\zeta_2$. In the case $\sigma_2 > 0$ the point $\zeta = (\zeta_1,\zeta_2)$ can not be real because of the normatization condition $2\ \mathrm{Im}\ \zeta_2 = \Phi^*\sigma_2\Phi > 0$. Let $c = (c_1,c_2)$ be a real point of the curve $f = 0$. We will construct a function $w(c,x,z)$ of the class $\Omega_f(\sigma_1,\sigma_2)$ which has only one singular point $z = x_1c_1+x_2c_2$. We will also find a corresponding pair of commuting Volterra operators which has a continuum set of joint invariant subspaces. Formally to a continuum set of joint invariant subspaces there corresponds a continuous analog of recursive formulas (66):

$$(c_1\sigma_2-c_2\sigma_1+\gamma(t))\Phi(t)\Phi^*(t) = 0 ,$$

$$\frac{d\gamma}{dt} = i(\sigma_1\Phi(t)\Phi^*(t)\sigma_2-\sigma_2\Phi(t)\Phi^*(t)\sigma_1) \qquad (67)$$

$$\gamma(0) = \gamma_0$$

Taking into account that $\sigma_2 > 0$ we can assume without loss of generality that $\sigma_2 = I$. Denoting $\sigma_1 = \sigma$, $\Phi(t)\Phi^*(t) = \Pi(t)$ we will write equations (67) in the following form

$$(c_1I-c_2\sigma+\gamma(t))\Pi(t) = 0$$

$$\frac{d\gamma}{dt} = i[\sigma,\Pi(t)] , \qquad (68)$$

$$\gamma(0) = \gamma_0$$

Equations of the form (66) or (68) will be called *linkage equations*. They play the main role in the construction of commuting operators [3,6,10]

THEOREM 13. *Let* $f(\lambda_1,\lambda_2) = 0$ *be a real irreducable curve of the form* $f(\lambda_1,\lambda_2) = \det(\lambda_1I-\lambda_2\sigma+\gamma_0)$, $(\sigma=\sigma^*,\ \gamma_0=\gamma_0^*)$ *and* $c = (c_1,c_2)$ *be a real regular point of this curve. Then there exist operator-functions* $\gamma(t) = \gamma^*(t)$, $\Pi(t) \geq 0$ *satisfying the linkage equations* (68), *and such that rank* $\Pi(t) = 1$, $\mathrm{tr}\Pi(t) = 1$. *The linkage equations* (68) *have the following remarkable family of integrals, depending on two complex parameters* λ_1, λ_2:

$$\det(\lambda_1I-\lambda_2\sigma+\gamma(t)) \equiv f(\lambda_1,\lambda_2) \qquad (69)$$

PROOF. Using an orthonormal basis in E, we will deal with corresponding matrices σ, γ, Π. Let $\varepsilon_1,\varepsilon_2\cdots$ be a sequence of positive numbers such that $\varepsilon_m \to 0$. Let us consider a sequence of sets $\{\zeta_m^{(1)},\ldots,\zeta_m^{(m)}\}$ of points $\zeta = (\zeta_1,\zeta_2)$

belonging to the region $\operatorname{Im} \zeta_2 > 0$ of the curve $f(\zeta_1, \zeta_2) = 0$ and satisfying the conditions $|\zeta_m^{(k)} - c| < \varepsilon_m$, $k=1,2,\ldots,m$. Let $2\Sigma \operatorname{Im} \zeta_{m2}^{(k)} = \ell$, where ℓ is a given positive number.

We will consider the following discrete linkage equations:

$$\gamma_m^{(1)} = \gamma_0 ,$$

$$(\bar{\zeta}_{m1}^{(k)} I - \bar{\zeta}_{m2}^{(k)} \sigma + \gamma_m^{(k)}) \Pi_m^{(k)} = 0 \tag{70}$$

$$\gamma^{(k+1)} = \gamma_m^{(k)} + i[\sigma, \Pi_m^{(k)}] , \quad k=1,2,\ldots,m .$$

We can assume that $\zeta_m^{(k)}$ are regular points of the cirve. Hence [2,6], rank $\Pi_m^{(k)} = 1$ and we can satisfy the normalization conditions $\operatorname{tr} \Pi_m^{(k)} = 2 \operatorname{Im} \zeta_{m2}^{(k)}$. Solving equations (70) step by step we obtain two sequences $\gamma_m^{(k)}$, $\Pi_m^{(k)}$ such that

$$\det(\lambda_1 I - \lambda_2 \sigma + \gamma_m^{(k)}) \equiv f(\lambda_1, \lambda_2) \tag{71}$$

Now we define the following functions of t, $0 \leqslant t \leqslant \ell$.

$$\Pi_m(t) = \frac{\Pi_m^{(k)}}{t_m^{(k)} - t_m^{(k-1)}} , \quad t_{k-1} \leqslant t < t_k \quad (k=1,2,\ldots,m), \tag{72}$$

where $t_m^{(0)} = 0$, $t_m^{(k)} = 2 \sum_{j=1}^{k} \operatorname{Im} \zeta_{m2}^{(j)}$

$$Q_m(t) = \begin{cases} \dfrac{t}{t_m^{(1)}} \Pi_m^{(1)} , & 0 \leqslant t \leqslant t_1 \\[2ex] \displaystyle\sum_{j=1}^{k-1} \Pi_m^{(j)} + \frac{t - t_m^{(k-1)}}{t_m^{(k)} - t_m^{(k-1)}} \Pi_m^{(k)} , & t_m^{(k-1)} \leqslant t \leqslant t_m^{(k)} \end{cases} \tag{73}$$

$$\gamma_m'(t) = \gamma_m^{(k)} + \frac{t - t_m^{(k-1)}}{t_m^{(k)} - t_m^{(k-1)}} (\gamma_m^{(k+1)} - \gamma_m^{(k)}) , \quad t_m^{(k-1)} \leqslant t \leqslant t_m^{(k)} \tag{74}$$

$$\gamma_m''(t) = \gamma_m^{(k)} , \quad t_m^{(k-1)} \leqslant t < t_m^{(k)} \tag{75}$$

Analogously we define step-functions (t), ζ (t). It is clear that $\gamma''_m(t) = \gamma'_m(t_m^{(k)})$, $t_m^{(k)} \leq t < t^{(k+1)}$. From (72), (73) it follows

$$Q_m(t) = \int_0^t \Pi_m(s)ds \ , \quad 0 \leq t \leq \ell \ , \tag{76}$$

$\operatorname{tr} Q_m(t) = t$, $\operatorname{tr} \Pi_m(t) = 1$, $Q_m(t+\Delta t) \geq Q_m(t)$ $(\Delta t > 0)$, $||Q_m(t+\Delta t)-Q_m(t)|| \leq |\Delta t|$.
Equations (70) can be written as follows:

$$(\zeta_{m1}(t)I-\zeta_{m2}(t)\sigma+\gamma''(t))\Pi_m(t) = 0 \ ,$$
$$\gamma'_m(t) = \gamma_0+i[\sigma,Q_m(t)] \tag{77}$$

and

$$\det(\lambda_1 I-\lambda_2\sigma+\gamma''_m(t)) \equiv f(\lambda_1,\lambda_2)$$

Taking a convergent subsequence $Q_m(t) \to Q(t)$, $0 \leq t \leq \ell$, $(\operatorname{tr}Q(t)=t)$ we obtain $\gamma'_{mj}(t) \to \gamma(t)$ and

$$\gamma(t) = \gamma_0 + i[\sigma,Q(t)] \tag{78}$$

Relations (78), (77) imply $||\gamma(t+\Delta t)-\gamma(t)|| \leq C\cdot|\Delta t|$, $||\gamma'_m(t+\Delta t)=\gamma(t)|| \leq C\cdot|\Delta t|$. Hence $\gamma''_m(t) \to \gamma(t)$. If (ζ_1,ζ_2) is a regular point of the curve $f=0$ and $f(\zeta_1,\zeta_2) = \det(\zeta_1 I-\zeta_2\sigma+\gamma)$ is a determinantal representation of this curve then the normalized solution Π of the equation

$$(\zeta_1 I-\zeta_2\sigma+\gamma)\Pi = 0$$

has the form

$$\Pi = pp^* \ , \quad p = \begin{pmatrix} K_{1j} \\ \vdots \\ K_{nj} \end{pmatrix} \left(\sum_{i=1}^{n} |K_{ij}|^2 \right)^{-\frac{1}{2}} \ , \tag{79}$$

where K_{ij} are the algebraic complements of the matrix $\zeta_1 I-\zeta_2\sigma+\gamma$. Hence Π depends continuously on elements of the matrix. Thus, if $m \to \infty$ we obtain

$$\begin{aligned} &[c_1 I-c_2\sigma+\gamma(t)]\Pi(t) = 0 \ , \\ &\gamma(t) = \gamma_0+i[\sigma,Q(t)] \ , \end{aligned} \qquad (0 \leq t \leq \ell) \tag{80}$$

$$Q(t) = \int_0^t \Pi(S)ds \ , \qquad\qquad (81)$$

$$\det(\lambda_1 I - \lambda_2\sigma + \gamma(t)) \equiv f(\lambda_1,\lambda_2)$$

The matrix $\Pi(t)$ has the form $\Pi(t) = P(t)P^*(t)$,

$$P(t) = \begin{pmatrix} p_1(t) \\ \vdots \\ p_n(t) \end{pmatrix} , \qquad \sum_{j=1}^n |p_j(t)|^2 = 1$$

To the solution $p(t)$ of the linkage equations there corresponds two Volterra operators in $L_2[0,\ell]$:

$$(A_j f)(t) = c_j f(t) + i \int_0^t p^*(t)\sigma_j p(s)ds \ , \quad j = 1,2, \qquad (82)$$

where $\sigma_1 = \sigma$, $\sigma_2 = I$.
It is easy to see that

$$\frac{1}{i}(A_j - A_j^*)f = \int_0^\ell p^*(t)\sigma_j p(s)ds \qquad (83)$$

and that $(A_1,A_2;\ \Phi,L_2,E,\sigma_1,\sigma_2)$ forms a colligation if $E = R_n$,

$$\Phi(f) = \int_0^\ell p(s)f(s)ds \ , \quad \Phi^*(u) = p^*(t)u \ , \qquad (84)$$

where

$$u = \begin{pmatrix} \xi_1 \\ \xi_r \end{pmatrix}$$

belongs to E.
<u>Operators</u> A_1, A_2 <u>are commutative</u>. Indeed, if $B_k = A_k - c_k I$ then

$$(B_1\ B_2)f = i \int_0^t P^*(t)\left[\int_r^t \sigma_1 p(s)p^*(s)\sigma_2\ ds\right]$$

and

$$(B_1B_2 - B_2B_1)f = \int_0^t p^*(t)\left[\int_r^t i[\sigma,\Pi(s)]ds\right]p(r)f(r)dr =$$

$$= \int_0^t p^*(t)\left(\int_r^t \frac{d\gamma}{ds}\ ds\right)p(r)f(r)dr = \int_0^t p^*(t)(\gamma(t)-\gamma(r)p(r)f(r)dr =$$

$$= \int_0^t p^*(t)(c_2\sigma_1 - c_1\sigma_2)f(r)dr - \int_0^t p^*(t)(c_2\sigma_1 - c_1\sigma_2)f(r)dr = 0$$

Let us check also the regularity condition

$$A_2\Phi^*\sigma_1 - A_1\Phi^*\sigma_2 = \Phi^*\gamma_0$$

Relations (82), (84) imply

$$A_2\Phi^*\sigma_1 - A_1\Phi^*\sigma_2 = c_2p^*(t)\sigma_1 - c_1p^*(t)\sigma_2 - i\int_0^t p^*(t)[\sigma,\Pi(s)]ds =$$

$$= p^*(t)(c_2\sigma_1 - c_1\sigma_2 + \gamma_0 - \gamma(t)) = p^*(t)\gamma_0 = \Phi^*\gamma_0$$

The CCF can be calculated by the standard procedure [1]:

$$w(c,\gamma_0,\ell,x,z) = \exp\int_0^\ell \left(\frac{i\Pi(s)ds}{z - x_1c_1 - x_2c_2}(x_1\sigma_1 + x_2\sigma_2)\right)$$

Hence we have obtained:

THEOREM 14. *The solution of the linkage equations exists and it is defined uniquely by the initial condition* $\gamma(0) = \gamma_0$.

The corresponding multiplicative integral $w(c,\gamma_0,\ell;\ x,x_1\lambda_1+x_2\lambda_2)$ maps $\ker(\lambda_1\sigma_2 - \lambda_2\sigma_1 + \gamma(0))$ onto $\ker(\lambda_1\sigma_2 - \lambda_2\sigma_1 + \gamma(\ell))$, $0 \leqslant \ell < \infty$. The functions $w(c,\gamma_0,\ell,x,z)$ all belong to the class $\Omega_f(\sigma_1,\sigma_2)$ and have an essential singularity at the point $z = x_1c_1 + x_2c_2$. It is interesting to note that the restriction

$$w(c,\gamma_0,\ell,\lambda) = w(c,\gamma_0,\ell,x,x_1\lambda_1+x_2\lambda_2)|\ker(\lambda_1\sigma_2 - \lambda_2\sigma_1 + \gamma_0)$$

does not depend on x_1, x_2.

REFERENCES

1. Brodskii, M.S. and Livšic, M.S.: Spectral analysis of nonselfadjoint operators and intermediate systems, Uspekhi Mat. Nauk 13 (1958), 3-85 [Engl. Transl. Amer. Math. Soc. Transl. (2), 13 (1960), 265-346].

2. Kravitsky, N.: Regular colligation for several operators in Banach space, Int. Equat. and Operator Theory, v.3/1 (1983).

3. Kravitsky, N.: On commuting integral operators, Operator Theory: Advances and Appl., v. 12, 1984.

4. Livšic, M.S. and Jancevich, A.A.: Theory of Operator colligations in Hilbert space, Engl. Transl., J. Wiley, N.Y., 1979.

5. Livšic, M.S.: The inverse problems for the characteristic functions of several commuting operators, Int. Equat. and Operator Theory, v. 2/2, 1979.

6. Livšic, M.S.: Cayley-Hamilton theorem, vector bundles and divisors of commuting operators, Int. Equat. and Operator Theory, v. 6 (1983).

7. Livšic, M.S.: On commuting nonselfadjoint operators, Int. Equat. and Operator Theory, to appear.

8. Nagy, B. Sz. and Foiaš, C.: Harmonic analysis of operators in Hilbert space, Amsterdam (1970).

9. Nikolskij, N.K.: Lectures on shift operators (Russian), Engl. Transl.: Springer-Verlag (to appear).

10. Waksman, L.L.: Harmonic analysis of multiparameter semigroups of contractions, Viniti 3991-80, Dep 1-166 (Russian), Engl. Transl: Springer Verlag (to appear).

M.S. Livšic
The David and Helena Zlotowski Chair
in Operator Theory and Systems
Department of Mathematics
Ben Gurion University of the Negev
Beer Sheva, Israel

BOUNDARY EIGENVALUE PROBLEMS

Reinhard Mennicken and Manfred Möller

0. INTRODUCTION

This paper deals with boundary eigenvalue problems of the following form

$$(0.1)\qquad \begin{cases} y'(x) - A(x,\lambda)y(x) = 0 & (x \in [a,b]), \\ \sum_{j=1}^{m} W^{(j)}(\lambda)y(a_j) + \int_a^b W(t,\lambda)y(t)dt = 0 & \end{cases}$$

where $a = a_1 < a_2 < \ldots < a_m = b$ and the coefficients are $n\times n$-matrices which are assumed to be sufficiently smooth in x and holomorphic in λ. The aim of this paper is to establish expansion theorems for these eigenvalue problems which are nonlinear in the parameter λ. This will be achieved in two steps: first by establishing the formal expansion and secondly by proving the convergence of the formal expansion.

Formal expansion is achieved by the application of a theorem of Keldyš [12], [13], cf. also Gohberg and Sigal [8]. The theorem states that the principal parts of the inverse of a holomorphic Fredholm-valued operator function having non-empty resolvent set can be expressed by root functions (eigenvectors and associated vectors) of the given operator function itself and its adjoint operator function. For this application the explicit form of the adjoint boundary eigenvalue problem is needed. The proof of the convergence of the formal expansion is based on criteria for regularity. The proofs of these criteria are obtained by making a careful analytic study of the Green's matrix associated with the given boundary eigenvalue problem.

Section 1 contains some basic functional analytic notations, definitions, and results. In section 2 we define the boundary eigenvalue operator function given by (0.1). Formal expansions are established by applying the theorem of Keldyš mentioned above and a modification of this theorem by the authors [17], cf. also Kaashoek [11]. Section 3 contains a theorem about the existence of a suitable asymptotic fundamental matrix of the differential system in (0.1) where we assume that the coefficient matrix $A(x,\lambda)$ has the following λ-asymptotic form

$$(0.2) \qquad A(x,\lambda) = \sum_{j=-1}^{k} \lambda^{-j} A_j(x) + O(\lambda^{-k-1}) \qquad (|\lambda| \to \infty).$$

The proof of this statement is similar to that of Langer [14]; cf. also Wasow [19] and Braaksma [2]. In view of the applications, we are interested in smoothness conditions with respect to x which are as weak as possible. In section 4 the Green's matrix for (0.1) is introduced and the inverse of the operator function defined by (0.1) is represented in terms of the Green's matrix. In section 5 we state a criterion for regularity. We consider boundary eigenvalue problems of type (0.1) whose differential equation has an asymptotic fundamental matrix of the form

$$(0.3) \qquad C(x,\lambda)\left\{\sum_{\kappa=0}^{k} \lambda^{-\kappa} P^{[\kappa]}(x) + o(\lambda^{-k})\right\}E(x,\lambda) \qquad (|\lambda| \to \infty)$$

where $C(x,\lambda)$ is an $n\times n$-matrix function which is a polynomial in λ and $E(x,\lambda)$ is a diagonal matrix with entries of the form $\exp\{\lambda R_i(x)\}$. We prove a geometric regularity criterion which goes back to Cole [3] for the special case that $C(x,\lambda)$ is the identity matrix. In section 6 we study boundary eigenvalue problems of the form (0.1) with

$$(0.4) \qquad A(x,\lambda) = A_0(x) + \lambda A_1(x)$$

where $A_1(x)$ is a diagonal matrix whose elements must fulfil certain additional conditions. In contrast to Cole [3] we allow $A_1(x)$ to have elements which are identical on $[a,b]$ and we also permit some of these elements to vanish identically on $[a,b]$. Thus we do not assume that $A_1(x)$ is invertible on $[a,b]$. We prove the expansibility of those vector functions having certain smoothness properties and fulfilling boundary conditions which, in contrast to the original boundary conditions in (0.1), are independent of λ. Boundary conditions of this type have already been introduced by Mennicken in [15].

In subsequent papers we will be concerned with boundary eigenvalue problems for n-th order differential equations. The boundary conditions will contain interface conditions and also some integral terms. The coefficients of the differential equation as well as of the boundary conditions will be allowed to be holomorphically or polynomially dependent on the eigenvalue parameter. In particular we shall consider boundary eigenvalue problems for differential equations of the form

$$\mathbb{K}\eta - \lambda \mathbb{H}\eta = 0 \tag{0.5}$$

with $\mathbb{K}$ and $\mathbb{H}$ being differential operators of order $n \geq 1$ and $0 \leq p \leq n-1$. Eigenvalue problems of this type for $p = 0$ and λ-independent boundary conditions have been studied by many authors, cf. e.g. Naimark [18]. The general case of (0.5) with arbitrary order p has been treated by Eberhard and Freiling [4], [5], [6], Freiling [7] and Heisecke [9], [10]. We shall extend their results and, for the proof of regularity criteria, make use of the generalization of Cole's regularity criterion, which is proved in section 5 of the present paper.

1. PRELIMINARIES

Let E and F be Banach spaces. $L(E,F)$ denotes the Banach space of all continuous linear operators on E to F. If $S \in L(E,F)$, $N(S)$ denotes its null-space and $R(S)$ its range. $S \in L(E,F)$ is called a Fredholm operator if both its nullity $\mathrm{nul}(S) := \dim N(S)$ and its deficiency $\mathrm{def}(S) := \mathrm{codim}\, R(S)$ are finite. $\Phi(E,F)$ denotes the set of all Fredholm operators on E to F. If $S \in \Phi(E,F)$, then $\mathrm{ind}(S) := \mathrm{nul}(S) - \mathrm{def}(S)$ is called the index of S.

E' and F' denote the dual Banach spaces of E or F respectively and $S^* \in L(F',E')$ the adjoint of $S \in L(E,F)$. We set

$$(y \otimes v)(w) := \langle w,v \rangle y \qquad (w \in F)$$

for $y \in E$ and $v \in F'$. Note that $y \otimes v \in L(F,E)$.

If U is an open subset of $\mathbb{C}$, $H(U,E)$ denotes the set of all holomorphic mappings defined on U with values in the Banach space E.

Let $T \in H(U,L(E,F))$. $\rho(T) := \{\lambda \in U : T(\lambda)$ is invertible$\}$ is called the resolvent set of T, $\sigma(T) := U \setminus \rho(T)$ its spectrum and $\sigma_p(T) := \{\lambda \in U : T(\lambda)$ is not injective$\}$ its point spectrum or the set of eigenvalues of T. We set $T^{-1}(\lambda) := T(\lambda)^{-1}$ for $\lambda \in \rho(T)$ and $T^*(\lambda) := T(\lambda)^*$ for $\lambda \in U$.

In the following we assume $\mu \in U$.

DEFINITION 1.1. $y \in H(U,E)$ is called a *root function* of T at μ if $y(\mu) \neq 0$ and $(Ty)(\mu) = 0$. $\nu(y)$ denotes the order of the zero of Ty at μ and is called the multiplicity of y.

For $i \in \mathbb{N}$ we set

(1.1) $L_i := \{y(\mu) : y$ is a root function of T at μ with $\nu(y) \geq i\} \cup \{0\}$.

This set L_i is a subspace of $N(T(\mu))$. From now on we assume that $\dim N(T(\mu)) = r < \infty$ and that, for some $s > 0$, $L_s \neq \{0\}$ and $L_{s+1} = \{0\}$. The last condition is fulfilled if T^{-1} has a pole of order s at μ. We define

(1.2) $m_j := \max\{i \in \mathbb{N} : \dim L_i \geq j\}$ $(j = 1,\dots,r)$

and state that $m_j \geq m_{j+1}$.

DEFINITION 1.2. A system $\{y_1,\dots,y_r\}$ of root functions of T at μ is called a *canonical system of root functions* (CSRF) if $y_1(\mu),\dots,y_r(\mu)$ are linearly independent (and thus form a basis of $N(T(\mu))$) and one of the equivalent conditions

i) $\nu(y_j) = \max\{\nu(y) : y$ is a root function of T at μ and $y(\mu) \notin \operatorname{span}\{y_1(\mu),\dots,y_{j-1}(\mu)\}\}$ $(j = 1,\dots,r)$,

ii) $\nu(y_j) = m_j$ $(j = 1,\dots,r)$,

iii) $\nu(y_j) \geq m_j$ $(j = 1,\dots,r)$

is fulfilled.

A root function corresponds to a chain of an eigenvector and associated vectors. A canonical system of root functions is related to a canonical system of eigenvectors and associated vectors. For more details see [16].

2. BOUNDARY EIGENVALUE OPERATOR FUNCTIONS

Let $-\infty < a < b < \infty$. We consider the Sobolev spaces

$$W^{j,p}(a,b) := \{y \in L_p(a,b) : y^{(i)} \in L_p(a,b),\ 1 \leq i \leq j\}$$

($j \in \mathbb{N}$, $1 \leq p \leq \infty$) where the derivative is the weak derivative in the sense of distribution theory. $W^{j,p}(a,b)$ is a Banach space with respect to the norm

$$|y|_{W^{j,p}(a,b)} := \sum_{i=0}^{j} |y^{(i)}|_{L_p(a,b)} \qquad (y \in W^{j,p}(a,b)),$$

cf. e.g. [1], Theorem 3.2. We write $H_j(a,b)$ instead of $W^{j,2}(a,b)$. $H_1(a,b)$ is continuously embedded in $C^0([a,b])$, cf. e.g. [1], p. 98.

For an arbitrary set G, $M_{n,m}(G)$ denotes the set of all $n\times m$-matrices with entries in G. If $m = n$, we briefly write $M_n(G)$.

Let $n \in \mathbb{N} \setminus \{0\}$ and $A \in H(\mathbb{C}, M_n(L_\infty(a,b)))$. We define

$$\text{(2.1)} \qquad T^D(\lambda)y := y' - A(\cdot,\lambda)y \qquad (y \in H_1^n(a,b),\ \lambda \in \mathbb{C})$$

where $H_1^n(a,b)$ is the n-fold product of $H_1(a,b)$. Let $m \geq 2$, $a = a_1 < a_2 < \ldots < a_m = b$, $W^{(j)} \in H(\mathbb{C}, M_n(\mathbb{C}))$ $(j = 1,\ldots,m)$ and $W \in H(\mathbb{C}, M_n(L_1(a,b)))$. For $y \in H_1^n(a,b)$ and $\lambda \in \mathbb{C}$ we set

$$\text{(2.2)} \qquad T^R(\lambda)y := \sum_{j=1}^{m} W^{(j)}(\lambda)y(a_j) + \int_a^b W(\xi,\lambda)y(\xi)d\xi$$

and

$$\text{(2.3)} \qquad T(\lambda)y := (T^D(\lambda)y, T^R(\lambda)y).$$

Since

(2.4) $T(\lambda)y = f$ $(f \in L_2^n(a,b)\times\mathbb{C}^n)$

is a boundary value problem, we call T the boundary eigenvalue operator function of (2.4).

From [17] and [15] we know

PROPOSITION 2.1. i) $T \in H(\mathbb{C},\Phi(H_1^n(a,b),L_2^n(a,b)\times\mathbb{C}^n))$.
ii) $\operatorname{ind} T(\lambda) = 0$ $(\lambda \in \mathbb{C})$.
iii) *If* $\rho(T) \neq \emptyset$, *then* $\sigma(T)$ *is a discrete subset of* $\mathbb{C}$, $\sigma(T) = \sigma_p(T)$, *and* T^{-1} *is a meromorphic function whose poles are the eigenvalues of* T.

The dual of $H_1^n(a,b)$ can be identified with

$$H_{-1}^n[a,b] := \{v_o - v_1' : v_o, v_1 \in L_2^n(\mathbb{R});\ \operatorname{supp}(v_o - v_1') \subset [a,b]\},$$

cf. [17], (4.2). The adjoint $T^* \in H(\mathbb{C},L(L_2^n(a,b)\times\mathbb{C}^n,H_{-1}^n[a,b]))$ of T has the form

$$(2.5)\qquad T^*(\lambda)(u,d) = -u' - A^t(\cdot,\lambda)u + \sum_{j=1}^{m} W^{(j)^t}(\lambda)d\delta_{a_j} + W^t(\cdot,\lambda)d$$

$(u \in L_2^n(a,b),\ d \in \mathbb{C}^n)$, where δ_{a_j} is the Dirac distribution with support at a_j and ^{t} denotes the transposition, cf. [17], section 4. We say that

(2.6) $T^*(\lambda)(u,d) = h$ $(h \in H_{-1}^n[a,b])$

is the "adjoint boundary eigenvalue problem" of (2.4).

From now on we assume that $\rho(T) \neq \emptyset$. An immediate consequence of a theorem of Keldyš, cf. e.g. [16], (2.1), is the following

THEOREM 2.2. *Let* $\mu \in \sigma(T)$ *and* $\{y_1,\ldots,y_r\}$ *be a* CSRF *of* T *at* μ.

Then there are polynomials $v_j : \mathbb{C} \to L_2^n(a,b)\times\mathbb{C}^n$ *of degree* $< m_j$ *such that*

$$D := T^{-1} - \sum_{j=1}^{r} (\cdot-\mu)^{-m_j} y_j \otimes v_j$$

is holomorphic at μ. *The* v_j *are uniquely determined by the system* $\{y_1,\ldots,y_r\}$. $\{v_1,\ldots,v_r\}$ *is a* CSRF *of* T^* *at* μ, $\nu(v_j) = m_j$, *and the biorthogonal relationships*

$$\frac{1}{l!}\frac{d^l}{d\lambda^l}\langle\eta_{ih},v_j\rangle(\mu) = \delta_{ij}\,\delta_{m_i-h,l} \tag{2.7}$$

$$(1 \leq h \leq m_i;\ 0 \leq l \leq m_j-1;\ i,j = 1,\ldots,r)$$

hold where $\eta_{ih} := (\cdot-\mu)^{-h}Ty_i$ *and* $\langle\ ,\ \rangle$ *is the canonical bilinear form on* $L_2^n(a,b)\times\mathbb{C}^n$.

The Theorem (2.10) in [17] leads to the following interesting modification of the foregoing theorem.

THEOREM 2.3. *Let* $k_1 \geq k_2 \geq \ldots \geq k_r$ *be positive natural numbers. Assume that* $\mu \in \sigma(T)$. *Let* $y_1,\ldots,y_r$ *be root functions of* T *at* μ *and* $v_1,\ldots,v_r$ *be root functions of* T^* *at* μ. *Assume that* $\nu(y_j) \geq k_j$ *and* $\nu(v_j) \geq k_j$ *for* $j \in \{1,\ldots,r\}$. *Set* $\eta_i := (\cdot-\mu)^{-k_i}Ty_i$ *and suppose that the biorthogonal relationships*

$$\frac{1}{l!}\frac{d^l}{d\lambda^l}\langle\eta_i,v_j\rangle(\mu) = \delta_{ij}\,\delta_{0l} \qquad (0 \leq l \leq k_j-1;\ i,j = 1,\ldots,r)$$

are fulfilled.

Then $\{y_1,\ldots,y_r\}$ *is a* CSRF *of* T *at* μ, $\{v_1,\ldots,v_r\}$ *is a* CSRF *of* T^* *at* μ, $\nu(y_j) = \nu(v_j) = k_j$ $(j = 1,\ldots,r)$ *and*

$$D := T^{-1} - \sum_{j=1}^{r} (\cdot-\mu)^{-k_j}\, y_j \otimes v_j$$

is holomorphic at μ.

Let $Y \in H(\mathbb{C},M_n(W^{1,\infty}(a,b)))$ be the fundamental matrix of $T^D y = 0$ with $Y(a,\lambda) = I_n$ for $\lambda \in \mathbb{C}$. The existence of such a fundamental matrix has been proved in [17], section 1. We set

$$M(\lambda) := T^R(\lambda)Y(\cdot,\lambda) \tag{2.8}$$

and call it the *characteristic matrix function* of (2.4). Obviously, $M \in H(\mathbb{C},M_n(\mathbb{C}))$ and $\sigma(M) = \sigma(T)$. In [17] the authors proved the

THEOREM 2.4. *Let* $\mu \in \sigma(M) = \sigma(T)$. *Let* $\{c_1,\dots,c_r\}$ *be a* CSRF *of* M *at* μ *and* $\{d_1,\dots,d_r\}$ *be a* CSRF *of* $M^* = M^t$ *at* μ. *Suppose that the biorthogonal relationships*

$$\frac{1}{l!}\frac{d^l}{d\lambda^l}\langle(\cdot-\mu)^{-m_i} Mc_i, d_j\rangle(\mu) = \delta_{ij}\,\delta_{0l} \tag{2.9}$$

$(0 \le l \le m_j - 1;\ i,j = 1,\dots,r)$ *hold where* m_i *is the multiplicity of* c_i. *We define*

$$y_i(\lambda) := Y(\cdot,\lambda)c_i(\lambda) \qquad (i = 1,\dots,r;\ \lambda \in \mathbb{C})$$

and

$$u_i(\lambda)(x) := Y^t(x,\lambda)^{-1}\left\{\sum_{j=1}^{m-1} Y^t(a_j,\lambda)W^{(j)^t}(\lambda)\chi_{(a_j,b)}(x) + \int_a^x Y^t(\xi,\lambda)W^t(\xi,\lambda)\,d\xi\right\}d_i(\lambda)$$

$(i = 1,\dots,r;\ \lambda \in \mathbb{C})$. *We set*

$$v_i := (u_i, d_i) \qquad (i = 1,\dots,r).$$

Then $\{y_1,\dots,y_r\}$ *is a* CSRF *of* T *at* μ, $\{v_1,\dots,v_r\}$ *is a* CSRF *of* T^* *at* μ, *the biorthogonal relationships*

$$\frac{1}{l!}\frac{d^l}{d\lambda^l}\langle(\cdot-\mu)^{-m_i} Ty_i, v_j\rangle(\mu) = \delta_{ij}\,\delta_{0l} \tag{2.10}$$

$(0 \le l \le m_j - 1;\ i,j = 1,\dots,r)$ *hold, and*

$$T^{-1} - \sum_{j=1}^{r} (\cdot-\mu)^{-m_j}\, y_j \otimes v_j$$

is holomorphic at μ.

Kaashoek [11] proved that T is globally equivalent to the canonical $L_2^n(a,b)$-extension of M.

3. ASYMPTOTIC FUNDAMENTAL MATRICES

In this section we consider the first order differential system

$$(3.1)\qquad y' - \widetilde{A}(\cdot,\lambda)y = 0$$

where, for some $k \in \mathbb{N}$ and $\gamma_0 > 0$,

$$(3.2)\qquad \widetilde{A}(\cdot,\lambda) = \sum_{j=-1}^{k} \lambda^{-j} A_{-j} + \lambda^{-k-1} A^k(\cdot,\lambda) \qquad (|\lambda| \geq \gamma_0).$$

We shall construct a fundamental matrix of (3.1) which has an appropriate asymptotic behaviour for $\lambda \to \infty$. For this purpose we make the following assumptions:

i) $A_1 \in M_n(W^{k,\infty}(a,b))$,

ii) $A_{-j} \in M_n(W^{k-j,\infty}(a,b)) \qquad (j = 0,\ldots,k)$,

iii) $A^k(\cdot,\lambda) \in M_n(L_\infty(a,b))$ for $|\lambda| \geq \gamma_0$ and is bounded in $M_n(L_\infty(a,b))$ as $\lambda \to \infty$,

iv) A_1 has the diagonal form

$$A_1 = \begin{bmatrix} A_0^1 & & & \\ & A_1^1 & & 0 \\ 0 & & \ddots & \\ & & & A_l^1 \end{bmatrix}$$

where

$$A_\nu^1 = r_\nu I_{n_\nu} \quad (\nu = 0,\ldots,l), \qquad \sum_{\nu=0}^{l} n_\nu = n.$$

For $\nu,\mu = 0,\ldots,l$ and $x \in [a,b]$

$$r_\nu(x) - r_\mu(x) = |r_\nu(x) - r_\mu(x)| e^{i\varphi_{\nu\mu}}$$

with some $\varphi_{\nu\mu} \in \mathbb{R}$ which are independent of x. Finally

$$(r_\nu - r_\mu)^{-1} \in L_\infty(a,b) \qquad (\nu \neq \mu).$$

We set

$$R_\nu(x) := \int_a^x r_\nu(\xi)\,d\xi \qquad (\nu = 0,\ldots,l;\ x \in [a,b]),$$

$$E_\nu(x,\lambda) := \exp(\lambda R_\nu(x)) I_{n_\nu} \qquad (\nu = 0,\ldots,l),$$

$$(3.3)\qquad E(x,\lambda) = \begin{bmatrix} E_0(x,\lambda) & & & \\ & E_1(x,\lambda) & & 0 \\ 0 & & \ddots & \\ & & & E_l(x,\lambda) \end{bmatrix}.$$

We shall often deal with functions having a special asymptotic behaviour when the parameter λ tends to infinity. For this we introduce some notations.

Let $|\;|$ be a fixed norm on $M_{i,j}(\mathbb{C})$. Let U be an unbounded subset of $\mathbb{C}$, f be a function on U with values in $M_{i,j}(\mathbb{C})$ and g be a $\mathbb{C}$-valued function on U. We write

$$f(\lambda) = O(g(\lambda)),$$

if there is a $C > 0$ such that $|f(\lambda)| \leq C|g(\lambda)|$ for $\lambda \in U$.

$$f(\lambda) = o(g(\lambda))$$

means that $|f(\lambda)||g(\lambda)|^{-1} \to 0$ as $\lambda \to \infty$ in U. Let $a \in M_{i,j}(\mathbb{C})$. We write

$$f(\lambda) = [a]$$

if $f(\lambda) - a = o(1)$.

Now let $1 \leq p \leq \infty$ and $h \in M_{i,j}(L_p(a,b))$. $|h|_p$ denotes the L_p-norm of the function $|h(\cdot)|$ where $|\;|$ is the norm on $M_{i,j}(\mathbb{C})$ chosen above. Let $f(\cdot,\lambda) \in M_{i,j}(L_p(a,b))$ for $\lambda \in U$ and, as above, g be defined on U with values in $\mathbb{C}$. We write

$$f(\cdot,\lambda) = \{O(g(\lambda))\}_p \quad\text{or}\quad f(\cdot,\lambda) = O(g(\lambda)) \quad\text{in } M_{i,j}(L_p(a,b))$$

if there is a $C > 0$ such that $|f(\cdot,\lambda)|_p \leq C|g(\lambda)|$ for $\lambda \in U$, and

$$f(\cdot,\lambda) = \{o(g(\lambda))\}_p \quad\text{or}\quad f(\cdot,\lambda) = o(g(\lambda)) \quad\text{in } M_{i,j}(L_p(a,b))$$

if $|f(\cdot,\lambda)|_p|g(\lambda)|^{-1} \to 0$ as $\lambda \to \infty$. If $h \in M_{i,j}(L_p(a,b))$, we write

$$f(\cdot,\lambda) = [h]_p$$

if $f(.,\lambda) - h = \{o(1)\}_p$.

PROPOSITION 3.1. *Let* $g \in L_p(a,b)$ *and* $r \in L_\infty(a,b)$. *Suppose that* $r(x) \geq 0$ *for all* $x \in [a,b]$ *and* $r^{-1} \in L_\infty(a,b)$. *Set*

$$R(x) := \int_a^x r(\xi)d\xi \ ,$$

$$F(x,y,\lambda) := \int_y^x \exp\{\lambda(R(x)-R(t))\}g(t)dt$$

and

$$v(\lambda) := \max\{|F(x,y,\lambda)| : x,y \in [a,b];\ \mathrm{Re}\lambda(x-y) \leq 0\}.$$

We assert that $v(\lambda) = o(1)$. *If* $g \in L_\infty(a,b)$, *then* $v(\lambda) = O((1+|\mathrm{Re}\lambda|)^{-1})$.

PROOF. Let $\varepsilon > 0$. Since $\frac{g}{r} \in L_1(a,b)$, there is a function $h \in C_0^\infty(a,b)$ such that

$$|\tfrac{g}{r} - h|_1 |r|_\infty \leq \frac{\varepsilon}{2} \ .$$

Obviously

$$\left|\int_y^x \exp\{\lambda(R(x)-R(t))\}(g(t)-h(t)r(t))dt\right| \leq |\tfrac{g}{r} - h|_1\, |r|_\infty \leq \frac{\varepsilon}{2} \ .$$

Integration by parts yields

$$\int_y^x \exp\{\lambda(R(x)-R(t))\}h(t)r(t)dt = \frac{1}{\lambda}\Big\{h(y)\exp\{\lambda(R(x)-R(y))\}$$

$$- h(x) + \int_y^x \exp\{\lambda(R(x)-R(t))\}h'(t)dt\Big\} = O(\tfrac{1}{\lambda}),$$

whence the first assertion is proved. If $g \in L_\infty(a,b)$, we obtain

$$|F(x,y,\lambda)| \leq |\tfrac{g}{r}|_\infty \left|\int_y^x r(t)\exp\{\mathrm{Re}\lambda(R(x)-R(t))\}dt\right|$$

$$= |\tfrac{g}{r}|_\infty |\mathrm{Re}\lambda|^{-1}(1 - \exp\{\mathrm{Re}\lambda(R(x)-R(y))\})$$

$$\leq \gamma |\tfrac{g}{r}|_\infty (1+|\mathrm{Re}\lambda|)^{-1}$$

with some $\gamma \geq 0$ which is independent of x, y and λ.

PROPOSITION 3.2. *Let* $s \in \mathbb{N} \setminus \{0\}$ *and* $C \in M_n(W^{s,\infty}(a,b))$. *Assume that* C *is invertible in the algebra* $M_n(L_\infty(a,b))$.

Then $C^{-1} \in M_n(W^{s,\infty}(a,b))$.

PROOF. Since $C^{-1}C' \in M_n(L_\infty(a,b))$, the differential system

$$Y' = -C^{-1}C'Y \tag{3.4}$$

has a fundamental solution $Y \in M_n(W^{1,\infty}(a,b))$ with $Y(a) = C^{-1}(a)$. Since $(CY)' = 0$ and $C(a)Y(a) = I_n$, we obtain $CY = I_n$ and hence $C^{-1} = Y$. Let $1 < l \leq s$ and suppose that $Y \in M_n(W^{l-1,\infty}(a,b))$. We conclude that $C^{-1}C'$ belongs to $M_n(W^{l-1,\infty}(a,b))$. The differential system (3.4) yields $Y' \in M_n(W^{l-1,\infty}(a,b))$ whence $Y \in M_n(W^{l,\infty}(a,b))$. By this induction the assertion is proved.

For the matrices A_j and the matrices $P^{[r]}$, defined below, we form the block matrices

$$A_j =: \left(A_{j,\nu\mu}\right)_{\nu,\mu=0}^{l} \quad \text{and} \quad P^{[r]} =: \left(P^{[r]}_{\nu\mu}\right)_{\nu,\mu=0}^{l}$$

according to the block structure of A_1.

THEOREM 3.3. A. *For* $r \in \{0,\dots,k\}$ *there are matrices* $P^{[r]} \in M_n(W^{k+1-r,\infty}(a,b))$ *such that the equations*

$$P^{[0]}A_1 - A_1P^{[0]} = 0, \quad P^{[0]}(a) = I_n, \tag{3.5}$$

$$P^{[r]\prime} - \sum_{j=0}^{r} A_{-j}P^{[r-j]} + P^{[r+1]}A_1 - A_1P^{[r+1]} = 0 \quad (r=0,\dots,k-1), \tag{3.6}$$

$$P^{[k]\prime}_{\nu\nu} - A_{0,\nu\nu}P^{[k]}_{\nu\nu} = \sum_{\substack{q=0\\ q\neq\nu}}^{l} A_{0,\nu q}P^{[k]}_{q\nu} + \sum_{j=1}^{k}\sum_{q=0}^{l} A_{-j,\nu q}P^{[k-j]}_{q\nu} \tag{3.7}$$

$$(\nu = 0,\dots,l)$$

hold.

B. *We set*

$$(3.8)\qquad \tau(\lambda) = \begin{cases} \displaystyle\max_{\substack{\nu,\mu=0\\ \nu\neq\mu}}^{l} (1+|\mathrm{Re}(\lambda e^{i\varphi_{\nu\mu}})|)^{-1} & \text{if } l>0 \\ |\lambda|^{-1} & \text{if } l=0 . \end{cases}$$

For $r \in \{0,\dots,k\}$ *let the matrices* $P^{[r]}$ *belong to* $M_n(W^{k+1-r,\infty}(a,b))$ *and fulfil the equations (3.5), (3.6) and (3.7). We assert that for sufficiently large* $\lambda \in \mathbb{C}$, *say* $|\lambda| \geq \Upsilon$, *there is a matrix function* $B_k(\cdot,\lambda) \in M_n(W^{1,\infty}(a,b))$ *with the following properties:*

i) *For* $|\lambda| \geq \Upsilon$

$$\tilde{Y}(\cdot,\lambda) := \left(\sum_{r=0}^{k} \lambda^{-r} P^{[r]} + \lambda^{-k} B_k(\cdot,\lambda)\right) E(\cdot,\lambda)$$

is a fundamental matrix of the differential system (3.1).

ii) *For large* λ *we have the asymptotic estimates*

$$(3.9)\qquad B_k(\cdot,\lambda) = \{o(1)\}_\infty ,$$

$$(3.10)\qquad B_k(\cdot,\lambda) = \{O(\tau(\lambda))\}_\infty ,$$

$$(3.11)\qquad \frac{1}{\lambda} B_k'(\cdot,\lambda) = \{o(1)\}_\infty ,$$

$$(3.12)\qquad \frac{1}{\lambda} B_k'(\cdot,\lambda) = \{O(\tau(\lambda))\}_\infty .$$

PROOF. A. The equations (3.5), (3.6) and (3.7) are fulfilled iff the block submatrices of $P^{[r]}$ satisfy the relationships

$$(3.13)\qquad P_{\nu\mu}^{[0]} = 0 \quad (\nu,\mu = 0,\dots,l;\ \nu \neq \mu),\quad P_{\nu\nu}^{[0]}(a) = I_{n_\nu} \quad (\nu=0,\dots,l),$$

$$(3.14)\qquad P_{\nu\nu}^{[r]\prime} - A_{0,\nu\nu}P_{\nu\nu}^{[r]} = \sum_{\substack{q=0\\ q\neq\nu}}^{l} A_{0,\nu q}P_{q\nu}^{[r]} + \sum_{j=1}^{r}\sum_{q=0}^{l} A_{-j,\nu q}P_{q\nu}^{[r-j]}$$

$$(\nu = 0,\dots,l;\ r = 0,\dots,k),$$

$$(3.15)\qquad P_{\nu\mu}^{[r+1]} = (r_\nu - r_\mu)^{-1}\left\{P_{\nu\mu}^{[r]\prime} - \sum_{j=0}^{r}\sum_{q=0}^{l} A_{-j,\nu q}P_{q\nu}^{[r-j]}\right.$$

$$(\nu,\mu = 0,\dots,l;\ \nu \neq \mu;\ r = 0,\dots,k-1).$$

We are going to solve these equations. For $\nu \in \{0,\ldots,l\}$ let $P^{[0]}_{\nu\nu} \in M_n(W^{1,\infty}(a,b))$ be the fundamental matrix of $y' - A_{0,\nu\nu}y = 0$ with $P^{[0]}_{\nu\nu}(a) = I_{n_\nu}$. If $\nu \neq \mu$ we set $P^{[0]}_{\nu\mu} = 0$. Then (3.13) and, for $r = 0$, (3.14) are valid. From the last equation we infer that $P^{[0]} \in M_n(W^{k+1,\infty}(a,b))$. For $r \in \{1,\ldots,k\}$ the matrices $P^{[r]}$ are defined recursively by (3.14) and (3.15). Since $(r_\nu - r_\mu) \in W^{k,\infty}(a,b)$ and $(r_\nu - r_\mu)^{-1} \in L_\infty(a,b)$, $(r_\nu - r_\mu)^{-1}$ belongs to $W^{k,\infty}(a,b)$ by Proposition 3.2. From this fact we conclude by induction that the matrices $P^{[r]}$ have the asserted smoothness properties.

B. Let $|\ |$ be a norm on $M_n(\mathbb{C})$ which makes it a Banach algebra. Then $M_n(L_\infty(a,b))$ is a Banach algebra with respect to $|\ |_\infty$. $P^{[0]}$ belongs to $M_n(W^{k+1,\infty}(a,b))$ and $P^{[0]}(x)$ is invertible for each $x \in [a,b]$. It follows that there is a $K > 0$ such that

$$|P^{[0]}|_\infty \leq \frac{K}{2}, \quad |P^{[0]^{-1}}|_\infty \leq \frac{K}{2}.$$

For $k > 0$ we define

$$P_k(\cdot,\lambda) := \sum_{r=0}^{k} \lambda^{-r} P^{[r]} \qquad (\lambda \in \mathbb{C} \setminus \{0\}). \tag{3.16}$$

For $k = 0$ we choose a matrix $\tilde{A}_0 = (\tilde{A}_{0,\nu\mu})^l_{\nu,\mu=0} \in M_n(L_\infty(a,b))$ such that

$$\begin{cases} \tilde{A}_{0,\nu\nu} = A_{0,\nu\nu} & (\nu = 0,\ldots,l), \\ (r_\nu - r_\mu)^{-1}\tilde{A}_{0,\nu\mu} \in M_n(C_0^\infty(a,b)) & (\nu,\mu = 0,\ldots,l;\ \nu \neq \mu), \\ |A_0 - \tilde{A}_0|_1 \leq (l+1)^{-2}K^{-8}. \end{cases} \tag{3.17}$$

We set

$$P^{[1]}_{\nu\nu} := 0 \qquad (\nu = 0,\ldots,l),$$

$$P^{[1]}_{\nu\mu} := (r_\mu - r_\nu)^{-1}\tilde{A}_{0,\nu\mu}P^{[0]}_{\mu\mu} \qquad (\nu,\mu = 0,\ldots,l;\ \nu \neq \mu)$$

which yields $P^{[1]} = (P^{[1]}_{\nu\mu})^l_{\nu,\mu=0} \in M_n(W^{1,\infty}(a,b))$ and the equation

(3.18) $P^{[0]'} - \tilde{A}_0 P^{[0]} + P^{[1]} A_1 - A_1 P^{[1]} = 0.$

We define

$$P_0(\cdot,\lambda) := P^{[0]} + \lambda^{-1} P^{[1]} \qquad (\lambda \in \mathbb{C} \setminus \{0\})$$

and then for $k \geq 0$

$$S_k := P_k E$$

where E is the matrix function defined in (3.3). Since $P^{[0]}$ is invertible in $M_n(L_\infty(a,b))$, there is a $\gamma > 0$ such that for $|\lambda| \geq \gamma$ $P_k(\cdot,\lambda)$ is invertible in the same matrix function space and

(3.19) $|P_k(\cdot,\lambda)|_\infty \leq K, \quad |P_k(\cdot,\lambda)^{-1}|_\infty \leq K.$

We may assume $\gamma \geq \gamma_0$. For $|\lambda| \geq \gamma$ we set

$$D_0(\cdot,\lambda) := \Big\{ (\tilde{A}_0 - A_0) P^{[0]} + \lambda^{-1} (P^{[1]'} - A_0 P^{[1]} - A^0(\cdot,\lambda) P_0(\cdot,\lambda)) \Big\} P_0(\cdot,\lambda)^{-1}$$

and, if $k \geq 1$,

$$D_k(\cdot,\lambda) := \Big\{ P^{[k]'} - \sum_{r=k}^{2k} \sum_{j=r-k}^{k} \lambda^{k-r} A_{-j} P^{[r-j]} - \lambda^{-1} A^k(\cdot,\lambda) P_k(\cdot,\lambda) \Big\} P_k(\cdot,\lambda)^{-1}.$$

Let $\kappa = \max\{1,k\}$. The equation $E'(\cdot,\lambda) = \lambda A_1 E(\cdot,\lambda)$ and the relationships (3.5), (3.6) and, if $k = 0$, (3.18) yield

$$S_k'(\cdot,\lambda) - \tilde{A}(\cdot,\lambda) S_k(\cdot,\lambda)$$

$$= P_k'(\cdot,\lambda) E(\cdot,\lambda) + P_k(\cdot,\lambda) E'(\cdot,\lambda) - \tilde{A}(\cdot,\lambda) P_k(\cdot,\lambda) E(\cdot,\lambda)$$

$$= \Big\{ \sum_{r=0}^{\kappa} \lambda^{-r} P^{[r]'} + \lambda \sum_{r=0}^{\kappa} \lambda^{-r} P^{[r]} A_1$$

$$- \Big(\sum_{j=-1}^{k} \lambda^{-j} A_{-j} + \lambda^{-k-1} A^k(\cdot,\lambda) \Big) \sum_{r=0}^{\kappa} \lambda^{-r} P^{[r]} \Big\} E(\cdot,\lambda)$$

$$= \Big\{ \sum_{r=0}^{\kappa} \lambda^{-r} P^{[r]\prime} + \sum_{r=0}^{\kappa-1} \lambda^{-r} P^{[r+1]} A_1 + \lambda P^{[0]} A_1$$

$$- \sum_{r=-1}^{k+\kappa} \sum_{j=\max\{-1,r-\kappa\}}^{\min\{k,r\}} \lambda^{-r} A_{-j} P^{[r-j]} - \lambda^{-k-1} A^k(\cdot,\lambda) P_k(\cdot,\lambda) \Big\} *$$

$$* E(\cdot,\lambda)$$

$$= \Big\{ \lambda\Big(P^{[0]}A_1 - A_1 P^{[0]}\Big) + \sum_{r=0}^{\kappa-1} \lambda^{-r}\Big(P^{[r]\prime} - \sum_{j=0}^{r} A_{-j} P^{[r-j]}$$

$$+ P^{[r+1]}A_1 - A_1 P^{[r+1]}\Big) + \lambda^{-\kappa}\Big(P^{[\kappa]\prime} - \sum_{j=0}^{k} A_{-j} P^{[\kappa-j]}\Big)$$

$$- \lambda^{-\kappa-1} \sum_{r=\kappa+1}^{k+\kappa} \sum_{j=r-\kappa}^{k} \lambda^{\kappa+1-r} A_{-j} P^{[r-j]} - \lambda^{-k-1} A^k(\cdot,\lambda) P_k(\cdot,\lambda) \Big\} *$$

$$* E(\cdot,\lambda)$$

$$= \lambda^{-k} D_k(\cdot,\lambda) S_k(\cdot,\lambda).$$

Let $I^{(\nu)}$ be the $n\times n$-matrix whose ν-th diagonal block is I_{n_ν} and whose components are zero elsewhere. For $\lambda \neq 0$ we set $x_{\nu\mu}(\lambda) = a$ if $\operatorname{Re}(\lambda e^{i\varphi_{\nu\mu}}) \leq 0$ and $x_{\nu\mu}(\lambda) = b$ if $\operatorname{Re}(\lambda e^{i\varphi_{\nu\mu}}) > 0$.

Next we prove that the integral equation

$$C(x,\lambda) = I_n - \lambda^{-k} \sum_{\nu,\mu=0}^{1} \int_{x_{\nu\mu}(\lambda)}^{x} S_k(x,\lambda) I^{(\nu)} S_k^{-1}(t,\lambda) D_k(t,\lambda) * \tag{3.20}$$

$$* C(t,\lambda) S_k(t,\lambda) I^{(\mu)} S_k^{-1}(x,\lambda)\,dt$$

has a unique solution in $M_n(L_\infty(a,b))$ for all sufficiently large λ. For this purpose we consider the bounded operator

$$T_\lambda : M_n(L_\infty(a,b)) \to M_n(L_\infty(a,b))$$

given by

$$T_\lambda(f)(x) := I_n - \lambda^{-k} \sum_{\nu,\mu=0}^{1} \int_{x_{\nu\mu}(\lambda)}^{x} S_k(x,\lambda) I^{(\nu)} S_k^{-1}(t,\lambda) *$$

$$* D_k(t,\lambda) f(t) S_k(t,\lambda) I^{(\mu)} S_k^{-1}(x,\lambda)\,dt$$

where $f \in M_n(L_\infty(a,b))$. An easy calculation yields

$$E(x,\lambda) I^{(\nu)} E(t,\lambda)^{-1} = \exp\{\lambda(R_\nu(x) - R_\nu(t))\} I^{(\nu)}.$$

By assumption

$$R_\nu(x) - R_\nu(t) + R_\mu(t) - R_\mu(x) = e^{i\varphi_{\nu\mu}} \int_t^x |r_\nu(\eta) - r_\mu(\eta)|\,d\eta$$

whence the equation

$$T_\lambda(f)(x) - T_\lambda(g)(x) = \lambda^{-k} \sum_{\nu,\mu=0}^{1} \int_{x_{\nu\mu}(\lambda)}^{x} \exp\Big\{\lambda e^{i\varphi_{\nu\mu}} *$$

$$* \int_t^x |r_\nu(\eta) - r_\mu(\eta)|\,d\eta\Big\} P_k(x,\lambda) I^{(\nu)} P_k^{-1}(t,\lambda) D_k(t,\lambda) *$$

$$* (g(t) - f(t)) P_k(t,\lambda) I^{(\mu)} P_k^{-1}(x,\lambda)\,dt$$

holds for $f,g \in M_n(L_\infty(a,b))$. By the choice of $x_{\nu\mu}(\lambda)$ we obtain

$$|\exp(\lambda e^{i\varphi_{\nu\mu}} \int_t^x |r_\nu(\eta) - r_\mu(\eta)|\,d\eta)|$$

$$= \exp(\mathrm{Re}(\lambda e^{i\varphi_{\nu\mu}}) \int_t^x |r_\nu(\eta) - r_\mu(\eta)|\,d\eta) \le 1$$

for t in the compact interval with the endpoints $x_{\nu\mu}(\lambda)$ and $x \in [a,b]$. We may assume that $|I^{(\nu)}|_\infty \le K$. Then

$$|T_\lambda| \le |\lambda|^{-k} (1+1)^2 K^7 |D_k(\cdot,\lambda) P_k(\cdot,\lambda)|_1.$$

Since $A^k(\cdot,\lambda)$ is bounded in $M_n(L_\infty(a,b))$ as $\lambda \to \infty$, there is a constant $M > 0$ such that for all sufficiently large λ

$$(3.21) \qquad |D_k(\cdot,\lambda) P_k(\cdot,\lambda)|_1 \le M$$

whence

$$|T_\lambda| \leq |\lambda|^{-k}(1+1)^2K^7M$$

follows. For $k=0$ we are able to improve this estimate. With a suitable $M>0$ we obtain

(3.21') $$|D_0(\cdot,\lambda)P_0(\cdot,\lambda)|_1 \leq |\tilde{A}_0 - A_0|_1\frac{K}{2} + |\lambda|^{-1}M$$

for all sufficiently large λ. With the aid of the estimate in (3.17) we infer

$$|T_\lambda| \leq \frac{1}{2} + |\lambda|^{-1}(1+1)^2K^7M.$$

We conclude that there is a positive $\delta<1$ such that $|T_\lambda| \leq \delta$ for $|\lambda| \geq \gamma$ if γ is sufficiently large. By Banach's fixed-point theorem there is a unique solution $C(\cdot,\lambda) \in M_n(L_\infty(a,b))$ of (3.20) for each $|\lambda| \geq \gamma$. In addition, the fixed-point theorem yields the estimate

(3.22) $$|C(\cdot,\lambda) - I_n|_\infty \leq (1-\delta)^{-1}|T_\lambda(I_n) - I_n|_\infty \qquad (|\lambda| \geq \gamma).$$

From (3.20) we see that $C(\cdot,\lambda) \in M_n(W^{1,\infty}(a,b))$ and

$$C'(x,\lambda) = -\lambda^{-k}D_k(x,\lambda)C(x,\lambda)$$

$$-\lambda^{-k}\sum_{\nu,\mu=0}^{1}\int_{x_{\nu\mu}(\lambda)}^{x} S_k'(x,\lambda)I^{(\nu)}S_k^{-1}(t,\lambda)D_k(t,\lambda)C(t,\lambda)S_k(t,\lambda)*$$

$$*\,I^{(\mu)}S_k^{-1}(x,\lambda)\,dt + \lambda^{-k}\sum_{\nu,\mu=0}^{1}\int_{x_{\nu\mu}(\lambda)}^{x} S_k(x,\lambda)I^{(\nu)}S_k^{-1}(t,\lambda)\,*$$

$$*\,D_k(t,\lambda)C(t,\lambda)S_k(t,\lambda)I^{(\mu)}S_k^{-1}(x,\lambda)S_k'(x,\lambda)S_k^{-1}(x,\lambda)\,dt$$

$$= -\lambda^{-k}D_k(x,\lambda)C(x,\lambda) + S_k'(x,\lambda)S_k^{-1}(x,\lambda)(C(x,\lambda) - I_n)$$

$$-(C(x,\lambda) - I_n)S_k'(x,\lambda)S_k^{-1}(x,\lambda)$$

$$= -\lambda^{-k}D_k(x,\lambda)C(x,\lambda) + S_k'(x,\lambda)S_k^{-1}(x,\lambda)C(x,\lambda)$$

$$-C(x,\lambda)S_k'(x,\lambda)S_k^{-1}(x,\lambda).$$

We define

(3.23) $\tilde{Y} := CS_k = CP_kE$

and infer that

$$\tilde{Y}' = C'S_k + CS_k' = \tilde{A}\,\tilde{Y}$$

by inserting the right sides of the differential system which we obtained above for S_k and C. We set

(3.24) $B_0(\cdot,\lambda) := \{C(\cdot,\lambda) - I_n\}P_0(\cdot,\lambda) + \lambda^{-1}P^{[1]}$

and, for $k > 0$,

(3.24') $B_k(\cdot,\lambda) := \lambda^k\{C(\cdot,\lambda) - I_n\}P_k(\cdot,\lambda)$

and state that

$$\tilde{Y}(\cdot,\lambda) = \left\{\sum_{r=0}^{k} \lambda^{-r}P^{[r]} + \lambda^{-k}B_k(\cdot,\lambda)\right\}E(\cdot,\lambda)$$

holds for $k \geq 0$.

It remains to prove the estimates for $B_k(\cdot,\lambda)$ as $\lambda \to \infty$. First we give the proof of (3.9) and (3.10). In view of (3.22) it is sufficient to estimate the matrix function $\lambda^k(T_\lambda(I_n) - I_n)$. We set

$$Q^{[0]} := (\tilde{A}_0 - A_0)P^{[0]}$$

and, for $k > 0$,

$$Q^{[k]} := P^{[k]'} - \sum_{j=0}^{k} A_{-j}P^{[k-j]}.$$

From (3.7) and (3.17) we conclude that the block diagonal of $Q^{[k]}$ is zero. Hence we obtain

$$\lambda^k(T_\lambda(I_n) - I_n)(x) = -\sum_{\substack{\nu,\mu=0\\ \nu\neq\mu}}^{1} \int_{x_{\nu\mu}(\lambda)}^{x} \exp\left\{\lambda e^{i\varphi_{\nu\mu}} *\right.$$

$$* \int_t^x |r_\nu(\eta) - r_\mu(\eta)|d\eta \Big\} P^{[0]}(x) I^{(\nu)} P^{[0]^{-1}}(t) Q^{[k]}(t) I^{(\mu)} *$$

$$* P^{[0]^{-1}}(x)\,dt + \frac{1}{\lambda} Q(x,\lambda)$$

where $Q(\cdot,\lambda)$ is bounded in $M_n(L_\infty(a,b))$ as $\lambda \to \infty$. The estimates (3.9) and (3.10) are obviously fulfilled if $l = 0$. For $l > 0$ we have to estimate terms of the form

$$\widetilde{F}(x,\lambda) = \int_{x_{\nu\mu}(\lambda)}^{x} \exp\Big\{\lambda e^{i\varphi_{\nu\mu}} \int_t^x |r_\nu(\eta) - r_\mu(\eta)|d\eta\Big\} g(t)\,dt$$

where $\nu \neq \mu$ and $g \in L_\infty(a,b)$. For $x \in [a,b]$ we set

$$r(x) := |r_\nu(x) - r_\mu(x)|.$$

Let F be defined as in Proposition 3.1. Then

$$\widetilde{F}(x,\lambda) = F(x, x_{\nu\mu}(\lambda), \lambda e^{i\varphi_{\nu\mu}}).$$

Since $\mathrm{Re}(\lambda e^{i\varphi_{\nu\mu}})(x - x_{\nu\mu}(\lambda)) \leq 0$, Proposition 3.1 yields $\widetilde{F}(\cdot,\lambda) = o(1)$ as $\lambda \to \infty$ and $\widetilde{F}(\cdot,\lambda) = O((1+|\mathrm{Re}\lambda e^{i\varphi_{\nu\mu}}|)^{-1})$ in $L_\infty(a,b)$.

Finally we shall prove the estimates (3.11) and (3.12). For $|\lambda| \geq \gamma$ we have

$$B_k(\cdot,\lambda) = \lambda^k \Big\{\widetilde{Y}(\cdot,\lambda) E^{-1}(\cdot,\lambda) - \sum_{j=0}^{k} \lambda^{-j} P^{[j]}\Big\}.$$

The differential equations fulfilled by $\widetilde{Y}(\cdot,\lambda)$ and $E(\cdot,\lambda)$ yield

$$B_k'(\cdot,\lambda) = \lambda^k \Big\{\widetilde{A}(\cdot,\lambda)\widetilde{Y}(\cdot,\lambda)E^{-1}(\cdot,\lambda) - \lambda\widetilde{Y}(\cdot,\lambda)E^{-1}(\cdot,\lambda)A_1 - \sum_{j=0}^{k} \lambda^{-j} P^{[j]'}\Big\}.$$

We insert the asymptotic expressions for $\widetilde{A}(\cdot,\lambda)$ and $\widetilde{Y}(\cdot,\lambda)E^{-1}(\cdot,\lambda)$ and obtain

$$B_k'(\cdot,\lambda) = \lambda^k\Big\{\lambda\Big(A_1P^{[0]} - P^{[0]}A_1\Big) - \sum_{r=0}^{k-1}\lambda^{-r}\Big(P^{[r]'}$$

$$- \sum_{j=0}^{r} A_{-j}P^{[r-j]} + P^{[r+1]}A_1 - A_1P^{[r+1]}\Big) - \lambda^{-k}P^{[k]'}$$

$$+ \sum_{r=k}^{2k}\sum_{j=r-k}^{k} \lambda^{-r}A_{-j}P^{[r-j]} + \lambda^{-k-1}A^k(\cdot,\lambda)\sum_{r=0}^{k}\lambda^{-r}P^{[r]}\Big\}$$

$$+ \widetilde{A}(\cdot,\lambda)B_k(\cdot,\lambda) - \lambda B_k(\cdot,\lambda)A_1.$$

Since the matrices $P^{[r]}$ fulfil the relationships (3.5) and (3.6), the equation

$$(3.25)\qquad B_k'(\cdot,\lambda) = -P^{[k]'} + \sum_{r=0}^{k}\sum_{j=r}^{k}\lambda^{-r}A_{-j}P^{[r+k-j]}$$

$$+ \lambda^{-1}A^k(\cdot,\lambda)\sum_{r=0}^{k}\lambda^{-r}P^{[r]} + \widetilde{A}(\cdot,\lambda)B_k(\cdot,\lambda) - \lambda B_k(\cdot,\lambda)A_1$$

holds. It follows that $\frac{1}{\lambda}B_k'(\cdot,\lambda)$ has the same asymptotic behaviour as $B_k(\cdot,\lambda)$.

COROLLARY 3.4. *Apart from (iv) we now assume that the following conditions hold for some* $\kappa \in \mathbb{N}$:

i') $A_1 \in M_n(W^{k+\kappa,\infty}(a,b))$,

ii') $A_{-j} \in M_n(W^{k+\kappa-j,\infty}(a,b))$ $(j = 0,\ldots,k)$,

iii') $A^k(\cdot,\lambda) \in M_n(W^{\kappa,\infty}(a,b))$ *for* $|\lambda| \geq \gamma_0$ *and is bounded in* $M_n(W^{\kappa,\infty}(a,b))$ *as* $\lambda \to \infty$.

We suppose that, for $r \in \{0,\ldots,k\}$, *the matrices* $P^{[r]}$ *belong to* $M_n(W^{k+1-r,\infty}(a,b))$ *and fulfil (3.5), (3.6) and (3.7). For* $|\lambda| \geq \gamma$ *let the matrix function* $B_k(\cdot,\lambda)$ *be defined as in part B of Theorem 3.3.*

We assert that $P^{[r]} \in M_n(W^{k+\kappa+1-r,\infty}(a,b))$ *for* $r \in \{0,\dots,k\}$ *and* $B_k(\cdot,\lambda) \in M_n(W^{\kappa+1,\infty}(a,b))$ *for* $|\lambda| \geq \gamma$. *We have*

(3.26) $$\frac{1}{\lambda^p} B_k^{(p)}(\cdot,\lambda) = \{o(1)\}_\infty$$

and

(3.27) $$\frac{1}{\lambda^p} B_k^{(p)}(\cdot,\lambda) = \{O(\tau(\lambda))\}_\infty$$

for $|\lambda| \geq \gamma$ *where* $p \in \{0,\dots,\kappa+1\}$ *and* τ *is the function defined in (3.8).*

PROOF. The first assertion follows from the equations (3.13), (3.14) and (3.15) by induction on r. From (3.25) we infer that the matrix function

$$B_k'(\cdot,\lambda) - \tilde{A}(\cdot,\lambda)B_k(\cdot,\lambda) + \lambda B_k(\cdot,\lambda)A_1$$

belongs to $M_n(W^{\kappa,\infty}(a,b))$ and is bounded in this space as $\lambda \to \infty$. By induction we obtain $B_k(\cdot,\lambda) \in M_n(W^{\kappa+1,\infty}(a,b))$ and

$$\frac{1}{\lambda^p} B_k^{(p)}(\cdot,\lambda) = \sum_{j=0}^{p-1} \binom{p-1}{j} \left\{ \frac{1}{\lambda^{p-j}} \tilde{A}^{(p-1-j)}(\cdot,\lambda) \frac{1}{\lambda^j} B_k^{(j)}(\cdot,\lambda) \right.$$

$$\left. - \frac{1}{\lambda^j} B_k^{(j)}(\cdot,\lambda) \frac{1}{\lambda^{p-1-j}} A_1^{(p-1-j)} \right\} + \{O(\frac{1}{\lambda^p})\}_\infty$$

for $p = 1,\dots,\kappa+1$. The estimates (3.26) and (3.27) follow from these equations again by induction on p.

4. THE INVERSE OF THE BOUNDARY EIGENVALUE OPERATOR FUNCTION

Let T be the boundary eigenvalue operator function defined by (2.1), (2.2) and (2.3). For $\lambda \in \rho(T)$ and $f_1 \in L_2^n(a,b)$, $f_2 \in \mathbb{C}^n$ we set

(4.1) $$R_1(\lambda)f_1 := T^{-1}(\lambda)(f_1,0),$$

(4.2) $$R_2(\lambda)f_2 := T^{-1}(\lambda)(0,f_2).$$

If $\lambda \in \rho(T)$ and $f \in H_1^n(a,b)$, we obtain

$$(4.3) \qquad f = T^{-1}(\lambda)T(\lambda)f = R_1(\lambda)T^D(\lambda)f + R_2(\lambda)T^R(\lambda)f.$$

We now give an analytic representation of $R_1(\lambda)$, $R_2(\lambda)$ and $T^{-1}(\lambda)$. For this purpose we set

$$(4.4) \qquad F(x,\lambda) := \sum_{\substack{j=1 \\ a_j<x}}^{m} W^{(j)}(\lambda) + \int_a^x W(t,\lambda)\,dt \qquad (a \le x < b),$$

$$(4.5) \qquad F(b,\lambda) := \sum_{j=1}^{m} W^{(j)}(\lambda) + \int_a^b W(t,\lambda)\,dt.$$

It follows that

$$(4.6) \qquad T^R(\lambda)y = \int_a^b dF(t,\lambda)y(t) \qquad (y \in H_1^n(a,b))$$

where the integral is the Riemann-Stieltjes integral of the vector function y with respect to the integrator $F(\cdot,\lambda)$. Let $Y \in H(\mathbb{C}, M_n(W^{1,\infty}(a,b))$ be the fundamental matrix of $T^D y = 0$ with $Y(a,\lambda) = I_n$ for $\lambda \in \mathbb{C}$. Let M be the characteristic matrix function defined in (2.8). For $\lambda \in \rho(T)$ the Green's matrix of T is defined by

$$(4.7) \qquad G(x,\xi,\lambda) := \begin{cases} \displaystyle\int_{t=a}^{\xi} Y(x,\lambda)M^{-1}(\lambda)d_tF(t,\lambda)Y(t,\lambda)Y^{-1}(\xi,\lambda) \\ \qquad\qquad (a \le \xi \le x \le b) \\ \displaystyle -\int_{t=\xi}^{b} Y(x,\lambda)M^{-1}(\lambda)d_tF(t,\lambda)Y(t,\lambda)Y^{-1}(\xi,\lambda) \\ \qquad\qquad (a \le x < \xi \le b) \end{cases}$$

where the integrator is $F(\cdot,\lambda)$ as in (4.6). G is of bounded variation with respect to ξ. Hence $G(x,\cdot,\lambda)$ belongs to $M_n(L_\infty(a,b))$. We set

$$(4.8) \qquad \hat{G}(x,\lambda) := Y(x,\lambda)M^{-1}(\lambda) \qquad (x \in [a,b],\ \lambda \in \rho(T))$$

and state

PROPOSITION 4.1. *For* $\lambda \in \rho(T)$ *and* $f_1 \in L_2^n(a,b)$, $f_2 \in \mathbb{C}^n$ *we have*

$$(4.9)\qquad (R_1(\lambda)f_1)(x) = \int_a^b G(x,\xi,\lambda)f_1(\xi)d\xi\ ,$$

$$(4.10)\qquad (R_2(\lambda)f_2)(x) = \hat{G}(x,\lambda)f_2\ ,$$

$$(4.11)\qquad (T^{-1}(\lambda)(f_1,f_2))(x) = \int_a^b G(x,\xi,\lambda)f_1(\xi)d\xi + \hat{G}(x,\lambda)f_2\ .$$

PROOF. The statement is clear if we prove (4.11). From

$$(4.12)\qquad \begin{cases} \int_a^b G(x,\xi,\lambda)f_1(\xi)d\xi \\ \\ = Y(x,\lambda)\int_a^x \int_{t=a}^{\xi} M^{-1}(\lambda)d_tF(t,\lambda)Y(t,\lambda)Y^{-1}(\xi,\lambda)f_1(\xi)d\xi \\ \\ - Y(x,\lambda)\int_x^b \int_{t=\xi}^{b} M^{-1}(\lambda)d_tF(t,\lambda)Y(t,\lambda)Y^{-1}(\xi,\lambda)f_1(\xi)d\xi \end{cases}$$

we infer that the function defined by $\int_a^b G(\cdot,\xi,\lambda)f_1(\xi)d\xi$ belongs to $H_1^n(a,b)$. On the right side of (4.12) we add and subtract the term

$$Y(x,\lambda)\int_a^x \int_{t=\xi}^{b} M^{-1}(\lambda)d_tF(t,\lambda)Y(t,\lambda)Y^{-1}(\xi,\lambda)f_1(\xi)d\xi\ .$$

We obtain

$$(4.13)\qquad \begin{cases} \int_a^b G(x,\xi,\lambda)f_1(\xi)d\xi = Y(x,\lambda)\int_a^x Y^{-1}(\xi,\lambda)f_1(\xi)d\xi \\ \\ - Y(x,\lambda)\int_a^b \int_{t=\xi}^{b} M^{-1}(\lambda)d_tF(t,\lambda)Y(t,\lambda)Y^{-1}(\xi,\lambda)f_1(\xi)d\xi \end{cases}$$

since, by (4.6),

$$M(\lambda) = \int_a^b dF(t,\lambda)Y(t,\lambda).$$

The equations (4.13) and $T^D(\cdot,\lambda)Y(\cdot,\lambda) = 0$ imply that

$$(4.14)\qquad T^D(\lambda)\left(\int_a^b G(\cdot,\xi,\lambda)f_1(\xi)d\xi + \hat{G}(\cdot,\lambda)f_2\right) = f_1.$$

Again from (4.13) we deduce that

$$T^R(\lambda)\int_a^b G(\cdot,\xi,\lambda)f_1(\xi)d\xi = \int_a^b d_xF(x,\lambda)Y(x,\lambda)\int_a^x Y^{-1}(\xi,\lambda)f_1(\xi)d\xi$$

$$-\int_a^b \int_{t=\xi}^b d_tF(t,\lambda)Y(t,\lambda)Y^{-1}(\xi,\lambda)f_1(\xi)d\xi = 0$$

whence

$$(4.15)\qquad T^R(\lambda)\left(\int_a^b G(\cdot,\xi,\lambda)f_1(\xi)d\xi + \hat{G}(\cdot,\lambda)f_2\right) = f_2.$$

and (4.11) is proved.

Suppose now that $\rho(T) \neq \emptyset$. By Proposition 2.1 the spectrum $\sigma(T)$ is a discrete subset of $\mathbb{C}$ and T^{-1} is meromorphic. Hence the following definition makes sense. For $\nu \in \mathbb{N}$ we fix closed Jordan curves $\Gamma_\nu \subset \rho(T)$ such that

$$0 \in \operatorname{int}\Gamma_\nu, \quad \overline{\operatorname{int}\Gamma_\nu} \subset \operatorname{int}\Gamma_{\nu+1}, \quad \bigcup_{\nu\in\mathbb{N}} \operatorname{int}\Gamma_\nu = \mathbb{C}.$$

DEFINITION 4.2. Let $|\ |$ be a continuous norm on $H_1^n(a,b)$ and let $p,p' \in \mathbb{Z}$.

The boundary eigenvalue problem (2.4) or the associated operator function T is called $|\ |$-regular of order (p,p') (with respect to the curves Γ_ν) if

$$\int_{\Gamma_\nu} |\lambda^{-p-1}R_1(\lambda)f_1||d\lambda| \to 0 \quad (\nu\to\infty)$$

and

$$\int_{\Gamma_\nu} |\lambda^{-p'-1}R_2(\lambda)f_2||d\lambda| \to 0 \quad (\nu\to\infty)$$

for all $f_1 \in L_2^n(a,b)$, $f_2 \in \mathbb{C}^n$.

In the following section we shall give sufficient conditions for regularity.

5. A REGULARITY CRITERION

We suppose as in section 2 that $A \in H(\mathbb{C}, M_n(L_\infty(a,b)))$, $W^{(j)} \in H(\mathbb{C}, M_n(\mathbb{C}))$ $(j = 1,\dots,m)$ and $W \in H(\mathbb{C}, M_n(L_1(a,b)))$ where $m \geq 2$. We consider the boundary eigenvalue operator function T defined by (2.1), (2.2) and (2.3).

In this section we make the following additional assumptions:

a) The matrix functions A, $W^{(j)}$ $(j = 1,\dots,m)$ and W are polynomials.

b) There are a matrix function $C \in H(\mathbb{C}, M_n(W^{1,\infty}(a,b)))$ and a positive real number γ_o which fulfil the following conditions:

b_1) C is a polynomial.

b_2) $C(\cdot,\lambda)$ is invertible in the algebra $M_n(W^{1,\infty}(a,b))$ if $|\lambda| \geq \gamma_o$.

b_3) For $|\lambda| \geq \gamma_o$ and $y \in H_1^n(a,b)$ the equation

$$C^{-1}(\cdot,\lambda)T^D(\lambda)C(\cdot,\lambda)y = y' - \widetilde{A}(\cdot,\lambda)y \tag{5.1}$$

holds where $\widetilde{A}$ is a matrix function of type (3.2), i.e.

$$\widetilde{A}(\cdot,\lambda) = \sum_{j=-1}^{k} \lambda^{-j} A_{-j} + \lambda^{-k-1} A^k(\cdot,\lambda) \qquad (|\lambda| \geq \gamma_o)$$

with some $k \in \mathbb{N}$. The coefficients A_{-j} $(j = -1,0,\dots,k)$ and A^k satisfy the conditions i), ii), iii) and iv) stated in section 3.

b_4) Let $r_o,\dots,r_l$ be the functions defined in condition iv) mentioned in b_3). For $\nu \in \{0,\dots,l\}$ and $x \in [a,b]$

$$r_\nu(x) = |r_\nu(x)| e^{i\varphi_\nu}$$

with some $\varphi_\nu \in \mathbb{R}$ which are independent of x. Finally either $r_\nu = 0$ or $r_\nu^{-1} \in L_\infty(a,b)$.

Condition b_2) is fulfilled if for $x \in [a,b]$ and $|\lambda| \geq \gamma_o$ the matrix $C(x,\lambda)$ is invertible: Since $C(\cdot,\lambda)$ is continuous in $[a,b]$, $C^{-1}(\cdot,\lambda)$ belongs to $M_n(L_\infty(a,b))$ if $|\lambda| \geq \gamma_o$. Proposition 3.2 yields $C^{-1}(\cdot,\lambda) \in M_n(W^{1,\infty}(a,b))$ for $|\lambda| \geq \gamma_o$.

If $s \in \mathbb{N}$, $\gamma_o > 0$ and f is a complex-valued function defined on $U = \{\lambda \in \mathbb{C} : |\lambda| \geq \gamma_o\}$, we call f an asymptotic polynomial of order s (with respect to $\frac{1}{\lambda}$) if there are $f_j \in \mathbb{C}$ $(j = 0,\dots,s)$ such that

$$f(\lambda) = \sum_{j=0}^{s} \lambda^{-j} f_j + \lambda^{-s} o(1)$$

as $\lambda \to \infty$. f is called an asymptotic polynomial if it is an asymptotic polynomial of some order s. Analogously asymptotic polynomials are defined for vector-valued functions.

PROPOSITION 5.1. *For* $|\lambda| \geq \gamma_o$

$$\det C^{-1}(\cdot,\lambda) = \lambda^{\tilde{q}}\,(c_o + \tfrac{1}{\lambda} c_1(\cdot,\lambda))$$

where $\tilde{q} \in \mathbb{Z}$, c_o *is an element in* $C^o([a,b])$ *which does not vanish in* $[a,b]$ *and* c_1 *is an asymptotic polynomial in* $C^o([a,b])$ *of arbitrary order. There is a* $\hat{q} \in \mathbb{Z}$ *such that*

$$C^{-1}(\cdot,\lambda) = O(\lambda^{\hat{q}}) \quad \textit{in} \quad C^o([a,b])$$

also for $|\lambda| \geq \gamma_o$.

PROOF. Since C is a polynomial with coefficients in $M_n(W^{1,\infty}(a,b))$, there is a natural number s such that

$$\det C(x,\lambda) = \sum_{j=0}^{s} a_j(x)\lambda^j$$

for $x \in [a,b]$ and $\lambda \in \mathbb{C}$ where $a_j \in W^{1,\infty}(a,b)$ for $j \in \{0,\dots,s\}$ and $a_s \neq 0$. If a $\delta > 0$ exists such that $|a_s(x)| \geq \delta$ for $x \in [a,b]$, then $\lambda^s(\det C(\cdot,\lambda))^{-1}$ is an asymptotic polynomial in $C^o([a,b])$ of arbitrary order which tends to a_s^{-1} as $\lambda \to \infty$. This proves the first assertion; the second assertion follows by Cramer's rule.

Let us suppose that there is no $\delta > 0$ with $|a_s(x)| \geq \delta$ for all x in $[a,b]$. Since a_s is not identically zero in $[a,b]$, there is a sequence $(x_\nu)_o^\infty$ in $[a,b]$ such that $a_s(x_\nu) \neq 0$ $(\nu \in \mathbb{N})$ and $a_s(x_\nu) \to 0$ $(\nu \to \infty)$. By choosing a subsequence if necessary we obtain an $\varepsilon > 0$ such that for $j \in \{0,\dots,s-1\}$ either $|a_j(x_\nu)| \geq \varepsilon$ $(\nu \in \mathbb{N})$ or $a_j(x_\nu) \to 0$ $(\nu \to \infty)$. Suppose that

$a_j(x_\nu) \to 0 \quad (\nu \to \infty)$ for all j. Then

$$\det C^{-1}(x_\nu,\lambda) = (\det C(x_\nu,\lambda))^{-1} \to \infty \qquad (\nu \to \infty)$$

for all $\lambda \in \mathbb{C}$ and this contradicts the fact that $C^{-1}(\cdot,\lambda)$ belongs to $M_n(W^{1,\infty}(a,b))$ if $|\lambda| \geq \gamma_0$. Consequently there is a $j_0 \in \{0,\ldots,s-1\}$ such that

$$(5.2) \qquad |a_{j_0}(x_\nu)| \geq \varepsilon \qquad (\nu \in \mathbb{N})$$

and $a_j(x_\nu) \to 0 \quad (\nu \to \infty)$ for $j \in \{j_0+1,\ldots,s\}$. We define

$$\rho := \begin{cases} \max\{|a_j(x)| : x \in [a,b],\ j \in \{0,\ldots,j_0-1\}\} & \text{if } j_0 > 0 \\ 0 & \text{if } j_0 = 0 \end{cases}$$

and $r := \max\{\frac{2j_0\rho}{\varepsilon}, \gamma_0, 1\}$. We choose some $\nu \in \mathbb{N}$ such that

$$(5.3) \qquad \sum_{j=j_0+1}^{s} |a_j(x_\nu)| r^{j-j_0} < \frac{\varepsilon}{2} .$$

The estimates (5.2), (5.3) and the definitions of ρ and r yield

$$|\sum_{\substack{j=0 \\ j \neq j_0}}^{s} a_j(x_\nu)\lambda^j| < \varepsilon |\lambda|^{j_0} \leq |a_{j_0}(x_\nu)\lambda^{j_0}|$$

for $|\lambda| = r$. By Rouché's theorem $\det C(x_\nu,\lambda)$ has exactly j_0 zeros in the open disk with center 0 and radius r. Since $j_0 < s$ and $a_s(x_\nu) \neq 0$, there is at least one λ with $|\lambda| > r$ and $\det C(x_\nu,\lambda) = 0$. This contradicts the assumption b_2) which implies that $C(x_\nu,\lambda)$ is invertible for $|\lambda| > r$.

In the following let $\tilde{Y}(\cdot,\lambda)$ be the fundamental matrix of (3.1) constructed in Theorem 3.3 for $|\lambda| \geq \gamma$. By condition b_3)

$$(5.4) \qquad \hat{Y}(\cdot,\lambda) := C(\cdot,\lambda)\tilde{Y}(\cdot,\lambda) \qquad (|\lambda| \geq \gamma)$$

is a fundamental matrix of $T^D(\lambda)y = 0$. We set

$$(5.5) \qquad \hat{M}(\lambda) := T^R(\lambda)\hat{Y}(\cdot,\lambda).$$

PROPOSITION 5.2. *Let* M *be the characteristic matrix function of (2.4) defined in (2.8). Then*

$$\det M(\lambda) = \lambda^{\tilde{q}} d(\lambda) \det \hat{M}(\lambda) \qquad (|\lambda| \geq \gamma)$$

where $\tilde{q} \in \mathbb{Z}$ *and* d *is an asymptotic polynomial of order* k *which has a non-zero limit as* $\lambda \to \infty$.

PROOF. As in section 2 let $Y \in H(\mathbb{C}, M_n(W^{1,\infty}(a,b)))$ be the fundamental matrix of $T^D(\lambda)y = 0$ which is uniquely determined by the initial condition $Y(a,\lambda) = I_n$ $(\lambda \in \mathbb{C})$. We infer that, for $|\lambda| \geq \gamma$,

$$(5.6) \qquad Y(\cdot,\lambda) = \hat{Y}(\cdot,\lambda)\hat{Y}^{-1}(a,\lambda)$$

whence

$$(5.7) \qquad M(\lambda) = \hat{M}(\lambda)\hat{Y}^{-1}(a,\lambda)$$

and thus

$$(5.8) \qquad \det M(\lambda) = \det \hat{M}(\lambda) \cdot \det \hat{Y}^{-1}(a,\lambda) .$$

By Theorem 3.3, $\tilde{Y}^{-1}(a,\lambda)$ is an asymptotic polynomial in $M_n(\mathbb{C})$ of order k and tends to I_n as $\lambda \to \infty$. By Proposition 5.1

$$\det C^{-1}(a,\lambda) = \lambda^{\tilde{q}} (c_o(a) + \frac{1}{\lambda} c_1(a,\lambda))$$

where $c_o(a) \neq 0$ and $c_1(a,\lambda)$ is an asymptotic polynomial in $\mathbb{C}$ of arbitrary order. By (5.4) the assertion holds.

In the sequel we shall study the asymptotic behaviour of $\hat{M}(\lambda)$ as $\lambda \to \infty$. Let q_o and q_2 be the orders of the polynomials $C(\cdot,\lambda)$ and $W(\cdot,\lambda)$, respectively, and q_1 be the maximum of the orders of the polynomials $W^{(j)}(\lambda)$ $(j = 1,\ldots,m)$. In the following we denote the diagonal elements of $A_1(x)$ by $\tilde{r}_\mu(x)$ $(\mu = 1,\ldots,n)$. We define

$$(5.9) \qquad \tilde{R}_\mu(x) := \int_a^x \tilde{r}_\mu(t)\,dt \qquad (\mu = 1,\ldots,n).$$

PROPOSITION 5.3. *Let* I *be a subinterval of* [a,b], $[a_I,b_I]$ *be the closure of* I *and set*

$$\hat{M}_I(\lambda) := \sum_{a_j\in I} W^{(j)}(\lambda)\hat{Y}(a_j,\lambda) + \int_I W(t,\lambda)\hat{Y}(t,\lambda)dt \tag{5.10}$$

for $|\lambda|\geq\gamma$ *where* $\hat{Y}(\cdot,\lambda)$ *is the fundamental matrix defined in (5.4). Define*

$$N(I) := \{a_j\in I : j\in\{1,\ldots,m\}\}\cup\{a_I,b_I\}. \tag{5.11}$$

We assert that

$$\hat{M}_I(\lambda) = \lambda^q\Big(\sum_{s\in N(I)} m_{\nu\mu}^{(s)}(a_I,b_I,\lambda)\exp(\lambda\tilde{R}_\mu(s))\Big)_{\nu,\mu=1}^n$$

where $q\in\mathbb{Z}$, $q\leq\max\{q_o+q_1,q_o+q_2\}$ *and the* $m_{\nu\mu}^{(s)}$ *are asymptotic polynomials in* $B([a,b]\times[a,b])$, *the space of bounded functions on* $[a,b]\times[a,b]$.

PROOF. By Theorem 3.3 there is a $q_1'\in\mathbb{Z}$, $q_1'\leq q_o+q_1$ such that the matrix functions

$$S_j(\lambda) := \lambda^{-q_1'}W^{(j)}(\lambda)C(a_j,\lambda)\Big(\sum_{r=0}^k \lambda^{-r}P^{[r]}(a_j) + \lambda^{-k}B_k(a_j,\lambda)\Big)$$

$(j=1,\ldots,m)$ are asymptotic polynomials in $M_n(\mathbb{C})$ of order $k_1' := k-(q_o+q_1-q_1')\geq 0$. We define $(s_{\nu\mu}^{(j)}(\lambda))_{\nu,\mu=1}^n := S_j(\lambda)$ for $j=1,\ldots,m$,

$$S_I(\lambda) := \sum_{a_j\in I} W^{(j)}(\lambda)\hat{Y}(a_j,\lambda)$$

and obtain

$$S_I(\lambda) = \lambda^{q_1'}\Big(\sum_{a_j\in I} s_{\nu\mu}^{(j)}(\lambda)\exp(\lambda\tilde{R}_\mu(a_j))\Big)_{\nu,\mu=1}^n \tag{5.12}$$

where the $s_{\nu\mu}^{(j)}$ are asymptotic polynomials of order k_1'.

Next we choose a $q_2'\in\mathbb{Z}$, $q_2'\leq q_o+q_2$ such that

$$Q(\cdot,\lambda) := \lambda^{-q_2'}W(\cdot,\lambda)C(\cdot,\lambda)\Big(\sum_{r=0}^k \lambda^{-r}P^{[r]} + \lambda^{-k}B_k(\cdot,\lambda)\Big)$$

is an asymptotic polynomial in $M_n(L_1(a,b))$ of order

$k_2' := k-(q_o+q_2-q_2') \geq 0$, i.e.

$$(5.13)\qquad Q(\cdot,\lambda) = \sum_{i=0}^{k_2'} \lambda^{-i} Q_i + \lambda^{-k_2'} \hat{Q}(\cdot,\lambda)$$

with $Q_i \in M_n(L_1(a,b))$ $(i=0,\dots,k_2')$ and $\hat{Q}(\cdot,\lambda) = \{o(1)\}_1$. It follows that

$$Q_I(\lambda) := \int_I W(t,\lambda)\hat{Y}(t,\lambda)dt = \lambda^{q_2'} \int_I Q(t,\lambda)E(t,\lambda)dt.$$

We set $(q_{\nu\mu}(t,\lambda))_{\nu,\mu=1}^n := Q(t,\lambda)$ and obtain

$$(5.14)\qquad Q_I(\lambda) = \lambda^{q_2'}\left(\int_{a_I}^{b_I} q_{\nu\mu}(t,\lambda)\exp(\lambda\tilde{R}_\mu(t))dt\right)_{\nu,\mu=1}^n$$

where the $q_{\nu\mu}$ are asymptotic polynomials in $L_1(a,b)$ of order k_2'. First let $\tilde{r}_\mu = 0$. We set

$$(5.15)\qquad q_{\nu\mu}^{(1)}(a_I,b_I,\lambda) := \int_{a_I}^{b_I} q_{\nu\mu}(t,\lambda)dt, \qquad q_{\nu\mu}^{(2)}(a_I,b_I,\lambda) := 0$$

for $\nu = 1,\dots,n$. The $q_{\nu\mu}^{(j)}$ are asymptotic polynomials in the space $C^o([a,b]\times[a,b])$ of order k_2' and, because $\tilde{R}_\mu = 0$,

$$(5.16)\qquad \int_{a_I}^{b_I} q_{\nu\mu}(t,\lambda)\exp(\lambda\tilde{R}_\mu(t))dt =$$

$$q_{\nu\mu}^{(1)}(a_I,b_I,\lambda)\exp(\lambda\tilde{R}_\mu(a_I)) + q_{\nu\mu}^{(2)}(a_I,b_I,\lambda)\exp(\lambda\tilde{R}_\mu(b_I)).$$

Now let $\tilde{r}_\mu \neq 0$. By assumption $b_4)$ $\tilde{r}_\mu$ has constant argument in $[a,b]$, i.e.

$$(5.17)\qquad \tilde{r}_\mu(x) = |\tilde{r}_\mu(x)|e^{i\tilde{\varphi}_\mu} \qquad (x\in[a,b])$$

with some $\tilde{\varphi}_\mu \in \mathbb{R}$. We set

$$(5.18)\qquad x_\mu(\lambda) := \begin{cases} a_I & \text{if } \operatorname{Re}(\lambda e^{i\tilde{\varphi}_\mu}) \leq 0 \\ b_I & \text{if } \operatorname{Re}(\lambda e^{i\tilde{\varphi}_\mu}) > 0 \end{cases}$$

and state that

$$(5.19)\qquad |\exp\{\lambda(\tilde{R}_\mu(t) - \tilde{R}_\mu(x_\mu(\lambda)))\}| \leq 1 \qquad (t\in[a_I,b_I]).$$

If λ is large, we choose some $l_o, \kappa \geq 0$ such that $W(\cdot,\lambda) \in M_n(W^{l_o,1}(a,b))$, $C(\cdot,\lambda) \in M_n(W^{l_o,\infty}(a,b))$ and the assumptions of Corollary 3.4 are fulfilled with respect to κ. We set

$$l_1 := \min\{l_o, \kappa, k_2'\}. \tag{5.20}$$

By Corollary 3.4 and the definition of Q we obtain that the functions $q_{\nu\mu}(\cdot,\lambda)$ $(\nu = 1,\dots,n)$ belong to $W^{l_1,1}(a,b)$. Since $\tilde{r}_\mu \in W^{\kappa,\infty}(a,b)$ and $\tilde{r}_\mu^{-1} \in L_\infty(a,b)$, Proposition 3.2 yields that $\tilde{r}_\mu^{-1} \in W^{\kappa,\infty}(a,b)$. Hence the functions

$$u_{\nu\mu}^{[0]}(\cdot,\lambda) := q_{\nu\mu}(\cdot,\lambda),$$

$$u_{\nu\mu}^{[j+1]}(\cdot,\lambda) := \left(\frac{u_{\nu\mu}^{[j]}}{\tilde{r}_\mu}\right)'(\cdot,\lambda) \qquad (j = 0,\dots,l_1-1)$$

are well-defined. From Corollary 3.4 we infer that the $u_{\nu\mu}^{[j]}$ are asymptotic polynomials in $M_n(L_1(a,b))$ of order $k_2'-j$. An integration by parts yields

$$\int_{a_I}^{b_I} u_{\nu\mu}^{[j]}(t,\lambda)\exp(\lambda\tilde{R}_\mu(t))\,dt = \frac{1}{\lambda}\frac{u_{\nu\mu}^{[j]}(t,\lambda)}{\tilde{r}_\mu(t)}\exp(\lambda\tilde{R}_\mu(t))\Big|_{t=a_I}^{b_I}$$
$$-\frac{1}{\lambda}\int_{a_I}^{b_I} u_{\nu\mu}^{[j+1]}(x,\lambda)\exp(\lambda\tilde{R}_\mu(t))\,dt$$

for $j = 0,\dots,l_1-1$. Hence we obtain

$$\int_{a_I}^{b_I} q_{\nu\mu}(t,\lambda)\exp(\lambda\tilde{R}_\mu(t))\,dt$$
$$= \sum_{j=0}^{l_1-1} (-1)^j\lambda^{-j-1}\frac{u_{\nu\mu}^{[j]}(t,\lambda)}{\tilde{r}_\mu(t)}\exp(\lambda\tilde{R}_\mu(t))\Big|_{t=a_I}^{b_I} + (-1)^{l_1}\lambda^{-l_1} *$$
$$* \int_{a_I}^{b_I} u_{\nu\mu}^{[l_1]}(t,\lambda)\exp\{\lambda(\tilde{R}_\mu(t)-\tilde{R}_\mu(x_\mu(\lambda)))\}dt\,\exp(\lambda\tilde{R}_\mu(x_\mu(\lambda)))$$

by a recursive application of the foregoing equation. We fix ν, μ and obtain

$$u_{\nu\mu}^{[l_1]}(\cdot,\lambda) = g + \overline{g}(\cdot,\lambda)$$

with $g \in L_1(a,b)$ and $\overline{g} = \{o(1)\}_1$. From Proposition 3.1 we conclude that

$$\int_{a_I}^{b_I} g(t)\exp\{\lambda(\tilde{R}_\mu(t)-\tilde{R}_\mu(x(\lambda)))\}dt$$

tends to zero in $C^o([a,b]\times[a,b])$ with respect to the variables a_I, b_I as $\lambda\to\infty$. The integral

$$\int_{a_I}^{b_I} \overline{g}(t,\lambda)\exp\{\lambda(\tilde{R}_\mu(t)-\tilde{R}_\mu(x(\lambda)))\}dt$$

has the same asymptotic behaviour since it can be estimated by

$$\int_{a_I}^{b_I} |\overline{g}(t,\lambda)|dt \leq \int_a^b |\overline{g}(t,\lambda)|dt = o(1)$$

because of (5.19). We conclude that a representation of the form (5.16) also holds if $\tilde{r}_\mu \neq 0$ but we have to observe that in this case the coefficients $q_{\nu\mu}^{(i)}$ are asymptotic polynomials in $C^o([a,b]\times[a,b])$ only of order l_1. Since

$$\hat{M}_I(\lambda) = S_I(\lambda) + Q_I(\lambda) ,$$

the assertion is proved by combining the relationships (5.12), (5.14) and (5.16).

We define

$$\text{(5.21)} \qquad E := \left\{ \sum_{l=1}^{n} \tilde{R}_l(a_{j(l)}) : j(1),\dots,j(n) \in \{1,\dots,m\} \right\}$$

and state the following

COROLLARY 5.4. *There are* $\nu_c \in \mathbb{Z}$ *and* $b_c \in \mathbb{C}$ *for* $c \in E$

such that

$$\det \hat{M}(\lambda) = \sum_{c\in E} \lambda^{\nu_c}[b_c]\exp(\lambda c). \tag{5.22}$$

PROOF. The foregoing proposition yields the assertion because $\hat{M}(\lambda) = \hat{M}_{[a,b]}(\lambda)$ and $N([a,b]) = \{a_1,\dots,a_m\}$. It is $\nu_c \le nq \le n\max\{q_o+q_1, q_o+q_2\}$ for all $c \in E$.

Let E be the set defined in (5.21) and P be the convex hull of E. For $K \subset [0,1]$ we set

$$E_K := \left\{ \sum_{l=1}^{n} \alpha_l \tilde{R}_l(b) : \alpha_l \in K \quad (l = 1,\dots,n) \right\}$$

and denote the convex hull of E_K by P_K.

PROPOSITION 5.5. $P = P_{\{0,1\}} = P_{[0,1]}$ *which means that* P *only depends on the values of the functions* $\tilde{R}_l$ *at the point* b.

PROOF. Since $\tilde{R}_l(a) = 0$ and $\tilde{R}_l(a_j)$ belongs to the line segment with the endpoints 0 and $\tilde{R}_l(b)$, the inclusions

$$E_{\{0,1\}} \subset E \subset E_{[0,1]}$$

hold. Thus we only have to prove that $P_{[0,1]} \subset P_{\{0,1\}}$. For $l \in \{1,\dots,n\}$ let $\alpha_l \in [0,1]$. Choose l_i $(i=1,\dots,n)$ such that $\{l_1,\dots,l_n\} = \{1,\dots,n\}$ and $\alpha_{l_1} \le \dots \le \alpha_{l_n}$. We obtain

$$\sum_{l=1}^{n} \alpha_l \tilde{R}_l(b) = \alpha_{l_1} \sum_{l=1}^{n} \tilde{R}_l(b) + \sum_{i=2}^{n} (\alpha_{l_i} - \alpha_{l_{i-1}}) \sum_{j=i}^{n} \tilde{R}_{l_j}(b) + (1-\alpha_{l_n})\cdot 0.$$

The sums

$$\sum_{l=1}^{n} \tilde{R}_l(b), \quad \sum_{j=i}^{n} \tilde{R}_{l_j}(b) \quad (i = 2,\dots,n)$$

and 0 belong to $E_{\{0,1\}}$ and thus to $P_{\{0,1\}}$. Furthermore

$$0 \le \alpha_{l_1} \le 1, \quad 0 \le \alpha_{l_i} - \alpha_{l_{i-1}} \le 1 \quad (i=2,\dots,n), \quad 0 \le 1 - \alpha_{l_n} \le 1$$

and

$$\alpha_{l_1} + \sum_{i=2}^{n} (\alpha_{l_i} - \alpha_{l_{i-1}}) + (1 - \alpha_{l_n}) = 1 .$$

Hence $P_{[0,1]} \subset P_{\{0,1\}}$ since $P_{\{0,1\}}$ is convex.

The boundary of P consists of a finite number of line segments P_s $(s = 1,\ldots,S)$. We have

$$P_s \cap E = \{c_s + \alpha_s t_{s,i} : i = 1,\ldots,m_s\}$$

where c_s is a fixed element in $P_s \cap E$, $\alpha_s \in \mathbb{C} \setminus \{0\}$ and $t_{s,i} \in \mathbb{R}$ $(i = 1,\ldots,m_s)$. For $c \in E$ let ν_c and $\tilde{b}_c(\lambda) := \lambda^{\nu_c}[b_c]$ be given by formula (5.22). If $c = c_s + \alpha_s t_{s,i}$, we set $\nu_{s,i} := \nu_c$. Let $\tilde{E}$ be the set of the vertices of P. For $s \in \{1,\ldots,S\}$ we define

$$\tilde{L}_s := \{(t_{s,i},\nu_{s,i}) : i \in \{1,\ldots,m_s\}, c_s+\alpha_s t_{s,i} \in \tilde{E} \text{ or } \tilde{b}_{c_s+\alpha_s t_{s,i}} \neq 0\}.$$

Let L_s be the set of those points $(t_{s,i},\nu_{s,i}) \in \tilde{L}_s$ for which there are no two different points (t_{s,i_1},ν_{s,i_1}), $(t_{s,i_2},\nu_{s,i_2}) \in \tilde{L}_s$ and no $\mu \in (0,1)$ such that

$$t_{s,i} = \mu t_{s,i_1} + (1-\mu)t_{s,i_2}, \quad \nu_{s,i} \leq \mu\nu_{s,i_1} + (1-\mu)\nu_{s,i_2} .$$

We define

$$\hat{E} := \{c_s+\alpha_s t_{s,i} : s \in \{1,\ldots,S\}, i \in \{1,\ldots,m_s\}, (t_{s,i},\alpha_{s,i}) \in L_s\}. \tag{5.23}$$

PROPOSITION 5.6. $\tilde{E}$ *is a subset of* $\hat{E}$. *If all* ν_c *are equal, then* $\tilde{E} = \hat{E}$.

PROOF. Let $c \in \tilde{E}$. Then c is the endpoint of some P_s. We infer that $c = c_s + \alpha_s t_{s,i_0}$ with some $i_0 \in \{1,\ldots,m_s\}$, $(t_{s,i_0},\nu_{s,i_0}) \in \tilde{L}_s$ and $t_{s,i} \geq t_{s,i_0}$ for all $i \in \{1,\ldots,m_s\}$ or $t_{s,i} \leq t_{s,i_0}$ also for all $i \in \{1,\ldots,m_s\}$. Hence there are no two different points (t_{s,i_1},ν_{s,i_1}), $(t_{s,i_2},\nu_{s,i_2}) \in \tilde{L}_s$ and no $\mu \in (0,1)$ such that

$$t_{s,i_0} = \mu t_{s,i_1} + (1-\mu)t_{s,i_2} .$$

Hence $(t_{s,i_0},\nu_{s,i_0}) \in L_s$ and $c \in \hat{E}$.

The second assertion is clear from the definition of L_s.

LEMMA 5.7. *Assume that* $b_c \neq 0$ *for all* $c \in \hat{E}$ *and set*

$g := \min\{\nu_c : c \in \hat{E}\}$.

Then there is an increasing sequence $(d_j)_0^\infty$ *of positive numbers which tends to* ∞ *as* $\lambda \to \infty$ *and a* $\delta > 0$ *with the following property: for all* $\lambda \in \mathbb{C}$ *with* $|\lambda| = d_j$ *for some* $j \in \mathbb{N}$ *there is a* $c(\lambda) \in \tilde{E}$ *such that*

$$\mathrm{Re}\{\lambda(c - c(\lambda))\} \leq 0 \qquad (c \in P) \tag{5.24}$$

and the estimate

$$|\lambda^{-g} \exp(-\lambda c(\lambda)) \det \hat{M}(\lambda)| \geq \delta \tag{5.25}$$

holds.

The foregoing statement stems from Langer, cf. [14], p. 176, but under the following stronger assumptions: The coefficients $\tilde{b}_c(\lambda)$ in the exponential sum $\det \hat{M}(\lambda)$ have the form

$$\tilde{b}_c(\lambda) = \lambda^{\nu_c} \sum_{k=0}^{\infty} b_{c,k} \lambda^{-k} \qquad (c \in E).$$

The $b_c = b_{c,0}$ are different from zero for all $c \in E$ and g is defined to be the minimum of all exponents ν_c. From an analysis of Langer's proof we conclude that his proof also goes through under the weaker assumptions stated above.

THEOREM 5.8. *Assume that* $b_c \neq 0$ *for all* $c \in \hat{E}$. *Then there are* $p, p' \in \mathbb{Z}$ *and an increasing sequence* $(d_j)_0^\infty$ *of positive numbers which tends to* ∞ *as* $\lambda \to \infty$ *such that* T *is* $|\ |_{L_2^n(a,b)}$*-regular of order* (p,p') *with respect to the curves* $\Gamma_j := \{\lambda \in \mathbb{C} : |\lambda| = d_j\}$ $(j \in \mathbb{N})$.

PROOF. In the following we suppose that λ is large. Let $f \in L_2^n(a,b)$. The definitions (4.7), (5.6) and the relationship (5.7) yield that

$$\int_a^b G(x,\xi,\lambda) f_1(\xi) d\xi =$$

$$= \hat{Y}(x,\lambda)\int_a^x \int_{t=a}^{\xi} \hat{M}^{-1}(\lambda)d_tF(t,\lambda)\hat{Y}(t,\lambda)\hat{Y}^{-1}(\xi,\lambda)f_1(\xi)d\xi$$

$$- \hat{Y}(x,\lambda)\int_x^b \int_{t=\xi}^{b} \hat{M}^{-1}(\lambda)d_tF(t,\lambda)\hat{Y}(t,\lambda)\hat{Y}^{-1}(\xi,\lambda)f_1(\xi)d\xi$$

$$= \hat{Y}(x,\lambda)\int_a^x \hat{M}^{-1}(\lambda)\hat{M}_{[a,\xi)}(\lambda)\hat{Y}^{-1}(\xi,\lambda)f_1(\xi)d\xi$$

$$- \hat{Y}(x,\lambda)\int_x^b \hat{M}^{-1}(\lambda)\hat{M}_{[\xi,b]}(\lambda)\hat{Y}^{-1}(\xi,\lambda)f_1(\xi)d\xi$$

where the second equation holds because of the definitions (4.4), (4.5) and (5.10). We define

$$(5.26)\qquad \hat{M}(x,\xi,\lambda) := \begin{cases} \hat{M}_{[a,\xi)}(\lambda) & \text{if } a \le \xi \le x \le b, \\ -\hat{M}_{[\xi,b]}(\lambda) & \text{if } a \le x < \xi \le b \end{cases}$$

and obtain

$$(5.27)\qquad \begin{cases} \int_a^b G(x,\xi,\lambda)f_1(\xi)d\xi \\ = \hat{Y}(x,\lambda)\int_a^b \hat{M}^{-1}(\lambda)\hat{M}(x,\xi,\lambda)\hat{Y}^{-1}(\xi,\lambda)f_1(\xi)d\xi \, . \end{cases}$$

We set

$$(5.28)\qquad \hat{N}(x,\xi) := \begin{cases} N([a,\xi)) & \text{if } a \le \xi \le x \le b, \\ N([\xi,b]) & \text{if } a \le x < \xi \le b \end{cases}$$

with $N(I)$ being defined in (5.11). From Proposition 5.3 and Definition (5.26) we infer that the μ-th column of $\hat{M}(x,\xi,\lambda)$ has the form

$$(5.29)\qquad \hat{m}_\mu(x,\xi,\lambda) = \lambda^q \sum_{s\in\hat{N}(x,\xi)} \hat{a}_\mu^{(s)}(x,\xi,\lambda)\exp(\lambda\tilde{R}_\mu(s))$$

where $q \in \mathbb{Z}$, $q \le \max\{q_o+q_1, q_o+q_2\}$ and the $\hat{a}_\mu^{(s)}$ are asymptotic polynomials in $L_\infty^n((a,b)\times(a,b))$.

For convenience we recall some well-known results from linear algebra. Let $M = (m_1,\dots,m_n) \in M_n(\mathbb{C})$ be invertible. Then

$$\det M \cdot M^{-1} = \left(\det M_{\nu\mu}\right)_{\nu,\mu=1}^{n}$$

where $M_{\nu\mu} := (m_1,\dots,m_{\nu-1},e_\mu,m_{\nu+1},\dots,m_n)$ and e_μ is the μ-th unit vector in $\mathbb{C}^n$, i.e. the ν-th column of M is substituted by the μ-th unit vector. Let $A = (a_1,\dots,a_n) \in M_n(\mathbb{C})$. Then

$$(5.30) \qquad (\det M)\, M^{-1} A = \left(\det M_{\nu\mu}^{A}\right)_{\nu,\mu=1}^{n}$$

where $M_{\nu\mu}^{A} := \det(m_1,\dots,m_{\nu-1},a_\mu,m_{\nu+1},\dots,m_n)$. For the proof of (5.30) let $A =: (a_{\alpha\mu})_{\alpha,\mu=1}^{n}$. We obtain

$$(\det M)\, M^{-1} A = \left(\sum_{\alpha=1}^{n} (\det M_{\nu\alpha})\, a_{\alpha\mu}\right)_{\nu,\mu=1}^{n}$$

$$= \left(\det(m_1,\dots,m_{\nu-1},\sum_{\alpha=1}^{n} a_{\alpha\mu} e_\alpha, m_{\nu+1},\dots,m_n)\right)_{\nu,\mu=1}^{n}$$

$$= \left(\det(m_1,\dots,m_{\nu-1},a_\mu,m_{\nu+1},\dots,m_n)\right)_{\nu,\mu=1}^{n} .$$

Let $\hat{m}_\mu(\lambda)$ be the μ-th column of $\hat{M}(\lambda)$. For $\nu,\mu \in \{1,\dots,n\}$ we set

$$h_{\nu\mu}(x,\xi,\lambda) := \exp\{\lambda(\tilde{R}_\nu(x)-\tilde{R}_\mu(\xi))\} \,*$$
$$*\, \det(\hat{m}_1(\lambda),\dots,\hat{m}_{\nu-1}(\lambda),\hat{m}_\mu(x,\xi,\lambda),\hat{m}_{\nu+1}(\lambda),\dots,\hat{m}_n(\lambda)).$$

The formula (5.30) yields that

$$(5.31) \qquad E(x,\lambda)(\det\hat{M}(\lambda))\hat{M}^{-1}(\lambda)\hat{M}(x,\xi,\lambda)E^{-1}(\xi,\lambda) = (h_{\nu\mu}(x,\xi,\lambda))_{\nu,\mu=1}^{n} .$$

By Proposition 5.3

$$\hat{m}_\mu(\lambda) = \lambda^q \sum_{j=1}^{m} \hat{m}_\mu^{(j)}(\lambda) \exp(\lambda \tilde{R}_\mu(a_j))$$

where q is the same integer as in (5.29) and the coefficients $\hat{m}_\mu^{(j)}$ are asymptotic polynomials in $\mathbb{C}^n$. We set

$$\Lambda_o(x,\xi) := \hat{N}(x,\xi)\times\{1,\dots,m\}^n.$$

If $\iota\in\Lambda_o(x,\xi)$ and $\iota=(s,j_1,\dots,j_n)$, we define

$$c_\nu^{(\iota)}(x,\xi) := \tilde{R}_\nu(x) - \tilde{R}_\nu(\xi) + \sum_{\substack{l=1\\ l\neq\nu}}^{n} \tilde{R}_l(a_{j_l}) + \tilde{R}_\nu(s)$$

and

$$h_\nu^{(\iota)}(x,\xi,\lambda) := \frac{1}{m}\det(\hat{m}_1^{(j_1)}(\lambda),\dots,\hat{m}_{\nu-1}^{(j_{\nu-1})}(\lambda),$$

$$\hat{a}_\nu^{(s)}(x,\xi,\lambda),\hat{m}_{\nu+1}^{(j_{\nu+1})}(\lambda),\dots,\hat{m}_n^{(j_n)}(\lambda))$$

where the $\hat{a}_\nu^{(s)}$ are asymptotic polynomials in $L_\infty^n((a,b)\times(a,b))$ defined by (5.29). An easy calculation shows that

$$h_{\nu\nu}(x,\xi,\lambda) = \lambda^{nq} \sum_{\iota\in\Lambda_o(x,\xi)} h_\nu^{(\iota)}(x,\xi,\lambda)\exp(\lambda c_\nu^{(\iota)}(x,\xi)) \tag{5.32}$$

for $\nu=1,\dots,n$. If $s\in\hat{N}(x,\xi)$, ξ is an element of the interval with the endpoints s and x and therefore

$$0 \le |\tilde{R}_\nu(x)| - |\tilde{R}_\nu(\xi)| + |\tilde{R}_\nu(s)| \le |\tilde{R}_\nu(b)|. \tag{5.33}$$

We set

$$x_\nu(\lambda) := \begin{cases} a & \text{if } \operatorname{Re}(\lambda e^{i\tilde{\varphi}_\nu}) \le 0 \\ b & \text{if } \operatorname{Re}(\lambda e^{i\tilde{\varphi}_\nu}) > 0 \end{cases} \tag{5.34}$$

and, for $\iota=(s,j_1,\dots,j_n)$

$$c_\nu^{(\iota)}(\lambda) := \tilde{R}_\nu(x_\nu(\lambda)) + \sum_{\substack{l=1\\ l\neq\nu}}^{n} \tilde{R}_l(a_{j_l}) .$$

From (5.33) and (5.34) we infer that

$$\begin{cases} \operatorname{Re}\{\lambda(c_\nu^{(\iota)}(x,\xi) - c_\nu^{(\iota)}(\lambda))\} \\ = \operatorname{Re}(\lambda e^{i\tilde{\varphi}_\nu})\left\{|\tilde{R}_\nu(x)| - |\tilde{R}_\nu(\xi)| + |\tilde{R}_\nu(s)| - |\tilde{R}_\nu(x_\nu(\lambda))|\right\} \le 0. \end{cases} \tag{5.35}$$

Obviously $c_\nu^{(\iota)}(\lambda)$ belongs to the set E defined in (5.21). For $c \in E$ we define

$$h_\nu^c(x,\xi,\lambda) := \sum_{\substack{\iota\in\Lambda_o(x,\xi)\\ c_\nu^{(\iota)}(\lambda)=c}} h_\nu^{(\iota)}(x,\xi,\lambda)\exp\{\lambda(c_\nu^{(\iota)}(x,\xi)-c_\nu^{(\iota)}(\lambda))\}$$

and conclude from (5.32) and (5.35) that

$$(5.36) \qquad h_{\nu\nu}(x,\xi,\lambda) = \lambda^{nq}\sum_{c\in E} h_\nu^c(x,\xi,\lambda)\exp(\lambda c)$$

where $h_\nu^c(\cdot,\cdot,\lambda) = \{0(1)\}_\infty$.

Now let $\nu \neq \mu$. We define

$$(5.37) \qquad \check{M}(x,\xi,\lambda) := \begin{cases} \hat{M}_{[\xi,b]}(\lambda) & \text{if } a \le \xi \le x \le b, \\ -\hat{M}_{[a,\xi)}(\lambda) & \text{if } a \le x < \xi \le b \end{cases}$$

and obtain

$$(5.38) \qquad \hat{M}(x,\xi,\lambda) + \check{M}(x,\xi,\lambda) = \begin{cases} \hat{M}(\lambda) & \text{if } a \le \xi \le x \le b, \\ -\hat{M}(\lambda) & \text{if } a \le x < \xi \le b. \end{cases}$$

As for $\hat{M}(x,\xi,\lambda)$ we conclude that the μ-th column of $\check{M}(x,\xi,\lambda)$ has the form

$$(5.39) \qquad \check{m}_\mu(x,\xi,\lambda) = \lambda^q \sum_{s\in\check{N}(x,\xi)} \check{a}_\mu^{(s)}(x,\xi,\lambda)\exp(\lambda\tilde{R}_\mu(s))$$

where

$$(5.40) \qquad \check{N}(x,\xi) := \begin{cases} N([\xi,b]) & \text{if } a \le \xi \le x \le b, \\ N([a,\xi)) & \text{if } a \le x < \xi \le b \end{cases}$$

and the $\check{a}_\mu^{(s)}$ are asymptotic polynomials in $L_\infty^n((a,b)\times(a,b))$. In the determinant defining $h_{\nu\mu}(x,\xi,\lambda)$ we subtract the ν-th column from the μ-th column if $\xi \le x$ and add the ν-th column to the μ-th column if $x < \xi$. These operations lead to

$$h_{\nu\mu}(x,\xi,\lambda) = \exp\{\lambda(\tilde{R}_\nu(x)-\tilde{R}_\mu(\xi))\} *$$

$$* \det(\hat{m}_1(\lambda),\ldots,\sigma\check{m}_\mu(x,\xi,\lambda),\ldots,\hat{m}_\mu(x,\xi,\lambda),\ldots,\hat{m}_n(\lambda))$$

where $\sigma=+1$ for $\xi \leq x$ and $\sigma=-1$ for $x<\xi$. We set

$$\Lambda(x,\xi) := \hat{N}(x,\xi) \times \check{N}(x,\xi) \times \{1,\ldots,m\}^n.$$

If $\iota \in \Lambda(x,\xi)$ and $\iota=(s,t,j_1,\ldots,j_n)$, we define

$$c_{\nu\mu}^{(\iota)}(x,\xi) := \tilde{R}_\nu(x) - \tilde{R}_\mu(\xi) + \sum_{\substack{l=1\\ l\neq\nu,\mu}}^{n} \tilde{R}_l(a_{j_l}) + \tilde{R}_\mu(s) + \tilde{R}_\mu(t)$$

and

$$h_{\nu\mu}^{(\iota)}(x,\xi,\lambda) := \frac{1}{m^2}\det(\hat{m}_1^{(j_1)}(\lambda),\ldots,\sigma\check{a}_\mu^{(t)}(x,\xi,\lambda),\ldots,$$
$$\hat{a}_\mu^{(s)}(x,\xi,\lambda),\ldots,\hat{m}_n^{(j_n)}(\lambda))$$

where the $\check{a}_\mu^{(t)}$ and $\hat{a}_\mu^{(s)}$ are the asymptotic polynomials defined in (5.29) and (5.39), respectively. It is easy to see that

$$(5.41)\qquad h_{\nu\mu}(x,\xi,\lambda) = \lambda^{nq} \sum_{\iota\in\Lambda(x,\xi)} h_{\nu\mu}^{(\iota)}(x,\xi,\lambda)\exp(\lambda c_{\nu\mu}^{(\iota)}(x,\xi)).$$

If $s \in \hat{N}(x,\xi)$ and $t \in \check{N}(x,\xi)$, then either $a \leq s \leq \xi \leq t \leq b$ or $a \leq t \leq \xi \leq s \leq b$ whence

$$(5.42)\qquad 0 \leq |\tilde{R}_\mu(s)| - |\tilde{R}_\mu(\xi)| + |\tilde{R}_\mu(t)| \leq |\tilde{R}_\mu(b)|.$$

Let $x_\nu(\lambda)$ be defined as in (5.34). Set

$$c_{\nu\mu}^{(\iota)}(\lambda) := \tilde{R}_\nu(x_\nu(\lambda)) + \tilde{R}_\mu(x_\mu(\lambda)) + \sum_{\substack{l=1\\ l\neq\nu,\mu}} \tilde{R}_l(a_{j_l})$$

for $\iota=(s,t,j_1,\ldots,j_n)$. From (5.34) and (5.42) we deduce that

$$(5.43)\qquad \begin{cases} \operatorname{Re}\{\lambda(c_{\nu\mu}^{(\iota)}(x,\xi)-c_{\nu\mu}^{(\iota)}(\lambda))\} = \operatorname{Re}(\lambda e^{i\tilde{\varphi}_\nu})\left\{|\tilde{R}_\nu(x)|-|\tilde{R}_\nu(x_\nu(\lambda))|\right\} \\ + \operatorname{Re}(\lambda e^{i\tilde{\varphi}_\mu})\left\{|\tilde{R}_\mu(s)|-|\tilde{R}_\mu(\xi)|+|\tilde{R}_\mu(t)|-|\tilde{R}_\mu(x_\mu(\lambda))|\right\} \leq 0 . \end{cases}$$

We have $c_{\nu\mu}^{(\iota)}(\lambda) \in E$ and define

$$h^c_{\nu\mu}(x,\xi,\lambda) := \sum_{\substack{\iota\in\Lambda(x,\xi)\\ c^{(\iota)}_{\nu\mu}(\lambda)=c}} h^{(\iota)}_{\nu\mu}(x,\xi,\lambda)\exp\{\lambda(c^{(\iota)}_{\nu\mu}(x,\xi)-c^{(\iota)}_{\nu\mu}(\lambda))\}$$

for $c\in E$. From (5.41) and (5.43) we infer that

$$h_{\nu\mu}(x,\xi,\lambda) = \lambda^{nq}\sum_{c\in E} h^c_{\nu\mu}(x,\xi,\lambda)\exp(\lambda c) \qquad (\nu \neq \mu) \tag{5.44}$$

where $h_{\nu\mu}(\cdot,\cdot,\lambda) = \{O(1)\}_\infty$.

By (5.4), (5.27) and (5.31)

$$\int_a^b G(x,\xi,\lambda)f_1(\xi)d\xi = (\det\hat{M}(\lambda))^{-1}C(x,\lambda)\tilde{Y}(x,\lambda)E^{-1}(x,\lambda) *$$

$$* \int_a^b (h_{\nu\mu}(x,\xi,\lambda))^n_{\nu,\mu=1}E(\xi,\lambda)\tilde{Y}^{-1}(\xi,\lambda)C^{-1}(\xi,\lambda)f_1(\xi)d\xi$$

By Theorem 3.3 the estimates

$$\tilde{Y}(\cdot,\lambda)E^{-1}(\cdot,\lambda) = \{O(1)\}_\infty, \quad E(\cdot,\lambda)\tilde{Y}^{-1}(\cdot,\lambda) = \{O(1)\}_\infty \tag{5.45}$$

hold. Furthermore we have

$$C(\cdot,\lambda) = \{O(\lambda^{q_o})\}_\infty \tag{5.46}$$

by assumption and

$$C^{-1}(\cdot,\lambda) = \{O(\lambda^{\hat{q}})\}_\infty \tag{5.47}$$

by Proposition 5.1. From these estimates and the representations (5.36) and (5.44) we obtain that

$$\int_a^b G(x,\xi,\lambda)f_1(\xi)d\xi = \lambda^{q'}(\det\hat{M}(\lambda))^{-1}\Big(\sum_{c\in E} k^c_{\nu\mu}(x,\lambda)\exp(\lambda c)\Big)^n_{\nu,\mu=1} \tag{5.48}$$

where q' is an integer (which is less or equal $q_o+\hat{q}+nq$), the $k^c_{\nu\mu}(\cdot,\lambda)$ belong to $L_\infty(a,b)$ and

$$k^c_{\nu\mu}(\cdot,\lambda) = \{O(1)\}_\infty . \tag{5.49}$$

Now choose $(d_j)_0^\infty$ and $\delta > 0$ according to Lemma 5.7. Fix some $j \in \mathbb{N}$ and let $|\lambda| = d_j$. Again by Lemma 5.7 choose $c(\lambda) \in \tilde{E}$ such that the estimates (5.24) and (5.25) hold. Then

$$(5.50) \qquad |(\det \hat{M}(\lambda))^{-1} \exp(\lambda c)| \leq \delta^{-1} |\lambda|^{-g} |\exp\{\lambda(c-c(\lambda))\}| \leq \delta^{-1} |\lambda|^{-g}$$

for all $c \in E$. This estimate, the relationships (4.9) and (5.48), and the statement (5.49) lead to

$$(5.51) \qquad \int_{\Gamma_j} |\lambda^{-q'+g-2} R_1(\lambda) f_1|_{L_2^n(a,b)} |d\lambda| \to 0 \qquad (j \to \infty)$$

where the Γ_j are the positively oriented circles with center 0 and radius d_j.

In the following we shall estimate the "boundary part" $R_2(\lambda) f_2$ of the resolvent of T. By (4.8) and (4.10) we have

$$(5.52) \qquad R_2(\lambda) f_2 = \hat{Y}(\cdot,\lambda) \hat{M}^{-1}(\lambda) f_2 \qquad (f_2 \in \mathbb{C}^n).$$

For $\nu, \mu \in \{1,\ldots,n\}$ we set

$$\hat{h}_{\nu\mu}(x,\lambda) := \exp(\lambda \tilde{R}_\nu(x)) * \\ * \det(\hat{m}_1(\lambda),\ldots,\hat{m}_{\nu-1}(\lambda), e_\mu, \hat{m}_{\nu+1}(\lambda),\ldots,\hat{m}_n(\lambda))$$

where e_μ denotes the μ-th unit column in $\mathbb{C}^n$. Hence

$$(5.53) \qquad E(\cdot,\lambda)(\det \hat{M}(\lambda)) \hat{M}^{-1}(\lambda) = (\hat{h}_{\nu\mu}(\cdot,\lambda))_{\nu,\mu=1}^n$$

by formula (5.30). We set

$$\hat{\Lambda} := \{1,\ldots,m\}^n.$$

If $\iota \in \hat{\Lambda}$ and $\iota = (j_1,\ldots,j_n)$, we define

$$\hat{c}_\nu^{(\iota)}(x) := \tilde{R}_\nu(x) + \sum_{\substack{l=1 \\ l \neq \nu}}^n \tilde{R}_l(a_{j_l})$$

and

$$\hat{h}_{\nu\mu}^{(\iota)}(\lambda) := \frac{1}{m}\det(\hat{m}_1^{(j_1)}(\lambda),\ldots,\hat{m}_{\nu-1}^{(j_{\nu-1})}(\lambda),e_\mu,$$
$$\hat{m}_{\nu+1}^{(j_{\nu+1})}(\lambda),\ldots,\hat{m}_n^{(j_n)}(\lambda)).$$

Obviously the $\hat{h}_{\nu\mu}^{(\iota)}$ are asymptotic polynomials and it is easy to see that

$$(5.54)\qquad \hat{h}_{\nu\mu}(x,\lambda) = \lambda^{(n-1)q}\sum_{\iota\in\hat{\Lambda}}\hat{h}_{\nu\mu}^{(\iota)}(\lambda)\exp(\lambda\hat{c}_\nu^{(\iota)}(x)).$$

With the $x_\nu(\lambda)$ defined in (5.34) we set

$$c_\nu^{(\iota)}(\lambda) := \tilde{R}_\nu(x_\nu(\lambda)) + \sum_{\substack{l=1\\ l\neq\nu}}^{n}\tilde{R}_l(a_{j_l})$$

for $\iota=(j_1,\ldots,j_n)$ and obtain

$$(5.55)\qquad \operatorname{Re}\{\lambda(\hat{c}_\nu^{(\iota)}(x)-c_\nu^{(\iota)}(\lambda)) = \operatorname{Re}(\lambda e^{i\tilde{\varphi}_\nu})\{|R_\nu(x)|-|R_\nu(x_\nu(\lambda))|\}\leq 0.$$

For $c\in E$ let

$$\hat{h}_{\nu\mu}^{c}(x,\lambda) := \sum_{\substack{\iota\in\hat{\Lambda}\\ c_\nu^{(\iota)}(\lambda)=c}}\hat{h}_{\nu\mu}^{(\iota)}(\lambda)\exp\{\lambda(\hat{c}_\nu^{(\iota)}(x)-c_\nu^{(\iota)}(\lambda))\}.$$

From (5.54) and (5.55) we infer that

$$(5.56)\qquad \hat{h}_{\nu\mu}(x,\lambda) = \lambda^{(n-1)q}\sum_{c\in E}\hat{h}_{\nu\mu}^{c}(x,\lambda)\exp(\lambda c)$$

where the $\hat{h}_{\nu\mu}^{c}(\cdot,\lambda)$ belong to $L_\infty(a,b)$ and $\hat{h}_{\nu\mu}^{c}(\cdot,\lambda)=\{O(1)\}_\infty$.

Let $f_2\in\mathbb{C}^n$. By (5.4), (5.52), (5.53) and (5.56) we obtain

$$(R_2(\lambda)f_2)(x)$$
$$= (\det\hat{M}(\lambda))^{-1}C(x,\lambda)\tilde{Y}(x,\lambda)E^{-1}(x,\lambda)(\hat{h}_{\nu\mu}(x,\lambda))_{\nu,\mu=1}^{n}f_2$$
$$= \lambda^{q''}(\det\hat{M}(\lambda))^{-1}\sum_{c\in E}\hat{k}_{\nu\mu}^{c}(x,\lambda)\exp(\lambda c)$$

where q'' is an integer (which is less or equal $q_o+(n-1)q$), the $\hat{k}^c_{\nu\mu}(\cdot,\lambda)$ are elements of $L_\infty(a,b)$ and

(5.57) $\hat{k}^c_{\nu\mu}(\cdot,\lambda) = \{O(1)\}_\infty.$

From (5.50) and (5.57) we conclude that

(5.58) $\int_{\Gamma_j} |\lambda^{-q''+g-2} R_2(\lambda) f_2|_{L_2^n(a,b)} |d\lambda| \to 0 \qquad (j\to\infty).$

The relationships (5.51) and (5.58) show that T is $|\,|_{L_2^n(a,b)}$-regular of order $(q'-g+1, q''-g+1)$.

Let q^* be the order of the polynomial A. From assumption b_3) we conclude that

$$A(\cdot,\lambda) = C(\cdot,\lambda)\tilde{A}(\cdot,\lambda)C^{-1}(\cdot,\lambda) + C'(\cdot,\lambda)C^{-1}(\cdot,\lambda)$$

for $|\lambda| \geq \gamma_0$. Hence $q^* \leq q_0 + \hat{q} + 1$ where q_0 is the order of the polynomial C and $\hat{q}$ is defined in Proposition 5.1.

PROPOSITION 5.9. *Assume that* T *is* $|\,|_{L_2^n(a,b)}$*-regular of order* (p,p') *with respect to the curves* Γ_j $(j\in\mathbb{N})$. *Suppose that*

$$\int_{\Gamma_j} |\lambda|^{-2} |d\lambda| \to 0 \qquad (j\to\infty).$$

Set $\hat{p} := \max\{1, p+q^*\}$ *and* $\hat{p}' := p'+q^*$.

Then T *is* $|\,|_{H_1^n(a,b)}$*-regular of order* $(\hat{p},\hat{p}')$ *with respect to the curves* Γ_j.

PROOF. Suppose that $f=(f_1,f_2)\in L_2^n(a,b)\times\mathbb{C}^n$ and let $\lambda\in\rho(T)$. The equation $T(\lambda)T^{-1}(\lambda)f = f$ and the definitions (2.1), (4.1) and (4.2) yield

(5.59) $T^D(\lambda)R_1(\lambda)f_1 = f_1, \quad T^D(\lambda)R_2(\lambda)f_2 = 0.$

The first equation implies that

(5.60) $(R_1(\lambda)f_1)' = f_1 + A(\cdot,\lambda)R_1(\lambda)f_1.$

Since $\hat{p}\geq p$, we obtain

$$\int_{\Gamma_j} |\lambda^{-\hat{p}-1} R_1(\lambda) f_1|_{L_2^n(a,b)} |d\lambda| \to 0 \qquad (j \to \infty).$$

Because $\bigcup_{j\in\mathbb{N}} \operatorname{int}\Gamma_j = \mathbb{C}$ and $\overline{\operatorname{int}\Gamma_j} \subset \operatorname{int}\Gamma_{j+1}$, we may assume that $|\lambda| \geq 1$ for all $\lambda \in \bigcup_{j\in\mathbb{N}} \Gamma_j$. This assumption, the equation (5.60) and the definition of $\hat{p}$ lead to the estimate

$$\int_{\Gamma_j} |\lambda^{-\hat{p}-1} (R_1(\lambda) f_1)'|_{L_2^n(a,b)} |d\lambda| \leq \int_{\Gamma_j} |\lambda|^{-2} |d\lambda| |f_1|_{L_2^n(a,b)}$$

$$+ \int_{\Gamma_j} |\lambda^{-p-1} \lambda^{-q^*} A(\cdot,\lambda) R_1(\lambda) f_1|_{L_2^n(a,b)} |d\lambda|.$$

The first term on the right side of this inequality tends to zero as $j \to \infty$ by assumption. Since $\lambda^{-q^*} A(\cdot,\lambda) = \{O(1)\}_\infty$, the second term can be estimated by

$$\rho \int_{\Gamma_j} |\lambda^{-p-1} R_1(\lambda) f_1|_{L_2^n(a,b)} |d\lambda|$$

with some constant $\rho \geq 0$, whence it also tends to zero as $j \to \infty$. Altogether it follows that

$$(5.61) \qquad \int_{\Gamma_j} |\lambda^{-\hat{p}-1} R_1(\lambda) f_1|_{H_1^n(a,b)} |d\lambda| \to 0 \qquad (j \to \infty).$$

The second equation in (5.59) yields

$$(R_2(\lambda) f_2)' = A(\cdot,\lambda) R_2(\lambda) f_2$$

and thus the statement

$$(5.62) \qquad \int_{\Gamma_j} |\lambda^{-\hat{p}'-1} R_2(\lambda) f_2|_{H_1^n(a,b)} |d\lambda| \to 0 \qquad (j \to \infty)$$

in a similar way as (5.61). The relationships (5.61) and (5.62) prove the assertion.

We close this section with the following remark: Let

$$A(\cdot,\lambda) = A_0 + \lambda A_1 \quad (\lambda \in \mathbb{C})$$

where A_0 and A_1 belong to $M_n(W^{k,\infty}(a,b))$ for some $k \in \mathbb{N}$. Suppose (as above) that the matrix functions $W^{(j)}$ $(j = 1,\dots,m)$ and W are polynomials. Then all the assumptions made at the beginning of this section are fulfilled if A_1 has the additional properties which are stated in condition iv) at the beginning of section 3 and in condition b_4) here in this section.

In fact, we only have to choose $\gamma_0 = 0$ and $C(x,\lambda) = I_n$, $A_{-j}(x) = 0$ $(j = 1,\dots,k)$, $A^k(x,\lambda) = 0$ for all $x \in (a,b)$ and $\lambda \in \mathbb{C}$.

6. AN EXPANSION THEOREM

In this section we also consider the boundary eigenvalue operator function $T = (T^D, T^R)$ defined by (2.1), (2.2) and (2.3). Here the coefficient matrix functions A, $W^{(j)}$ $(j = 1,\dots,m)$ and W of T^D and T^R, respectively, have to fulfil the following assumptions:

I) $A(\cdot,\lambda) = A_0 + \lambda A_1$ for $\lambda \in \mathbb{C}$ where A_0 and A_1 belong to $M_n(W^{k,\infty}(a,b))$ for some $k \in \mathbb{N}$. A_1 is a block matrix with respect to the decomposition $\mathbb{C}^n = \mathbb{C}^p \times \mathbb{C}^l$ and has the form

$$A_1 = \begin{bmatrix} 0 & 0 \\ 0 & \Omega_1 \end{bmatrix} \tag{6.1}$$

where $1 \le l \le n$ and Ω_1 is invertible in $M_n(L_\infty(a,b))$ (and thus in $M_n(W^{k,\infty}(a,b))$ by Proposition 3.2).

II) The $W^{(j)}$ $(j = 1,\dots,m)$ and W are polynomials.

The assumptions with respect to A_1 are fulfilled if $A_1 = \operatorname{diag}(\tilde{r}_1,\dots,\tilde{r}_n)$ where $\tilde{r}_1 = \dots = \tilde{r}_p = 0$ for some $p \in \{0,\dots,n-1\}$, $\tilde{r}_{p+1},\dots,\tilde{r}_n$ belong to $W^{k,\infty}(a,b)$ and $\tilde{r}_{p+1}^{-1},\dots,\tilde{r}_n^{-1}$ are elements of $L_\infty(a,b)$. In fact, set $l := n-p$ and $\Omega_1 := \operatorname{diag}(\tilde{r}_{p+1},\dots,\tilde{r}_n)$.

We also write A_0 as a block matrix with respect to $\mathbb{C}^p \times \mathbb{C}^l$

$$A_0 =: \begin{bmatrix} A_{11}^{[0]} & A_{12}^{[0]} \\ A_{21}^{[0]} & A_{22}^{[0]} \end{bmatrix}. \tag{6.2}$$

PROPOSITION 6.1. *Let* $f \in H_{k+1}^n(a,b)$. *We set*

$$f =: f^{[0]} =: \begin{bmatrix} y^{[0]} \\ z^{[0]} \end{bmatrix} \in H_{k+1}^p(a,b) \times H_{k+1}^1(a,b)$$

and assume that $y^{[0]\prime} - A_{11}^{[0]} y^{[0]} - A_{12}^{[0]} z^{[0]} = 0$.

Then there are $f^{[j]} \in H_{k+1-j}^n(a,b)$ *for* $j \in \{1,\dots,k+1\}$ *such that*

$$f^{[j]\prime} - A_0 f^{[j]} - A_1 f^{[j+1]} = 0 \qquad (j = 0,\dots,k). \tag{6.3}$$

PROOF. According to the block structure of A_0 and A_1 we decompose

$$f^{[j]} = \begin{bmatrix} y^{[j]} \\ z^{[j]} \end{bmatrix}.$$

We state that the $f^{[j]}$ are solutions of (6.3) iff the $y^{[j]}$ and $z^{[j]}$ fulfil the equations

$$\begin{cases} y^{[j]\prime} - A_{11}^{[0]} y^{[j]} - A_{12}^{[0]} z^{[j]} = 0, \\ z^{[j+1]} = \Omega_1^{-1}\left\{z^{[j]\prime} - A_{21}^{[0]} y^{[j]} - A_{22}^{[0]} z^{[j]}\right\} \end{cases} \tag{6.4}$$

for $j = 0,\dots,k$. By assumption the first equation holds for $j = 0$. The second equation for $j = 0$ uniquely defines $z^{[1]}$ and it is easy to see that $z^{[1]} \in H_k^1(a,b)$. Then we define $y^{[1]}$ to be a solution of the first equation for $j = 1$ and it follows that $y^{[1]} \in H_{k+1}^p(a,b)$. In this manner we continue until we have defined $y^{[k]}$ and $z^{[k+1]}$. Finally we set $y^{[k+1]} := 0$.

By assumption II), T^R is a polynomial, say of order q, i.e.

$$(6.5)\qquad T^R(\lambda) = \sum_{r=0}^{q} \lambda^r T_r^R$$

with $T_r^R \in L(H_1^n(a,b), \mathbb{C}^n)$ for $r = 0,\dots,q$.

THEOREM 6.2. *Let* $|\ |$ *be a continuous norm on* $H_1^n(a,b)$. *Assume that* T *is* $|\ |$*-regular of order* (p,p') *with respect to the curves* Γ_ν $(\nu \in \mathbb{N})$. *Suppose that* $k \geq \max\{p,p'\}$ *where* k *is the natural number given in the assumption I). Let* $f \in H_{k+1}^n(a,b)$. *We set*

$$f =: f^{[0]} =: \begin{bmatrix} y^{[0]} \\ z^{[0]} \end{bmatrix} \in H_{k+1}^p(a,b) \times H_{k+1}^1(a,b)$$

and assume that $y^{[0]\prime} - A_{11}^{[0]} y^{[0]} - A_{12}^{[0]} z^{[0]} = 0$. *For* $j \in \{1,\dots,k+1\}$ *choose functions* $f^{[j]} \in H_{k+1-j}^n(a,b)$ *such that the equations (6.3) hold. Suppose that the "boundary conditions"*

$$(6.6)\qquad \sum_{r=0}^{\min\{q,k-j+1\}} T_r^R f^{[j+r-1]} = 0 \qquad (j = 1,\dots,p')$$

are satisfied.

Under these assumptions the expansion

$$(6.7)\qquad \begin{cases} f = \lim\limits_{\nu\to\infty} \sum\limits_{\mu\in\sigma(T)\cap \operatorname{int}\Gamma_\nu} \Big\{-(\operatorname{res}_\mu R_1)A_1 f \\ \qquad + \operatorname{res}_{\lambda=\mu}\Big(R_2(\lambda) \sum\limits_{j=0}^{q-1} \lambda^j \sum\limits_{r=j+1}^{\min\{q,k+j+1\}} T_r^R f^{[r-j-1]}\Big)\Big\} \end{cases}$$

holds in $(H_1^n(a,b), |\ |)$.

PROOF. The relationship (4.3) and Proposition 6.1 lead to

$$R_1(\lambda)A_1 f^{[j]} = -\frac{1}{\lambda} f^{[j]} + \frac{1}{\lambda} R_1(\lambda)A_1 f^{[j+1]} + \frac{1}{\lambda} R_2(\lambda) T^R(\lambda) f^{[j]}$$

for $j = 0,\dots,k$. A recursive substitution for $j = 0,\dots,k$ yields

$$R_1(\lambda)A_1 f = -\sum_{j=0}^{k} \lambda^{-j-1} f^{[j]} + \lambda^{-k-1} R_1(\lambda)A_1 f^{[k+1]} +$$

$$+\sum_{j=0}^{k}\lambda^{-j-1}R_2(\lambda)T^R(\lambda)f^{[j]}.$$

We integrate along the curves Γ_ν and obtain

$$(6.8)\quad \begin{cases} f = -\sum\limits_{\mu\in\sigma(T)\cap\mathrm{int}\Gamma_\nu}(\mathrm{res}_\mu R_1)A_1 f + \frac{1}{2\pi i}\int\limits_{\Gamma_\nu}\sum\limits_{j=0}^{k}\lambda^{-j-1}R_2(\lambda) * \\ \quad * T^R(\lambda)f^{[j]}d\lambda + \frac{1}{2\pi i}\int\limits_{\Gamma_\nu}\lambda^{-k-1}R_1(\lambda)A_1 f^{[k+1]}d\lambda. \end{cases}$$

An easy calculation yields

$$\sum_{j=0}^{k}\lambda^{-j-1}T^R(\lambda)f^{[j]} = \sum_{j=0}^{k}\sum_{r=0}^{q}\lambda^{r-j-1}T_r^R f^{[j]}$$

$$= \sum_{r=0}^{q}\sum_{j-r+1=-r+1}^{k-r+1}\lambda^{r-j-1}T_r^R f^{[j]}$$

$$= \sum_{r=0}^{q}\sum_{j=-r+1}^{k-r+1}\lambda^{-j}T_r^R f^{[j+r-1]}$$

$$= \sum_{j=-q+1}^{k+1}\lambda^{-j}\sum_{r=\max\{0,-j+1\}}^{\min\{q,k-j+1\}}T_r^R f^{[j+r-1]}$$

$$= \sum_{j=-q+1}^{0}\lambda^{-j}\sum_{r=-j+1}^{\min\{q,k-j+1\}}T_r^R f^{[j+r-1]}$$

$$+\sum_{j=1}^{k+1}\lambda^{-j}\sum_{r=0}^{\min\{q,k-j+1\}}T_r^R f^{[j+r-1]}\ .$$

This equation, the conditions (6.6) and the relationship (6.8) lead to

$$f = \sum_{\mu\in\sigma(T)\cap\mathrm{int}\Gamma_\nu}\Big\{-(\mathrm{res}_\mu R_1)A_1 f + \mathrm{res}_{\lambda=\mu}\Big(R_2(\lambda)\sum_{j=0}^{q-1}\lambda^j *$$

$$* \sum_{r=j+1}^{\min\{q,k+j+1\}}T_r^R f^{[r-j-1]}\Big)\Big\} + \frac{1}{2\pi i}\int_{\Gamma_\nu}\sum_{j=p'+1}^{k+1}\lambda^{-j}R_2(\lambda) *$$

$$* \sum_{r=0}^{\min\{q,k-j+1\}} T_r^R f^{[j+r-1]} d\lambda + \frac{1}{2\pi i} \int_{\Gamma_\nu} \lambda^{-k-1} R_1(\lambda) A_1 f^{[k+1]} d\lambda .$$

Thus the regularity of T proves the theorem.

REMARK 6.3. Let $\max\{q,q+p'\} \leq k+1$. Then (6.6) and (6.7) read

$$\sum_{r=0}^{q} T_r^R f^{[j+r-1]} = 0 \qquad (j = 1,\dots,p') \tag{6.6'}$$

and

$$\begin{cases} f = \lim\limits_{\nu\to\infty} \sum\limits_{\mu\in\sigma(T)\cap\mathrm{int}\Gamma_\nu} \Big\{-(\mathrm{res}_\mu R_1) A_1 f \\ \qquad + \mathrm{res}_{\lambda=\mu}\Big(R_2(\lambda) \sum\limits_{j=0}^{q-1} \lambda^j \sum\limits_{r=j+1}^{q} T_r^R f^{[r-j-1]}\Big)\Big\}. \end{cases} \tag{6.7'}$$

By Theorem 2.2 or Theorem 2.4 the residues in (6.7) and (6.7') can be expressed in terms of canonical systems of root functions (eigenvectors and associated vectors) of T and the adjoint operator function T^*. We omit the details and leave the exact formulation to the reader, cf. also [16], section 3.

REFERENCES

1. Adams, R.A.: Sobolev spaces, Academic Press, New York - San Francisco - London, 1975.

2. Braaksma, B.L.J.: Global reduction of linear differential systems involving a small singular parameter, SIAM J. Math. Anal. 2 (1971), 149-165.

3. Cole, R.H.: General boundary conditions for an ordinary linear differential system, Trans. Amer. Math. Soc. 111 (1964), 521-550.

4. Eberhard, W. and Freiling, G.: Nicht S-hermitische Rand- und Eigenwertprobleme, Math. Z. 133 (1973), 187-202.

5. Eberhard, W. and Freiling, G.: Das Verhalten der Greenschen Matrix und der Entwicklungen nach Eigenfunktionen N-regulärer Eigenwertprobleme, Math. Z. 136 (1974), 13-30.

6. Eberhard, W. and Freiling, G.: Stone-reguläre Eigenwertprobleme, Math. Z. 160 (1978), 139-161.

7. Freiling, G.: Reguläre Eigenwertprobleme mit Mehrpunkt-Integral-Randbedingungen, Math. Z. 171 (1980), 113-131.

8. Gohberg, I.C. and Sigal, E.I.: An operator generalization of the logarithmic residue theorem and Rouché's theorem (Russian), Mat. Sb. (N.S.) 84(126) (1971), 607-629, Engl. transl.: Math. USSR-Sb. 13 (1971), 603-625.

9. Heisecke, G.: Rand-Eigenwertprobleme $N(y)=\lambda P(y)$ bei λ-abhängigen Randbedingungen, Mitt. Math. Sem. Gießen No. 145 (1980), 1-74.

10. Heisecke, G.: Rand-Eigenwertprobleme $N(y)=\lambda P(y)$ bei λ-abhängigen Randbedingungen, ZAMM 61 (1981), T242-T244.

11. Kaashoek, M.A.: Analytic equivalence of the boundary eigenvalue operator function and its characteristic matrix function, to appear in: Integral Equations Operator Theory.

12. Keldyš, M.V.: On the eigenvalues and eigenfunctions of certain classes of nonselfadjoint equations (Russian), Dokl. Akad. Nauk SSSR 77 (1951), 11-14.

13. Keldyš, M.V.: On the completeness of eigenfunctions of some classes of nonselfadjoint linear equations (Russian), Uspehi Mat. Nauk 26 (1971), 15-41, Engl. transl.: Russian Math. Surveys 26 (1971), 15-44.

14. Langer. R.E.: The boundary problem of an ordinary linear differential system in the complex domain, Trans. Amer. Math. Soc. 46(1939), 151-190.

15. Mennicken, R.: Spectral theory for certain operator polynomials, in: Functional Analysis, Holomorphy and Approximation Theory II, ed.: G. Zapata, North-Holland Math. Studies 86 (Notas de Matemática 92) (1984), 203-243.

16. Mennicken, R. and Möller, M.: Root functions, eigenvectors, associated vectors and the inverse of a holomorphic operator function, Arch. Math. (Basel) 42 (1984), 455-463.

17. Mennicken, R. and Möller, M.: Root functions of boundary eigenvalue operator functions, to appear in: Integral Equations Operator Theory.

18. Naimark, M.A.: Linear differential operators, part I: Elementary theory of linear differential operators,

Ungar, New York, 1967.

19. Wasow, W.: Asymptotic expansions for ordinary differential equations, Interscience Publishers, New York - London - Sydney, 1965.

Reinhard Mennicken and Manfred Möller
NWF I - Mathematik
Universität Regensburg
Universitätsstr. 31
8400 Regensburg
West Germany

Operator Theory:
Advances and Applications, Vol.19

SYSTEM THEORY FOR LOSSLESS WAVE SCATTERING*

C.V.K. Prabhakara Rao and P. Dewilde

In this paper, a system theory is developed for discrete time systems which preserve the L^2 norm of input and output signals. In many applications such as low sensitivity filter design and least squares estimation of stochastic processes, a central problem called the "Lossless Inverse Scattering" (LIS) problem arises. Using system theory, solutions are defined directly in terms of paramcters derived from the data. In the course of the paper some classical system theoretical concepts, like the canonical construction of a state space from the I/O map, and pertinent information concerning scattering theory are reviewed so as to make the paper selfcontained.

1. INTRODUCTION

In this paper, we study discrete-time systems, which preserve the L^2-norms of the input and output signals. Moreover, the systems considered will be analogous to transmission-lines with forward and backward going waves. Such systems arise, for example, in the obvious context of discretized physical wave scattering systems [1, 2], as well as in low sensitivity digital filter design [3, 4], interpolation problems [5] and least squares estimation of stochastic processes [6].

Interestingly enough, in all the above mentioned cases, the relevant properties of the lossless system are best brought out by analyzing, not the given system itself, but a transformed version of it, namely a J-lossless system, where J is a signature matrix of the form $I_p \oplus -I_q$, p,q >0. In this publication, we shall restrict ourselves to a class of J-lossless systems, namely those whose transfer function is analytic and bounded in the unit disc of the complex plane. (This class is sufficiently large to cover many practical situations). After reviewing certain well known but important properties of such transfer functions, we shall develop their state space theory. Following the Nerode definition, we characterize the state and null spaces as subspaces of input and output signal spaces of vector valued analytic functions. This approach leads to a description of the state space, bringing out the essential parameters of the system and clearly exhibits the applicability of lossless systems to the above mentioned contexts, e.g. interpolation problems.

A problem that is common and central to most applications is what has been termed in recent literature as the Lossless Inverse Scattering (LIS) problem [6, 7, 8, 9, 10]. Here one seeks a lossless system that produces a wave scattering behavior which

* This research has been partially supported by Z.W.O. (the Netherlands Foundation for Fundamental Research) under contract no. R69-56.

approximates/matches a given scattering function. It turns out that using the theory developed, the state space of the lossless solution can be determined directly in terms of the given scattering data. In the next step, the solution itself can be retrieved in one of many ways: 1) RKHS method, 2) Systems Realization Method, and 3) Recursive Method. Each equivalent but different approach sheds a new light on the situation thereby providing useful connections and identities.

2. LOSSLESS WAVE SCATTERING SYSTEMS

2.1 Signal Spaces

Let a(t) be a mx1 vector valued discrete time signal for $-\infty \leqslant t \leqslant \infty$ and let $A(e^{i\theta})$ be its formal Fourier representation defined below:

$$A(e^{i\theta}) := \sum_t a(t)e^{it\theta} \tag{1}$$

Further, define the *energy* of the signal a(t) formally as:

$$E\{a\} := \sum_{-\infty}^{\infty} a^*(t)a(t) = \int_{-\pi}^{\pi} A^*(e^{i\theta})A(e^{i\theta})\, d\theta/2\pi \tag{2}$$

where the superscript * denotes conjugate transposition. Throughout, we will only consider signals with *finite energy*. Each component of such a vectorial signal a(t) is a scalar finite energy signal whose Fourier transform $A(e^{i\theta})$ is known to be well defined. Furthermore, the space of such signals is l^2 (in time domain) or L^2 (in Fourier domain). Accordingly, we shall refer to the space of finite energy mx1 signals as l_m^2 or L_m^2.

Two important subspaces of L_m^2 are H_m^2 and K_m^2. H_m^2 is the space of Fourier representations of time signals that are zero for t<0, and K_m^2 is the space of Fourier representations of time signals that are zero for t⩾0. The scalar components of a signal $A(e^{i\theta})$ in H_m^2 belong to the well known Hardy space H^2, so that we call H_m^2 the mx1 vectorial Hardy space. Such functions can be extended analytically into the unit disc D={z: |z| < 1} of the complex plane as:

$$A(z) = \sum_0^{\infty} a(t)z^t \quad \text{and} \quad \lim_{r\to 1} A(re^{i\theta}) = A(e^{i\theta}) \ \ a.e. \tag{3}$$

where a.e. stands for "almost everywhere".

Similarly, we shall call K_m^2 the mx1 vectorial *conjugate* Hardy space and functions in K_m^2 can be extended analytically outside the unit disc E ={z: |z| > 1} as:

$$A(z) = \sum_{-\infty}^{-1} a(t)z^t \quad \text{and} \quad \lim_{r\to 1} A(re^{i\theta}) = A(e^{i\theta}) \ \ a.e. \tag{4}$$

Note that L_m^2, H_m^2 and K_m^2 are Hilbert spaces with respect to the following innerproduct: for f(z) and g(z) in L_m^2,

$$\langle f\ , g \rangle := \int_{-\pi}^{\pi} g^*(e^{i\theta})f(e^{i\theta})\, d\theta/2\pi \tag{5}$$

The theory of these spaces can be found in [11] for the scalar functions, and in [12] for vectorial functions.

2.2 Lossless Systems

Abstractly, a system is a map between an input signal space and an output signal space. Consider a system S that maps L_m^2 into L_n^2 with n$\geq$m. S is said to be a *Lossless Map* if for any input signal in L_m^2, the energy of the input signal equals the energy of the corresponding output signal. Moreover, the system is said to be a *causal map* if for all t, and for all input-output pairs {u(t),y(t)}, the following inequality holds:

$$\sum_{-\infty}^{t} u^*(\tau)u(\tau) \geq \sum_{-\infty}^{t} y^*(\tau)y(\tau) \tag{6}$$

Due to a theorem of Bochner and Chandrasekharan [13], such a system can be described by an nxm *matrix transfer function* $\Sigma(z)$, such that

1. $\Sigma(z)$ is analytic in D and the radial limits $\Sigma(e^{i\theta}) = \lim_{r\to 1} \Sigma(re^{i\theta})$ exist for almost all θ.
2. $\Sigma(z)$ is contractive in D, i.e.

$$I_m - \Sigma^*(z)\Sigma(z) \geq 0 \ \ in\ D \tag{7a}$$

3. $\Sigma(z)$ is unitary on the unit circle, i.e.

$$I_m - \Sigma^*(e^{i\theta})\Sigma(e^{i\theta}) = 0 \ \ a.e.\ on\ T \tag{7b}$$

By convention, a (semi)positive definite Hermitian matrix A($=A^*$) is written as (A$\geq$0) A$>$0.
$\Sigma(z)$ is said to be an *all-pass* function or a *phase function* in Electrical Network theory and an *isometry* in Complex Function theory. If m=n, then $\Sigma(z)$ has square dimensions and is called an *inner* function. In this case, we also have that

$$I_m - \Sigma(z)\Sigma^*(z) \geq 0 \qquad |z|<1 \tag{8a}$$

$$= 0 \ \ a.e.\ \ |z|=1 \tag{8b}$$

In Complex Function theory, $\Sigma(z)$ is said to belong to H^{∞}_{mxm}, namely the space of mxm matrices whose elements belong to the Hardy space H^{∞}, i.e. are bounded analytic functions in D [11, 14, 15]. The elements can be rational or nonrational depending on the system map. Strictly speaking, the transfer function $\Sigma(z)$ is defined only in D and a.e. on T of the complex plane. We can use the following extension of $\Sigma(z)$ into E. Firstly, define the so called para-Hermitian conjugate of $\Sigma(z)$ as:

$$\Sigma_*(z) := \Sigma^*(1/z^*) \tag{9}$$

$\Sigma_*(z)$ is well defined and analytic in E and possesses radial limits a.e. on T. Now we define $\Sigma'(z) = \Sigma_*^{-1}(z)$ as the extension of $\Sigma(z)$ into E. Since $\Sigma'(z)$ is meromorphic in E and its radial limits equal those of $\Sigma(z)$ a.e. on T, thanks to (7b), $\Sigma'(z)$ is said to be the pseudo-meromorphic extension of $\Sigma(z)$. If $\Sigma(z)$ is a rational function, then it is analytic a.e. in the complex plane as also the extension $\Sigma'(z)$ and they are equal to each other a.e. in the complex plane.

Substituting $[\Sigma']^{-1}$ for Σ^* in (7b), and using the identity theorem for analytic functions, [16], we get

$$I_m - \Sigma_*(z)\Sigma(z) = 0 \ \ a.e. \tag{10}$$

Using identical arguments, we also get

$$I_m - \Sigma(z)\Sigma_*(z) = 0 \ \ a.e. \tag{11}$$

2.3 Lossless Wave Scattering Systems

In this section, we will consider discrete time systems, which are analogous to lossless transmission lines. See Fig-1. We shall call them Lossless Wave Scattering Systems (LWSSs).

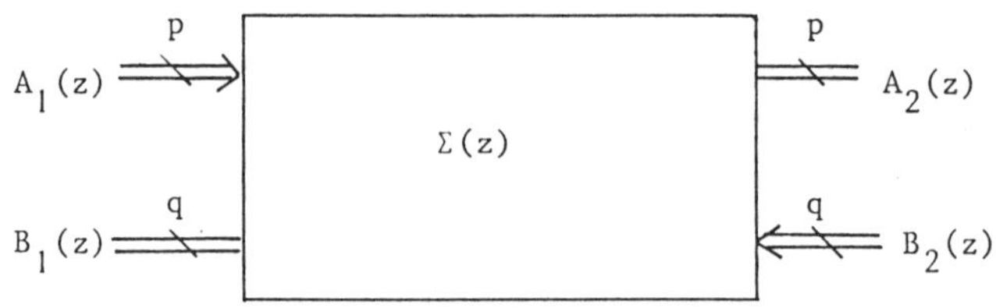

Figure 1. Lossless Wave Scattering System

The system can be viewed as processing a right to left going wave $A_1 \rightarrow A_2$, and a left to right going wave $B_2 \rightarrow B_1$. With reference to the system, A_1, B_2 are said to be the incident waves and A_2,B_1 are said to be the scattered waves. We shall refer to A_1 & B_1 as belonging to port-1 and A_2 & B_2 as belonging to port-2.
Let the inputs and outputs be related as follows:

$$\begin{bmatrix} A_2(z) \\ B_1(z) \end{bmatrix} = \Sigma(z) \begin{bmatrix} A_1(z) \\ B_2(z) \end{bmatrix} \tag{12}$$

so that $\Sigma(z)$ is an mxm (m=p+q) matrix transfer function called *Scattering Matrix* (SM). We assume throughout this paper that Σ is an inner or a lossless function, satisfying the strict inequalities in (7a) and (8a), that is:

$$I_m - \Sigma^*(z)\Sigma(z) > 0 \text{ for } |z|<1 \tag{13a}$$

$$= 0 \text{ for } |z|=1 \text{ } a.e. \tag{13b}$$

and consequently,

$$I_m - \Sigma(z)\Sigma^*(z) > 0 \text{ for } |z|<1 \tag{14a}$$

$$= 0 \text{ for } |z|=1 \text{ } a.e. \tag{14b}$$

Remark: This is no serious restriction because, if the strict inequality is not satisfied in (13a), then there exists a point α in D so that

$$rank \left[I_m - \Sigma^*(\alpha)\Sigma(\alpha) \right] = k < m$$

Now due to the maximum modulus theorem for matrix valued analytic functions [17, 18], it follows that:

$$rank \left[I_m - \Sigma^*(z)\Sigma(z) \right] = k < m \text{ for } all \ |z|<1$$

Now it follows that $\Sigma(z)$ can be decomposed as:

$$\Sigma(z) = U \begin{bmatrix} \Sigma_k(z) & \\ & I_{m-k} \end{bmatrix} V$$

where U and V are unitary matrices and $\Sigma_k(z)$ satisfies (13). Thus it suffices to consider $\Sigma(z)$ satisfying the strict inequality as in (13a).

The wave scattering system $\Sigma(z)$ can be characterized by four partial transfers, obtained by partitioning $\Sigma(z)$ as follows:

$$\Sigma(z) = \begin{bmatrix} T(z) & S_c(z) \\ S(z) & T_c(z) \end{bmatrix} \tag{15}$$

See Fig-2.

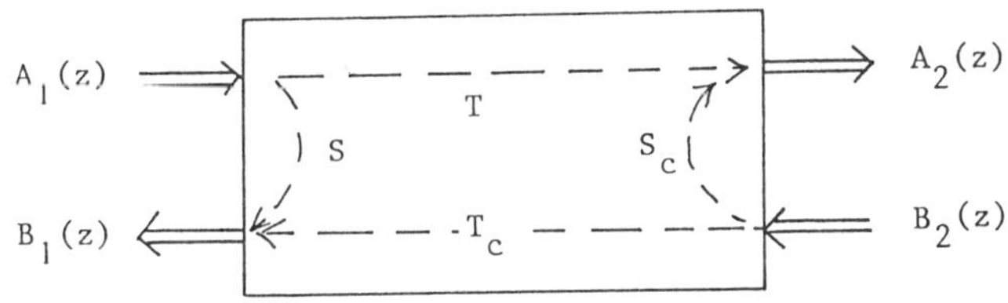

Figure 2. Partial Transfers of a Wave Scattering System

T(z) & T_c(z) are called *Transmission* and *complementary Transmission* functions respectively, while S(z) & S_c(z) are called *Scattering* and *complementary Scattering* functions respectively. Substituting (15) in (13), it follows easily that:

$$I - f^*(z)f(z) > 0 \text{ for } |z| \leqslant 1 \tag{16}$$

where f(z) can be S,S_c,T or T_c and I indicates unit matrix of appropriate dimensions. An analytic function f(z) satisfying (16) belongs to H^∞ and is said to be a *contractive* or a *Schur* function [19]. Such functions occur profusely in Electrical Network Theory [20].

Since $\Sigma(z)$ is a lossless system, the four partial transfers are tightly coupled together. This is best revealed by substituting (15) into (10) and (11), which yields:

$$TT_* + S_cS_{c*} = I_p ; T_cT_{c*} + SS_* = I_q ; TS_* + S_cT_{c*} = 0_{pxq} \tag{17a}$$

$$T_*T + S_*S = I_p ; T_{c*}T_c + S_{c*}S_c = I_q ; T_*S_c + S_*T_c = 0_{pxq} \tag{17b}$$

The following property on the location of the zeros of T and T_c will be of use later. Outer matrix functions have been defined in complete generality in the literature, see e.g. [12]. For our purposes, it will be enough to characterize outer functions as square rational functions which are analytic in D with no zeros in D. (Zeros in T are allowed.) Such functions are also called *minimum phase functions*. In the rational case, a maximum phase function may be defined as analytic in D with no zeros in E.

Proposition 1: If T_c(z) is a minimum phase function, then T(z) is a maximum phase function.

Proof: Suppose the proposition is not true. Then there exists a point α in D and a vector $\xi \epsilon C^m$ such that

$$T(1/\alpha^*)\xi = 0 \quad \xi \neq 0$$

Evaluating the identity $S_{c*}T + T_{c*}S = 0$ at $z = 1/\alpha^*$ and noting that $S_c(\alpha)$ and $T_c(\alpha)$ are finite as $|\alpha|<1$, we obtain

$$S(1/\alpha^*)\xi = 0$$

Substituting the above two equations in $T_* T\xi + S_* S\xi = \xi$ and noting that $T(\alpha)$ and $S(\alpha)$ are finite for $|\alpha|<1$, we obtain the absurd relation $0=\xi$. This disproves the supposition and proves the proposition.

[]

The converse is not generally true, as the example $\Sigma(z) = zI_2$ shows.
The zeros of T(z) are defined to be the *Transmission Zeros* (TZs).

2.4 Chain Scattering Matrix Description

It will be of use to describe the LWSS alternately by the so called chain scattering matrix (CSM). Formally, CSM relates the incident and scattered waves in Fig-1 as follows:

$$\begin{bmatrix} A_2(z) \\ B_2(z) \end{bmatrix} = \Theta(z) \begin{bmatrix} A_1(z) \\ B_1(z) \end{bmatrix} \tag{18}$$

$\Theta(z)$ can be thought of as describing a *partial inverse* of $\Sigma(z)$ and is related to it as:

$$\Theta(z) = \begin{bmatrix} I & S_c \\ & I \end{bmatrix} \begin{bmatrix} T & \\ & T_c^{-1} \end{bmatrix} \begin{bmatrix} I & \\ -S & I \end{bmatrix} \tag{19}$$

It can be easily verified that the losslessness of $\Sigma(z)$ given by (13) and (14) translates as follows to $\Theta(z)$:

$$Let \;\; J = \begin{bmatrix} I_p & \\ & -I_q \end{bmatrix} \tag{20}$$

Then

$$J - \Theta^*(z)J\Theta(z) > 0 \quad |z|<1 \tag{21}$$
$$= 0 \quad |z|=1$$

and

$$J - \Theta(z)J\Theta^*(z) > 0 \quad |z|<1 \tag{22}$$
$$= 0 \quad |z|=1$$

Such functions are called *J-lossless* or *J-inner* functions.

Note however that due to the presence of T_c^{-1} in (19) $\Theta(z)$ need not be analytic in D. From now on, we assume that $T_c(z)$ has no zeros in D or on T. This makes $\Theta(z)$ analytic in D with no poles on T. Although T_c is a bounded (H^∞) function, its inverse is not bounded in general. (In fact, we can only say that it belongs to the Nevanlinna class of analytic functions [15,p.16,Th.2.1]). Thus $\Theta(z)$ is not a bounded function in general. However, if $\Theta(z)$ is a rational function, then it is analytic almost everywhere in the complex plane and hence automatically bounded in a neighbourhood of $D \cup T$.

2.5 Chain Matrix Description

When the size of the A-waves equals the size of the B-waves in the LWSS depicted in Fig-1, i.e. p=q, another useful description comes about by transforming the input and output variables. Consider the following constant transformation matrix M:

$$M = \frac{1}{2^{1/2}} \begin{bmatrix} I_p & I_p \\ I_p & -I_p \end{bmatrix} \tag{23}$$

Define the Chain Matrix (CM) U(z) as follows:

$$U(z) = M\Theta(z)M \tag{24}$$

It is easily verified that the J-lossless property of Θ(z) translates to the J_1-losslessness for U(z) as:

$$\begin{aligned} J_1 - U^*(z)J_1U(z) &> 0 \quad |z|<1 \\ &= 0 \quad |z|=1 \end{aligned} \tag{25}$$

and

$$\begin{aligned} J_1 - U(z)J_1U^*(z) &> 0 \quad |z|<1 \\ &= 0 \quad |z|=1 \end{aligned} \tag{26}$$

$$where \quad J_1 = MJM = \begin{bmatrix} 0 & I_p \\ I_p & 0 \end{bmatrix} \tag{27}$$

Since M is a constant matrix, the analytic properties of U(z) are the same as those of Θ(z). Thus assuming that T_c has no zeros in D or on T renders U(z) analytic in D with no poles on T, while if T_c is also rational then U(z) is analytic in a neighbourhood of the closed unit disc $D \cup T$.

2.6 Scalar Example

Before proceeding further, we consider the case when the partial transfers T, T_c, S, S_c are scalar rational functions. The general form of a 2x2 lossless Scattering Matrix (SM) is well known for the continuous time case in Electrical Network theory [18,p.277]. It is easy to transport this result to the discrete time case, which we give below. Also the corresponding forms of Θ(z) and U(z) defined via (19) and (24) are given, all parametrized in a form convenient for our needs.

$$\Sigma(z) = \frac{1}{D+Q} \begin{bmatrix} 2P^\# & Q^\# - D^\# \\ D-Q & 2P \end{bmatrix} \tag{28a}$$

$$\Theta(z) = \frac{1}{2P} \begin{bmatrix} D^\# + Q^\# & D^\# - Q^\# \\ D-Q & D+Q \end{bmatrix} \tag{28b}$$

$$U(z) = \frac{1}{2P} \begin{bmatrix} Q^\# + Q & D^\# - D \\ Q^\# - Q & D^\# + D \end{bmatrix} \tag{28c}$$

In the above formulas, P,Q,D are polynomials, such that D+Q has roots in E (making Σ(z) analytic in $D \cup T$). The superscript # is defined below, eg for P(z):

$$P^\#(z) = z^N P_*(z) = z^N P^*(1/z^*)$$

where N is an integer greater than or equal to the highest degree of P,Q,D and the P,Q,D polynomials satisfy the Lyapunov equation:

$$DQ^{\#} + QD^{\#} = 2\,PP^{\#} \tag{29}$$

The most general form of a lossless 2x2 SM is obtained by multiplying $\Sigma(z)$ in (28a) from the left and right by constant diagonal unitary matrices. In future, we use the forms given in (28) with P,Q,D all of degree N to illustrate various ideas clearly. From our assumption on the zeros of T_c, it follows that P(z) can be represented as:

$$P(z) = \prod_{k=1}^{N} (1 - \omega_k^* z) \quad |\omega_k| < 1 \tag{30}$$

3. PROPERTIES AND APPLICATIONS OF LWSSs

In this section, we highlight the most important properties of LWSSs, and suggest applications. This will lead us to the Lossless Inverse Scattering problem which forms the content of Section-4 onwards.

3.1 Interpolation

Consider the LWSS in Fig-2 terminated by at the left port by a Schur/ contractive function (sometimes referred to as a *passive load*) $S_L(z)$ as shown in Fig-3.

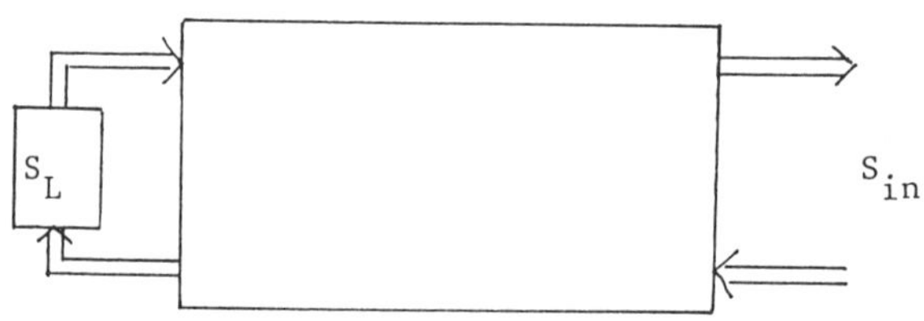

Figure 3. LWSS terminated in a Passive Load

The transfer at the right port from B_2 to A_2 is easily calculated as:

$$S_{in} = S_c + T\,[I - S_L S]^{-1} S_L\, T_c \tag{31}$$

$S_{in}(z)$ is clearly a Schur function on physical grounds, but can also be directly verified. We will now show that $S_c(z)$ belonging to the LWSS interpolates $S_{in}(z)$ in the following sense.

Under the assumption that $T_c(z)$ has its zeros in E, it follows from Prop-1 that all zeros of T lie in D. Consider the inner-outer factorization of the pxp T(z) [11, p.370].

$$T(z) = V(z)\,G(z) \tag{32}$$

where V(z) is an inner function whose zeros are the zeros of T(z) (in D) and G(z) is an outer function which, by definition, has no zeros in D. Moreover, it has no zeros on T as well, due to the fact that T itself has no zeros on T.

Define now the so called *interpolation space* h :

$$h = H_p^2 \ominus VH_p^2 \tag{33}$$

where H_p^2 denotes the px1 vectorial Hardy space, and $\ominus$ denotes orthogonal complementation with respect to the $L^2(d\theta)$ innerproduct. (see (5)).

Interpolation Theorem: For all passive loads $S_L(z)$ in (31),

$$P S_{in}^* f = P S_c^* f \quad \text{for } all \ f \in h \tag{34}$$

where P denotes orthogonal projection from L_q^2 onto H_q^2 with respect to the $L^2(d\theta)$ innerproduct.

Proof: From (31), we have

$$P S_{in}^* f = P S_c^* f + P X^* f$$

where $X = T\,[I-S_L S]^{-1} S_L T_c$. We will prove that for all f in h, $X^* f$ is orthogonal to H_q^2. Indeed, let $g(z) \in H_q^2$:

$$< g \ , X^* f > \ = \ < Xg \ , f > \ = \ < Vh \ , f > \ = 0 \ \textit{from}\ (33)$$

where $h = G\,[I-S_L S]^{-1} S_L T_c$ using (32). Note that h belongs to H_p^2 because the inverse is well defined due to the contractivity of S and S_L.

[]

This abstract idea of interpolation becomes concrete by considering a simple scalar example.

Case 1

Assume that the points ω_k in (30) are distinct. Then, the inner-outer factorization (32) yields

$$T = \frac{2P^\#}{D+Q} = \frac{P^\#}{P}\,\frac{2P}{D+Q}$$

so that $V = P^\# / P$ and

$$h = span\,\{1/(1-\omega_k^* z): \ k = 1..N\}$$

Substituting in the interpolation equation (34) leads to

$$S_{in}(z) = S_c(z) \quad z = \omega_1 \cdots \omega_N \tag{35}$$

[]

Case 2

Assume that all the points ω_k in (30) coalesce at 0. Then (32) becomes

$$T(z) = z^N \, \frac{2}{D+Q}$$

so that $h = span\,\{1, z, ..., z^{N-1}\}$. Evaluating the interpolation equation gives:

$$S_{in}(z) - S_c(z) = z^N \, S'(z) \tag{36}$$

where S' (z) is an analytic function at z=0, indicating that the first N terms in the Taylor's series of S_{in} and S_c match.

[]

This type of interpolation is called Nevanlinna-Pick (NP) interpolation and has been studied extensively by a number of researchers. [21, 22, 23].

3.2 Spectral Factorization

Here, we assume that p=q so that T, S, T_c, S_c are all square functions. Consider the Cayley transform of S(z) defined below:

$$C(z) = [I+S][I-S]^{-1} \tag{37}$$

Note that C(z) is analytic in D (and belongs to the Nevanlinna class). Moreover, the contractivity of S(z) in D makes the real part of C(z) positive there:

$$\tfrac{1}{2}[C(z) + C^*(z)] = [I-S]^{-1}[I-SS^*][I-S]^{-1*}$$
$$> 0 \quad |z|<1 \tag{38}$$

Thus C(z) is called a p.r. (positive real) or a Caratheodory function. It is well known in complex function theory [11, 14] that C(z) has the following integral representation due to Herglotz:

$$C(z) = \int_{-\pi}^{\pi} \frac{e^{i\theta}+z}{e^{i\theta}-z} \frac{d\mu}{2\pi} + i \ \mathrm{Im}\{C(0)\} \tag{39}$$

where μ is a finite positive measure on T and Im indicates the imaginary part. Moreover, it can be argued * that in the present context μ is absolutely continuous, so that

$$d\mu = W(\theta)\, d\theta \tag{40}$$

with $W(\theta)$ Hermitian and semi-positive definite. Furthermore, $W(\theta)$ can be obtained as:

$$W(\theta) = \lim_{r\to 1} \tfrac{1}{2}\left[C(re^{i\theta}) + C^*(re^{i\theta})\right] \tag{41}$$

Note that $W(\theta)$ can be considered as the power spectral density of a stationary stochastic sequence. An important problem in linear least squares estimation of stochastic sequences is the spectral factorization of $W(\theta)$. These spectral factors can be retrieved from the partial transfers of the LWSS thanks to the losslessness property.

$$\text{Let} \ \ F(z) = [I-S]^{-1}T_c, \ \ R(z) = T[I-S]^{-1} \tag{42}$$

It is easily verified that

$$W(\theta) = F(e^{i\theta})F^*(e^{i\theta}) = R^*(e^{i\theta})R(e^{i\theta}) \tag{43}$$

Noting from (42) that F and R are analytic in D and have their zeros in E and D respectively (due to the corresponding properties of T_c and T respectively), F and R are called the minimum and maximum phase spectral factors of $W(\theta)$.

Similarly, we can determine the minimum phase spectral factor R_c and the maximum phase spectral factor F_c of $W_c(\theta)$ obtained from $C_c(z) = [I+S_c][I-S_c]^{-1}$, as follows:

* This follows from the assumption that T_c and hence T due to Prop.1 have no zeros on T. Thus from (17) S(z) is strictly contractive on T so that C(z) has no pole on T. This guarantees that μ is absolutely continuous.

$$F_c(z) = T_c\,[I - S_c]^{-1},\ R_c(z) = [I - S_c]^{-1}\,T \tag{44}$$

$$\text{and}\ \ W_c(\theta) = F_c^* F_c = R_c R_c^* \tag{45}$$

These factors can all be retrieved from the LWSS using appropriate boundary conditions. See Fig-4.

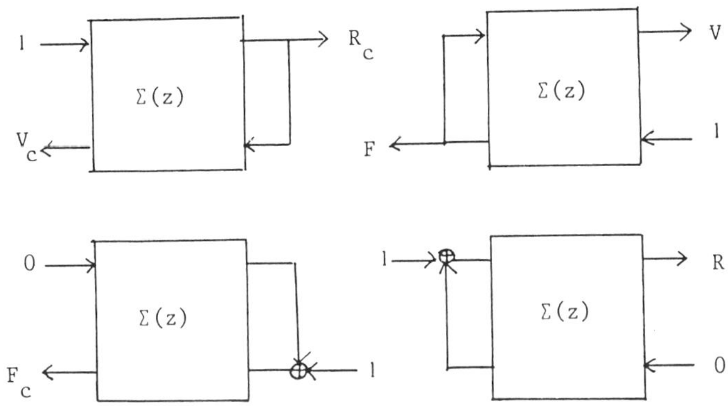

Figure 4. Spectral Factors

3.3 Maximum Entropy

Consider the situation of Fig-3, where the LWSS is terminated by a passive load $S_L(z)$. Let $C_{in} = [I+S_{in}][I-S_{in}]^{-1}$ be the Caratheodory function associated with $S_{in}(z)$ and let $W_{in}(\theta)$ be the corresponding spectral density function related as in (39) and (40). Consider the spectral density function $W_c(\theta)$ defined as in (45) and (44) and produced by the LWSS alone. It can be shown that $W_c(\theta)$ maximizes a certain entropy integral given below:

$$\int_{-\pi}^{\pi} \log \det [W_c(\theta)] d\theta \geqslant \int_{-\pi}^{\pi} \log \det [W_{in}(\theta)] d\theta \tag{46}$$

for all $W_{in}(\theta)$ obtained by different passive loads $S_L(z)$. For a proof, see [7].

The use of the above inequality is the following: Assume that the LWSS is a rational function, while $S_L(z)$ is possibly a nonrational function. The spectral density function $W_{in}(\theta)$ is in general a nonrational function, to which $W_c(\theta)$ (produced by the LWSS alone) can be considered as a rational approximant. The sense of approximation is as in (46) and is known to be a statistically desirable criterion [24].

Lossless Inverse Scattering Problem

From the above discussion, the following problem, named the Lossless Inverse Scattering (LIS) problem emerges as the main problem to be solved. See Fig-5.

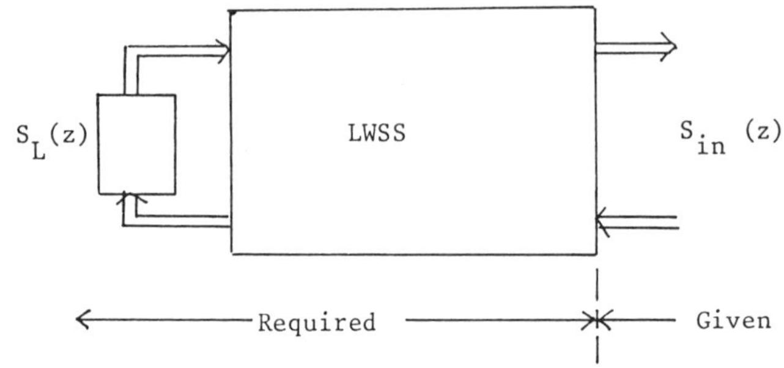

Figure 5. Lossless Inverse Scattering Problem (LIS Problem)

Here $S_{in}(z)$ is a given Schur function (possibly nonrational) and the problem is to find and parametrize all possible representations of $S_{in}(z)$ as a LWSS terminated with a passive load $S_L(z)$. This is equivalent to finding and parametrizing all possible representations for $S_{in}(z)$ as in (31). Having solved such a problem, one can then use the LWSS obtained to produce Nevanlinna-Pick interpolants for $S_{in}(z)$ or maximum entropy approximants to the spectral density $W_{in}(\theta)$ associated with $S_{in}(z)$. Note that for these applications, the exact form of the load $S_L(z)$ is unimportant, so long as its existence is guaranteed. By contrast, the spectral factorization application described is a deterministic problem where we require the load to be 1 as shown in Fig-4. Another example of such a deterministic application is the synthesis of the so called orthogonal digital filters. See [3, 4].

In the rest of the paper, we solve the LIS problem. From the discussion on interpolation it will not be surprising to find that the solutions of are parametrized in terms of the interpolation space defined as in (33). We solve the LIS problem following a system theoretic approach. Accordingly, we begin by characterizing the state space of the CSM of a LWSS.

4. STATE SPACE ANALYSIS OF A J-LOSSLESS SYSTEM

4.1 Definitions and Explicit Formulae

Let $\Theta(z)$ be an mxm (m=p+q) J-lossless function satisfying (21) and parametrized as in (19). Assume that $\Theta(z)$ is a rational function with poles in E, so that it represents a bounded map from L_m^2 to L_m^2. Consider the restricted Input-Output map H taking signals in the strict past (K_m^2) into signals in the future (H_m^2) according to the following rule:

$$f \in K_m^2 \rightarrow Hf = P\,\Theta f \in H_m^2 \tag{47}$$

where P denotes orthogonal projection from L_m^2 into H_m^2 with respect to the $L^2(d\theta)$ innerproduct. See Fig-6.

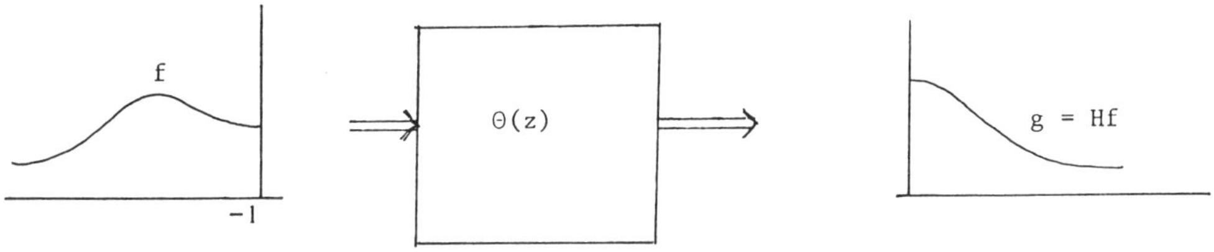

Figure 6. Restricted Input-Output Map

In terms of H, we define the state and null spaces of the system as follows:

$$\textit{Input Nullspace}: \; N = \textit{Kernel } H = \left\{ f \in K_m^2 : \; P\,\Theta f = 0 \right\} \tag{48a}$$

$$\textit{Output Statespace}: \; H^\circ = \textit{Range } H = \left\{ P\,\Theta f : f \in K_m^2 \right\} = P\,\Theta K_m^2 \tag{48b}$$

$$\textit{Input Statespace}: \; H := K_m^2 \ominus J N \tag{48c}$$

$$\textit{Output Nullspace}: \; N^\circ := H_m^2 \ominus J H^\circ \tag{48d}$$

Notes

1. The definitions of N and H° are physically motivated. Signals in N bring the system $\Theta(z)$ at t=0 into the "null state", characterized by the zero output response from t=0 onwards. Similarly, each output response for t≥0 defines a state of the system and their collection forms the "output statespace".
2. The definitions of N° and H are not unique, and our choice (including a J-factor) matches with the fact that $\Theta(z)$ is a J-lossless function. Its consequences are dramatic as will be seen soon.
3. Since $\Theta(z)$ is assumed to be rational, the range of H is a finite dimensional space and hence is automatically closed. In general, one would have to indicate closures in the definition of H°.

We have the following properties of the four spaces:

1) N° and H° are respectively forward and backward shift invariant subspaces in H_m^2, i.e.

$$z N^\circ \subset N^\circ \quad \text{and} \quad P z^{-1} H^\circ \subset H^\circ \tag{49}$$

where "$\subset$" denotes set inclusion.

2) N and H are backward and forward shift invariant subspaces in K_m^2, i.e.

$$z^{-1} N \subset N \quad \text{and} \quad Q z H \subset H \tag{50}$$

where Q denotes orthogonal projection from L_m^2 onto K_m^2 with respect to the $L^2(d\theta)$ innerproduct.

3) Define, what may be called as the generalized backward shift operator in H_m^2, R_ω as:

$$f \in H_m^2, \quad R_\omega f = [f - f(\omega)]/[z-\omega] \in H_m^2 \tag{51}$$

Note that the backward shift operator Pz^{-1} defined in (49) is in fact R_0. Furthermore, it can be easily proved from (49) that

$$R_\omega H^\circ \subset H^\circ \quad \text{for } |\omega|<1 \tag{52}$$

4) The output state space H° can be also expressed as:

$$H^\circ = \left\{ R_0^k \Theta\xi, \ \xi \in C^m, \ k \geqslant 1 \right\} \tag{53}$$

where $R_0^k f = (R_0(R_0..(R_0 f)..))$

Since $\Theta(z)$ is a J-lossless system, the input nullspace and the output statespace defined in (48) can be expressed as follows:

Theorem 1:

$$N = \Theta^{-1}K_m^2, \quad H^\circ = H_m^2 \ominus J\Theta H_m^2 \tag{54}$$

Proof: By definition, if $f \in N$, then $P\,\Theta f = 0$, so that $\Theta f \in K_m^2$ or $f \in \Theta^{-1}K_m^2$. Conversely, if $f \in \Theta^{-1}g$ with $g \in K_m^2$, clearly, $P\,\Theta f = 0$.

As for H°, if $f \in K_m^2$, clearly $P\,\Theta f \in H_m^2$. Moreover,

$$< P\,\Theta f \ , J\Theta H_m^2> \ = \ <\Theta f \ , J\Theta H_m^2> \ = \ <f \ , H_m^2> \ = 0$$

so that $P\,\Theta f$ subset $H_m^2 \ominus J\Theta H_m^2$. Conversely, if $f \in H_m^2 \ominus J\Theta H_m^2$, it is easy to check that $u = \Theta^{-1}f$ belongs to K_m^2, so that $f = P\,\Theta u$ which belongs to H°.

[]

Formulas for the statespaces can be derived in terms of the partial transfers of the LWSS.

Theorem 2: Let $\Theta(z)$ be a J-lossless function satisfying (19) and let it be rational with poles in E. The output state space is given by

$$H^\circ = \begin{bmatrix} I \\ P S_c^* \end{bmatrix} h \tag{55}$$

where h is the interpolation space defined in (33), i.e.

$$h = H_p^2 \ominus VH_p^2$$

and $T = VG$ is the inner-outer factorization of $T(z)$.

Proof: We first prove that the right hand side of (55) is contained in $H_m^2 \ominus J\Theta H_m^2$. It suffices to show that, if $f \in h$,

$$\begin{bmatrix} I \\ S_c^* \end{bmatrix} f \ \perp \ J\Theta \begin{bmatrix} g \\ h \end{bmatrix}, \quad g, h \in H_q^2$$

Indeed,

$$< \begin{bmatrix} I \\ S_c^* \end{bmatrix} f \ , J\Theta \begin{bmatrix} g \\ h \end{bmatrix} > \ = \ < f \ , [I \ S_c] J\Theta \begin{bmatrix} g \\ h \end{bmatrix} >$$

$$= < f \ , Tg > \ = \ < f \ , VGg > \ = \ 0 \ \textit{since} \ f \perp VH_q^2$$

Conversely, consider f $\in K_p^2$, and g $\in K_m^2$, and let

$$\begin{bmatrix} a \\ b \end{bmatrix} = \Theta \begin{bmatrix} f \\ g \end{bmatrix}$$

so that

$$a = S_c T_c^{-1} h + (T - S_c T_c^{-1} S) g$$

$$= S_c T_c^{-1} h + T_*^{-1} g \ \textit{ from } (17)$$

$$b = T_c^{-1} (h - Sg)$$

Now P a belongs to H because VH^2 = TH^2 and

$$<a \ , TH^2> \ = \ <S_c T_c^{-1} h + T^{-1*} g \ , TH^2> \ = \ <-S^* h + g \ , H^2> \ \textit{ from } (17)$$

which vanishes due to Cauchy's theorem.

Moreover, P b = $P \ S_c^* \ P$ a = $P \ S_c^*$a because

$$b - S_c^* a = \left| T_c^{-1} (h - Sg) \right| - S_c^* \left| S_c T_c^{-1} (h - Sg) + Tg \right|$$

$$= T_c^* [h - Sg] - S_c^* Tg = T_c^* h \in K_m^2$$

[]

Remark: Similarly, the input state space can be shown to be

$$H = \begin{bmatrix} I \\ Q\,S \end{bmatrix} h \tag{56}$$

with $h = K_p^2 \ominus V^{-1} K_p^2$ and V is obtained from the outer-inner factorization of T(z), T = GV where V is inner and and G is outer. Q denotes orthogonal projection with respect to the $L^2(d\theta)$ innerproduct from L_q^2 into K_q^2.

[]

4.2 State Space as a RKHS

Define an indefinite innerproduct, namely the J-innerproduct, on $L_m^2(d\theta)$ as follows: For f,g$\in L_m^2$,

$$[f \ , g] = < f \ , Jg > \tag{57}$$

Note that [f,f] need not be positive due to the presence of the J-factor. Such spaces are called Krein spaces [25]. However, for the output state space H° of Θ(z), we find that it is positive:

$$\textit{Let } \ F = \begin{vmatrix} f \\ P S_c^* f \end{vmatrix} \in H^\circ$$

$$\textit{Then } \ [F \ , F] = \| f \|^2 - \| P S_c^* f \|^2 \geqslant \| f \|^2 [1 - \| P S_c^* \|_o^2]$$

where $\|.\|_o$ indicates operator norm. Similarly,

$$<F\,,F> = \|f\|^2 + \|PS_c^*f\|^2 \leqslant 2\|f\|^2$$

$$\textit{Thus}\ [F\,,F] \geqslant \mu <F\,,F>,\ \ \mu = ½(1 - \|PS_c^*\|_o^2) > 0 \tag{58}$$

That $\mu > 0$ is justified as follows. From our assumption that T_c has no zeros on T, it follows from (17) that the L^∞-norm of S_c^*, is strictly smaller than one. Since the L^∞-norm equals the operator norm and since P is a contraction, it follows that $\mu > 0$.

H^o is said to be a J-uniformly positive subspace of the Krein space H_m^2, and allows us to rewrite (48) as:

$$H^o = H_m^2[-]\,\Theta H_m^2,\ \ N^o = \Theta H_m^2 \tag{59}$$

(In general if a subspace is not J-uniformly positive, the decomposition of a Krein space into a J-orthogonal sum of subpaces is not trivial. See [25] for a full discussion.) Moreover, in our case, H^o is finite dimensional because $\Theta(z)$ is a rational function, so that it is a genuine Hilbert space in the J-innerproduct. Furthermore, it possesses an mxm function $K_\omega(z)$ called the Reproducing Kernel (RK) which satisfies, by definition, the two properties:

1. The columns of $K_\omega(z)$ belong to H^o,

$$K_\omega(z)\xi \in H^o \ \ \text{for}\ \xi\epsilon C^m, |\omega| < 1,\ |z| < 1 \tag{60a}$$

2. For $f \epsilon H^o$, the reproducing property holds:

$$[f\,,K_\omega\xi] = \xi^* f(\omega)\ \ \text{for}\ \xi\epsilon C^m,\ |\omega| < 1 \tag{60b}$$

Note that the innerproduct is computed with respect to the variable z. Such a space is referred to as a Reproducing Kernel Hilbert Space (RKHS).

Theorem 3: The RK of H^o with respect to the J-innerproduct is given by:

$$K_\omega(z) = \frac{J - \Theta(z)J\Theta^*(\omega)}{1-\omega^* z}\quad |\omega| < 1,\ |z| < 1 \tag{61}$$

Proof: To prove (60a), consider $K_\omega(z)\xi$ which clearly belongs to H_m^2. Moreover, it can be easily verified using Cauchy's theorem, that it is J-orthogonal to ΘH_m^2 so that it belongs to H^o. To prove the reproducing property, consider f(z) in H^o. Then,

$$[f\,,K_\omega\xi] = \xi^* f(\omega) - [f\,,\Theta g]\ \textit{with}\ g = \frac{J\Theta^*(\omega)\xi}{1-\omega^* z} \in H_m^2$$

$$= \xi^* f(\omega)\ \ \textit{since}\ f \perp J\Theta H_m^2$$

[]

Similar arguments hold to show that the input state space H is also an RKKS.

4.3 Other Connections

The output nullspace N^o is a forward shift invariant subspace in the Krein space H_m^2 and has been represented as a J-inner function times H_m^2 (see (59)). This form reminds one of the well known representation of shift invariant subspaces in the Hilbert space H_m^2 due to Beurling and Lax [11]. The generalization of the Beurling-Lax theorem to the Krein space setting has been done by Ball and Helton [26], and of course the representation of N^o agrees with it.

The output state space H^o is characterized by the fact that it is a backwards shift invariant subspace in H_m^2 and that it is a RKHS with respect to the J-innerproduct, with the RK having the form in (61). De Branges obtained the necessary and sufficient conditions for a RKHS in H_m^2 (with respect to *any* innerproduct) to have its RK of the form in (61). For a statement of the theorem and applications, see [9, 10]. We simply remark that the state space H^o meets all the requirements of the de Branges characterization.

5. LOSSLESS INVERSE SCATTERING SOLUTIONS

In this section, we obtain and parametrize a class of solutions to the LIS problem depicted and introduced in Sec 3.4. Let us recapitulate the problem.
Referring to Fig-5, we are given a pxq (possibly nonrational) Schur function $S_{in}(z)$ analytic in D and bounded on T, satisfying

$$I_q - S_{in}(z)S_{in}^*(z) \geqslant 0 \quad in \ D \cup T \tag{62}$$

The LIS-problem is to determine a Lossless Wave Scattering System (LWSS) described by a Chain Scattering Matrix (CSM) $\Theta(z)$ and a Schur function $S_L(z)$ such that

$$S_{in}(z) = T_\Theta[S_L] = [\Theta_{11}S_L + \Theta_{12}]\,[\Theta_{21}S_L + \Theta_{22}]^{-1} \tag{63}$$

where Θ_{ij} are block entries of $\Theta(z)$ partitioned such that Θ_{11}, Θ_{12}, Θ_{21} and Θ_{22} are pxp, pxq, qxp, and qxq respectively. The class of solutions to be obtained now will be described by a $\Theta(z)$ which is a rational function, with poles in E. We shall call such solutions rational *Schur type solutions.* The procedure to be followed will be seen to reflect the properties of the (output) state space of a J-lossless function, as discovered in Sec-4. Briefly, it is as follows:

We first choose an interpolation space h , of the type occuring in (33), but otherwise arbitrary, which in fact parametrizes the LIS-solution. We construct a so called Resolution space H^o, of the type occuring in (55), and determine a J-lossless function $\Theta(z)$ from it such that the resolution space is the output state space of $\Theta(z)$. We determine $\Theta(z)$ in three illuminating ways, namely the Reproducing Kernel approach, the Systems realization approach and a recursive approach. Finally, we prove the existence of a Schur function $S_L(z)$ such that (63) holds.
From now on, we shall use the following short hand notation:

$$\rho_\omega(z) = 1 - \omega^* z \tag{64}$$

5.1 Basic Constructions

The interpolation space h is chosen as follows:
Let ω_i $(1 \leqslant i \leqslant n < \infty)$ be distinct points in D, called interpolation points. Associated with each point ω_i, let ξ_{ij} be m_i vectors in C^p for $1 \leqslant j \leqslant m_i \leqslant p$, and let ν_{ij} be positive integers called multiplicity of $\{\omega_i, \xi_{ij}\}$. Define

$$h = span\left\{ h_{ij}(z) = \frac{z^{k-1}\xi_{ij}}{\rho_{\omega_i}^k(z)},\ 1 \leqslant i \leqslant n,\ 1 \leqslant j \leqslant m_i,\ 1 \leqslant k \leqslant \nu_{ij} \right\} \tag{65}$$

and let $\nu = \sum_{i,j} \nu_{ij}$.

It is easy to see that h is a ν-dimensional RKHS with respect to the $L^2(d\theta)$ innerproduct. Moreover, it is a backwards shift invariant subspace in H_q^2, and in fact,

$$R_{\omega} h \subset h \quad \text{for } |\omega|<1 \tag{66}$$

where R_{ω} was defined in (51). Assume that the given Schur function $S_{in}(z)$ satisfies the following inequality:

$$\text{if } f \in h\,, \ \| P S_{in}^{*} f \| < \| f \| \tag{67}$$

where the norms are computed in the $L^2(d\theta)$ metric. Note that this is a weaker condition than assuming that $S_{in}(z)$ is strictly contractive on T. In fact, this condition is equivalent to assuming the positivity of a Nevanlinna Pick (NP) interpolation matrix [21]. This is most easily seen in the case where the interpolation points are simple:
Let

$$h = span \ \{ f_k(z) = \xi_k / \rho_{\omega_k}(z),\ 1 \leqslant k \leqslant n \} \tag{68}$$

with ω_k being distinct points in D. An elementary computation shows that

$$P S_{in}^{*} f_k = S_{in}^{*}(\omega_k) \xi_k / \rho_{\omega_k}(z) \tag{69}$$

Requiring that the inequality holds for all linear combinations of f_k leads to the condition that the following matrix is positive definite:

$$0 < P = \left[p_{ij} = \langle f_j , f_i \rangle = \frac{\xi_i^{*} [I - S_{in}(\omega_i) S_{in}^{*}(\omega_j)] \xi}{1 - \omega_j^{*} \omega_i},\ 1 \leqslant i,j \leqslant n \right] \tag{70}$$

"P" is nothing but the well known Nevanlinna Pick interpolation matrix [27].

[]

Now, we build the *Resolution space* as follows:

$$H^{o} = \begin{bmatrix} I \\ P S_{in}^{*} \end{bmatrix} h \tag{71}$$

Clearly, H^{o} is a ν-dimensional subspace of H_m^2, m=p+q. Moreover, it is a RKHS in the J-innerproduct [.,.] defined in (57), thanks to the assumption (67). Since H^{o} is a finite dimensional space, the Reproducing Kernel (RK), say $K_{\omega}(z)$ is a rational function. This can be seen readily from the well known formula for the RK:

$$K_{\omega}(z) = \sum_{j=1}^{\nu} \phi_j(z) \phi_j^{*}(\omega) \tag{72}$$

where $\{\phi_j\}$ are an J-orthonormal basis for H^{o}. Observe also that $K_{\omega}(z)$ is in fact well defined a.e. in the complex plane and in particular, in a neighbourhood of the closed unit disc. Finally, R_{ω}-invariance of h as in (66) induces the R_{ω}-invariance of H^{o}:

$$R_{\omega} H^{o} \subset H^{o} \tag{73}$$

Indeed, consider

$$\begin{bmatrix} f \\ P S_{in}^{*} f \end{bmatrix} \in H^{o}.$$

Clearly $R_{\omega} f \in h$, by construction, while the following computation completes the argument to prove (73):
For all $\xi \in C^p$, and $\beta \in D$,

$$\xi^{*} \left[R_{\omega} P S_{in}^{*} f \right](\beta) = \frac{1}{\beta - \omega} \langle S_{in}^{*} f , \frac{\xi}{\rho_{\beta}} - \frac{\xi}{\rho_{\omega}} \rangle = \xi^{*} \left[P S_{in}^{*} (R_{\omega} f) \right](\beta)$$

5.2 Determination of $\Theta(z)$

From H°, we determine a J-lossless function $\Theta(z)$ is n the following three different ways:

RK Approach

Choose a normalization point α on the unit circle, and define the following mxm rational function in terms of the RK of H°,

$$\Theta(z) = I - \rho_{\alpha}(z)\, K_{\alpha}(z)\, J \tag{74}$$

Recall that $K_{\alpha}(z)$ is well defined so that $\Theta(z)$ is rational and analytic in a neighbourhood of the closed unit disc. Observe also that $\Theta(z)$ is normalized as $\Theta(\alpha)=I$.

Proposition 2: $\Theta(z)$ is J-unitary on the unit circle:

$$J - \Theta^{*}(z) J\, \Theta(z) = 0 \ \ a.e.\ on\ T \tag{75a}$$

$$J - \Theta(z) J\, \Theta^{*}(z) = 0 \ \ a.e.\ on\ T \tag{75b}$$

Proof: It suffices to show that

$$[\xi\, ,\eta] = [\Theta\xi\, ,\Theta\eta] \ \text{for } all\ \xi,\eta\epsilon C^{m}$$

Substituting for $\Theta(z)$ in the right hand side and using the following relations leads to the identity in a straightforward way.

$$[\xi\, , \rho_{\alpha}K_{\alpha}J\eta] = \eta^{*}JK_{\alpha}^{*}(0)J\xi$$

$$[\rho_{\alpha}K_{\alpha}J\xi\, ,\eta] = \eta^{*}JK_{\alpha}(0)J\xi$$

$$[\rho_{\alpha}K_{\alpha}J\xi\, , \rho_{\alpha}K_{\alpha}J\eta] = \eta^{*}J\,[K_{\alpha}(0) + K_{\alpha}^{*}(0)]\,J\xi$$

[]

Theorem 4: The output state space of $\Theta(z)$ is H°, that is,

$$P\ \Theta K_{m}^{2} = H^{\circ} \tag{76}$$

Proof: Firstly note that

$$K_{m}^{2} = span \left\{ \xi/(z-\omega),\ \xi\epsilon C^{m},\ \omega\epsilon D \right\}$$

Substituting for Θ (z) from (74), we can easily show that

$$P\ \Theta\frac{\xi}{z-\omega} = \alpha^{*}K_{\alpha}(z)J\xi - \rho_{\alpha}(\omega)\,[R_{\omega}K_{\alpha}](z)J\,\xi\ \epsilon H^{\circ}$$

To prove the converse, first note that

$$P\ \Theta K_{m}^{2} = H_{m}^{2}\,[-]\,\Theta H_{m}^{2} \ \text{ and}$$

$$H_{m}^{2} = span \left\{ \frac{\xi}{1-\omega^{*}z},\ \xi\epsilon C^{m},\ \omega\epsilon D \right\}$$

Since H° is already in H_{m}^{2}, the following computation completes the proof:

$$\text{for } f\ \epsilon H^{\circ},\ [f\ ,\Theta\frac{\xi}{z-\omega}] = \xi^{*}\ [f(\omega) - f(\alpha) + (\alpha-\omega)(R_{\omega}f\ )(\alpha)] = 0$$

[]

From Sec-4, it follows that the RK of H° has the form in (61). This can now establish the J-contractivity of $\Theta(z)$ in D:

$$\xi^* [J - \Theta^*(\omega)J\Theta(\omega)]\,\xi = (1-|\omega|^2)\,\xi^* K_\omega(\omega)\xi$$
$$= (1-|\omega|^2)\,[K_\omega\xi\,,K_\omega\xi] > 0 \text{ for } |\omega|<1$$

This makes $\Theta(z)$ a J-inner function satisfying (22).

System Realization Approach:

An alternative way to construct $\Theta(z)$ proceeds by determining a set of realization matrices {A,B,C,D} from H° such that the associated transfer function is the desired $\Theta(z)$.
Let A be an operator mapping H° into H° and C be an operator mapping H° into C^m, as defined below:

$$\text{if } f \in H^{\circ},\ Af := R_0 f\ , Cf := f(0) \tag{77}$$

Note that $R_0 f \in H^{\circ}$ due to R_0-invariance of H°, see (71). Choose a set of orthonormal basis vectors with respect to the J-innerproduct in H°, and in C^m. With respect to those bases, the operators A & C become constant matrices A & C. They satisfy the identity,

$$A^*A + C^*JC = I_\nu \tag{78}$$

Indeed, if ξ,η are any two vectors in C^ν, and f(z), g(z) are the corresponding functions in H° (with respect to the chosen basis), we have

$$\eta^*\xi = [f\ ,g] = [zR_0 f + f(0)\,,zR_0 g + g(0)]$$
$$= [R_0 f\ ,R_0 g] + g^*(0)Jf(0) = \eta^*\ [A^*A + C^*JC]\,\xi$$

Proposition 3: The eigenvalues of A (and A) are in D.

Proof: Let $Af = R_0 f = \lambda f$ for some f in H°. Substituting for $R_0 f$, this means $f = f(0)/(1-\lambda z)$. But f(z) is in H_m^2, implying $\lambda \in D$.

[]

This enables us to define the B and D matrices:

$$B = (I-A)(I-A)^{-1*}C^*J \tag{79}$$

$$D = I - C(I-A)^{-1}B \tag{80}$$

The matrices {A,B,C,D} form a J-orthogonal realization, that is,

$$\begin{bmatrix} A & B \\ C & D \end{bmatrix} \begin{bmatrix} I & \\ & J \end{bmatrix} \begin{bmatrix} A & B \\ C & D \end{bmatrix}^* = \begin{bmatrix} I & \\ & J \end{bmatrix} \tag{81}$$

(The verification is straightforward.) This means that the associated transfer function

$$\Theta(z) = D + zC(I-zA)^{-1}B \tag{82}$$

is a J-lossless function. Indeed,

$$J - \Theta^*J\Theta = B^*(I-zA)^{-1*}\,(1-|z|^2)\,(I-zA)^{-1}B$$

which is positive in D and zero on T.

Define now the J-weighted controllability Grammian as follows:

$$G_c = \sum_{k=0}^{\infty} A^k BJB^* A^{k*} \tag{83}$$

G_c satisfies the identity,

$$G_c = BJB^* + AG_cA^*$$

Using the fact that $AA^*+BJB^* = I$ which comes about from (81) by interchanging the order of multiplication on the left hand side, the above equation leads to $G_c = I$. In particular, this means that the controllability matrix

$$W_c = [B, AB, A^2B, \cdots] \tag{84}$$

has full column rank.

Theorem 5: The output state space of $\Theta(z)$ is H^o,

$$H^o = P\, \Theta K_m^2 \tag{85}$$

Proof: By choosing the natural basis $\{z^{\pm k}\}$ in H_m^2 and K_m^2, we can write

$$P\, \Theta K_m^2 = [1, z, z^2, ...]\ Range\ W_o W_c$$
$$= [1, z, z^2, ...]\ Range\ W_o$$

where $W_o^* = [C^*, A^*C^*, A^{2*}C^* \ ...]$ and where we used the fact that W_c has full column rank.

The following identity holds for functions in H_m^2:

$$f(z) = \sum_0^{\infty} z^k\, [R_o^k f\,](0)$$

If $f(z) \in H^o$, we have $R_o^k f = CA^k f$, so that

$$f(z) = \sum_0^{\infty} z^k\, CA^k \xi$$

with ξ being the vector representation of f(z) with respect to the chosen basis in H^o. It follows from (86) that H^o is contained in $P\, \Theta K_m^2$. They are in fact equal because the dimension of the right hand side can not be greater than ν, as can be seen from (86).

[]

Recursive Approach

Without impairing generality, we assume that each interpolation point ω_k in the definition of h (65) has only one associated interpolation direction ξ_k and multiplicity one. Thus H^o in (71) can be written down as follows with the aid of (69):

$$H^o = span\left\{ f_k(z) = \frac{u_k}{1-\omega_k^* z},\ 1 \leqslant k \leqslant \nu \right\},\ u_k = \begin{bmatrix} \xi_k \\ S_{in}^*(\omega_k)\xi_k \end{bmatrix} \tag{87}$$

Now we will recursively J-orthogonalize $\{f_k\}$ to produce $\{\phi_k\}$, generating at each step a degree-1 J-lossless function $\theta_k(z)$ whose product will be the desired $\Theta(z)$. We shall use the following basic form of a degree-1 J-lossless function:

$$\theta_k(z) = I + (\zeta_k - 1)\, V_k V_k^* J / V_k^* J V_k \tag{88}$$

where $\zeta_k = (z-\omega_k)/(1-\omega_k^* z)$ and V_k is an mx1 vector such that $V_k^* J V_k \neq 0$. The J-lossless

properties of $\theta_k(z)$ are easy to verify, while the (Mcmillan) degree of $\theta_k(z)$ is one because V_k is a mx1 vector.
The following properties of $\theta_k(z)$ will be of use later:

$$1)\ V_k^* \theta_k(\omega_k)\xi = 0 \quad \text{for all } \xi \epsilon C^m \tag{89}$$

$$2)\ \lim_{z \to 1/\omega_k^*} (1-\omega_k^* z)\,\theta_k(z)\xi = \alpha V_k \tag{90}$$

for some constant α.

Recursive Algorithm

Initialization: Let $\phi_1(z) = f_1(z)$ and $V_1 = u_1$ as defined in (87), so that $V_1^* J V_1 > 0$. Let $\theta_1(z)$ be a degree-1 J-lossless factor as defined in (88).

Recursion $1 < k \leqslant \nu$: Assume that we have (k-1) J-orthogonal vectors $\{\phi_j\}$ and associated J-lossless factors $\{\theta_j\}$ such that, for $1<j<k$,

$$1)\ V_j = \left[\theta_1 \cdots \theta_{j-1}\right]^{-1}(1/\omega_j^*)\, u_j \tag{91}$$

$$2)\ \phi_j(z) = \frac{1}{1-\omega_j^* z}\left[\theta_1 \cdots \theta_{j-1}\right](z)\, V_j \tag{92}$$

where θ_j are as defined in (89). Note that θ_i^{-1} has a pole at ω_i and a zero (=pole of θ_i) at $1/\omega_i^*$, so that V_j in (91) is finite and nonzero.
The next J-orthogonal vector $\phi_k(z)$ is given by:

$$\phi_k(z) = \frac{1}{1-\omega_k^* z}\left[\theta_1 \cdots \theta_{k-1}\right]^{-1}(1/\omega_k^*)\, u_k \tag{93a}$$

$$\text{with}\ \ V_k = \left[\theta_1 \cdots \theta_{k-1}\right]^{-1}(1/\omega_k^*)\, u_k \tag{93b}$$

[]

We now justify the algorithm. Firstly, using (89) and (90), we can write

$$\phi_k(z) = \sum_{j=1}^{k-1} \alpha_j f_j(z) + f_k(z)$$

where α_j are some constants. Thus $\phi_k \epsilon H^\circ$ and moreover is nonzero, since f_j are linearly independent. Using (89), it further follows easily that ϕ_k is J-orthogonal to $\phi_1 \dots \phi_{k-1}$. Finally,

$$[\phi_k\ ,\phi_k] = \frac{V_k^* J V_k}{1-|\omega_k|^2} > 0$$

because $0 \neq \phi_k\ \epsilon H^\circ$, a J-positive space. Thus $\theta_k(z)$ is a well defined J-lossless factor.

[]

$$\text{Let}\quad \Theta(z) = \theta_1(z)\ \cdots\ \theta_\nu(z) \tag{94}$$

It follows easily that $\Theta(z)$ is a J-lossless function, because $\{\theta_j\}$ are. Furthermore, the following theorem proves that the output state space of $\Theta(z)$ is H°.

Theorem 6:

$$H^{\circ} = H_m^2 [-] \Theta H_m^2 \tag{95}$$

Proof: Using (89), it follows that $\phi_j(z)$ is J-orthogonal to ΘH_m^2. Since $\{\phi_j\}$ span H°, this implies that H° is contained in the right hand side of (95). Since the dimension of the right hand side cannot be greater than ν, the theorem is proved.

[]

5.3 LIS-Theorem

Having obtained a J-lossless function $\Theta(z)$ starting from $S_{in}(z)$ and h via H° such that the output state space of $\Theta(z)$ is H°, it remains to prove the existence of a passive load (Schur function) $S_L(z)$ such that $S_{in}(z)$ admits a representation given in (63). The proof we give uses the interpolation property of $S_{in}(z)$ and the complementary scattering function associated with $\Theta(z)$. To simplify the proof, we assume that $S_{in}(z)$ satisfies,

$$\| P S_{in}^{*} f \| < \| f \| \text{ for } all \ f \in H_p^2 \tag{96}$$

Note that this is a stronger assumption than the one in (67). A proof which does not need such an extended assumption is possible, and uses the notion of admissibility and geometry of Krein spaces [9, 28].

Interpolation

Consider the complementary scattering function ($=\Theta_{11}\Theta_{22}^{-1}$) associated with $\Theta(z)$ obtained in the last section, and call it $S_0(z)$. From Sec-4, we know that the output state space of $\Theta(z)$ is given by:

$$\begin{bmatrix} I \\ P S_0^{*} \end{bmatrix} h' \ \textit{with} \ h' = H_p^2 \ominus V H_p^2 \tag{97}$$

and V(z) is obtained from the inner-outer factorization of T(z) [=V(z)G(z)]. Since this space equals H°, we have from (71) that

$$h' = h \text{ , and } P S_0^{*} h = P S_{in}^{*} h \tag{98}$$

As discussed in Sec-3, this means that S_0 and S_{in} interpolate each other in the interpolation space h .

Construction of $S_{Lj}(z)$

Now we will construct a sequence of Schur functions $S_{Lj}(z)$ such that $S_j = T_{\Theta}[S_{Lj}]$ interpolates $S_{in}(z)$ on increasingly bigger interpolation spaces h_j. Define a sequence of nested interpolation spaces h, subset h_1 subset h_2 subset ...etc., as:

$$h_j = h + span\left\{ e_i \rho_{\alpha_k}, 1 \leqslant i \leqslant p, 1 \leqslant k \leqslant j \right\} \tag{99}$$

where α_k are distinct points in D and distinct from $\{\omega_k\}$ in (65). $\{e_i\}$ are the natural basis for C^p, i.e. e_i is a px1 vector with 1 in the i-th place and zeros elsewhere. Construct the spaces

$$[H^{\circ}]_j = \begin{bmatrix} I \\ P S_{in}^{*} \end{bmatrix} h_j \tag{100}$$

Note that these spaces are nested as h_j are. Due to the assumption (96), $[H^{\circ}]_j$ is an

RKHS with respect to the J-innerproduct. Following any of the procedures in the last section, let us construct a J-lossless function $\Theta_j(z)$ such that the output state space of it equals $[H^\circ]_j$. If S_j is the complementary scattering function associated with $\Theta_j(z)$, we again have the interpolation relation:

$$P S_{in}^* h_j = P S_j^* h_j \tag{101}$$

Proposition 4: Under the present condition,

$$\Theta_j(z) = \Theta(z)\, \Theta_{rj}(z) \tag{102}$$

where $\Theta(z)$ and $\Theta_j(z)$ are obtained from H° and $[H^\circ]_j$ respectively, and $\Theta_{rj}(z)$ is a J-lossless function.

Proof: Let $K_\omega(z)$ and $K_\omega^j(z)$ be the reproducing kernels of H° and $[H^\circ]_j$ respectively. For all ω in D and ξ in C^m, we obtain by applying Schwarz's inequality in $[H^\circ]_j$,

$$[K_\omega \xi \,,\, K_\omega^j \xi] \leqslant \|K_\omega \xi\| \,.\, \|K_\omega^j \xi\|$$

Evaluating both sides using the reproducing property of the kernels leads to

$$\xi^* K_\omega(\omega)\xi \leqslant \xi^* K_\omega^j(\omega)\xi$$

Using the form of the RKs from (61), we obtain easily

$$\xi^* \Theta(\omega)\, [J - \Theta_{rj}(\omega) J \Theta_{rj}^*(\omega)]\, \Theta^*(\omega)\xi \geqslant 0 \quad \omega \epsilon D\,,\, \xi \epsilon C^m$$

where $\Theta_{rj} = \Theta^{-1}\Theta_j$. This establishes the J-contractivity on T, while the J-unitarity on T falls out at once from the J-unitarity of Θ and Θ_j.

[]

This means that $S_j(z)$ can be expressed as

$$S_j(z) = T_\Theta[S_{Lj}] \quad or \quad S_{Lj}(z) = T_{\Theta^{-1}}[S_j] \tag{103}$$

where S_{Lj} is the complementary scattering function associated with $\Theta_{rj}(z)$. See Fig-7.

Convergence of $S_{Lj}(z)$

Firstly we use the classical argument of Schur [19] to prove that $S_j(z)$ converges to $S_{in}(z)$, provided the extra interpolation points $\{\alpha_k\}$ in (99) are chosen so that

$$\sum_1^\infty [1 - |\alpha_k|] = \infty \tag{104}$$

From the interpolation relation (101), and from our construction of $h_j(z)$, it follows that

$$[S_{in} - S_j]/b_j(z), \quad with \quad b_j(z) = \prod_{i=1}^{j} \frac{z-\alpha_i}{1-\alpha_i^* z}$$

is analytic in D. Applying Schwarz's lemma [11, p.271], it follows that in D,

$$|\eta^* [S_{in} - S_j]\xi| \leqslant 2|b_j| \quad for\ all\ \xi \epsilon C^q,\ \eta \epsilon C^p$$

By choosing $\{\alpha_j\}$ to satisfy (104), the right hand side converges to zero, so that $S_j(z)$ converges to $S_{in}(z)$. Now from (103), it follows that the sequence of Schur functions

$S_{L_j}(z)$ converge to a Schur function $S_L(z)$ a.e. in D because $\Theta^{-1}(z)$ is bounded a.e. in D.

6. CONCLUDING REMARKS

The theory developed in this paper can be extended in several ways. Firstly, the CSMs considered might have poles on the unit circle (so called "Brune points"- see [8, 9, 10].) The theory as given here will break down because the elements of the relevant $\Theta(z)$ matrix are no longer in L^2. This can be fixed but at the cost of some unappealing limiting procedures. (Preliminary discussions can be found in [28].) Secondly, infinite dimensional interpolation and resolution spaces can be handled as well, resulting in nonrational LIS-solutions. A reference for this is [28]. A further extension is to Partial Lossless Inverse Scattering (PLIS) problems where only incomplete data on the input scattering function is given. Although there are preliminary results [29] this theory is largely undeveloped. This problem has potentially powerful applications to α-stationary stochastic processes.

References

1. P. Dewilde, J.T. Fokkema, and I. Widya, "Inverse Scattering and Linear Prediction, The Time Continuous Case," *Stochastic Systems: The Mathematics of Filtering and Identification and Applications*, pp. 351-382 D. Riedel Publ. House, (1981).

2. A. Bruckstein, B. Levy, and T. Kailath, "Differential Methods in Inverse Scattering," *SIAM J. Appl. Math.* 45(2) pp. 312-335 (April 1985).

3. E.F. Deprettere and P. Dewilde, "Orthogonal Cascade Realization of Real Multiport Digital Filters," *Intl. J. Circuit Theory and Applns.* 8 pp. 245-272 (1980).

4. C.V.K. PrabhakaraRao, E. Deprettere, and P. Dewilde, *Cascade Orthogonal Digital Filters: A Tutorial*, to be published 1985.

5. Ph. Delsarte, Y. Genin, and Y. Kamp, "On the Role of the Nevanlinna-Pick Problem in Circuit and System Theory," *Intl. J. Circuit Theory and Applns.* 9 pp. 177-187 (1981).

6. P. Dewilde and H. Dym, "Schur Recursions, Error Formulas and Convergence of Rational Estimators for Stationary Stochastic Sequences," *IEEE Trans. Info. Theory* IT-27(4) pp. 446-461 (Jul 1981).

7. P. Dewilde and H. Dym, "Lossless Chain Scattering Matrices and Optimum Linear Prediction: The Vector Case," *Intl. J. Circuit Theory and Appln.* 9 pp. 135-175 (1981).

8. P. Dewilde and H. Dym, "Lossless Inverse Scattering, Estimation Theory and Digital Filters," *IEEE Trans. Info. Theory* IT-30(4) pp. 644-662 (July 1984).

9. D. Alpay and H. Dym, "Hilbert Spaces of Analytic Functions, Inverse Scattering and Operator Models. I," *Integral Equations and Operator Theory* 7 pp. 589-641 (1984).

10. D. Alpay and H. Dym, "Hilbert Spaces of Analytic Functions, Inverse Scattering and Operator Models. II," *Integral Equations and Operator Theory* 8 pp. 145-180 (1985).

11. W. Rudin, *Real & Complex Analysis*, McGraw Hill Inc., New York, USA (1974). Second Edition

12. H. Helson, *Lectures on Invariant Subspaces*, Acad. Press, NY, USA (1964).

13. S. Bochner and K. Chandrasekharan, *Fourier Transforms*, Princeton Univ. Press (1949).

14. P. Koosis, *Introduction to Hilbert Spaces*, Cambridge Univ. Press (1980). London Math. Soc. Lecture Note Series 40

15. P.L. Duren, *Theory of* H^p*-Spaces*, Acad. Press, NY, USA (1970).

16. K. Knopp, *Theory of Functions*, Dover Publications (1945).

17. V. Potapov, "The Multiplicative Structure of J-Contractive Matrix Functions," *Transl. AMS, Series 2* **15** pp. 131-243 (1960).

18. V. Belevitch, *Classical Network Theory*, Holden Day, San Fransisco (1968).

19. J. Schur, "Uber Potenzreihen, die im Innern des Einheitskreises beschrankt sind," *J. fur die Riene und Angewandte Math.* **147** pp. 205-232 (Sept 1917). (in German)

20. D.C. Youla, L.J. Castriota, and H.J. Carlin, "Bounded Real Scattering Matrices and the Foundations of Linear Passive Network Theory," *IRE Trans.* **CT-4**(1) pp. 102 - 124 (March 1959).

21. N.I. Akhiezer, *The Classical Moment Problem*, Oliver & Boyd, Edinburgh (1965).

22. V.M. Adamjan, D.Z. Arov, and M.G. Krein, "Analytic Properties of Schmidt pairs for a Hankel Operator and the Schur-Takagi Problem," *Math. USSR Sbornik* **15**(1) pp. 31-73 (1971).

23. D. Sarason, "Generalized Interpolation in H^∞," , pp. 179-203 (May 1967).

24. J.P. Burg, "Maximum Entropy Spectral Analysis," *Ph.D. Thesis, Stanford Univ., USA*, (1975).

25. M.G. Krein, "Introduction to the Geometry of Indefinite J-Spaces and to the Theory of Operators on those spaces," *Amer. Math. Soc. Transl., Series 2*, **93** pp. 103-176 (1970).

26. J. Ball and J.W. Helton, "A Beurling-Lax Theorem for the Lie Group U(m,n) which contains most classical Interpolation Theory," *J. Operator Theory* **9** pp. 107-142 (1983).

27. Ph. Delsarte, Y. Genin, and Y. Kamp, "The Nevanlinna Pick Problem for Matrix Valued Functions," *SIAM J. Appl. Math.* **36**(1) pp. 47-61 (Feb 1979).

28. C.V.K. PrabhakaraRao, "A Systems Approach to the Multiport Lossless Inverse Scattering Problem," Ph.D. Thesis, Delft University of Technology, Delft, The Netherlands (Dec 1985).

29. D. Alpay, P. Dewilde, and H. Dym, "On Partial Lossless Inverse Sacttering and Alpha-Stationary Estimation," *to be presented at ISCAS'86, San Jose*, (1986).

C.V.K. Prabhakara Rao
Department of Electrical
and Computer Engineering
College of Engineering
Drexel University
Philadelphia PA19104
U.S.A.

P. Dewilde
Department of Electrical
Engineering
Delft University of Technology
P.O. Box 5031
2600 GA Delft
The Netherlands

Operator Theory:
Advances and Applications, Vol. 19

A HADAMARD FACTORIZATION THEOREM FOR ENTIRE MATRIX VALUED FUNCTIONS

Carl L. Prather[1)] and André C.M. Ran[2)]

INTRODUCTION

The Hadamard factorization theorem for scalar valued entire functions is well known. It allows one to factor a scalar valued entire function of finite order ρ into a canonical product times the exponential of a polynomial having degree at most ρ. Standard proofs of this result may be found in many books on function theory (see, e.g., [2] or [10]). A beautiful yet elementary proof which bypasses the standard approach can be found in [7].

In the present paper the authors will give an analogue of the Hadamard factorization theorem for a certain class of matrix valued entire functions. Our approach is completely different from any of the approaches to the scalar case. Recently, a version of the Weierstrass factorization theorem has been proved for the case of matrix valued complex functions (see [6]). The result we obtain in this paper is closer to the classical result than the theorem in [6]. However, we have to restrict ourselves to a smaller class of matrix valued functions.

---

1)The first author was partially supported by NSF grant nr. RII-8410705.

2)The second author was supported by a grant from the Niels Stensen Stichting at Amsterdam.

The paper consists of two sections. In the first section we study a model originally developed in [9,4] in more detail. This model, combined with a factorization method taken from [1], is used to establish the main theorem in the second section. We restrict ourselves to regular entire matrix valued functions A(z), i.e. such that $\det A(z)\not\equiv 0$. Further, we assume that A(z) is transcendental, for the polynomial case compare [3], Theorem 3.21.

1. A MODEL FOR ENTIRE MATRIX VALUED FUNCTIONS

Let A(z) be an mxm matrix valued entire function. We say that A(z) has order ρ whenever for any $\varepsilon>0$ there exist positive constants B and C such that

$$\|A(z)\| \leqslant B\exp(C|z|^{\rho+\varepsilon})$$

for all sufficiently large $|z|$, and ρ is the smallest number for which this holds. Here, as usual

$$\|A(z)\| = \max_{\substack{\|x\|=1\\ x\in\mathbb{C}^m}} \|A(z)x\|.$$

Next, we introduce a model for entire matrix valued functions as developed in [9], Section II.5 (see also [4]). Let $A(z)=\sum_{n=0}^{\infty} A_n z^n$ be a transcendental entire mxm matrix function. Define

$$\alpha_k = 2\cdot\sup_{n\geqslant k} \|A_n\|^{\frac{1}{n}},$$

$$\beta_0 = 1,\ \beta_k = \beta_{k-1}\cdot\alpha_k.$$

Then $\alpha_1\geqslant\alpha_2\geqslant\ldots>0$ and $\beta_j\to 0$ as $j\to\infty$. Put $X=\ell^2(\mathbb{C}^m)$, and define operators $w: X\to\mathbb{C}^m$, $\tau: \mathbb{C}^m\to X$, $V: X\to X$ and $M: X\to X$ by

$$w(y_0,y_1y_2\ldots) = y_0,\quad \tau y = (y,0,0\ldots),$$

$$V(y_0,y_1,y_2\ldots) = (0,\alpha_1 y_0,\alpha_2 y_1,\alpha_3 y_2\ldots),$$

$$M(y_0,y_1,y_2\ldots) = \Big(\sum_{j=0}^{\infty}\beta_j^{-1}A_j y_j,0,0\ldots\Big).$$

The operators w,τ and V are clearly bounded, the boundedness of M follows from

$$\|A_j\| = (\|A_j\|^{\frac{1}{j}})^j \leqslant 2^{-j}\alpha_j^j \leqslant 2^{-j}\beta_j.$$

Further one easily checks that I-zV is invertible for all $z\in\mathbb{C}$. A direct computation then shows that

$$A(z) = I + w(M-I)(I-zV)^{-1}\tau.$$

Since A(z) is assumed to be regular we can assume without loss of generality that $A_0=I$. Then, for $z\neq 0$

$$A(\tfrac{1}{z}) = I + w(M-I)\tau + w(M-I)V(z-V)^{-1}\tau.$$

Since $A(0)=I$ we get that $I+w(M-I)\tau=I$, i.e. $w(M-I)\tau=0$. Consequently

$$A(\tfrac{1}{z}) = I + w(M-I)V(z-V)^{-1}\tau. \tag{1.1}$$

As is well known (see [1])

$$A(\tfrac{1}{z})^{-1} = I - w(M-I)V(z-V^{x})^{-1}\tau,$$

where the operator V^{x} is defined by

$$V^{x} = V - \tau w(M-I)V = [I-\tau w(M-I)]V. \tag{1.2}$$

By Theorem 2.6 of [1], V^{x} is a linearization of an extension of $A(\frac{1}{z})$.

The operator V^{x} can be represented in matrix form as follows:

$$V^{x} = \begin{bmatrix} -\beta_0^{-1}A_1 & -\beta_1^{-1}A_2 & -\beta_2^{-1}A_3 & \cdot & \cdot & \cdot & \cdot \\ \alpha_1 I_m & & & & & & \\ & \alpha_2 I_m & & & & & \\ & & \alpha_3 I_m & & & & \\ & & & \cdot & & & \\ & & & & \cdot & & \\ & & & & & \cdot & \\ & & & & & & \cdot \end{bmatrix},$$

where I_m denotes the mxm identity matrix.

The remainder of this section is devoted to the study of properties of the operators V and V^{x}.

THEOREM 1.1. The following hold

(i) $\operatorname{Ker} V = (0)$, $\operatorname{Ker} V^{x} = (0)$,

(ii) V and V^{x} are compact.

PROOF. (i) Since $A(z)$ is transcendental we have $\alpha_j\neq 0$ for all j and hence $\operatorname{Ker} V=(0)$. Next, suppose $V^{x}x=0$. Then $Vx\in\operatorname{Ker}(I-\tau w(M-I))$, hence $Vx\in\operatorname{Im}\tau$ and so $x=0$.

(ii) First we show that V is compact. Define projections P_n on $\ell^2(\mathbb{C}^m)$ by

$$P_n(x_0,x_1,x_2\ldots) = (x_0,x_1\ldots,x_n,0,0,\ldots).$$

Consider $(I-P_n)V$. We have that

$$((I-P_n)Vx)_i = \begin{cases} 0 & \text{for } 0\leqslant i\leqslant n+1, \\ \alpha_i x_i & \text{for } i>n+1. \end{cases}$$

As a result $\|(I-P_n)V\|\leqslant \sup_{i\geqslant n+2} \alpha_i=\alpha_{n+2}$, since the α's decrease monotonically. Thus

$\|V-P_nV\|\to 0$ as $n\to\infty$. Since P_n is a finite rank operator so is P_nV, and therefore V is compact.

From the compactness of V and formula (1.2) we obtain that also V^x is compact.□

Recall (see [5]) that a compact operator A is said to belong to the Schatten p-class S_p if the sequence of eigenvalues $\{s_j\}_{j=1}^{\infty}$ of $(A^*A)^{\frac{1}{2}}$ lies in ℓ^p, that is $\sum_{j=1}^{\infty} |s_j|^p<\infty$.

Next, we investigate conditions which guarantee that V and V^x lie in S_p. From formula (1.2) one sees that if $V\in S_p$ for some p, then also $V^x\in S_p$, since S_p forms an ideal in the algebra of bounded operators. The fact that $V^x\in S_p$ implies also $V\in S_p$ is shown in the following theorem.

THEOREM 1.2. The following are equivalent:

(i) $V\in S_p$,

(ii) $V^x\in S_p$,

(iii) the sequence $\{\alpha_j\}_{j=1}^{\infty}$ is in ℓ^p.

PROOF. To establish the equivalence of (i) and (ii) it suffices to show that $I-\tau w(M-I)$ is invertible, by virtue of (1.2). Suppose $x\in \mathrm{Ker}(I-\tau w(M-I))$. Then $0=x_1=x_2=\ldots$, and $x_0=w(M-I)x = \sum_{j=1}^{\infty} \beta_j^{-1}A_jx_j=0$. Hence $x=0$ and $\mathrm{Ker}(I-\tau w(M-I))=(0)$. Next, let $y=(y_0,y_1,y_2\ldots)\in\ell^2(\mathbb{C}^m)$, and put

$$x = \left(\sum_{j=0}^{\infty} \beta_j^{-1}A_jy_j,y_1,y_2\ldots\right).$$

One easily checks that $(I-\tau w(M-I))x=y$, so $\mathrm{Im}(I-\tau w(M-I))=\ell^2(\mathbb{C}^m)$. Therefore, $I-\tau w(M-I)$ is invertible.

It remains to show that (i) and (iii) are equivalent. Consider $V^*Vx=(\alpha_1^2x_0,\alpha_2^2x_1,\ldots)$. So the eigenvalues of $(V^*V)^{\frac{1}{2}}$ are α_i, $i=1,2,\ldots$. As a result $V\in S_p$ if and only if $\{\alpha_i\}_{i=1}^{\infty}\in\ell^p$.□

We complete this section with the following important theorem.

THEOREM 1.3. If the entire matrix function $A(z)$ has order ρ, then $\{\alpha_j\}_{j=1}^{\infty} \in \ell^{\rho+\varepsilon}$ for every $\varepsilon>0$.

PROOF. Take $A(z)=\sum_{j=0}^{\infty} A_j z^j$. The order of growth of the entire function $A(z)$ produces a certain growth on the coefficients A_j. In a form convenient for our purpose it can be expressed as

$$\lim_{j\to\infty}(j!\ \|A_j\|^{\rho+\varepsilon})^{\frac{1}{j}}=0, \text{ for all } \varepsilon>0.$$

For a proof of this in the scalar case see [8], Chapter I; this proof goes through also in the case of a matrix valued function.

From the formula above it follows that $j!\|A_j\|^{\rho+\varepsilon}<1$ for large j and $\varepsilon>0$, that is

$$\|A_j\| < \left(\frac{1}{j!}\right)^{\frac{1}{\rho+\varepsilon}}$$

for large j and $\varepsilon>0$. It follows that for $\varepsilon>0$ and large j

$$\alpha_j<2\sup_{n\geq j}\left(\frac{1}{n!}\right)^{\frac{1}{\rho+\varepsilon}} = 2\left(\frac{1}{j!}\right)^{\frac{1}{\rho+\varepsilon}}.$$

Therefore $(\alpha_j)^{\rho+\varepsilon}<2^{\rho+\varepsilon}\frac{1}{j!}$ for $\varepsilon>0$ and large j. Consequently $\{\alpha_j\}_{j=1}^{\infty}\in\ell^{\rho+\varepsilon}$ for $\varepsilon>0$.□

2. THE MAIN THEOREM

In this section we state and prove the main factorization theorem. Before stating the theorem we need some preliminary definitions.

Let $A(z)$ be an entire mxm-matrix valued function. A complex number λ is said to be a zero of $A(z)$ if $\det A(\lambda)=0$. We shall assume that $A(z)$ has only linear elementary divisors, that is for each zero λ of $A(z)$ $\dim\operatorname{Ker}A(\lambda)$ is equal to the order of λ as a zero of $\det A(z)$. Further we assume $A(z)$ is regular, i.e. $\det A(z)\not\equiv 0$. In this case we number the zeros $\{z_j\}_{j=1}^{\infty}$ as follows: $|z_1|\leqslant|z_2|\leqslant\ldots$. We do not exclude the case that there is a finite number of zeros.

Put $\lambda_j=\frac{1}{z_j}$, and define

$$\sigma = \inf\{\alpha \mid \sum_{j=1}^{\infty} |\lambda_j|^{\alpha}<\infty\}.$$

The number σ is called the <u>exponent of convergence</u> of the sequence $\{\lambda_j\}_{j=1}^{\infty}$. Further, let $p=[\sigma]$ if σ is not an integer, $p=\sigma-1$ if σ is an integer and $\sum_{j=1}^{\infty}|\lambda_j|^{\sigma}<\infty$ and $p=\sigma$ if σ is an integer and $\sum_{j=1}^{\infty}|\lambda_j|^{\sigma}=\infty$. The number p is called the <u>rank of the function</u> A(z). If A(z) is an entire function of finite order ρ it follows from Theorems 1.3 and 1.2 that the operator V^X associated with A(z) is in $S_{\rho+\varepsilon}$ for every $\varepsilon>0$. From this it follows that $\sum_{j=1}^{\infty}|\lambda_j|^{\rho+\varepsilon}<\infty$ (see[5], Section III.7). One obtains the following inequalities:

(2.1) $$p \leqslant \sigma \leqslant \rho.$$

We are now in a position to state our main result.

THEOREM 2.1. Let A(z) be an entire mxm-matrix valued function of finite order ρ with only linear elementary divisors. Then

$$A(z) = B(z)\prod_{j=1}^{\infty}E_p(zZ_j),$$

where B(z) is an entire everywhere invertible mxm-matrix function having order at most ρ, and

$$E_p(zZ_j) = (I-zZ_j)\exp\Big(\sum_{k=1}^{p}\frac{1}{k}(zZ_j)^k\Big).$$

Here p is the rank of the function A(z), and Z_j is a normal mxm matrix such that $\det E_p(zZ_j)=0$ if and only if $z=z_j$, the j-th zero of A(z) and rank $Z_j=\dim\operatorname{Ker}A(z_j)$. Further for every zero z_i of A(z) we have $\operatorname{Ker}A(z_i)=\operatorname{Ker}\prod_{j=1}^{\infty}E_p(z_iZ_j)$.

Moreover, $\sum_{j=1}^{\infty}E_p(zZ_j)$ has order at most σ, the exponent of convergence of the sequence $\{\frac{1}{z_j}\}_{j=1}^{\infty}$. Finally, the order of multiplication in the infinite product is understood to be from right to left.

The Hadamard theorem for the scalar case states that the everywhere invertible factor B(z) can be written as the exponential of a polynomial having degree at most ρ. However, in the matrix valued case the theorem above is the best result obtainable, since not every function which is everywhere invertible can be written as the exponential of a polynomial. This is shown by the following example: $B(z) = \begin{bmatrix} 1 & 0 \\ e^{z} & 1 \end{bmatrix}$, which is an everywhere invertible function of finite order. Now B(z) can not be written as the exponent of a polynomial, which can easily be seen by letting $z \to -\infty$.

The proof of Theorem 2.1 will be given in several steps. The first step is to combine the model developed in Section 1 with the factorization method from [1] to show that corresponding to every zero of A(z) there is a linear right divisor of A(z). Our second step will be to show that the infinite product appearing in Theorem 2.1 converges uniformly on compact subsets of $\mathbb{C}$. Finally we shall prove the statements about the order of B(z) and the infinite product.

THEOREM 2.2. Let $\frac{1}{\lambda}$ be a zero of A(z). Then one can factorize

$$A(z) = B_1(z)(I - zZ_\lambda),$$

where Z_λ is a normal mxm matrix with eigenvalues 0 and λ and rank $Z_\lambda = \dim \operatorname{Ker} A(\frac{1}{\lambda})$. Further $\det B_1(\frac{1}{\lambda}) \neq 0$.

PROOF. First note that if $\frac{1}{\lambda}$ is a zero of A(z) then λ is an eigenvalue of V^x. We compute $\operatorname{Ker}(V^x - \lambda I)$. Let $x \in \operatorname{Ker}(V^x - \lambda I)$, then $\alpha_i x_{i-1} = \lambda x_i$ for $i \geq 1$ and $-\sum_{j=0}^{\infty} \beta_j^{-1} A_{j+1} x_j = \lambda x_0$. It follows that $x_i = \beta_i x_0 \lambda^{-i}$ so that

$$-\sum_{j=0}^{\infty} \beta_j^{-1} A_{j+1} x_j = -\sum_{j=0}^{\infty} A_{j+1} x_0 \lambda^{-j} = \lambda x_0,$$

in other words $A(\frac{1}{\lambda}) x_0 = 0$. Next $\sum_{j=0}^{\infty} (\beta_j \lambda^{-j})^2 \|x_0\|^2 < \infty$. Indeed, $(\beta_i / \lambda^i)^2 (\lambda^{i-1} / \beta_{i-1})^2 = \alpha_i^2 / \lambda^2 \to 0$ as $i \to \infty$. Hence $(x_0, \beta_1 x_0 / \lambda, \beta_2 x_0 / \lambda^2, \ldots) \in \ell^2(\mathbb{C}^m)$. So

$$\operatorname{Ker}(V^x - \lambda I) = \{x = (x_0, x_1, x_2, \ldots) \mid x_0 \in \operatorname{Ker} A(\tfrac{1}{\lambda}),\ x_i = \frac{\beta_i}{\lambda^i} x_0,\ i \geq 1\}$$

Now, set $M_0 = \operatorname{Ker} A(\frac{1}{\lambda})^{\perp}$ and put $M = \{y \in \ell^2(\mathbb{C}^m) \mid y_0 \in M_0\}$. Then M is a closed V-invariant subspace of $\ell^2(\mathbb{C}^m)$. We next show that

$$\ell^2(\mathbb{C}^m) = M \dotplus \operatorname{Ker}(V^x - \lambda I).$$

Indeed, suppose $y \in M \cap \operatorname{Ker}(V^x - \lambda I)$. Then $y_0 \in M_0 \cap \operatorname{Ker} A(\frac{1}{\lambda})$ and $y_i = \beta_i \lambda^{-i} y_0$ for $i \geq 1$. Hence $y = 0$. Next, let $z \in \ell^2(\mathbb{C}^m)$ and write $z_0 = x_0 + y_0$, with

$x_0\in \mathrm{Ker}A(\frac{1}{\lambda})$ and $y_0\in M_0$. Take $y_i=z_i-\beta_i\lambda^{-i}x_0$ for $i\geq 1$. Then $y=(y_0,y_1,\ldots)\in \ell^2(\mathbb{C}^m)$ since

$$\sum_{i=0}^{\infty} \|y_i\|^2 = \|y_0\|^2 + \sum_{i=1}^{\infty} \|z_i-\beta_i\lambda^{-i}x_0\|^2 \leq$$

$$\leq \|y_0\|^2 + 2(\sum_{i=1}^{\infty} \|z_i\|^2+ \sum_{i=1}^{\infty} \beta_i\lambda^{-i}\|x_0\|^2) < \infty.$$

Then we have $z=x+y$ with $x=(x_0,\beta_1\lambda^{-1}x_0,\beta_2\lambda^{-2}x_0,\ldots)\in \mathrm{Ker}(V^X-\lambda I)$ and $y=(y_0,y_1,y_2,\ldots)\in M$.

Let π be the projection onto $\mathrm{Ker}(V^X-\lambda I)$ along M. According to [1], Section 1.1 we have the factorization

$$A(\tfrac{1}{z}) = [I+w(M-I)V(z-V)^{-1}(I-\pi)\tau][I+w(M-I)V\pi(z-V)^{-1}\tau].$$

Consider the right factor $I+w(M-I)V\pi(z-V)^{-1}\tau$, which is equal to $I+w(M-I)V\pi(z-\pi V\pi)^{-1}\pi\tau$. Its inverse is $I-w(M-I)V\pi(z-\pi V^X\pi)^{-1}\pi\tau$. Define the space $Y=\mathrm{Ker}(V^X-\lambda I)\dot{+}\tau M_0$. Then $\dim Y=m$. If we write $\mathbb{C}^m=\mathrm{Ker}A(\frac{1}{\lambda})\dot{+}M_0$ the isomorphism S between $\mathbb{C}^m$ and Y is given by the matrix $S=\begin{bmatrix}\pi\tau & 0\\ 0 & \tau\end{bmatrix}$, and its inverse S^{-1} is given by $w\mid Y$. We consider the operator $\pi V^X\pi$ as an operator acting on Y, $\pi\tau$: $\mathbb{C}^m\to Y$, $w(M-I)V\pi$: $Y\to\mathbb{C}^m$. Then

$$I-w(M-I)V\pi(z-\pi V^X\pi)^{-1}\pi\tau =$$

$$= I - w(M-I)V\pi S(z-S^{-1}\pi V^X\pi S)^{-1}S^{-1}\pi\tau.$$

Now $S^{-1}\pi V^X\pi S$, $S^{-1}\pi\tau$ and $w(M-I)V\pi S$ map $\mathbb{C}^m$ to $\mathbb{C}^m$. We determine their matrices with respect to the decomposition $\mathbb{C}^m=\mathrm{Ker}A(\frac{1}{\lambda})\dot{+}M_0$. For $x\in\tau M_0$ we have $w(M-I)V\pi x=0$ and for $x\in \mathrm{Ker}(V^X-\lambda I)$ one has

$$w(M-I)V\pi x = w(M-I)Vx =$$

$$= w(\sum_{j=1}^{\infty} A_jx_0\lambda^{-j+1}, -\beta_1x_0,-\beta_2x_0\lambda^{-1},\ldots) =$$

$$= \sum_{j=1}^{\infty} A_jx_0\lambda^{-j+1} = \lambda(A(\tfrac{1}{\lambda})-I)x_0 = -\lambda x_0 = -\lambda wx.$$

Thus the matrix of $w(M-I)V\pi S$ is $\begin{bmatrix}-\lambda I & 0\\ 0 & 0\end{bmatrix}$. The matrix of $S^{-1}\pi\tau$ is easily seen to be $\begin{bmatrix}I & 0\\ 0 & 0\end{bmatrix}$ and the matrix of $S^{-1}\pi V^X\pi S$ is $\begin{bmatrix}\lambda I & 0\\ 0 & 0\end{bmatrix}$. It follows that

$$I - w(M-I)V\pi S(z-S^{-1}\pi V^X\pi S)^{-1}S^{-1}\pi\tau =$$

$$= \begin{bmatrix}I & 0\\ 0 & I\end{bmatrix} - \begin{bmatrix}-\lambda I & 0\\ 0 & 0\end{bmatrix}(z-\begin{bmatrix}\lambda I & 0\\ 0 & 0\end{bmatrix})^{-1}\begin{bmatrix}I & 0\\ 0 & 0\end{bmatrix} =$$

$$= \begin{bmatrix} I + \frac{\lambda}{z-\lambda} I & 0 \\ 0 & I \end{bmatrix} = \begin{bmatrix} \frac{z}{z-\lambda} I & 0 \\ 0 & I \end{bmatrix}$$

For its inverse we have

$$I + w(M-I)V\pi(z-V)^{-1}\tau = I - \frac{1}{z}\begin{bmatrix} \lambda I & 0 \\ 0 & 0 \end{bmatrix}.$$

Now let $Z_\lambda = \begin{bmatrix} \lambda I_{\mathrm{ker}A(\frac{1}{\lambda})} & 0 \\ 0 & 0 \end{bmatrix}$. We have shown that

$$A(\tfrac{1}{z}) = B_1(\tfrac{1}{z})(I - \tfrac{1}{z} Z_\lambda)$$

where rank $Z_\lambda = \dim \mathrm{Ker} A(\frac{1}{\lambda})$ and the eigenvalues of Z_λ are exactly λ and 0. Clearly Z_λ is normal. Therefore

$$A(z) = B_1(z)(I - zZ_\lambda),$$

where $B_1(z)$ is an entire matrix valued function. Since all the partial multiplicities of $A(z)$ at $\frac{1}{\lambda}$ are one, $B_1(z)$ has no zero at $\frac{1}{\lambda}$, and of course has partial multiplicities equal to one at its zeros.□

Of course we could also have proved Theorem 2.2 directly by putting $Z_\lambda = \begin{bmatrix} \lambda I_{\mathrm{ker}A(\frac{1}{\lambda})} & 0 \\ 0 & 0 \end{bmatrix}$, and showing that $B_1(z) = A(z)(I - zZ_\lambda)^{-1}$ is entire and det $B_1(\frac{1}{\lambda}) \neq 0$. Our approach shows that this factorization can be seen as an application of the factorization method in [1].

Continuing the argument of Theorem 2.2 by induction one gets

$$A(z) = B_n(z) \prod_{j=1}^{n} (I - zZ_j)$$

where the order of multiplication is understood to be from right to left, $\sigma(Z_j) = \{\lambda_j, 0\}$, rank $Z_j = \dim \mathrm{Ker}\, A(z_j)$ and $\det B_n(z_j) \neq 0$ for $j = 1,2,\ldots,n$, and $B_n(z)$ is entire.

Before proving the convergence of the infinite product occuring in Theorem 2.1 we establish two lemmas.

LEMMA 2.3. Let B_j be mxm matrices and suppose that $\sum_{j=1}^{\infty} \|B_j\| < \infty$. Then $\prod_{j=1}^{\infty} (I + B_j)$ converges, where the product is understood to be taken from right to left.

PROOF. Let $\varepsilon > 0$. For $k \geq n$ we have that

$$\|\prod_{j=1}^{n}(I+B_j) - \prod_{j=1}^{k}(I+B_j)\| =$$

$$= \|[I-\prod_{j=n+1}^{k}(I+B_j)]\prod_{j=1}^{n}(I+B_j)\| =$$

$$= \|\{\sum_{j=n+1}^{k} B_j+\ldots+B_k\cdots B_{n+1}\}\prod_{j=1}^{n}(I+B_j)\| \leqslant$$

$$\leqslant \{\sum_{j=n+1}^{k}\|B_j\|+\ldots+\|B_k\|\cdots\|B_{n+1}\|\}\prod_{j=1}^{n}(1+\|B_j\|)$$

$$= |1-\prod_{j=n+1}^{k}(1+\|B_j\|)|\prod_{j=1}^{n}(1+\|B_j\|) =$$

$$|\prod_{j=1}^{n}(1+\|B_j\|) - \prod_{j=1}^{k}(1+\|B_j\|)|.$$

Since $\prod_{j=1}^{\infty}(1+\|B_j\|)$ converges (see [2], Corollary VII 5.6) there is an N such that for $n,k \geqslant N$ we have

$$|\prod_{j=1}^{n}(1+\|B_j\|) - \prod_{j=1}^{k}(1+\|B_j\|)| < \varepsilon.$$

Hence $\{\prod_{j=1}^{n}(I+B_j)\}_{n=1}^{\infty}$ is a Cauchy sequence and thus it converges.□

LEMMA 2.4. Let $E_p(z)=(1-z)\exp(\sum_{j=1}^{p}\frac{1}{j}z_j)$. Then for any normal matrix B with $\|B\|<1$ one has

$$\|E_p(B)-I\| \leqslant \|B\|^{p+1}.$$

PROOF. It is well known that if $|z|\leqslant 1$ then $|E_p(z)-1|\leqslant|z|^{p+1}$ (see e.g. [2], Lemma VII 5.11). Now if $|f(z)|\leqslant|g(z)|$ and f and g are defined and analytic in a neighborhood of $\sigma(B)$ then $\|f(B)\|\leqslant\|g(B)\|$, since $\|f(B)\|=\max\{|f(\lambda)| \mid \lambda\in\sigma(B)\}$ and $\|g(B)\|=\max\{|g(\lambda)| \mid \lambda\in\sigma(B)\}$. Putting $f(z)=E_p(z)-1$ and $g(z)=z^{p+1}$ we obtain the desired inequality.□

We are now ready for our next result.

THEOREM 2.5. Let p, σ and ρ be as defined above, $E_p(z)=(1-z)\exp(\sum_{j=1}^{p}\frac{1}{j}z^j)$, and let Z_j be any normal matrix with eigenvalues 0 and λ_j and rank $Z_j=\dim\operatorname{Ker}A(z_j)$, where z_j

is the j-th zero of A(z) and $\lambda_j = \frac{1}{z_j}$. Then the infinite product $\prod_{j=1}^{\infty} E_p(zZ_j)$, taken from right to left, converges uniformly on compact subsets of $\mathbb{C}$. Moreover, the infinite product is an entire matrix valued function of order at most σ.

PROOF. Choose a compact set $K \subset \mathbb{C}$ and let r be such that $k \subset \{z \mid |z| \leqslant r\}$. As before, we order the λ_j's so that $|\lambda_1| \geqslant |\lambda_2| \geqslant \ldots$. Choose N so large that $r < |z_N| = \frac{1}{|\lambda_N|}$. Then for $j \geqslant N$ we have $|z\lambda_j| = \|zZ_j\| < 1$ for $z \in K$. By Lemma 2.4 we obtain

$$\|E_p(zZ_j) - I\| \leqslant (|z| \; \|Z_j\|)^{p+1} \leqslant r^{p+1} |\lambda_j|^{p+1}$$

for $j \geqslant N$ and $z \in K$. Now note that $\sum_{j=1}^{\infty} |\lambda_j|^{p+1} < \infty$ by the definition of p and σ. Setting $B_j = E_p(zZ_j) - I$ in Lemma 2.3 one sees that the product $\prod_{j=N}^{\infty} E_p(zZ_j)$ converges. Hence also $\prod_{j=1}^{\infty} E_p(zZ_j)$ converges. As $\|B_j\| \leqslant r^{p+1} |\lambda_j|^{p+1}$ for $z \in K$, the convergence is uniform on K. It follows that $\prod_{j=1}^{\infty} E_p(zZ_j)$ in an entire function.

Next we prove the statement about the order of the infinite product. Set

$$P(z) = (p_{ij}(z))_{i,j=1}^{m} = \prod_{j=1}^{\infty} E_p(zZ_j).$$

Since Z_j is normal, it is diagonal with respect to some orthonormal basis. We let

$$S_j^{-1} Z_j S_j = \begin{bmatrix} \lambda_j & & & & \\ & \ddots & & & \\ & & \lambda_j & & \\ & & & 0 & \\ & & & & \ddots \\ & & & & & 0 \end{bmatrix} \Big\} m_j.$$

Then

$$S_j^{-1} E_p(zZ_j) S_j = E_p(S_j^{-1} z Z_j S_j) =$$

$$= \begin{bmatrix} E_p(z\lambda_j) & & & & \\ & \ddots & & & \\ & & E_p(z\lambda_j) & & \\ & & & 1 & \\ & & & & \ddots \\ & & & & & 1 \end{bmatrix} \Big\} m_j.$$

Therefore

$$E_p(zZ_j) = S_j \begin{bmatrix} E_p(z\lambda_j) & & & & \\ & \ddots & & & \\ & & E_p(z\lambda_j) & & \\ & & & 1 & \\ & & & & \ddots \\ & & & & & 1 \end{bmatrix} S_j^{-1}.$$

It now follows that the entries $p_{ij}(z)$ of $P(z)$ are either constants or multiples of canonical products, which by the scalar theorem ([10], Section 8.2.5) have order at most σ. Then for every $\varepsilon>0$ there exist positive B and C such that

$$\|P(z)\|^2 \leqslant \sum_{i=1}^{m}\sum_{j=1}^{m} |p_{ij}(z)|^2 \leqslant B \exp C|z|^{\sigma+\varepsilon}$$

for $|z|$ sufficiently large. So it follows that $P(z)$ has order at most σ.□

Finally we give the remaining parts of the proof of Theorem 2.1.

PROOF OF THEOREM 2.1. To obtain the matrices Z_j and the function

$$B(z) = A(z)(\prod_{j=1}^{\infty} E_p(zZ_j))^{-1}$$

we define a sequence $B_n(z)$ of entire functions as follows: apply the argument of Theorem 2.2 to $A(z)$ with its zero z_1, and let

$$B_1(z) = A(z)E_p(zZ_1)^{-1}.$$

Then $B_1(z)$ is entire since $\exp \sum_{j=1}^{p} \frac{1}{j}(zZ_1)^j$ is everywhere invertible, and $\det B_1(z_1)\neq 0$. Next, apply again Theorem 2.2, now to $B_1(z)$ with its zero z_2 to obtain the entire function

$$B_2(z) = B_1(z)E_p(zZ_2)^{-1} = A(z)(\prod_{j=1}^{2} E_p(zZ_j))^{-1}.$$

Continue this process by induction. Then

$$B(z) = B_n(z)(\prod_{j=n+1}^{\infty} E_p(zZ_j))^{-1},$$

so $B(z)$ is analytic at $\mathbb{C}\setminus\{z_{n+1},z_{n+2},\ldots\}$ for all n. Hence $B(z)$ is an entire function.

Further, $\det A(z)=\det B(z)\prod_{j=1}^{\infty}\det E_p(zZ_j)$, and $\det A(z)$ and $\det E_p(zZ_j)$ have z_j as common zero with the same multiplicity. It follows that $B(z)$ is everywhere invertible. One sees from this that $\operatorname{Ker} A(z_i)=\operatorname{Ker}\prod_{j=1}^{\infty} E_p(z_iZ_j)$ for every zero z_i of $A(z)$. This can also be seen directly from our construction.

It remains to show that $B(z)$ has order at most ρ. First we show that each entry of $B(z)$ has order at most ρ. We need the following elementary facts from the theory of complex valued entire functions: if f(z) and g(z) are entire functions having order ρ_1 and ρ_2, respectively, then f(z)+g(z), f(z)-g(z) and f(z)/g(z) (provided the latter function is entire) have order at most $\max\{\rho_1,\rho_2\}$. Now by Cramer's rule

$$\left[\prod_{j=1}^{\infty} E_p(zZ_j)\right]^{-1} = \frac{\operatorname{Adj}\left[\prod_{j=1}^{\infty} E_p(zZ_j)\right]}{\det \prod_{j=1}^{\infty} E_p(zZ_j))}.$$

By the remarks made above each entry of this matrix has order at most $\sigma \leqslant \rho$. Therefore each entry of

$$B(z) = A(z)\left[\prod_{j=1}^{\infty} E_p(zZ_j)\right]^{-1}$$

has order at most ρ. Now, by the argument given in the last part of the proof of Theorem 2.5, we see that B(z) has order at most ρ. This completes the proof.□

REFERENCES

1. Bart, H., Gohberg, I., and Kaashoek, M.A.: Minimal Factorization of Matrix and Operator Functions, Birkhäuser Verlag, Basel, 1979.

2. Conway, J.B.: Functions of One Complex Variable, 2nd edition, Springer Verlag, New York, 1978.

3. Gohberg, I., Lancaster P., and Rodman, L.: Matrix Polynomials, Academic Press, New York, 1982.

4. Gohberg, I., Kaashoek, M.A. and Lay, D.C.: Equivalence, linearization and decompositions of holomorphic operator functions, J. Funct. Anal. 28(1978), 102-144.

5. Gohberg, I., and Krein, M.G.: Introduction to the Theory of Linear Nonselfadjoint Operators, Translations of Mathematical Monographs, Vol.18, Amer. Math. Soc., Providence, R.I., 1969.

6. Gohberg, I., and Rodman, L.: Analytic matrix functions with prescribed local data, J. de Analyse Mathematique 40 (1981), 90-128.

7. Pólya, G.: Neuer beweis für die Produktdarstellung der ganzen transzendenten Funktionen endlicher Ordnung, S.B. Bayer Akad. Wiss.(1921), 29-40 = Collected Works, Vol.1, Singularities of Analytic Functions, ed. R.P. Boas, M.I.T. Press, Cambridge, Mass., 1974.

8. Pringsheim, A.: Elementare theorie der ganzen transzendenten Funktionen von endlicher Ordnung, Math. Annalen 58 (1904), 257-342.

9. van der Mee, C.V.M.: Realization and linearization, Rapport nr.109, Wiskundig

Seminarium der Vrije Universiteit, Amsterdam, 1979.

10. Titchmarsh, E.C.: The Theory of Functions, Oxford Univ. Press, 1939.

Carl L. Prather
Department of Mathematics
Virginia Polytechnic Institute
and State University,
Blacksburg, VA 24061,
USA

André C.M. Ran
Subfaculteit Wiskunde
en Informatica,
Vrije Universiteit,
Amsterdam,
The Netherlands

Operator Theory:
Advances and Applications, Vol. 19

THE ABSTRACT TIME-DEPENDENT TRANSPORT EQUATION

V. Protopopescu

The abstract time-dependent transport equation is considered as an evolution (initial-boundary value) problem in the Banach spaces L_p, $1 \leqslant p < \infty$. By using the trajectory method, the basic issues of existence, uniqueness, positivity, and dissipativity are settled for very general, possibly time-dependent domains, operators, and boundary conditions.

1. EXAMPLES OF TRANSPORT PROCESSES

Almost every physical, chemical, biological or ecological phenomenon is directly accompanied by, if not overwhelmingly dominated by, the transport of different quantities such as mass, electric charge, number of particles, energy, momentum, intensity, information, etc. pertinent for describing the problem at hand. Examples abound in every conceivable field and range from neutron transport in reactors, polarized light transfer in atmospheres, exciton hopping in semiconductors, and reactant diffusion in solutions to vehicle traffic on highways, population migration in environmental situs, and oil flows through porous rocks. Transport phenomena are described by transport equations which can be classified according to several criteria: linear or nonlinear, stationary or time-dependent, kinetic or hydrodynamic, deterministic or stochastic. We shall present here an abstract version of the linear, time-dependent, deterministic, kinetic equation together with some general results about its solution. The results have been obtained jointly with R. Beals [1]. For a more detailed account on the physical basis of the transport equation, the setting of the abstract problem, related results and more detailed proofs, we refer to the monograph written jointly with W. Greenberg and C. V. M. van der Mee [2].

2. A TYPICAL REALIZATION

A typical realization of the transport problem is provided by the transport of a dilute gas of classical particles in a host medium. The unknown is the scalar function $u = u(t,x,\xi)$, called the one-particle distribution function, which describes the number density of gas particles having the position $x = (x_1,x_2,x_3)$ and the velocity $\xi = (\xi_1,\xi_2,\xi_3)$ at time t. In general, the volume accessible to the gas is a finite domain $\Omega \subset R^3$, the velocity space is the whole R^3 (the theory is non-relativistic), and the time t takes values in R_+. The transport equation is a balance (rate) equation for the distribution function u and models the two main processes which are supposed to take place at the kinetic level: collisions between the molecules of the gas and the host medium and, between collisions, streaming of the molecules of the gas according to the laws of dynamics. Thus in a reasonably standard notation one writes:

$$\frac{\partial u}{\partial t}(t,x,\xi)+\xi\cdot\frac{\partial u}{\partial x}(t,x,\xi)+\left(a\cdot\frac{\partial u}{\partial \xi}\right)(t,x,\xi)+(hu)(t,x,\xi)$$

$$-\int\sigma(t,x,\xi'\to\xi)u(t,x,\xi')d\xi'=f(t,x,\xi) \ . \tag{1}$$

The first three terms give the total (Stokes) derivative of u with respect to t, a is the acceleration produced by the external fields, f is an autonomous source, $-h$ and the integral operator describe the collisional loss and gain, respectively, and the dot means the usual scalar product. In order to be uniquely solvable, the transport equation (1) has to be supplemented with an initial condition

$$u(t\downarrow 0,x,\xi)=u_0(x,\xi) \tag{2}$$

and with boundary conditions describing the processes which take place at the walls of the domain Ω. If $n(x)$ is the outward normal at the point x of the surface $\partial\Omega$, we shall denote

$$D_{\mp}=\left\{(x,\xi)\in\partial\Omega\times R^3 : n(x)\cdot\xi\lessgtr 0\right\} \ . \tag{3}$$

Particles at $(x,\xi)\in D_-$ (respectively D_+) are incoming (respectively outgoing) particles. The boundary condition establishes a relationship between them, in the form

$$u|_{D_-}=K(u|_{D_+})+g \ , \tag{4}$$

where K is an appropriate boundary operator and g an autonomous boundary source.

The typical form of the boundary condition in gas dynamics reads:

$$u(t,x,\xi)=\int_{n(x)\cdot\xi'>0} k(\xi'\to\xi)|n(x)\cdot\xi'|u(t,x,\xi')d\xi', \ x\in\partial\Omega, \quad n(x)\cdot\xi<0 \ , \tag{5}$$

where the positive kernel $k(\xi'\to\xi)$ satisfies the condition

$$\int_{n(x)\cdot\xi>0} k(\xi'\to\xi)|n(x)\cdot\xi|d\xi \leq 1 \ , \tag{6}$$

implying that at the boundary the current is either conserved or absorbed. In order to account for specular or reverse reflections, the kernels k can contain δ-functions. A smooth kernel k models a diffuse reflection, while $k\equiv 0$ describes absorption.

3. THE ABSTRACT VERSION

The abstract version of the transport equation (1) can be essentially written as

$$\tau\cdot\frac{\partial u}{\partial\zeta} + Au = f \ , \tag{7}$$

where $\tau\cdot\dfrac{\partial}{\partial\zeta}$ is a real vector field over an open set $\Sigma\subset R^{2m+1}$. By abuse we shall also call τ a vector field. In order to reinforce the connection with the time-dependent transport problem, we shall identify τ_0 with 1, $\tau_1, \ldots,\tau_m$ with $\xi_1,\ldots,\xi_m$, and $\tau_{m+1},\ldots,\tau_{2m}$ with $a_1,\ldots,a_m$ (and, accordingly, ζ_0 with t, $\zeta_1,\ldots,\zeta_m$ with $x_1,\ldots,x_m$, and $\zeta_{m+1},\ldots,\zeta_{2m}$ with $\xi_1,\ldots,\xi_m$), but in principle the abstract treatment can be carried out irrespective of these particular identifications. The set Σ is called the phase space of the problem; it is the domain where the independent variables take values; $\zeta = (\zeta_0,\ldots, \zeta_{2m})\in\Sigma$. In our previous example, it had the form $\Sigma = (0, T)\times\Omega\times R^3$, but in general the direct product structure is not needed. In fact, when the space and velocity domains vary in time, this structure cannot be assumed. In the following, we shall only impose that $\zeta_0\in(0,T)$, $T<\infty$. If the vector τ has no components identically equal to zero, Σ will be endowed with the ordinary Lebesgue measure $d\zeta = dtdxd\xi$ corresponding to the independent variables of the problem.

Remark

When $a\equiv 0$, the vector field acts only on the variables t and x, but ξ remains as an independent variable. This situation creates a certain freedom in choosing the measure $d\zeta$, which in this particular case can be taken as $d\zeta = dtdxd\rho(\xi)$ with $d\rho(\xi)$ chosen according to the physics of the collisions.

Equation (7) is an evolution equation for the distribution function u to be sought in a certain functional space. In general, u may be a vector with a finite number of components, each of them belonging to an appropriate Banach space. The definition of the field $\tau \cdot \frac{\partial}{\partial \zeta}$ and of the "collision" operator A is generalized in an obvious way as to include these cases too. To fix ideas, we shall limit this presentation only to scalar functions u to be sought in the Banach spaces $L_p(\Sigma; d\zeta)$, $1 \leqslant p < \infty$.

The boundary of the phase space, $\partial\Sigma$, can be partitioned in the following way:

$$\partial\Sigma = D^- \cup D^+ \cup D^0 \cup \widetilde{D} \ , \tag{8}$$

where

$$D^{\overset{\mp}{0}} = \left\{ \zeta \in \partial\Sigma : \tau(\zeta) \cdot n(\zeta) \lesseqqgtr 0 \right\} \ , \tag{9}$$

$$n(\zeta) = (n_0, \ldots, n_{2m}) = \text{the outer normal at } \zeta \in \partial\Sigma \ , \tag{10}$$

and

$$\widetilde{D} = \left\{ \zeta \in \partial\Sigma : n(\zeta) \text{ is not defined} \right\} \ . \tag{11}$$

In the following, we shall consider that $\partial\Sigma$ is piecewise C^1; therefore, $\widetilde{D}$ will be of measure zero with respect to $\partial\Sigma$.

In order to determine the evolution problem we must add boundary conditions in the form

$$u^- = \mathbb{K} u^+ + g \ , \tag{12}$$

where $u^{\mp}$ are the traces of u on the boundaries $D^{\mp}$ and, as before, g is the autonomous boundary source and $\mathbb{K}$ is the boundary operator

$$\mathbb{K} : L_p(D^+; d\nu^+) \to L_p(D^-; d\nu^-) \ . \tag{13}$$

Here we adopt the following definition of traces. If u and $\tau\cdot\frac{\partial u}{\partial \zeta}$ belong to $L_p(\Sigma;d\zeta)$, we say that $u^{\mp}\in L_p(D^{\mp};d\nu^{\mp})$ are traces for u if the extended Green's identity

$$\langle \tau\cdot\frac{\partial u}{\partial \zeta},w\rangle + \langle u,\tau\cdot\frac{\partial w}{\partial \zeta}\rangle = \int_{D^+} u^+ w d\nu^+ - \int_{D^-} u^- w d\nu^- \tag{14}$$

is valid for all w belonging to an appropriate test function space.

Proving the existence of the appropriate boundary measures $d\nu^{\mp}$, as well as the existence and uniqueness of the traces $u^{\mp}$ which make (12) meaningful, is one of the important ingredients of the abstract transport theory [1, 2, 3, 4]. However, we shall not insist on this subject here.

A vector $u\in L_p$ is a good candidate for a solution of (1), (12) if it belongs to the set

$$E_p = \left\{u\in L_p : \tau\cdot\frac{\partial u}{\partial \zeta} + Au\in L_p, u^{\mp}\in L_p(D^{\mp};d\nu^{\mp}) \ , \ u^- = \mathbb{K}u^+ + g\right\} . \tag{15}$$

We note that the relation (12) is in fact an *initial*-boundary condition. Since ζ_0 has been given the interpretation of time, $D^{\mp}$ are naturally and unequivocally split into

$$D^- = D_0 \cup D_-, \quad D^+ = D_T \cup D_+ \ , \tag{16}$$

where

$$D_0 := \partial\Sigma \cap \left\{\zeta_0 = 0\right\}, D_T := \partial\Sigma\cap\left\{\zeta_0 = T\right\} .$$

Accordingly, the boundary operator $\mathbb{K}$ and the source g must have the form

$$\mathbb{K} = (0,K) \quad K : L_p(D_+;d\nu^+|_{D_+}) \rightarrow L_p(D_-;d\nu^-|_{D_-}) \ , \tag{17}$$

$$g = (g_0;g_-) \ . \tag{18}$$

In general, K, D_-, D_+ will be dependent on time.

4. MAIN EXISTENCE RESULT

We shall assume that the vector field τ is Lipschitz continuous with respect to ζ and that it is divergent free; i.e., $\frac{\partial}{\partial \zeta}\cdot\tau = 0$. With our identifications, this condition reduces to

$\sum_{i=1}^{m} \frac{\partial a_i}{\partial \xi_i} = 0$. The Lipschitz condition ensures the existence and uniqueness of the integral curves as a consequence of Picard's theorem. The divergence-free condition is reminiscent of the original physical foundation of the problem and is related to the conservation of the measure $d\zeta$ under the dynamical flow in the absence of collisions. (Liouville's theorem). Yet, it is not a necessary condition from the abstract point of view and relaxing it can be done without special problems [5].

Moreover, we shall assume that the integral curves of the field $\tau \cdot \frac{\partial}{\partial \zeta}$ remain in a bounded region of Σ. Then, we may state:

THEOREM 1. *i.) Let* $f \in L_p(\Sigma; d\zeta), g \in L_p(D^-; d\nu^-)$, $\|A\| < \infty, \|\mathbb{K}\| < 1, 1 \leqslant p < \infty$. *Then the abstract transport problem (1), (12) is well posed and has a unique solution* $u \in E_p$ *and there exists* $C = C(p, A, \mathbb{K})$ *such that*

$$\|u\| \leqslant C\left(\|f\| + \|g\|\right) . \tag{19}$$

ii.) Let the conditions in i.) be satisfied, let $f \geqslant 0, g \geqslant 0$, *and suppose that* $\mathbb{K}$ *and* $\lambda_0 I - A$ *(for some* $\lambda_0 \in R$*) are positive operators in the cone-sense, i.e., they transform nonnegative functions into nonnegative functions. Then* $u \geqslant 0$.

iii.) Let the conditions in i.) be satisfied, let $f = 0, g = (g_0; 0)$ *and suppose*

$$\int_\Sigma (sgn u)|u|^{p-1} A u \, d\zeta \geqslant 0 . \tag{20}$$

Then

$$\|u\| \leqslant \|g_0\| .$$

The proof of the main result relies on the following three important propositions:

PROPOSITION 1. (Existence of Traces and Green's Identity) *i.) Let* $u \in L_p, \tau \cdot \frac{\partial u}{\partial \zeta} \in L_p$. *Then there exist unique traces* $u^{\mp}$.

ii.) If $u^+ \in L_p(D^+; d\nu^+)$, *then* $u^- \in L_p(D^-; d\nu^-)$ *and vice versa. Morever, Green's identity is valid in the strong form*

$$p \int_\Sigma (sgn u) \, |u|^{p-1} \tau \cdot \frac{\partial u}{\partial \zeta} d\zeta = \int_{D^+} |u^+|^p d\nu^+ - \int_{D^-} |u^-|^p d\nu^- . \tag{21}$$

PROPOSITION 2. (The Unperturbed Problem) *The problem*

$$\tau \cdot \frac{\partial u}{\partial \zeta} = f \quad \text{in } \Sigma \quad f \in L_p(\Sigma; d\zeta) \tag{22}$$

$$u^- = g \quad \text{on } D^- \quad g \in L_p(D^-; d\nu^-) \tag{23}$$

has a unique solution $u \in E_p$. *For* $f \geqslant 0, g \geqslant 0$ *we have* $u \geqslant 0$.

PROPOSITION 3. (The Modified Problem) *The unperturbed problem (22)-(23) has a unique solution* $u \in E_p$ *if and only if the problem*

$$\left(\tau \cdot \frac{\partial}{\partial \zeta} + \lambda \right) u_\lambda = f_\lambda \quad \text{in } \Sigma \quad f_\lambda \in L_p(\Sigma; d\zeta) \tag{24}$$

$$u_\lambda^- = g_\lambda \quad \text{on } D^- \quad g_\lambda \in L_p(D^-; d\nu^-) \tag{25}$$

has a unique solution $u_\lambda \in E_p$. *The relationship between the two problems is given by*

$$u(\zeta) = e^{\lambda \zeta_0} u_\lambda(\zeta), \ f(\zeta) = e^{\lambda \zeta_0} f_\lambda(\zeta), \ g(\zeta) = e^{\lambda \zeta_0} g_\lambda(\zeta) \ . \tag{26}$$

PROOF OF THEOREM 1. The idea of the proof is perturbative and it goes in three steps.

1. Since by Proposition 2 the problem (22)-(23) is uniquely solvable, then the problem (24)-(25) is uniquely solvable. We shall write its solution as

$$u_\lambda = S_\lambda(f_\lambda, g_\lambda) = S_\lambda(f_\lambda, 0) + S_\lambda(0, g_\lambda) \ , \tag{27}$$

where we used the linearity. The solution u_λ belongs to E_p and Green's identity (21) applied to u_λ reads

$$\|u_\lambda^+\|^p + \lambda \|u_\lambda\|^p \leqslant \|g_\lambda\|^p + \lambda^{1-p} \|f_\lambda\|^p \ , \tag{28}$$

yielding the useful estimates

$$\|S_\lambda(f_\lambda, 0)\| \leqslant \lambda^{-1} \|f_\lambda\| \ , \tag{29}$$

$$\|S_\lambda(f_\lambda,0)^+\| \leqslant \lambda^{-1+1/p}\|f_\lambda\| \ , \tag{30}$$

$$\|S_\lambda(0,g_\lambda)\| \leqslant \lambda^{-1/p}\|g_\lambda\| \ , \tag{31}$$

$$\|S_\lambda(0,g_\lambda)^+\| \leqslant \|g_\lambda\| \ . \tag{32}$$

2. We introduce now the bounded perturbation A and solve

$$\left(\tau\cdot\frac{\partial}{\partial\zeta}+A+\lambda\right)u_\lambda=f_\lambda \tag{33}$$

$$u_\lambda^- = g_\lambda \tag{34}$$

for $\lambda > \|A\|$. We look for a solution to the form

$$u_\lambda = S_\lambda(f_\lambda^*, g_\lambda) \ , \tag{35}$$

where f_λ^* is to be found. Forcing u_λ given by (35) to satisfy (33) yields an equation for f_λ^*:

$$(1+L_\lambda)f_\lambda^* = f - AS_\lambda(0,g_\lambda) \ , \tag{36}$$

where

$$L_\lambda f_\lambda^* = AS_\lambda(f_\lambda^*,0) \ . \tag{37}$$

By (29), $\|L_\lambda\| \leqslant \lambda^{-1}\|A\| < 1$ and we can write

$$f_\lambda^* = \sum_{m=0}^{\infty}(-L_\lambda)^m(f_\lambda - AS_\lambda(0,g_\lambda)) \ . \tag{38}$$

Introducing (38) in (35) gives

$$u_\lambda = T_\lambda(f_\lambda,g_\lambda) \ , \tag{39}$$

where T_λ is now a well defined solution operator. As before, we obtain estimates for $T_\lambda(f_\lambda,g_\lambda) = T_\lambda(f_\lambda,0) + T_\lambda(0,g_\lambda)$:

$$\|T_\lambda(f_\lambda,0)\| = \|S_\lambda(f^*_\lambda,0)\| \leqslant \frac{\lambda^{-1}\|f_\lambda\|}{1-\|L_\lambda\|} \tag{40}$$

$$\|T_\lambda(0,g_\lambda)\| \leqslant \|S_\lambda(f^*_\lambda,g_\lambda)\| \leqslant \lambda^{-1/P}\|g_\lambda\| + \frac{\lambda^{-1}\|f_\lambda\|}{1-\|L_\lambda\|} \tag{41}$$

$$\|T_\lambda(0,g_\lambda)^+\| = \|S_\lambda(f^*,g_\lambda)^+\| = \|S_\lambda(0,g_\lambda)^+ + S_\lambda(f^*_\lambda,0)^+\| \leqslant \left[1+(\lambda-\|A\|)^{-1}\|A\|\right]\|g_\lambda\| \tag{42}$$

3. Perturb the boundary operator:

$$\left(\tau\cdot\frac{\partial u}{\partial \zeta} + Au + \lambda\right)u_\lambda = f_\lambda \tag{43}$$

$$u_\lambda^- = \mathbb{K}u_\lambda^+ + g_\lambda \ . \tag{44}$$

We look now for a solution

$$u_\lambda = T_\lambda(f_\lambda, g^*_\lambda) \ , \tag{45}$$

where g^*_λ is to be found. Forcing u_λ given by (45) to satisfy (44) yields an equation for g^*_λ

$$(1-M_\lambda)g^*_\lambda = g_\lambda + \mathbb{K}T_\lambda(f_\lambda,0)^+ \tag{46}$$

$$M_\lambda g^*_\lambda = \mathbb{K}T_\lambda(0,g^*_\lambda)^+ \tag{47}$$

with

$$\|M_\lambda\| < (1+\left(\lambda-\|A\|\right)^{-1}\|A\|)\|\mathbb{K}\| < 1 \ . \tag{48}$$

Then g^*_λ can be found uniquely:

$$g^*_\lambda = \sum_{m=0}^{\infty} M^m_\lambda(g_\lambda + \mathbb{K}T_\lambda(f_\lambda,0)^+) \ . \tag{49}$$

The transformation $u = u_\lambda e^{\lambda\zeta_0} := U_\lambda(f,g)$ eventually gives the solution of the original problem (1),(12) and defines the solution operator U_λ. The bound (19) is straightforwardly obtained by extending to U_λ the estimates previously obtained for S_λ and T_λ.

ii.) By Proposition 2, the solution $u = e^{\lambda\zeta_0} S_\lambda(f_\lambda, g_\lambda)$ of the problem (22)-(23) is non-negative whenever $f \geq 0, g \geq 0$. All the operations yielding f^*, T_λ, g^*, and U_λ are positively preserving; therefore $u = e^{\lambda\zeta_0} U_\lambda(f,g) \geq 0$.

iii.) Inequality (20) is obtained from Green's identity (21) applied to the problem (22)-(23) with $f=0, g_-=0$. We get

$$\int_{D_T} |u_T|^p d\nu_T - \int_{D_0} |u_0|^p d\nu_0 = -\int_{D_+} |u_+|^p d\nu_+ + \int_{D_-} |Ku_+|^p d\nu_- -$$

$$-p\int (sgnu)|u|^{p-1} Au d\zeta < 0 \ , \tag{50}$$

implying $\|u_T\| \leq \|u_0\|$. The inequality remains valid if instead of $t=T$ we take a section in the phase space at the actual time $t=t$; therefore (20) is proved.

Remarks

1. Inequality (20) emphasizes the role of the time and of the "time-boundaries" D_0, D_T for our problem, which is described by a hyperbolic equation with a dissipation mechanism. If the external sources f, g_- are absent, the initial distribution $u_0 = g_0$ cannot but shrink in the appropriate norm. The dissipation is related to the irreversible character of the transport equations. Yet, the irreversibility is to be understood in a different sense than the irreversibility, say, of the diffusion equation. In general, the transport problem is well posed even for $t \in R_-$, but the corresponding evolution is not positively preserving; therefore, it cannot describe the evolution of a physical distribution function.

2. The whole idea of the proof is a perturbative one and therefore it imposes restrictions on the operator A. For simplicity we took A to be bounded. The proof can be carried out if A can be written as $A = -h + A_1$, where A_1 is bounded and h is a measurable, locally integrable, positive function [1,6]. Extending A to the class of second-order differential operators is yet to be done: the perturbative approach is certainly not applicable and the existence of traces is difficult to prove.

3. Besides the Banach spaces L_p, problems (22)-(23) may be considered in different other spaces – chosen for their technical or physical appropriateness. Results are available for Sobolev spaces [4,7] and for bounded Borel measures [1].

5. THE CONSERVATIVE CASE $\|\mathbb{K}\|=1$

We consider now the problem (1)-(12) with conservative boundaries. We recall that since $\mathbb{K}=(0,K)$, then $\|K\|=1$. The temporal part of the boundary operator is always zero, so imposed by the very nature of the *initial*-boundary value problem we deal with. The main result stated before for $\|K\|<1$ does not and cannot have an exact counterpart in the conservative case. Uniqueness, though, is easy to prove.

PROPOSITION 4. *Let us us consider the problem (1), (12) with $\|K\|=1$. If the problem has a solution $u\in E_p$, then the solution is unique.*

PROOF. Indeed, it suffices to prove that for $f=g=0, u=0$, Green's identity applied to the modified problem

$$\left(\tau\cdot\frac{\partial}{\partial\zeta}+A+\lambda\right)u_\lambda=0\ ,u_\lambda^-=\mathbb{K}u_\lambda^+ \tag{51}$$

yields

$$\lambda p\|u_\lambda\|^p\leqslant\|Ku_{\lambda,+}\|^p-\|u_{\lambda,+}\|^p+p\int_\Sigma(sgnu_\lambda)|u_\lambda|^{p-1}Au_\lambda d\zeta$$

$$\leqslant p\|A\|\|u_\lambda\|^p\ . \tag{52}$$

For $\lambda>\|A\|$, (52) forces $u_\lambda=0$. Consequently $u=u_\lambda e^{\lambda\zeta_0}=0$.

In general, Green's identity gives

$$\|u_{\lambda,T}\|^p+\lambda p\|u_\lambda\|^p-p\|A\|\|u_\lambda\|^p\leqslant\|g_{\lambda,0}\|^p+p\|f_\lambda\|\|u_\lambda\|^{p-1}+$$

$$+\|Ku_{\lambda,+}\|^p+\|g_{\lambda,-}\|^p=\|u_{\lambda,+}\|^p\leqslant$$

$$\|g_\lambda\|^p+(p-1)p^{p/(p-1)}\|u_\lambda\|^p+\|f_\lambda\|^p \tag{53}$$

Taking $t=T$ and λ big enough, we get

$$\|u\|\leqslant C_{\lambda,p}\left(\|g\|+\|f\|\right)\ . \tag{54}$$

However, since $\|K\|=1$, the inequality does not give global evaluations for the traces $u_{\mp}$. Therefore, it may happen that $u^{\mp}$ satisfy the boundary condition but do not belong to $L_p(D^{\mp};d\nu^{\mp})$. That this can actually happen was shown by Voigt [3].

An existence result can be obtained if we strengthen the assumptions on the operators A and K and we weaken the requirement on the solution. Namely, we impose that $K \geqslant 0, \lambda_0 I - A \geqslant 0$ for some $\lambda_0 \in R$ and, based on positivity, we extend, via positive sequences, the definition of K to junctions which may not belong to $L_p(D_+;d\nu_+)$.

Indeed, suppose v_m is a nondecreasing sequence of non-negative functions in $L_p(D_+;d\nu_+)$ with positive limit v. Then Kv_m is also nondecreasing and has limit depending only on v, so we may extend K to such functions by setting

$$Kv = \lim_m Kv_m \ . \tag{54}$$

A natural procedure in trying to solve (1),(12) for $\|K\|=1$ is to replace K by $K_m, m=1,2, \cdots$

$$K_m := \alpha_m K \ . \tag{55}$$

where α_m is an increasing sequence of non-negative scalars with limit 1. Then $\|K_m\| = \alpha_m < 1$, so the problem (1)-(12) with K replaced by K_m has a unique solution $u_m \in E_p$.

Let us now state the main result for the conservative case.

THEOREM 2. *Suppose* $\|A\| < \infty, K > 0$ *and* $\|K\| = 1$. *Suppose* $f \in L_p(\Sigma, d\zeta)$ $g \in L_p(D_-;d\nu_-)$ *are non-negative. Let* $K_m := \alpha_m K, 0 < \alpha_m < 1$, *let* α_m *increase with limit 1, and let* u_m *be the solution of (1), (12) with K replaced by* K_m.

Then the sequence u_m *is nondecreasing and, independently of* α_m, *converges in norm to a function* $u \in L_p(\Sigma; d\zeta)$, *which satisfies the transport equation () and which has traces* $u^{\mp}$ *which satisfy the boundary condition (12) with K extended.*

PROOF. Let us consider (1),(12) for K replaced by $K_{m_0} = \alpha_{m_0} K$. By Theorem 1, the solution is given by $U_{m_0}(f,g)$. For any $m > m_0$ we look for a solution in the form

$$u_m = U_{m_0}(f, g^*) \ . \tag{57}$$

Since $K_m - K_{m_0}$ is positive, we may repeat the argument used in Theorem 1 and conclude that $g^* > g$. Since U_{m_0} is positive, this gives $u_m \geqslant u_{m_0}$.

Since the sequence $\{u_m\}$ is nondecreasing and positive, to show that the positive limit belongs to L_p and is the norm-limit of the sequence, it is enough to obtain a uniform L_p bound for the u_m. Replacing A by $\lambda I + A$ with $\lambda > \|A\|$ and using Green's identity, we get

$$\|u_{\lambda,m}\| \leqslant C_{\lambda,p,m}(\|g_\lambda\| + \|f_\lambda\|) \ . \tag{58}$$

Transforming as before, we obtain (58) with $\lambda = 0$ as well. Green's inequality does not give a uniform bound for $(u_m)_+$; therefore convergence in E_p cannot be proved. Yet $\tau \cdot \frac{\partial u_m}{\partial \zeta}$ converges in $L_p(\Sigma; d\zeta)$; therefore (1) is satisfied. Moreover, (58) gives a uniform bound for $\{u_m\}_T$, which then converges to u_T in norm in $L_p(D_T; d\nu_T)$. Finally, the pointwise limits satisfy

$$u_- = \lim K(u_m)_- = \lim K_m(u_m)_+ = \lim \alpha_m K(u_m)_+ = \lim K(u_m)_+ \ . \tag{59}$$

Now, if $\{\alpha_m\}$, $\{\alpha_m'\}$ are two nondecreasing sequences of positive scalars with limit 1, and u and u' are the limits of the corresponding solutions $\{u_m\}$, $\{u_m'\}$, we construct the sequence $\{\alpha_m''\}$ by taking $\alpha_m'' =$ max $\{\alpha_m, \alpha_m'\}$ and conclude that $u = u'$. Hence the solution is independent of the choice of the sequence $\{\alpha_m\}$.

6. CONCLUDING REMARKS

1. When the space and velocity domains as well as vector field τ and the operators A, K do not depend on time, the abstract transport problem (1),(12) can be solved by semigroup techniques. The solution operator is then given by a uniparametric semigroup which, depending on the case, can be contractive and/or positive. In the Banach spaces L_p, $1 \leqslant p < \infty$, the semigroup is strongly continuous. For $p = \infty$ this is not true; nor is it for the space of bounded measures because the domain of the transport operator is not dense in the underlying spaces.

2. In the semigroup framework, the time-interval is naturally extended to the whole R_+. Determining the asymptotic behavior of the solution becomes one of the important questions of the theory. The answer to this question is related to the spectral mapping theorem for the exponential function and a general result is not available. For a detailed account on the semigroup method as applied to transport problems and on the available results, see References [1-8] and references therein.

ACKNOWLEDGEMENTS

I thank the organizers of the Workshop for having given me the opportunity to present these results. The support of the Engineering Physics and Mathematics Division during the preparation of the manuscript is gratefully acknowledged.

LITERATURE

1. R. Beals, V. Protopopescu: Abstract time-dependent transport equations, preprint, 1984; to appear in J. Math. Anal. Appl.
2. W. Greenberg, C. V. M van der Mee, V. Protopopescu: Boundary value problems in abstract kinetic theory, Boston & Basel, Birkhäuser Verlag, 1986.
3. J. Voigt: Functional analytic treatment of initial boundary value problems for collisionless gases, Habilitationsschrift, Universität München, 1980.
4. S. Ukai: Solutions of the Boltzmann equation, preprint, 1985.
5. C. Bardos: Boundary value problems for first order partial differential equations with real coefficients; approximation theorems; application to transport equation, Ann. Scient. Ec. Norm. Sup. 4 Série, 3 185-233, 1970 (in French).
6. J. Voigt: On substochastic C_0 - semigroups and their generators, in Semesterberich Funktional Analysis Wintersemester 84/85, Tübingen, p. 71-86.
7. Y. Shizuta: On the classical solutions of the Boltzmann equation, Comm. Pure Appl. Math. 36, 705-754, 1983.
8. J. Hejtmanek: Time dependent linear transport equations, in Kinetic theories and the Boltzmann equation, ed. C. Cercignani, L. N. M. no - 1048, Berlin, Springer Verlag, 1984, p. 1-59.

V. Protopopescu
Engineering Physics and Mathematics Division
Oak Ridge National Laboratory
Oak Ridge TN 37831
USA

Operator Theory:
Advances and Applications, Vol. 19
© 1986 Birkhäuser Verlag Basel

THE RIEMANN BOUNDARY PROBLEM AND COMPLETE INTEGRABLE SYSTEMS

Yu.L. Rodin

In these lectures I want to describe the relations between modern methods of nonlinear mathematical physics and some ideas of the complex analysis close to Wiener-Hopf equations and Toeplitz operators.

Consider the one-dimensional Schrödinger equation

$$L(t)f = -\frac{d^2f}{dx^2} + u(x,t)f = \lambda f \tag{1}$$

with the real potential satisfying the condition

$$\int_{-\infty}^{\infty} (1+|x|)u(x,t)dx < \infty. \tag{2}$$

Let the spectrum of L be independent of the time. Then the Lax condition

$$\dot{L} = [A,L] = AL - LA \tag{3}$$

is valid. Here A is an anti-symmetric operator. For any integer $q \geq 0$ there exists a differential operator A_q of order $2q+1$ such that $[A_q,L]$ is an operator of multiplication by a function. In this case the Eq. (3) is the famous Korteweg-de Vries equation for $q = 1$ and a higher KdV equation for $q > 1$.

As it is known, the continuous spectrum of the operator L(t) is the positive real semiaxis and the discrete spectrum is finite and negative.

Let $f_j(x,\lambda)$ $(j = 1,2)$ be the so-called Jost functions, i.e. the solutions of the Eq. (1) determined by the asymptotic behaviour

$$f_j(x,\lambda) = \exp(-i\sqrt{\lambda}\,x) + o(1), \qquad x \to (-1)^j\infty. \tag{4}$$

These functions are connected by the relation

$$\begin{aligned} &f_1(x,\lambda) = a(\lambda)f_2(x,\lambda) + b(\lambda)f_2(x,\lambda), \\ &|a(\lambda)|^2 - |b(\lambda)|^2 = 1, \qquad \operatorname{Im}\lambda = 0, \qquad \operatorname{Re}\lambda \geq 0. \end{aligned} \tag{5}$$

The value $a(\lambda)$ is named the scattering amplitude, which is analytic continued into the complex λ-plane cutted along the real positive semiaxis and has simple poles at the discrete spectrum points.

Consider the Riemann surface E of the function $w = \sqrt{z}$ consisting of two sheets $E_{\pm}$ determined by the sign of $\operatorname{Im}\sqrt{z}$. The functions $f_j(x,\lambda)$ are analytic continued into the sheets of this surface, E_+ (for $j = 1$) and E_- (for $j = 2$). $j = 2$). Therefore, the vector-functions

$$g_+(x,\lambda) = (f_1(x,\lambda), \overline{f_2(x,\bar{\lambda})})$$

$$g_-(x,\lambda) = (f_2(x,\lambda), \overline{f_1(x,\bar{\lambda})})$$

are analytic in the domains $E_{\pm}$, satisfy the boundary condition

$$g_+(x,\lambda) = g_-(x,\lambda)\frac{1}{\overline{a(\lambda)}}\begin{pmatrix} 1 & -\overline{b(\lambda)} \\ b(\lambda) & 1 \end{pmatrix}, \tag{6}$$

$$\lambda \in \Gamma\{\operatorname{Re}\lambda \geq 0,\ \operatorname{Im}\lambda = 0\},$$

and have the asymptotic behaviour

$$f_j(x,\lambda) = \exp(-i\sqrt{\lambda}\,x) + o(1), \qquad \lambda \to \infty. \tag{7}$$

The branch of $\sqrt{\lambda}$ is chosen on the corresponding sheet of E. Note that

$$\dot{b}(\lambda) = 8i(\sqrt{\lambda})^3 b(\lambda), \qquad \lambda \to \infty. \tag{8}$$

Therefore, we obtain the matrix Riemann boundary problem on the Riemann surface E. The method of the reduction of the inverse scattering problem to the Riemann boundary problem was proposed by V.E. Zakharov and A.B. Shabat.

In the case $b(\lambda) = 0$ (a reflectionless potential $u(x,t)$) to solve the inverse scattering problem it is necessary to construct analytic matrix functions with prescribed singularities. Corresponding potentials are called solitons.

If the condition (2) is invalid, the problem becomes complicated. In particular, analogs of reflectionless fastdecreasing potentials-periodic and almost periodic potentials lead to the problem of the construction of an analytic function with prescribed singularities on a hyperelliptic Riemann surface of finite or infinite genus (N.I. Akhiezer, B.A. Dubrovin, A.R. Its, I.M. Krichever, P. Lax, V.B. Matveev, V.A. Marchenko, I.V. Ostrovski - the finite-zones integration method; B.M. Levitan, H.P. McKean, E. Trubowitz - the infinite-zones case).

Here we outline the reduction of the inverse scattering problem to the Riemann boundary problem on Riemann surfaces proposed by the author.

We consider the class of potentials such that the operator L(0) possesses a finite-zones continuous spectrum and an arbitrary finite discrete spectrum. This class contains, in particular, periodic and almost periodic finite-zones potentials disturbed by fast-decreasing ones. In this case the discrete spectrum may be located in the lacunas of the continuous spectrum. If a disturbance is integrable but satisfies not the condition (2), then the discrete spectrum may be located in zones as well as in lacunas.

Let $\|\xi_{ij}(\tau)\|$ be the spectral matrix of the operator family L(t) and

(9) $$M_{ij}(\lambda) = \int_{-\infty}^{\infty} \frac{d\xi_{ij}(\tau)}{\lambda - \tau}, \qquad i,j = 1,2.$$

The Weil-Titchmarsh functions are defined by the relations

$$m_j(\lambda) = \frac{(-1)^{j+1}\frac{1}{2} + M_{12}(\lambda)}{M_{11}(\lambda)}, \qquad j = 1,2$$

and the Weil solutions are

(10) $$f_j(x,\lambda) = \Theta(x,\lambda) + m_j(\lambda)\varphi(x,\lambda), \qquad j = 1,2.$$

Here Θ,φ are solutions of Eq.(1) determined by the initial conditions

$$\Theta(0,\lambda) = \varphi'(0,\lambda) = 1,$$

$$\Theta'(0,\lambda) = \varphi(0,\lambda) = 0.$$

For the case of periodic finite-zones potentials the Weil functions coincide with the Floke solutions.

Let $\Gamma = \cup_{j=0}^{n-1} (\lambda_{2j},\lambda_{2j+1}) \cup (\lambda_{2n},\infty)$ be the continuous spectrum of the family $\{L(t)\}$, $R_{2n+1}(\lambda) = \Pi_{j=0}^{2n} (\lambda - \lambda_j)$ and $E_n = E_+ \cup E_-$ be the Riemann surface of the function $w^2 = R(\lambda)$. From Eq.(9),(10) it follows that the functions $f_j(x,\lambda)$ are analytic on the planes E correspondingly and have the asymptotic behaviour

(11) $$f_j(x,\lambda) = \exp\{-i\sqrt{\lambda - \lambda_{2n}}\,x\} + o(1), \qquad \lambda \to \infty.$$

In this case functions $g_\pm(x,\lambda)$ satisfy the boundary condition (6) on the hyperelliptic surface E_n of the genus n. Note that the function $a(\lambda)$ is analytic on E_+ and has zeros at discrete spectrum points.

The adducing construction is suitable for a wide class of operators. It may be Zakharov-Shabat operators

$$Lf = -\frac{df}{dx} + (\lambda J + B(x))f, \tag{12}$$

where J is a diagonal matrix and B is an off-diagonal one. These equations describe in particular well-known non-linear operators sin-Gordon, sh-Gordon, non-linear Schrödinger equation and so on. The same method is applicable for the system

$$\frac{df}{dx} - i \sum_{j=1}^{3} w_j(\lambda) S_j(x,t)\sigma_j = 0, \tag{13}$$

where σ_j $(j = 1,2,3)$ are Pauli matrices, $|\vec{S}| = 1$ and $w_j(\lambda)$ are some elliptic functions. This equation is connected with a non-linear Landau-Lifschitz equation describing spin waves in a ferromagnet. For fast-decreasing potential $\vec{S}(x,t)$ Eq.(13) is reduced to the Riemann problem on a torus (A.V. Mikhailov, Yu.L. Rodin, E.K. Sklyanin). For the class of potentials decribed above we obtain the Riemann problem on a two-sheet surface over the torus.

Note that in all cases the boundary problem has the form (6) where $b(\lambda)/a(\lambda)$ is the reflection coefficient, $a(\lambda)$ is the scattering amplitude and Γ is the continuous spectrum of the problem.

In conclusion we describe so-called D-BAR method of solving of the inverse scattering problem close to the Riemann problem method (M.J. Ablowitz, D. Bar-Yaacob, R. Beals, R.R. Coifman, A.S. Fokas, S.V. Manakov, V.E. Zakharov). This method is especially useful for the case of multidimsional scattering.

Consider the equation

$$\tilde{\theta}_\lambda f = Bf, \qquad \tilde{\theta}_\lambda = \frac{d}{dx} - \lambda \,\mathrm{ad}\, J. \tag{14}$$

Since the operator $\tilde{\theta}_\lambda$ is an analytical function of λ, $\tilde{\theta}_\lambda$ commutes with the operator $\partial/\partial\bar{\lambda}$ and hence a solution of Eq.(14) satisfies the equation

$$\frac{\partial f}{\partial \bar{\lambda}} = fa \qquad \text{where } \tilde{\theta}_\lambda a = 0. \tag{15}$$

In more complicated cases we obtain the Carleman-Bers-Vekua equation

(16) $$\frac{\partial f}{\partial \bar{\lambda}} = fa + \bar{f}b$$

on a Riemann surface.

The following lecture is devoted to the Riemann boundary problem generated by the inverse scattering problem.

Yu.L. Rodin
6 Institute Prospect, Apt.79
Chernogolovka
Moscow District
142432 USSR

Operator Theory:
Advances and Applications, Vol. 19

THE RIEMANN BOUNDARY PROBLEM ON RIEMANN SURFACES

Yu.L. Rodin

Let M be a compact Riemann surface of the genus g, let Γ be a Ljapunov contour consisting of a finite number of closed curves without self-intersections, let G(p) be an $n \times n$-matrix which satisfies the Hölder condition on Γ and is non-degenerate at each point of Γ. The Riemann problem is to find an analytic vector or matrix-valued function φ(q) in the domain M which is continuous up to the boundary and satisfies the boundary condition.

(1) $\qquad \varphi_+(p) = \varphi_-(p)G(p), \qquad p \in \Gamma.$

All restrictions on G(p), Γ and φ(q) listed above may be relaxed to a considerable extent.

There is a voluminous literature on this problem on the complex plane (J. Plem lj, F.D. Gakhov, I.C. Gohberg, A. Grothendieck, M.G. Krein, N.I. Muskhelishvili). For the case $g > 0$ this problem was intensively studied in the sixties-seventies by R.N. Abdulaev, L.I. Chibrikova, W. Koppelman, Yu.L. Rodin, H. Röhrl, E.I. Zverovich and other authors. They ascertained the equivalence of the Riemann problem, the central facts of the theory of algebraic functions of one variable (the Riemann-Roch, Abel theorems, the Jacobi inversion problem, the Riemann theta-function) and complex vector bundles over Riemann surfaces.

In the first half of the lecture we discuss these connections and methods of solving the problems described in the first lecture (see the preceding paper). Note that there are many important physical problems leading to the Riemann problem and vector bundles which we are forced to omit here (for example, the M. Atyah's theory of the auto-self-dual Yang Mills fields, the instantons theory and so on). In the second half of the lecture we consider the case $g = \infty$. It corresponds to the infinite-zones case of the inverse scattering problem. Here we use one approach to the study of noncompact surfaces proposed by the author in the seventies.

Associated to the problem (1) is the sheaf Ω_G of germs of holomorphic solutions of (1) and the vector bundle B_G over M with fiber $\mathbb{C}^n$ and transistion matrices

$$h_{jk} = (\psi_{+,k})^{-1}\psi_{+,j} = (\psi_{-,k})^{-1}\psi_{-,j}^{-1} \qquad \text{in } U_j \cap U_k,$$

where $\{\psi_{\pm,j}\}$ is a 0-cochain of degenerate holomorphic $n \times n$ matrices satisfying condition (1) in the domains U_j and $M\{U_j, j \in I\}$ is a fixed covering of M. The matrix solution X(q) of (1) is called *canonical* if at each point of M the order of det X is equal to the sum of orders of the strings.

THEOREM 1. *Let* $B \to M$ *be a complex vector bundle with fiber* $\mathbb{C}^n$ *and group* $GL(n,\mathbb{C})$. *Then there is a matrix-valued function* G(p) *on* Γ *such that* B_G *is equivalent to* B.

THEOREM 2. *A necessary and sufficient condition for problem* (1) *to have a canonical solution is that the bundle* B_G *decomposes into a direct sum of line bundles.*

The problem

$$\Psi_+(p) = \Psi_-(p)G^{*-1}(p) \tag{2}$$

for vector (matrix)-valued differentials is dual to (1). If $L_G = H^0(M,B_G)$ and $H_G = H^1(M,B_G)$ are the spaces of solutions of the problems (1) and (2) correspondingly, then the index expression

$$\dim L_G - \dim H_G = \operatorname{ind}_\Gamma \det G - ng + n \tag{3}$$

is equivalent to the Riemann-Roch theorem for M. This relation can be obtained from the corresponding system of singular integral equations or from the Riemann-Roch theorem directly.

It is clear that in the general case effective solving of the Riemann problem is impossible. But it is possible for many important particular cases. E.g., the problem (6), (7) of the first lecture

$$g_+(\lambda) = g_-(\lambda)\frac{1}{a(\lambda)}\begin{pmatrix} 1 & -\overline{b(\lambda)} \\ b(\lambda) & 1 \end{pmatrix} \tag{4}$$

on the surface E_n can be solved if $|b(\lambda)| \neq 1$ everywhere on Γ.

Let, for simplicity, $|b(\lambda)| < 1$ on Γ. Then

(5) $$\begin{pmatrix} 1 & -\overline{b(\lambda)} \\ b(\lambda) & 1 \end{pmatrix} = \psi_-^{-1}(\lambda)\psi_+(\lambda), \qquad \lambda \in \Gamma,$$

where $\psi_\pm(\lambda)$ are analytic in E and $\psi_\pm(\infty) = 1$. The matrices $\psi_\pm(\lambda)$ may be calculated by the series

(6) $$\psi_\pm(\lambda) = 1 + \psi_\pm^1(\lambda) + \psi_\pm^2(\lambda) + \dots ,$$

where $\psi_\pm^n(\lambda)$ are determined by the formulas

(7) $$\psi_\pm^n(\lambda) = \pm\tfrac{1}{2}\psi^{n-1}(\tau)\beta(\tau) + \frac{1}{2\pi i}\int_\Gamma \psi_-^{n-1}(\theta)\beta(\theta)K(\theta,\tau)d\theta;$$
$$\beta(\lambda) = \begin{pmatrix} 0 & -\overline{b(\lambda)} \\ b(\lambda) & 0 \end{pmatrix}, \qquad \tau \in \Gamma.$$

Here $K(\theta,\lambda)$ is a Cauchy type integral kernel. Such kernels were constructed by several authors (H. Behnke and K. Stein, S.Ja. Gusman and Yu.L. Rodin and others). With respect to the variable λ the kernel $K(\theta,\lambda)$ has a pole at $\lambda = \theta$ and g poles described by the some divisor $\delta \in \Gamma$, $\deg\delta = g$, $\dim\delta = 1$. Therefore, the function $\psi_\pm(\lambda)$ has g poles at the points of δ. Then the problem (4) is reduced to the scalar problem

(8) $$g_+(\lambda)\psi_+^{-1}(\lambda) = \frac{1}{\overline{a}(\lambda)}\, g_-(\lambda)\psi_-^{-1}(\lambda), \qquad \lambda \in \Gamma,$$

in the class of the analytic functions with prescribed poles at points of δ and points of the discrete spectrum and asymptotic behaviour at infinity (see Eq.(7) of Lect. 1). This problem can be easily solved by the Riemann theta-function.

In the case $g = \infty$ we assume that the surface M has a zero boundary and consider only the scalar case $n = 1$. The contour Γ is assumed to be compact. As is clear from Eq.(3), the index of the problem can be infinite. Moreover, in the noncompact case a solution can possess on infinite number of zeros and poles and this case is most interesting. In dictates the statement of the problem described below. In this statement we combine two ideas. A solution has to have a finite Dirichlet integral (the usual restriction in the noncompact surfaces theory) and an infinite number of poles and zeros.

We introduce some notions and designations. Let J be the sheaf of integers, A and Ω be the sheaves of germs of smooth and holomorphic functions and $\Omega^*(B)$ be the multiplicative sheaf of holomorphic functions different from zero. For the sheaf A and its subsheaves the D-norms of the 0-cochain $\{f_j\}$ and the 1-cochain $\{f_{jk}\}$ are different and given by the formulas

$$(9)\qquad \begin{aligned} \|\{f_i\}\| &= \Big(\sum_{i\in I}\int_{U_i} |df_i|^2\Big)^{\frac{1}{2}}, \\ \|\{f_{ij}\}\| &= \Big(\sum_{i,j\in I}\int_{U_i\cap U_j\neq\emptyset} |df_{ij}|^2\Big)^{\frac{1}{2}}. \end{aligned}$$

For the sheaf Ω^*

$$(10)\qquad \|\{f_i\}\| := \|\{d\ln f_i\}\|, \qquad \|\{f_{ij}\}\| := \|\{d\ln f_{ij}\}\|.$$

If P is one of the sheaves listed above, $\zeta^k_D(P)$ (k=0;1) are the groups of k-cochains with values in P with finite D-norms and $H^k_D(P)$ and $Z^k_D(P)$ are corresponding cohomology and cocycles groups.

Let $\{h_j\}$ be some 0-cochain such that $h_j(p)$ are meromorphic functions in U_j satisfying on the contours $\Gamma \cap U_j$ the condition

$$(11)\qquad h^+(p) = \frac{1}{G(p)}\, h^-(p),$$

the functions $h_{jk} = h_j(q)/h_k(q)$ are holomorphic and different from zero in the intersections $U_j \cap U_k$ and $\{h_{jk}\} \in Z^1_D(\Omega^*)$. The holomorphic cochain $\{\varphi_j\} \in \zeta^0_D(\Omega)$ such that $\varphi_j/\varphi_k = h_{jk}$ and different from zero on the ideal boundary (it means that there exists $\varepsilon > 0$ such that $|\varphi_j| > \varepsilon$ outside a compact part of M) is called a *strong solution* of the described Riemann problem (problem R).

Let $I_0 \subset I$, $\{U_j, j \in I_0\}$ be a triangulation of M and $\Sigma_j\, \alpha_j \equiv 1$ be a corresponding partition of unit. The Cauchy index of the problem R is defined by the formula

$$(12)\qquad x = \int_M \bar\partial\tau_j = \sum_{j\in I_0}\int_{U_j} \bar\partial\tau_j,$$

where

$$(13)\qquad \tau_j = \frac{1}{2\pi i}\sum_{k\in I} \alpha_k\, \partial \ln h_{jk}.$$

For an infinite cochain $\{h_j\}$ the value κ can be finite or infinite.

THEOREM 3. *Let the Cauchy index of the problem* R *be equal to* $\kappa < \infty$. *Then a strong solution of this problem has exactly* κ *zeros.*

Let $\{\psi_j\} \in \zeta^0_D(\Omega)$ be a finite positive divisor of the degree κ and $\{\eta_j\} \in \zeta^0_D(\Omega^*)$ such that

(14) $h_{jk} = \eta_j \psi_{jk} \eta_k^{-1}, \qquad \psi_{jk} = \psi_j/\psi_k.$

Then the cochain $\{\varphi_j\}$, $\varphi_j = \eta_j \psi_j$ is called a *weak solution* of the problem R.

Let $\{\varphi_j\}$ be a strong or weak solution of the problem R, the finite cochain $\{\psi_i\} \in \zeta_D^0(\Omega)$ determines the divisor of zeros of the solution and h_G and h_ψ be the cohomology classes of the group $H_D^1(\Omega^*)$ determined by the cocycles $\{h_{jk}\}$ and $\{\psi_{jk}\}$. We have the exact sequence

(15) $H^1(J) \to H_D^1(\Omega) \xrightarrow{e^*} H_D^1(\Omega\ ) \to H^2(J),$

where $e(f) = \exp 2\pi i\, f(q)$.

THEOREM 4. *For the weak solvability of the problem* R *of the index* $\kappa < \infty$ *it is necessary and sufficient that there exist a finite positive divisor* $\{\psi_j\} \in \zeta_D^0(\Omega)$ *of the degree* κ *such that the cohomology classes* $e^{*-1}(h_G)$ *and* $e^{*-1}(h_\psi)$ *coincide in the group* $H_D^1(\Omega)/i^*H^1(J)$. *The same condition is necessary for the strong solvability of the problem* R.

It is naturally to designate

(16) $J_D = H_D^1(\Omega)/i^*H^1(J)$

by the D-jacobian of the surface M. There exists an infinite-dimensional torus $J_S \subset J_D$ analogous to the Jacobi torus of a compact surface. For such Riemann problems of which the solvability is described by the torus J_S the solvability conditions can be written in the form of an infinite system of the equations analogous the Jacobi inverse problem.

If the cochain $\{h_j\}$ is finite, the problem R can be solved in closed form by the Cauchy type integrals. The index (finite or infinite) of this problem can be expressed using the same terms. The Riemann theta-function has been constructed for $g = \infty$ only for a hyperelliptic surface (H.P. McKean and E. Trubowitz).

Yu.L. Rodin
6 Institute Prospect, Apt 79
Chernogolavka
Moscow District
142432 USSR.

Operator Theory:
Advances and Applications, Vol. 19

STABLE INVARIANT SUBSPACES MODULO A SUBSPACE

Leiba Rodman[1)]

A general result concerning stability of subspaces in a finite dimensional linear space which are invariant for a given linear transformation modulo a subspace, is proved. This result encompasses the previously known results concerning stability of invariant subspaces and stability of (A,B)-invariant subspaces when the pair (A,B) is controllable.

1. INTRODUCTION

Let $A : C^n \to C^n$ be a linear transformation, and let $V \subset C^n$ be a subspace (C^n stands for the linear space of all n-dimensional column vectors with complex coordinates). A subspace $M \subset C^n$ is called A-<u>invariant modulo</u> V if $Ax \in M + V$ for every $x \in M$. Thus, the notion of an A-invariant subspace is a generalization of the familiar notion of an A-invariant subspace (the latter is a particular case of the former when $V = \{0\}$).

Invariant subspaces modulo a subspace play a key role in solving many problems in the linear system theory (see, e.g. [8]), where they often appear as (A,B)-invariant subspaces. Given linear transformations $A : C^n \to C^n$ and $B : C^m \to C^n$, a subspace $M \subset C^n$ is called (A,B)-<u>invariant</u> if there is a transformation $F : C^n \to C^m$ such that M is $(A + BF)$-invariant. It is well-known (and not difficult to verify) that M is (A,B)-invariant if and only if M is A-invariant modulo ImB.

---

1) Partially supported by NSF grant.

The following notion of a stable invariant subspace modulo a subspace is important in connection with computational aspects. We shall give the definition in the framework of (A,B)-invariant subspaces. Roughly speaking, an (A,B)-invariant subspace is stable if it is perturbed as little as we wish provided the perturbations of A and B are sufficiently small. We shall measure the closeness of subspaces in C^n in terms of the gap

$$\theta(M_1, M_2) = \| P_{M_1} - P_{M_2} \|$$

between subspaces M_1 and M_2 in C^n, where P_N is the orthogonal projector on the subspace $N \subset C^n$. (Everywhere in this paper we use the operator norm induced by the euclidean norm in C^n.) Thus, an (A,B)-invariant subspace M is called stable if for $\varepsilon > 0$ there exists $\delta > 0$ such that for any pair of linear transformations $A' : C^n \to C^n$ and $B' : C^m \to C^n$ satisfying

$$\|A - A'\| + \|B - B'\| < \delta$$

there exists an (A',B')-invariant subspace M' such that $\theta(M, M') < \varepsilon$.

The similar notion of a stable A-invariant subspace was studied extensively in the recent years. Given a linear transformation $A : C^n \to C^n$, an A-invariant subspace M is called stable if for every $\varepsilon > 0$ there esists $\delta > 0$ such that every linear transformation $A' : C^n \to C^n$ with $\|A-A'\| < \delta$ has an invariant subspace M' with the property that $\theta(M,M') < \varepsilon$. We shall see later that an A-invariant subspace M is stable if and only if it is stable as an (A,0)-invariant subspace, where $0 : C^n \to C^n$ is the zero transformation. The stable invariant subspaces have been described in [1,3] and in [2], where the framework of finite dimensional linear spaces over the reals was considered as well. For the comprehensive treatment of stability of invariant subspaces see [5]. Many applications of stable invariant subspaces are found in [2,4,5].

It turns out that generally not all A-invariant subspaces are stable. As proved in [1,3], an A-invariant subspace M is stable if and only if for every eigenvalue λ_0 of A with dim Ker $(\lambda_0 I-A) > 1$ either $M \supset R_{\lambda_0}(A)$ or $M \cap R_{\lambda_0}(A) = \{0\}$ holds. Here

$$R_{\lambda_0}(A) = \text{Ker }(\lambda_0 I-A)^n$$

is the *root subspace* corresponding to eigenvalue λ_0. In contrast, it was proved in [6] that every (A,B)-invariant subspace is stable provided the pair (A,B) is *controllable*, i.e.

$$\sum_{j=0}^{\infty} \text{Im}(A^j B) = C^n.$$

This fact is important in the investigation of stability of cascade decompositions of linear systems (see [7,6]) .

The main purpose of this paper is to prove general results on stability of invariant subspaces modulo a subspace which incorporate as particular cases the known results in the extreme situations (B = 0 and (A,B) is controllable).

2. MAIN RESULT

Let $A : C^n \to C^n$ and $B : C^m \to C^n$ be linear transformations. The subspace

$$J(A,B) = \sum_{j=0}^{\infty} \text{Im}(A^j B)$$

will play an important role in the sequel; it is called the *controllable subspace* of (A,B) and is widely used in the linear systems theory. It is easily seen that actually

$$J(A,B) = \sum_{j=0}^{n-1} \text{Im}(A^j B)$$

and that $J(A,B)$ is the smallest A-invariant subspace which contains ImB.

We state now the main result of this paper.

THEOREM 2.1. *Let K be a direct complement to $J(A,B)$ in C^n, and let $\Pi_K : C^n \to K$ be the projector on K along $J(A,B)$. Assume that an (A,B)-invariant subspace is such that the following condition holds:* (α) *for every eigenvalue λ_0 of the* transformation $\Pi_K A|_K : K \to K$ with dim Ker $(\lambda_0 I - \Pi_K A|_K) > 1$ either

$$\Pi_K M \supset R_{\lambda_0}(\Pi_K A|_K) \text{ or } \Pi_K M \cap R_{\lambda_0}(\Pi_K A|_K) = \{0\}$$

holds.

Then M is a stable (A,B)-invariant subspace.

Conversely, if M is a stable (A,B)-invariant subspace such that $M \cap J(A,B) = \{0\}$, then the condition (α) holds.

Before we proceed farther, some remarks regarding Theorem 2.1 are in order.

Remark 1. The subspace $\Pi_K M$ is $\Pi_K A|_K$-invariant. Indeed, let $x \in \Pi_K M$; then $x = y + z$ for some $y \in M$, $z \in J(A,B)$. By definition of $J(A,B)$ write

$$z = \sum_{j=0}^{n-1} A^j B q_j, \quad q_j \in C^m.$$

On the other hand, M is $(A+BF)$-invariant for some linear transformation $F : C^n \to C^m$. So we have $(A+BF)y \in M$ and

$$Ax = (A+BF)y - BFy + \sum_{j=0}^{n-1} A^{j+1} B q_j \in M + J(A,B).$$

Hence $\Pi_K Ax \in \Pi_K M$, and $\Pi_K M$ is $\Pi_K A|_K$-invariant. Comparing Theorem 2.1 with the description of stable invariant subspaces, we see that Theorem 2.1 can be alternatively formulated as follows.

THEOREM 2.2. Let K and Π_K be as in Theorem 2.1, and let M be an (A,B)-invariant subspace. If $\Pi_K M$ is stable as a $\Pi_K A|_K$-invariant subspace, then M is a stable (A,B)-invariant subspace. Conversely, if M is a stable (A,B)-invariant subspace and $M \cap J(A,B) = \{0\}$, then $\Pi_K M$ is a stable $\Pi_K A|_K$-invariant subspace.

Remark 2. The statement of Theorem 2.1 (or of Theorem 2.2) does not depend on the choice of K. To verify this, let K' be another direct complement to $J(A,B)$ in C^n. Write the identity transformation I in a 2×2 block matrix form

$$I = \begin{bmatrix} I & * \\ 0 & S \end{bmatrix} : J(A,B) \dotplus K \to J(A,B) \dotplus K'.$$

Then it is easily seen that $\Pi_{K'} = S\Pi_K$ and that $\Pi_{K'}A|_{K'} \cdot S = S \cdot \Pi_K A|_K$. Thus, $S : K \to K'$ is a similarity transformation between $\Pi_K A|_K$ and $\Pi_{K'} A|_{K'}$ such that $S(\Pi_K M) = \Pi_{K'} M$, and hence the statement of Theorem 2.1 is independent of the choice of K.

Remark 3. If $J(A,B) = C^n$, i.e. if (A,B) is controllable, then Theorem 2.1 implies that every (A,B)-invariant subspace is stable, thereby recovering the result proved in [6]. Also in the case when dim $J(A,B) = n - 1$ every (A,B)-invariant subspace is stable.

Remark 4. If $B = 0$, we obtain from Theorem 2.1 that an (A,0)-invariant subspace M is stable if and only if it is stable as an A-invariant subspace (a fact mentioned in the introduction).

Remark 5. It will be seen from the proof that Theorem 2.2 holds also in the framework of linear transformation and their invariant subspaces in finite dimensional real vector spaces. A description of stable invariant subspaces for real transformations (given in [2]) allows one to restate Theorem 2.2 in terms similar to Theorem 2.1. We will leave this restatement to the interested reader.

We conclude this section with an example which shows, in particular, that stability of the (A,B)-invariant subspace M does not generally imply stability of the $\Pi_K A|_K$-invariant subspace $\Pi_K M$ (in such case necessarily $M \cap J(A,B) \neq \{0\}$ in view of Theorem 2.1).

EXAMPLE 2.1. Let A be the 3 × 3 zero matrix, and let

$$B = \begin{bmatrix} 1 \\ 0 \\ 0 \end{bmatrix} .$$

Every subspace in C^3 is obviously (A,B)-invairant. It turns out that the (A,B)-invariant subspace

$$M = \text{Span} \left\{ \begin{bmatrix} 1 \\ 0 \\ 0 \end{bmatrix} , \begin{bmatrix} 0 \\ 1 \\ 0 \end{bmatrix} \right\}$$

is stable.

Indeed, let $\{A_p\}_{p=1}^{\infty}$ and $\{B_p\}_{p=1}^{\infty}$ be sequences of matrices of sizes 3×3 and 3×1, respectively, such that

$$\lim_{p\to\infty} A_p = 0; \quad \lim_{p\to\infty} B_p = B.$$

Denote by a_{ijp} the (i,j)-entry in A_p and by b_{jp} the j-th entry in B_p $(i,j=1,2,3)$. Choose sequences $\{x_p\}_{p=1}^{\infty}$, $\{y_p\}_{p=1}^{\infty}$, $\{u_p\}_{p=1}^{\infty}$ and $\{v_p\}_{p=1}^{\infty}$ of (complex) numbers which tend to zero and are such that for $p = 1, 2, \ldots$ the inequalities

$$v_p - y_p u_p \neq 0,$$

$$[x_p, 1] \begin{bmatrix} 1 & u_p \\ y_p & v_p \end{bmatrix}^{-1} \begin{bmatrix} b_{2p} \\ b_{3p} \end{bmatrix} - b_{1p} \neq 0 \quad (2.1)$$

hold (such a choice is obviously possible). Determine numbers $\{f_{jp}\}_{p=1}^{\infty}$, $j = 1,2,3$ from the equalities

$$d_p[f_{1p}, f_{2p}, f_{3p}] =$$

$$[a_{11p}, a_{12p}, a_{13p}] - [x_p, 1] \begin{bmatrix} 1 & u_p \\ y_p & v_p \end{bmatrix}^{-1} \begin{bmatrix} a_{21p} & a_{22p} & a_{23p} \\ a_{31p} & a_{32p} & a_{33p} \end{bmatrix}$$

for $p = 1,2,\ldots$, where d_p is the left-hand side of (2.1). Let

$$Q_p = \begin{bmatrix} 1 & u_p \\ y_p & v_p \end{bmatrix}^{-1} \begin{bmatrix} a_{21p}+b_{2p}f_{1p} & a_{22p}+b_{2p}f_{2p} & a_{23p}+b_{2p}f_{3p} \\ a_{31p}+b_{3p}f_{1p} & a_{32p}+b_{3p}f_{2p} & a_{33p}+b_{3p}f_{3p} \end{bmatrix} \begin{bmatrix} x_p & 1 \\ 1 & u_p \\ y_p & v_p \end{bmatrix}.$$

We have then the following equalities for $p = 1, 2, \ldots$:

$$(A_p + B_p[f_{1p}, f_{2p}, f_{3p}]) \begin{bmatrix} x_p & 1 \\ 1 & u_p \\ y_p & v_p \end{bmatrix} = \begin{bmatrix} x_p & 1 \\ 1 & u_p \\ y_p & v_p \end{bmatrix} Q_p,$$

which imply that for each p the subspace

$$_p = \text{Span} \left\{ \begin{bmatrix} x_p \\ 1 \\ y_p \end{bmatrix}, \begin{bmatrix} 1 \\ u_p \\ v_p \end{bmatrix} \right\}$$

is $(A_p+B_p[f_{1p}, f_{2p}, f_{3p}])$-invariant; hence M_p is (A_p, B_p)-

invariant. As obviously

$$\lim_{p\to\infty} \theta(\mathcal{M}_p, \mathcal{M}) = 0,$$

the stability of the (A,B)-invariant subspace $\mathcal{M}$ follows.

We have here

$$\Pi_{\mathcal{K}} A|_{\mathcal{K}} = 0,$$

and since $\mathcal{K}$ is 2-dimensional, the linear transformation $\Pi_{\mathcal{K}} A|_{\mathcal{K}}$ has no non-trivial (i.e. different from {0} and $\mathcal{K}$) stable invariant subspaces. In particular, $\Pi_{\mathcal{K}}$ is not a stable invariant subspace for $\Pi_{\mathcal{K}} A|_{\mathcal{K}}$. □

It is instructive to observe in Example 2.1 that the (A,B)-invariant subspace

$$\mathcal{N} = \mathrm{Span} \begin{bmatrix} 0 \\ 1 \\ 0 \end{bmatrix}$$

is not stable (as follows from Theorem 2.1). One can also see this directly: define

$$A_p = \begin{bmatrix} 0 & 0 & 0 \\ 0 & 0 & 0 \\ 0 & p^{-1} & 0 \end{bmatrix}, \quad p = 1, 2, \ldots, \quad B_p = B = \begin{bmatrix} 1 \\ 0 \\ 0 \end{bmatrix}.$$

Then there is no sequence of subspace $\{\mathcal{N}_p\}_{p=1}^{\infty}$ such that $\lim_{p\to\infty} \theta(\mathcal{N}_p, \mathcal{N}) = 0$ and for each p the subspace $\mathcal{N}_p$ is (A_p, B_p)-invariant.

3. PROOF OF THE MAIN RESULT.

According to Remark 1, we have to prove Theorem 2.2 only. We need some preliminaries for its proof.

Two pairs of transformations (A_1, B_1) and (A_2, B_2), where $A_j : \mathbb{C}^n \to \mathbb{C}^n$, $B_j : \mathbb{C}^m \to \mathbb{C}^n$ for $j = 1, 2$, are called <u>feedback equivalent</u> if there exist invertible linear transformations $M : C^n \to C^n$ and $N : C^m \to C^n$ and a linear transformation $F : C^n \to C^m$ such that

$$A_1 = M(A_2 + B_2F)M^{-1}, \quad B_1 = MB_2N. \tag{3.1}$$

The feedback equivalence is widely used in linear systems theory. It is not difficult to see that (A,B)-invariant subspaces behave well under feedback equivalence: if (3.1) holds, then a subspace L is (A_2, B_2)-invariant if and only if ML is (A_1, B_1)-invariant.

PROPOSITION 3.1. Assume (3.1) holds. Then (A_2, B_2)-invariant subspace L is stable if and only if (A_1, B_1)-invariant subspace ML is stable.

Proof. Assume L is a stable (A_2, B_2)-invariant subspace. Use the well-known property that for any subspace $L' \subset C^n$ we have

$$\theta(ML, ML') \le \| Q_{ML} - Q_{ML'} \|,$$

where $Q_{ML}(Q_{ML'})$ is a projector on $L(L')$, not necessarily orthogonal (see, e.g. [9]). It follows that

$$\theta(ML, ML') \le C_1 \| P_L - P_{L'} \|, \tag{3.2}$$

where $C_1 = \|M\| \, \|M^{-1}\|$. On the other hand, given linear transformations $A_1' : C^n \to C^n$ and $B_1' : C^m \to C^n$, define

$$A_2' = M^{-1}A_1'M - B_2'F, \quad B_2' = M^{-1}B_1'N^{-1}.$$

Then

$$\|A_2' - A_2\| + \|B_2' - B_2\| \le C_2 (\|A_1' - A_1\| + \|B_1' - B_1\|), \tag{3.3}$$

where

$$C_2 = \max(\|M^{-1}\| \cdot \|M\|, \ \|M^{-1}\| (\|N^{-1}F\| + \|N^{-1}\|)).$$

Now given $\varepsilon > 0$, choose $\delta > 0$ corresponding to εC_1^{-1} according to the definition of the stable (A_2, B_2)-invariant subspace L. Then in view of (3.2) and (3.3), as long as

$$\|A_1' - A_1\| + \|B_1' - B_1\| < \delta C_2^{-1},$$

we have $\theta(ML, ML') < \varepsilon$, i.e. ML is a stable (A_1, B_1)-invariant subspace.

Analogously one proves the converse statement. □

Now we take up a particular case of Theorem 2.2, namely, stability of the controllable subspace.

LEMMA 3.2. *The subspace* $\mathcal{J}(A,B)$ *is stable (as an* (A,B)*-invariant subspace). Moreover, for every* $\eta > 0$ *there exists a linear transformation* $F : C^n \to C^m$ *(which depends on* A, B *and* η *only) with the properties that* $\|F\| < \eta$ *and for every* $\varepsilon > 0$ *there exists* $\delta > 0$ *such that given any pair of linear transformations* $A' : C^n \to C^n$, $B' : C^n \to C^m$ *with*

$$\|A'-A\| + \|B'-B\| < \delta ,$$

there is an $(A'+B'F)$*-invariant subspace* $\mathcal{J}'$ *satisfying*

$$\theta(\mathcal{J}', \mathcal{J}(A,B)) < \varepsilon .$$

Proof. Let $\Pi : C^n \to \mathcal{J}(A,B)$ be a projector whose image is $\mathcal{J}(A,B)$. Then the pair $(A|_{\mathcal{J}(A,B)}, \Pi B)$ is controllable, i.e.

$$\sum_{j=0}^{\infty} \mathrm{Im}(A^j \Pi B) = \mathcal{J}(A,B).$$

By the spectral assignment theorem (see, e.g., [8]) there exists a linear transformation $F : C^n \to C^m$ such that the $(A+BF)$-invariant subspace $\mathcal{J}(A,B)$ is spectral, which means that

$$\sigma((A+BF)|_{\mathcal{J}(A,B)}) \cap \sigma((I-\Pi)(A+BF)|_{\mathrm{Ker}\Pi}) = \emptyset \qquad (3.4)$$

More exactly, let $\lambda_1, \ldots, \lambda_k$ be all the eigenvalues of the linear transformation $(I-\Pi)A|_{\mathrm{Ker}\Pi}$ acting on Ker Π, and observe that for every $G : C^n \to C^m$ we have

$$(I-\Pi)(A+BG)|_{\mathrm{Ker}\,\Pi} = (I-\Pi)A|_{\mathrm{Ker}\,\Pi}.$$

So the $(A+BF)$-invariant subspace $\mathcal{J}(A,B)$ is spectral if and only if

$$\det(\lambda_i I - (A+BF)|_{\mathcal{J}(A,B)}) \neq 0, \quad i = 1, \ldots, k. \qquad (3.5)$$

Clearly, for any given $\eta > 0$ we can find an F with $\|F\| < \eta$ satisfying (3.5). Now Lemma 3.2 follows from the stability of spectral invariant subspaces. □

Proof of Theorem 2.2. In view of Proposition 3.1 we can assume without loss of generality that the given (A,B)-invariant subspace M is actually A-invariant. Further, we can assume that the direct complement K to $\mathcal{J}(A,B)$ in C^n is such that

$$(M \cap \mathcal{J}(A,B)) + \Pi_K M = M . \qquad (3.6)$$

Let J_M be a direct complement to $M \cap J(A,B)$ in $J(A,B)$, and let K_M be a direct complement to $\Pi_K(M)$ in K. Then with respect to the direct sum decomposition

$$C^n = (M \cap J(A,B)) + J_M + \Pi_K(M) + K_M$$

the transformations A and B have the following form:

$$A = \begin{bmatrix} A_{11} & A_{12} & A_{13} & A_{14} \\ 0 & A_{22} & A_{23} & A_{24} \\ 0 & 0 & A_{33} & A_{34} \\ 0 & 0 & 0 & A_{44} \end{bmatrix}, \quad B = \begin{bmatrix} B_1 \\ B_2 \\ 0 \\ 0 \end{bmatrix}$$

It is easily seen that the pair (A_{22}, B_2) is controllable:

$$\sum_{j=0}^{\infty} \operatorname{Im}(A_{22}^j, B_2) = J_M .$$

By the spectral assignment theorem, there exists a linear transformation $G : J_M \to C^m$ such that $\sigma(A_{22}+B_2G) \cap \sigma(A_{11}) = \emptyset$. Using Proposition 3.1 again, we can assume that actually

$$\sigma(A_{22}) \cap \sigma(A_{11}) = \emptyset. \tag{3.7}$$

Now let $\Pi_K M$ be stable as $\Pi_K A|_K$-invariant subspace. Arguing by contradiction, suppose that M is not stable (as an (A,B)-invariant subspace). So there exist $\varepsilon_0 > 0$ and sequences of transformations $\{A_p\}_{p=1}^{\infty}$, $\{B_p\}_{p=1}^{\infty}$ such that

$$\lim_{p\to\infty} \{\|A_p - A\| + \|B_p - B\|\} = 0$$

and every (A_p, B_p)-invariant subspace N satisfies

$$\theta(M,N) \geq \varepsilon_0 \tag{3.8}$$

Using Lemma 3.2 and passing if necessary to subsequences of $\{A_p\}_{p=1}^{\infty}$ and $\{B_p\}_{p=1}^{\infty}$, we find a sequence of linear transformations $F_p : C^n \to C^m$, $p = 1,2,\ldots,$ such that

$$\lim_{p\to\infty} \theta(\ (A,B),\ _p) = 0, \quad \lim_{p\to\infty} F_p = 0, \tag{3.9}$$

for some sequence $\{J_p\}_{p=1}^{\infty}$, where J_p is a $(A_p+B_pF_p)$-invariant subspace for $p = 1,2,\ldots$. Consider the restrictions $A|_{J(A,B)}$ and $(A_p+B_pF_p)|_{J_p}$. Let $x_1, \ldots, x_k$ be a basis in $J(A,B)$. It follows from (3.9) that for $p = 1,2,\ldots$ there exists a basis $x_1^{(p)},\ldots, x_k^{(p)}$ in J_p such that

$$\lim_{p\to\infty} x_j^{(p)} = x_j, \quad j = 1, \ldots, k.$$

So in the matrix representation of $(A_p+B_pF_p)|J_p$ with respect to the basis $\{x_j^{(p)}\}_{j=1}^{k}$ and in the matrix representation of $A|_{J(A,B)}$ with respect to the basis $\{x_j\}_{j=1}^{k}$ we have

$$\lim_{p\to\infty} (A_p+B_pF_p)|J_p = A|_{J(A,B)}.$$

Because of (3.7), $M \cap J(A,B)$ is a spectral and hence stable $A|_{J(A,B)}$-invariant subspace. So for every $p=1,2,\ldots,$ there exists an $(A_p+B_pF_p)$-invariant subspace $M_p \subset J_p$ such that

$$\lim_{p\to\infty} \theta(M_p, M\cap J(A,B)) = 0. \tag{3.10}$$

On the other hand, let Π_p be the projector on K along J_p (observe that because of (3.9) J_p is a direct complement to K, at least for p large enough (see, e.g., [4])). Again in view of (3.9) we have (by choosing suitable bases in J_p and $J(A,B)$) that

$$\lim_{p\to\infty} \Pi_p(A_p+B_pF_p)|K = \Pi_K A|K.$$

As $\Pi_K M$ is a stable $\Pi_K A|_K$-invariant subspace, there exists a $\Pi_p(A_p+B_pF_p)|_K$-invariant subspace $N_p \subset K$ such that

$$\lim_{p\to\infty} \theta(N_p,\Pi_K M) = 0. \tag{3.11}$$

Now (3.10) and (3.11) imply

$$\lim_{p\to\infty} \theta(M_p+N_p,(M \quad J(A,B)) + \Pi_K M) = 0$$

or (by (3.6))

$$\lim_{p\to\infty} \theta(M_p+N_p, M) = 0. \tag{3.12}$$

Indeed, let

$$\lim_{k\to\infty} \theta(M_{p_k}+N_{p_k}, L) = 0$$

for some subspace L and some subsequence $\{p_k\}_{k=1}^{\infty}$. Using (3.10), (3.11) and the property that L consists exactly of those vectors $x \in C^n$ for which there is a sequence $\{x_{p_k}\}_{k=1}^{\infty}$ tending to x such that $x_{p_k} \in M_{p_k} + N_{p_k}$ for $k=1,2,\ldots$ (see, e.g. [4]), it is not difficult to verify that $L \subset M$. The consideration of dimensions shows that actually $L = M$. It remains to use the compactness of the set of all subspaces in C^n in the topology induced by the metric $\theta(\cdot,\cdot)$, and (3.12) follows.

The subspace $M_p + N_p$ is clearly $(A_p+B_pF_p)$-invariant, hence (A_p,B_p)-invariant. In view of (3.12) we obtain a contradiction with (3.8).

Conversely, assume M is a stable (A,B)-invariant subspace such that $M \cap J(A,B) = \{0\}$. We shall prove that $\Pi_K M$ is a stable $\Pi_K A|_K$-invariant subspace. Let $C_p : K \to K$, $p=1,2,\ldots$ be a sequence of linear transformations such that $\lim_{p\to\infty} C_p = \Pi_K A|_K$. Consider the sequence of linear transformations $\{D_p\}_{p=1}^{\infty}$, $D_p : C^n \to C^n$ defined by the properties that $D_p x = Ax$ for $x \in J(A,B)$; $D_p x = (I-\Pi_K)Ax + C_p x$ for $x \in K$. Clearly, $\lim_{p\to\infty} D_p = A$. By the stability of M as an (A,B)-invariant subspace, there exists a sequence of linear transformations $\{G_p\}_{p=1}^{\infty}$, $G_p: C^n \to C^m$ such

$$\lim_{p\to\infty} \theta(N_p, M) = 0 \tag{3.13}$$

where N_p is some (D_p+BG_p)-invariant subspace for $p=1,2,\ldots$. Note that $J(A,B)$ is an invariant subspace for $D_p + BG_p$ $(p=1,2,\ldots)$. Because of condition (3.6) (which now takes the form $M \subset K$) it is easily seen that there exists a sequence of subspaces $\{K_p\}_{p=1}^{\infty}$ such that

$$\lim_{p\to\infty} \theta(K_p, K) = 0 \tag{3.14}$$

and $N_p \subset K_p$ for all p. In particular, K_p is a direct complement to $J(A,B)$ for p sufficiently large. Let Π_p be the projector on K_p along $J(A,B)$, and let

$$C'_p = \Pi_p(D_p+BG_p)|_{K_p} : K_p \to K_p.$$

The subspace N_p is C'_p-invariant.

On the other hand,

$$C_p = \Pi_K(D_p+BG_p)|_K : K \to K.$$

So, on writing

$$I = \begin{bmatrix} I & S_p \\ 0 & T_p \end{bmatrix} : \quad J \dotplus K_p \to J \dotplus K,$$

we have $T_pC'_p = C_pT_p$ (p=1,2,...). As T_p is obviously invertible, the transformations C_p and C'_p are similar and the subspace T_pN_p is C_p-invariant. It remains to observe that because of (3.13) and (3.14) we have

$$\lim_{p\to\infty} \theta(T_pN_p, M) = 0,$$

and the stability of $\Pi_K M$ as an $\Pi_K A|_K$-invariant subspace is proved.

4. STABLE SUBSPACES MODULO A SUBSPACE.

In this section we shall restate the main result (Theorem 2.1) in the framework of invariant subspaces modulo a subspace.

Let $A : C^n \to C^n$ be a linear transformation and let $V \subset C^n$ be a subspace. We say that a subspace $M \subset C^n$ which is A-invariant modulo V is <u>stable</u> if for every $\varepsilon > 0$ there exists $\delta > 0$ such that for any linear transformation $A' : C^n \to C^n$ and any subspace $V' \subset C^n$ such that

$$\|A'-A\| + \theta(V',V) < \delta \tag{4.1}$$

there exists a subspace $M' \subset C^n$ which is A'-invariant modulo V' and satisfies the inequality

$$\theta(M,M') < \varepsilon. \tag{4.2}$$

If for every $\varepsilon > 0$ one can find $\delta > 0$ such that $\|A'-A\| < \delta$ implies $\theta(M,M') < \varepsilon$ for some M' which is A'-invariant modulo V, then the A-invariant modulo V subspace M will be called <u>stable with fixed</u> V.

THEOREM 4.1. <u>Let</u> $A : C^n \to C^n$ <u>be a linear transformation, and let</u> M <u>be an</u> A-<u>invariant subspace modulo a subspace</u> $V \subset C^n$. <u>Assume that condition (α) of Theorem 1.1 holds, where</u> K <u>is some direct complement in</u> C^n <u>to the minimal</u> A-<u>invariant subspace</u> $J(A,V)$ <u>which contains</u> V. <u>Then</u> M <u>is a stable</u> A-<u>invariant subspace modulo</u> V.

<u>Conversely, if</u> M <u>is stable with fixed</u> V <u>and</u> $M \cap J(A,V) = \{0\}$, <u>then the condition (α) of Theorem 1.1 holds</u>.

PROOF. The first part of the theorem follows from Theorem 1.1 when B is the orthogonal projector on V. The second part of the theorem follows from the proof of the second part of Theorem 1.1 when B is any (fixed) linear transformation whose image is V. □

Again, Theorem 4.1 is valid also in the framework of real linear transformations and real invariant subspaces.

REFERENCES

1. Bart, H., Gohberg, I., Kaashoek, M.A.: Stable factorization of monic matrix polynomials and stable invariant subspaces, Integral Equations and Operator Theory 1 (1978), 496-517.

2. Bart, H., Gohberg, I., Kaashoek, M.A.: Minimal factorization of matrix and operator functions, Birkhauser, Bäsel, 1979.

3. Campbell, S., Daughtry, J.: The stable solutions of quadratic matrix equations, Proc. AMS 74 (1979), 19-23.

4. Gohberg, I., Lancaster, P., Rodman, L.: Matrix polynomials, Academic Press, New York etc., 1982.

5. Gohberg, I., Lancaster, P., Rodman, L.: Invariant subspaces of matrices with applications, John Wiley, to appear.

6. Gohberg, I., Rubinstein, S.: Stability of minimal fractional decompositions of ration matrix functions, in press.

7. Helton, J. W., Ball, J. A.: The cascade decompositions of a given system vs the linear fractional decompositions of its transfer function, Integral Equations and Operator Theory 5 (1982), 341-385.

8. Wonham, W. M.: Linear multivariable control: a geometric approach, Springer-Verlag, Berlin, 1979.

Leiba Rodman,
School of Mathematical Sciences,
Tel-Aviv University,
Tel-Aviv, Israel
Department of Mathematics,
Arizona State University,
Tempe, Arizona 85287

Operator Theory:
Advances and Applications, Vol. 19

SELF-ADJOINT DETERMINANTAL REPRESENTATIONS OF REAL IRREDUCIBLE CUBICS

Victor Vinnikov

We obtain a complete classification of self-adjoint determinantal representations of real irreducible cubics. The case of real rational cubics have been treated previously[13], so here we treat mainly the case of real elliptic cubics. The complete set of canonical representations is built; it is topologically equivalent either to an punctured circle or to a disjoint union of a circle and a punctured circle. We investigate elementary transformations [7,8,9] which allow us to obtain from a given self-adjoint determinantal representation all the non-equivalent representations within the same component. Finally we introduce and study a certain notion of positivity, or definiteness; it turns out that the non-compact component (the punctured circle) consists of indefinite representations, while the compact component (the circle) consists of definite representations.

1. INTRODUCTION

Let $F(x_0,x_1,x_2)$ be a complex (algebraic) curve of degree n. A *determinantal representation* $U = U(x_0,x_1,x_2)$ of F is a matrix of order n whose entries are linear in x_0,x_1,x_2:

$$U(x_0,x_1,x_2) = x_1A + x_2B + x_0C \tag{1}$$

(A,B,C-complex matrices of order n)
satisfying $\det U(x) \equiv cF(x)$ $(c \in C,\ c \neq 0)$.

Determinantal representations have been previously investigated in [3,4,12]. It is shown there that a *general complex curve possesses a determinantal representation.*

Suppose now F is a real curve (i.e. all its coefficients are real). A *self-adjoint determinantal representation* $U(x_0,x_1,x_2) = x_1A + x_2B + x_0C$ of F is a determinantal representation satisfying $U^*(x) = U(x)$ for $x_0,x_1,x_2 \in R$ (i.e. A,B,C are complex hermitian matrices) and so that

$\det U(x) = cF(x)$ with $c > 0$.

Two self-adjoint determinantal representations U_1, U_2 of a real curve F of degree n are called *equivalent* if there exists a complex matrix $P \in GL(n,C)$ so that:

$$U_2 = PU_1P^* \tag{2}$$

Study and classification of self-adjoint determinantal representations of a real curve is an interesting problem by itself, it is equivalent to the algebraic problem of the simultaneous classification of triples of hermitian matrices, and as shown by Livšic[9], it has a deep connection with the theory of colligations for pairs of commuting operators in a Hilbert space. In this paper we investigate and classify self-adjoint determinantal representations of real irreducible cubics. A survey of the results appears in ref. 14.

We introduce an additional general concept. Let $U(x)$ be an arbitrary (not necessary self-adjoint) determinantal representation of a complex curve F of degree n, we define the *corresponding vector bundle* $\varepsilon(x)$ along the curve by

$$\varepsilon(x) = \ker U(x) \tag{3}$$

It is rather easy to see[3,9,13] that *if* $\tilde{x}$ *is a regular point of* F, *then the vector bundle at it is one-dimensional:* $\dim \varepsilon(\tilde{x}) = 1$. If the curve possesses singular points, the dimension of the vector bundle there may either remain 1 or grow up; see[13] for examples.

Let now $U_1(x)$, $U_2(x)$ be self-adjoint determinantal representations of a real curve F of degree n, the corresponding vector bundles $\varepsilon_1(x)$, $\varepsilon_2(x)$ are called *equivalent* if there exists a complex matrix $T \in GL(n,C)$ so that:

$$\varepsilon_2(x) = \varepsilon_1(x)T \tag{4}$$

Clearly, the equivalence of self-adjoint determinantal representations implies the equivalence of the corresponding vector bundles.

We proceed now to recall some facts about real irreducible cubics (see ref. 2,5,11,15).

2. REAL IRREDUCIBLE CUBICS

A complex irreducible curve $f(x,y)$ is called *rational* if there exist complex rational functions $\varphi(t)$, $\psi(t)$ satisfying:

a) For almost all $t_0 \in C$ (i.e., all but a finite number) the point $(\varphi(t_0),\psi(t_0))$ is on the curve: $f(\varphi(t_0), \psi(t_0)) = 0$;

b) For almost all points (x_0,y_0) on the curve there exists a unique $t_0 \in C$, so that $x_0 = \varphi(t_0)$, $y_0 = \psi(t_0)$.

For rational curves the genus $g = 0$, and the Riemann surface of the curve is homeomorphic to the closed complex plane, i.e. to the sphere.

We remind that an elliptic function is a doubly periodic meromorphic function (see ref. 1). A complex irreducible curve $f(x,y)$ is called *elliptic* if there exist elliptic functions $\varphi(u)$, $\psi(u)$ (with the same periods) satisfying:

a) For almost all u_0 in the period parallelogram, the point $(\varphi(u_0),\psi(u_0))$ is on the curve: $f(\varphi(u_0),\psi(u_0)) = 0$;

b) For almost all points (x_0,y_0) on the curve there exists a unique u_0 in the period parallelogram, so that $x_0 = \varphi(t_0)$, $y_0 = \psi(t_0)$.

For elliptic curves the genus $g = 1$, and the Riemann surface of the curve is homeomorphic to the torus.

Now, an irreducible cubic, in particular a real irreducible cubic, is either non-singular (smooth), in which case it is elliptic, or it has a unique singular point of multiplicity 2, in which case it is rational. We have presented a complete classification of self-adjoint determinantal representations of real rational cubics in ref. 13; in this paper (except for the last section) we are going to concentrate on real elliptic (i.e. smooth) cubics.

We remind, that complex curves can be brought into different forms by a projective change of coordinates in the complex (projective) plane. Since we are dealing with real curves, we shall consider only real changes of coordinates - which have real coefficients and leave the real part of the complex plane invariant; they are given by

$$[x_0',x_1',x_2'] = [x_0,x_1,x_2]A \tag{5}$$

where $A \in GL(3,R)$,

Suppose that the change of coordinates (5) transforms a real curve $F(x_0,x_1,x_2)$ into a real curve $F'(x_0',x_1',x_2')$. Then it clearly induces a one-to-one real linear correspondence between self-adjoint determinantal representations of F and those of F'. It is also clear that the equivalence of self-adjoint determinantal representations is preserved under such a correspondence. Hence instead of studying self-adjoint determinantal representations of the original curve F, we may study those of F'.

It can be shown that any real smooth cubic can be brought, by a real projective change of coordinates, to the following normal form:

$$y^2 = g(x) \tag{6}$$

where $g(x)$ is a (real) polynomial of degree 3 without multiple roots. $g(x)$ must have at least one real root and by performing an additional real shift in coordinates we may assume that this real root is 0. Thus our final normal form for real smooth cubics (under the real projective change of coordinates) is:

$$y^2 = x(x+\theta_1)(x+\theta_2) \tag{7}$$

where θ_1, θ_2 are either distinct real numbers, different from 0, or complex conjugates. In projective coordinates:

$$x_2^2x_0 - x_1(x_1+\theta_1x_0)(x_1+\theta_2x_0) = 0 \tag{8}$$

We shall denote this cubic by $F_{\theta_1\theta_2}(x_0,x_1,x_2)$. By the remarks of the previous paragraph we may limit ourselves now to the investigation of self-adjoint determinantal representations of $F_{\theta_1\theta_2}$.

Before proceeding further, we want to write down explicitly the parametrization of $F_{\theta_1\theta_2}$ by elliptic functions. To this end we recall the Weierstrass normal form for smooth cubics:

$$\tilde{y}^2-4(\tilde{x}-e_1)(\tilde{x}-e_2)(\tilde{x}-e_3) = 0 , \quad e_1+e_2+e_3 = 0 \tag{9}$$

Or:

$$\tilde{y}^2 - 4\tilde{x}^3 - g_2\tilde{x} - g_3 = 0 \qquad (10)$$

where $g_2 = -4(e_1e_2+e_1e_3+e_2e_3)$, $g_3 = 4e_1e_2e_3$. The parametrization of the Weiestrass normal form is given by

$$\begin{cases} \tilde{x} = \wp(u) \\ \tilde{y} = \wp'(u) \end{cases} \qquad (11)$$

where $\wp(u) = \wp(u;g_2,g_3)$ is the Weierstrass elliptic function with invariants g_2, g_3. We shall denote the corresponding periods (here and afterwards) by 2ω, $2\omega'$. We remind that the Weierstrass elliptic function is an elliptic function that possesses double poles with residue 0 at the points of the period lattice:

$$\wp(u) = \frac{1}{u^2} + \sum_{\substack{m,m' \\ 2m\omega+2m'\omega'\neq 0}} \left[\frac{1}{(u-2m\omega-2m'\omega')^2} - \frac{1}{(2m\omega+2m'\omega')^2}\right] \qquad (12)$$

It is immediately seen that the following (real) change of coordinates transforms our normal form (7) into the Weierstrass normal form (9):

$$\begin{cases} x = \sqrt[3]{4}\,\tilde{x} - \frac{\theta_1+\theta_2}{3} \\ y = \tilde{y} \end{cases} \qquad (13)$$

(the roots in the Weierstrass form are $e_1 = \frac{\theta_1+\theta_2}{3\sqrt[3]{4}}$, $e_2 = \frac{\theta_2-2\theta_1}{3\sqrt[3]{4}}$, $e_3 = \frac{\theta_1-2\theta_2}{3\sqrt[3]{4}}$). From (11) we obtain now the parametrization of $F_{\theta_1\theta_2}$ by elliptic functions:

$$\begin{cases} x = \sqrt[3]{4}\,\wp(u) - \frac{\theta_1+\theta_2}{3} \\ y = \wp'(u) \end{cases} \qquad (14)$$

Of course, since $F_{\theta_1\theta_2}$ is a real cubic, the invariants g_2, $g_3 \in \mathbb{R}$. The precise structure of the period parallelogram depends on whether the roots θ_1, θ_2 are real or complex conjugate, as follows.

a) $\theta_1, \theta_2 \in R$: Then in the Weierstrass normal form (9) the discriminant $g_2^3 - 27g_3^2 > 0$. In this case one of the periods (say 2ω) is real, while the other one is purely imaginary; the period parallelogram is thus a rectangle (Figure 1).

b) $\theta_1 = \bar{\theta}_2 \notin R$. Then in the Weierstrass normal form (9) the discriminant $g_2^3 - 27g_3^2 < 0$. In this case the periods are complex conjugate, hence the period parallelogram is a rhombus (Figure 2).

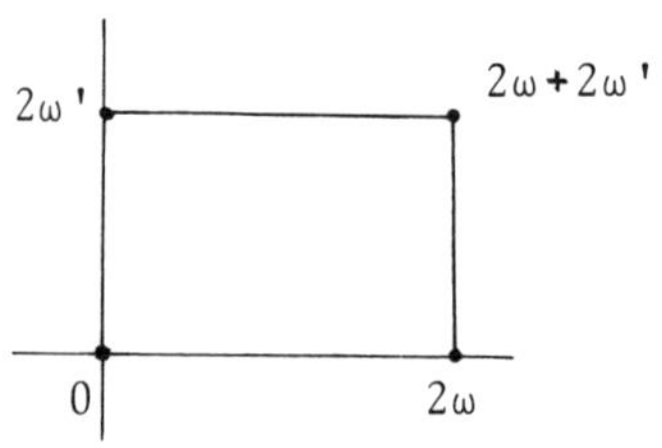

$\omega \in R$, $\omega' \in iR$

$\theta_1, \theta_2 \in R$

Figure 1

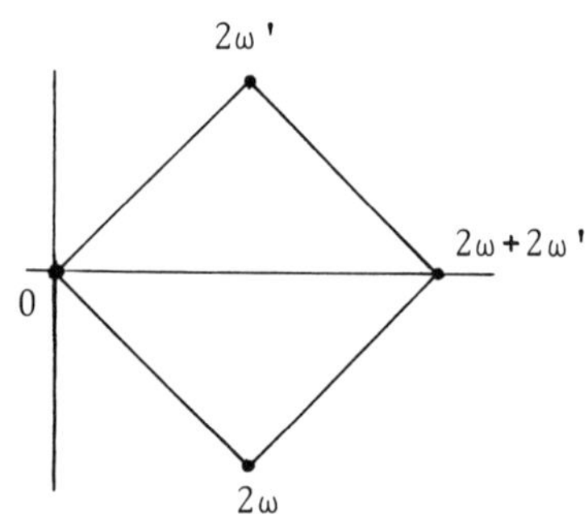

$\omega' = \bar{\omega}$, $\theta_1 = \bar{\theta}_2 \notin R$

Figure 2

3. CANONICAL FORMS FOR DETERMINANTAL REPRESENTATIONS

A first step in the classification of self-adjoint determinantal representations is to obtain a complete set of canonical forms for them, so that each self-adjoint determinantal representation of $F_{\theta_1\theta_2}$ is equivalent to exactly one canonical representation. To this end we shall use an algebraic approach (as we did in ref. 13 for real rational cubics).

Given any self-adjoint determinantal representation of $F_{\theta_1\theta_2}$:

$$\det(x_1A+x_2B+x_0C) = r\, F_{\theta_1\theta_2}(x_0,x_1,x_2) \qquad (r > 0) \tag{15}$$

we want to reduce it to the canonical form by the group action:

$$x_1A + x_2B + x_0C \to P(x_1A+x_2B+x_0C)P^* , \quad P \in GL(3,C) \qquad (16)$$

Alternatively, we may consider the group action:

$$(A,B,C) \to (PAP^*,PBP^*,PCP^*) , \quad P \in GL(3,C) \qquad (17)$$

on triples (A,B,C) of hermitian matrices of order 3 satisfying (15).

We proceed as follows: first, using the theory of indefinite inner-product spaces[6,10], we reduce simultaneously the above matrices A, B to the canonical form (by the transformation $A \to PAP^*$, $B \to PBP^*$); then we find a matrix P preserving this canonical form for A, B, and reduce the matrix C to the canonical forms by the transformation $C \to PCP^*$.

We note that from (15) $\det A = -r \neq 0$. We define now an indefinite metric in C^3 by

$$[h,g] = (hA,g) \qquad (18)$$

where $(\cdot, \cdot)$ denotes the usual scalar product. Since A is hermitian and non-singular, this indefinite metric is hermitian and non-degenerate. Clearly, A is the matrix of the metric (Gram matrix) in the standard basis.

A (complex) vector space endowed with an (indefinite) hermitian non-degenerate metric is called a *pseudo-unitary space*. Let V be an arbitrary pseudo-unitary space and denote the metric in it by $[\cdot, \cdot]$. A linear transformation K in V is called *self-adjoint* if

$$[hK,g] = [h,gK] \qquad (19)$$

Choose a basis in V, let A be the matrix of the metric in this basis and let K be the matrix of the linear transformation K; then K is self-adjoint when

$$KA = AK^* \qquad (20)$$

When a change of basis with a non-singular matrix P is performed, the matrix A of the metric is transformed into PAP^* (congruence), while the matrix K of the linear transformation K is transformed into PKP^{-1} (similarity). The following theorem describes the canonical form for a matrix of a self-adjoint transformation in a pseudo-unitary space.

THEOREM 1 (ref. 6,10): Let K be a self-adjoint linear transformation of the pseudo-unitary space V. Then there exists an orthogonal direct sum decomposition:

$$V = V_1 \oplus V_2 \oplus \dots \oplus V_n \tag{21}$$

so that each V_i is K-invariant, and one of the following holds:

a) There exists a basis of V_i in which the matrix of K is an elementary Jordan matrix with an elementary divisor $(\lambda-\alpha)^m$, $\alpha \in R$, and the matrix of the metric is

$$\varepsilon \begin{bmatrix} & & & 1 \\ & & 1 & \\ & \cdot^{\cdot^{\cdot}} & & \\ 1 & & & \end{bmatrix}, \quad \varepsilon = \pm 1 \tag{22}$$

(We shall say that the real elementary divisor $(\lambda-\alpha)^m$ of K has sign ε; the system of elementary divisors of K in which each real elementary divisor has been this assigned a sign will be called a *signed system of elementary divisors*; it may be shown that the signs in a signed system of elementary divisors don't depend on the particular way of decomposition in (21)).

b) There exists a basis of V_i in which the matrix of K is a direct sum of two elementary Jordan matrices corresponding to conjugate elementary divisors $(\lambda-\alpha)^m$, $(\lambda-\bar{\alpha})^m$, $\alpha \notin R$, and the matrix of the metric is:

$$\begin{bmatrix} & & & 1 \\ & & 1 & \\ & \cdot^{\cdot^{\cdot}} & & \\ 1 & & & \end{bmatrix} \quad \square \tag{23}$$

We return now to the coefficient matrices A, B, C of the self-adjoint determinantal representation (15) of $F_{\theta_1\theta_2}$ and to our indefinite metric (18) in C^3. We define $K=BA^{-1}$, i.e. $B=KA$. Since B is hermitian, it is easily seen that K is self-adjoint with respect to the indefinite metric (18) (i.e. (20) holds). The transformation $A \to PAP^*$, $K \to PKP^{-1}$ induces the following transformation on B:

$$B = KA \to (PKP^{-1})(PAP^*) = P(KA)P^* = PBP^* \tag{24}$$

It follows, by Theorem 1, that the canonical form for the pair of matrices A,B - under the desired transformation $A \to PAP^*$, $B \to PBP^*$ - is determined by the signed system of elementary divisors of K.

From (15) it can be seen that the signed system of elementary divisors of K is $+\lambda^3$. The canonical form for A, K and B is thus:

$$A = \begin{bmatrix} 0 & 0 & 1 \\ 0 & 1 & 0 \\ 1 & 0 & 0 \end{bmatrix}, \quad K = \begin{bmatrix} 0 & 1 & 0 \\ 0 & 0 & 1 \\ 0 & 0 & 0 \end{bmatrix}, \quad B=KA= \begin{bmatrix} 0 & 1 & 0 \\ 1 & 0 & 0 \\ 0 & 0 & 0 \end{bmatrix} \tag{25}$$

Until the end of this section we assume that A, B are in their canonical form (25).

We have to find now a matrix P preserving A, B:

$$\begin{cases} PAP^* = A \\ PBP^* = B \end{cases} \tag{26}$$

It can be shown that P is of the form

$$P = a\begin{bmatrix} 1 & q & s \\ 0 & 1 & q \\ 0 & 0 & 1 \end{bmatrix} \tag{27}$$

where $|a| = 1$, $q+\bar{q} = 0$, $\operatorname{Re} s = -\frac{|q|^2}{2}$. We assume until the end of the section that P is as above.

We want to reduce the matrix C to the canonical forms by the transformation $C \to PCP^*$. Using (15) we can show that C can be brought to the form

$$C = \begin{bmatrix} t & d & c \\ -d & b & 0 \\ c & 0 & -1 \end{bmatrix} \tag{28}$$

where $t, b, c \in R$ and d is purely imaginary. A direct calculation (using (15) again) shows that then:

$$c = \frac{\theta_1+\theta_2-b}{2}, \quad t = \theta_1\theta_2 + \frac{b-\theta_1-\theta_2}{2} \cdot \frac{\theta_1+\theta_2+3b}{2},$$
$$d = \pm i\sqrt{b(b-\theta_1)(b-\theta_2)} \tag{29}$$

It can be also seen that if C, C' are two different matrices of the form (28), then $PCP^* = C'$ is impossible. Hence (28)-(29) gives us a complete set of canonical forms for the matrix C (under the transformation $C \to PCP^*$). Thus:

THEOREM 2: The complete set of canonical forms for self-adjoint determinantal representations of $F_{\theta_1\theta_2}$ is given explicitly by:

$$x_1 \begin{bmatrix} 0 & 0 & 1 \\ 0 & 1 & 0 \\ 1 & 0 & 0 \end{bmatrix} + x_2 \begin{bmatrix} 0 & 1 & 0 \\ 1 & 0 & 0 \\ 0 & 0 & 0 \end{bmatrix} + x_0 \begin{bmatrix} t & d & c \\ -d & b & 0 \\ c & 0 & -1 \end{bmatrix} =$$
$$= \begin{bmatrix} tx_0 & dx_0+x_2 & cx_0+x_1 \\ -dx_0+x_2 & bx_0+x_1 & 0 \\ cx_0+x_1 & 0 & -x_0 \end{bmatrix} \tag{30}$$

where b is a real parameter and

$$c = \frac{\theta_1+\theta_2-b}{2}, \quad t = \theta_1\theta_2 + \frac{b-\theta_1-\theta_2}{2} \cdot \frac{\theta_1+\theta_2+3b}{2},$$

$d = \pm i\sqrt{b(b-\theta_1)(b-\theta_2)}$, d - purely imaginary □

Of course, since d is purely imaginary, b must take its values in certain intervals on the real axis; the precise structure of these intervals depends on whether θ_1, θ_2 are real or complex conjugate.

4. TOPOLOGICAL STRUCTURE OF THE SET OF CANONICAL REPRESENTATIONS

Our objective now is to clarify the topological structure of the set of canonical representations (30), i.e. the structure of the set of self-adjoint determinantal representations of $F_{\theta_1\theta_2}$ modulo the equivalence relation.

The canonical representation (30) is determined by two

numbers b, d ($b \in R$, $d \in iR$) satisfying

$$(-d)^2 = -b(-b+\theta_1)(-b+\theta_2) \tag{31}$$

Consider the point $(-b, -d) \in C^2$. It is an (affine) point, whose first coordinate is real and the second one is purely imaginary, and it is situated on the curve:

$$y^2 = x(x+\theta_1)(x+\theta_2) \tag{32}$$

i.e. on $F_{\theta_1\theta_2}$. We get thus a natural homeomorphism between the set of canonical representations (30) and the set of all affine points on $F_{\theta_1\theta_2}$ whose first coordinate is real and the second one is purely imaginary.

The set of these points on $F_{\theta_1\theta_2}$ looks differently according to whether the roots θ_1, θ_2 are real or complex conjugate. The precise picture is given by Figures 3-4, where, of course, the x-axis is real, while the y-axis is purely imaginary.

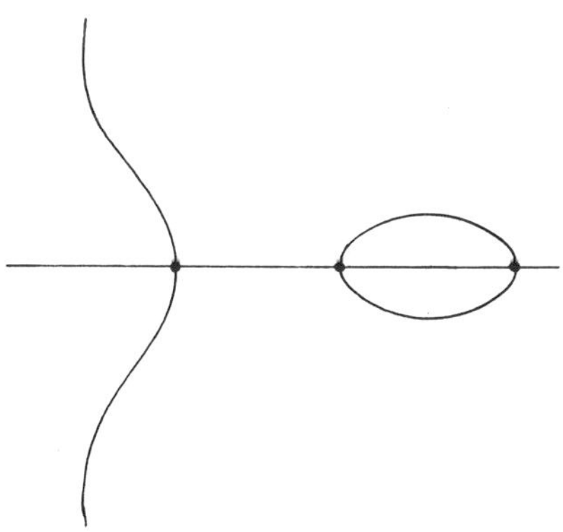

$\theta_1, \theta_2 \in R$

Figure 3

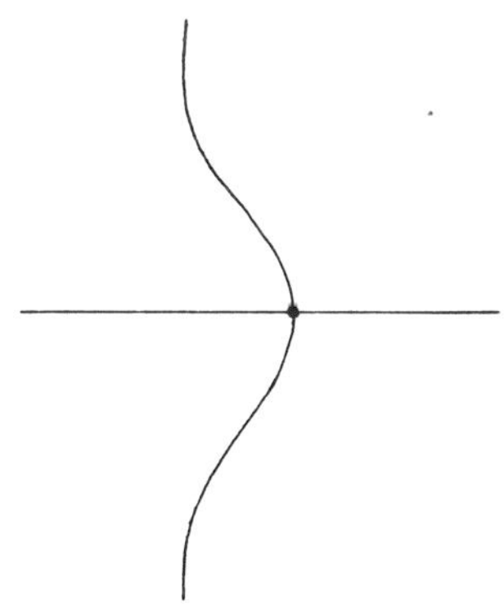

$\theta_1 = \bar{\theta}_2 \notin R$

Figure 4

We have thus

THEOREM 3: There is a natural homeomorphism between the set of canonical forms (30) for self-adjoint determinantal representations of $F_{\theta_1\theta_2}$ and the set of affine points on $F_{\theta_1\theta_2}$ whose first coordinate is real and the second one is purely imaginary. Consequently, the set of canonical representations (30) is topologically equivalent to:

a) $\theta_1, \theta_2 \in R$: a disjoint union of a punctured circle and a circle;

b) $\theta_1 = \bar{\theta}_2 \notin R$: a punctured circle □

We see that the set of canonical representations of $F_{\theta_1\theta_2}$ contains in any case a non-compact component; if $\theta_1, \theta_2 \in R$ it contains an additional component which is compact. The meaning of these two components will become clear later.

The topological structure in Theorem 3 and Figures 3-4 can be also understood through the parametrization (14) of $F_{\theta_1\theta_2}$ by elliptic functions. Indeed, (-b,-d) is a point on $F_{\theta_1\theta_2}$, hence there corresponds to it a certain value v in the period parallelogram. Since (-b,-d) is an affine point:

$$v \not\equiv 0 \pmod{2\omega, 2\omega'} \tag{33}$$

(the vertices of the period parallelogram correspond, under the parametrization (14), to the point of $F_{\theta_1\theta_2}$ at infinity, i.e. in projective coordinates to (0,0,1)). The condition that b is real and d - purely imaginary, means, according to (14), that

$$\wp(v) \in R, \quad \wp'(v) \in iR \tag{34}$$

This clearly holds on the imaginary axis (since the invariants $g_2, g_3 \in R$). The precise picture is as follows.

a) $\theta_1, \theta_2 \in R$: We have seen in section 2 that in this case the period parallelogram is a rectangle ($\omega \in R$, $\omega' \in iR$). (34) holds when

$$\text{Re } v \equiv 0 \pmod{2 \text{ Re } \omega, 2 \text{ Re } \omega'} \tag{35}$$

(this is the imaginary axis) and also when

$$\text{Re } v \equiv \omega \pmod{2 \text{ Re } \omega, 2 \text{ Re } \omega'} \tag{36}$$

(35) describes, taking into account (33), an open interval in the period parallelogram (homeomorphic to a punctured circle), and (36) describes a closed interval whose endpoints must be identified (homeomorphic to a circle). See Figure 5.

b) $\theta_1 = \bar{\theta}_2 \notin R$: In this case the period parallelogram is a rhombus ($\omega' = \bar{\omega}$). (34) holds only when

$$\text{Re } v \equiv 0 \pmod{2 \text{ Re } \omega, 2 \text{ Re } \omega'} \tag{35}$$

(on the imaginary axis). As we noted above, this equation describes, taking into account (33), an open interval in the period parallelogram. See Figure 6.

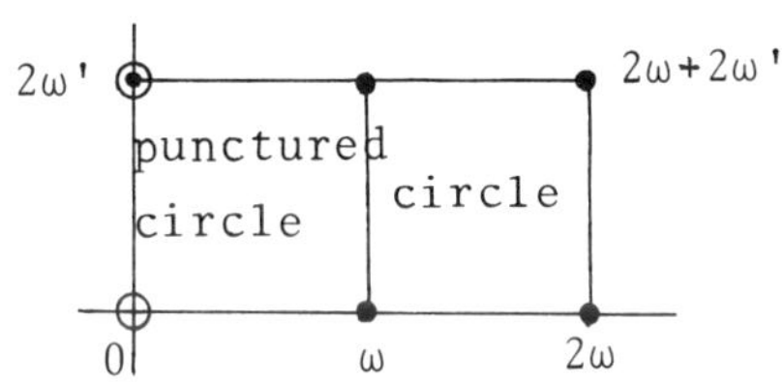

$\theta_1, \theta_2 \in R$

Figure 5

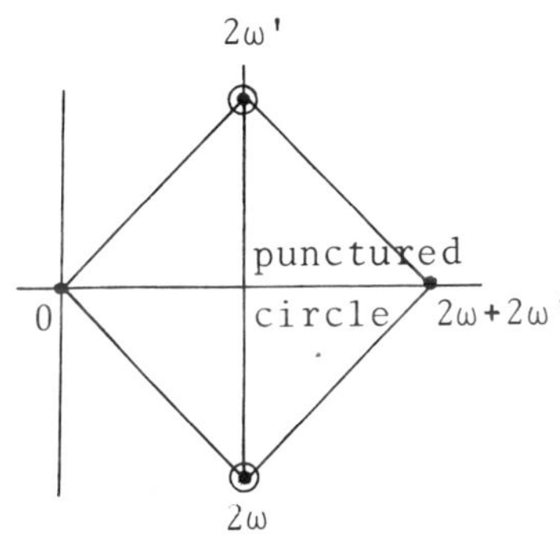

$\theta_1 = \bar{\theta}_2 \notin R$

Figure 6

5. CLASS OF DIVISORS OF THE VECTOR BUNDLE AND COMPLETE SYSTEM OF INVARIANTS

To complete the classification of self-adjoint determinantal representations of $F_{\theta_1\theta_2}$ we have to find a complete system of invariants for them, so that given two representations we will be able to tell whether they are equivalent or not. To this end we shall use the notion of the divisor (see ref. 2,5,11, 15).

First let $R(x_0, x_1, x_2)$ be a rational function in $x_0, x_1,$

x_2 (whose numerator and denominator are homogeneous of the same degree). We remind that we can assign multiplicities to its zeroes and its poles along the curve $F_{\theta_1\theta_2}$ (this can be done in several equivalent ways: we can use general methods of algebraic geometry, or we can consider the rational function along the curve as a meromorphic function on the corresponding Riemann surface, or we can use the fact that $F_{\theta_1\theta_2}$ is an elliptic curve and substitute into $R(x)$ the parametrization of $F_{\theta_1\theta_2}$ by elliptic functions).

Let now $U(x)$ be a self-adjoint determinantal representation of $F_{\theta_1\theta_2}$, and let $\varepsilon(x)$ be the corresponding vector bundle. Let $u(x)$ be a rational section of $\varepsilon(x)$ (i.e. a section (a representative) $u(x) \in \varepsilon(x)$ whose entries are rational functions of x_0, x_1, x_2; a rational section can always be constructed using the algebraic complements of a column in $U(x)$). We consider zeroes and poles of $u(x)$ along $F_{\theta_1\theta_2}$ and define their multiplicities in the natural way: the multiplicity of T as a zero (resp. as a pole) of $u(x)$ is defined as the minimal (resp. as the maximal) between the multiplicities of T as a zero (resp. as a pole) of the entries of $u(x)$.

Denote by $Z_1, \ldots, Z_k$ the zeroes and by $P_1, \ldots, P_\ell$ the poles of $u(x)$ along $F_{\theta_1\theta_2}$, and let $n_1, \ldots, n_k$ and $m_1, \ldots, m_\ell$ be the corresponding multiplicities. The divisor of $u(x)$ is by definition the following formal sum:

$$\operatorname{div}(u(x)) = \sum_i n_i Z_i - \sum_j m_j P_j \tag{37}$$

The divisor of a rational function $R(x)$ is defined similarly.

In general, a *divisor* D on $F_{\theta_1\theta_2}$ is any (finite) formal sum of points T_i on $F_{\theta_1\theta_2}$ with integer coefficients r_i: $D = \sum_i r_i T_i$. Clearly, the divisors on $F_{\theta_1\theta_2}$ form an (additively written) abelian group. The sum of the coefficients of the divisor D is called the *degree* of D: $\deg D = \sum_i r_i$.

A divisor D on $F_{\theta_1\theta_2}$ is said to be equivalent to 0: $D \equiv 0$, if there exists a rational function $R(x)$ so that $D = \mathrm{div}(R(x))$. A necessary condition for this is $\deg D = 0$, since the rational function $R(x)$ has an equal number of zeroes and of poles along $F_{\theta_1\theta_2}$, counting multiplicities (substituting into $R(x)$ the parametrization of $F_{\theta_1\theta_2}$ by elliptic functions, this turns out to be the familiar theorem that an elliptic function has the same number of zeroes and of poles, counting multiplicities, in the period parallelogram); it is not a sufficient conditon, since the elliptic cubic $F_{\theta_1\theta_2}$ has genus $g=1$, and its Riemann surface is homeomorphic to the torus, and not just to the closed complex plane. Two divisors D_1, D_2 on $F_{\theta_1\theta_2}$ are called (linearly) equivalent: $D_1 \equiv D_2$, if $D_1 - D_2 \equiv 0$.

We return now to our vector bundle $\varepsilon(x)$ and consider two rational sections $u(x)$, $v(x)$ of it. Since $F_{\theta_1\theta_2}$ is smooth, $\varepsilon(x)$ is one-dimensional, thus there exists a rational function $R(x)$ so that (along $F_{\theta_1\theta_2}$):

$$u(x) = R(x)v(x) \tag{38}$$

Hence:

$$\mathrm{div}(u(x)) = \mathrm{div}(R(x)) + \mathrm{div}(v(x)) \equiv \mathrm{div}(v(x)) \tag{39}$$

We conclude that the vector bundle $\varepsilon(x)$ determines uniquely a class of equivalent divisors on $F_{\theta_1\theta_2}$, that we shall denote by $C\ell\ \mathrm{div}(\varepsilon(x))$.

Let now $\varepsilon_1(x)$, $\varepsilon_2(x)$ be equivalent vector bundles (corresponding to some self-adjoint determinantal representations of $F_{\theta_1\theta_2}$), i.e. $\varepsilon_2(x) = \varepsilon_1(x)T$, $T \in GL(3,C)$. If $u_1(x)$ is a rational section of $\varepsilon_1(x)$, then

$$u_2(x) = u_1(x)T \tag{40}$$

is a rational section of $\varepsilon_2(x)$. However, since T is a constant, non-singular matrix, $u_1(x)$, $u_2(x)$ have the same zeroes and poles along $F_{\theta_1\theta_2}$, including multiplicities. Thus $\mathrm{div}(u_1(x)) = \mathrm{div}(u_2(x))$ and

$$C\ell\ \mathrm{div}(\varepsilon_1(x)) = C\ell\ \mathrm{div}(\varepsilon_2(x)) \tag{41}$$

Hence the class of divisors is invariant under the equivalence of vector bundles.

Consequently we have found an invariant for self-adjoint determinantal representations of $F_{\theta_1\theta_2}$. We note that our considerations were completely general: for self-adjoint determinantal representations of any real irreducible curve, we can define the class of divisors of the corresponding vector bundle, and it is invariant under the equivalence of self-adjoint determinantal representations.

For self-adjoint determinantal representations of $F_{\theta_1\theta_2}$ we claim that the class of divisors of the corresponding vector bundle forms the complete system of invariants. To prove this, we show that for different canonical representations (30), the corresponding vector bundles possess different classes of divisors. We consider a canonical representation:

$$U = \begin{bmatrix} t & d+y & c+x \\ -d+y & b+x & 0 \\ c+x & 0 & -1 \end{bmatrix}$$

(we are using now affine coordinates), and taking algebraic complements of the first column we obtain a rational section $f(x)$ of the vector bundle $\ker U$:

$$f(x) = [-b-x,\ d+y,\ -bc-(b+c)x-x^2] \tag{42}$$

The entries of $f(x)$ have, along $F_{\theta_1\theta_2}$, a common zero at $(-b,-d)$ (which turns out to be simple) and a common pole at infinity (which has multiplicity 4). Hence (in projective coordinates):

$$\mathrm{div}(f(x)) = (1,-b,-d) - 4(0,0,1) \tag{43}$$

Let now U, U' be two different canonical representations (30), and let $f(x)$, $f'(x)$ be the rational sections (42) of the corresponding vector bundles. Suppose that $\mathrm{div}(f(x)) \equiv \mathrm{div}(f'(x))$, i.e.

$$(1,-b,-d) - 4(0,0,1) \equiv (1,-b',-d') - 4(0,0,1) ,$$
$$(1,-b,-d) - (1,-b',-d') \equiv 0 \qquad (44)$$

The last equation is, however, impossible, since for any two distinct points A, B on $F_{\theta_1\theta_2}$, $A-B \not\equiv 0$ (this follows from the fact that there is no elliptic function whose only singularity in the period parallelogram is a simple pole).

We thus have:

THEOREM 4: U being a self-adjoint determinantal representation of $F_{\theta_1\theta_2}$, the class of divisors of the corresponding vector bundle forms the complete system of invariants for U (under the equivalence of self-adjoint determinantal representations). □

We obtain also the following Corollary, which is a particular case of the result obtained in ref. 3.

COROLLARY: If $u(x)$ is a rational section of the vector bundle corresponding to a self-adjoint determinantal representation of $F_{\theta_1\theta_2}$, then $\deg \operatorname{div}(u(x)) = -3$. □

Indeed, there exists a (unique) rational section $f(x)$ of the form (42), so that $\operatorname{div}(u(x)) \equiv \operatorname{div}(f(x))$. Hence $\deg \operatorname{div}(u(x)) = \deg \operatorname{div}(f(x)) = -3$.

In order to make an explicit use of Theorem 4, we must have explicit conditions for the equivalence of divisors on $F_{\theta_1\theta_2}$. These are based on Abel's theorem for elliptic curves (ref. 1,2).

THEOREM 5: $\{u_1,...,u_n\}$ and $\{v_1,...,v_n\}$ form the zero set and the pole set respectively of some elliptic function in the period parallelogram (each zero and each pole is written down a number of times according to its multiplicity) if and only if:

$$\Sigma u_i \equiv \Sigma v_i \pmod{2\omega, 2\omega'} \qquad □ \qquad (45)$$

We shall write $T \leftrightarrow u$ to denote that the point T on $F_{\theta_1\theta_2}$ corresponds to the value u in the period parallelogram under the parametrization (14) of $F_{\theta_1\theta_2}$ by elliptic functions.

We have then:

COROLLARY: Let $D_1 = \Sigma\, n_i\, T_i$, $D_2 = \Sigma\, m_j\, P_j$ be two divisors on $F_{\theta_1\theta_2}$, and let $T_i \leftrightarrow u_i$, $P_j \leftrightarrow v_j$. Then $D_1 \equiv D_2$ if and only if:

a) $\deg D_1 = \deg D_2$;

b) $\Sigma\, n_i u_i \equiv \Sigma\, m_j v_j \pmod{2\omega,\ 2\omega'}$. □

6. ELEMENTARY TRANSFORMATIONS

In this section we consider an elementary transformation, first suggested by Livšic in ref. 9 and Kravitsky in ref. 7, 8, which allows to obtain, for any real curve, infinitely many, in general non-equivalent, self-adjoint determinantal representations from a given one. We show then that given a self-adjoint determinantal representation of $F_{\theta_1\theta_2}$, we can obtain from it by elementary transformations all the non-equivalent representations within the same component (recall the structure given by Theorem 3).

Let $f(x_1,x_2)$ be a real irreducible curve of degree n and take a self-adjoint determinantal representation of it:

$$\det(x_1\sigma_2 - x_2\sigma_1 + \gamma) = cf(x_1,x_2) \qquad (c > 0) \tag{46}$$

(we shall use in this section affine coordinates and denote them by (x_1,x_2) instead of (x,y)). In addition to the corresponding vector bundle $\varepsilon(x) = \ker(x_1\sigma_2 - x_2\sigma_1 + \gamma)$, we shall consider now also the vector bundle $\varphi(x) = \operatorname{coker}(x_1\sigma_2 - x_2\sigma_1 + \gamma)$. Both $\varepsilon(x)$ and $\varphi(x)$ are one-dimensional, except perhaps at the singular points of f.

Let $\lambda = (\lambda_1,\lambda_2)$, $\bar\lambda = (\bar\lambda_1,\bar\lambda_2)$ be two conjugate affine points on f, and suppose they are non-singular and different (i.e. $\lambda,\bar\lambda \notin R^2$). Take $\hat v \in \varepsilon(\lambda)$, $\hat u \in \varphi(\bar\lambda)$:

$$\begin{cases} \hat v(\lambda_1\sigma_2 - \lambda_2\sigma_1 + \gamma) = 0 \\ (\bar\lambda_1\sigma_2 - \bar\lambda_2\sigma_1 + \gamma)\hat u = 0 \end{cases} \tag{47}$$

Since the representation $x_1\sigma_2 - x_2\sigma_1 + \gamma$ is self-adjoint, of course:

$$\hat u = q\hat v^* \quad (q \in C,\ q \neq 0) \tag{48}$$

From (47) it follows that:

$$(\lambda_1-\bar{\lambda}_1)\hat{v}\sigma_2\hat{u} = (\lambda_2-\bar{\lambda}_2)\hat{v}\sigma_1\hat{u} \tag{49}$$

Unless $\hat{v}\sigma_1\hat{u} = \hat{v}\sigma_2\hat{u} = 0$, it follows from (49) that we can normalize $\hat{v}$, $\hat{u}$ so that

$$\begin{cases}\lambda_1-\bar{\lambda}_1 = \hat{v}\sigma_1\hat{u}\\ \lambda_2-\bar{\lambda}_2 = \hat{v}\sigma_2\hat{u}\end{cases} \tag{50}$$

The pair of conjugate points λ, $\bar{\lambda}$ on f is called *admissible* with respect to the representation $x_1\sigma_2-x_2\sigma_1+\gamma$ if $\hat{v}\sigma_1\hat{u}$, $\hat{v}\sigma_2\hat{u}$ are not both 0. From (47), (48) we conclude that the pair λ, $\bar{\lambda}$ is unadmissible when:

$$\hat{v}\sigma_1\hat{v}^* = \hat{v}\sigma_2\hat{v}^* = \hat{v}\gamma\hat{v}^* = 0 \tag{51}$$

i.e. when $\hat{v} \in \varepsilon(\lambda)$ is a common self-orthogonal vector of σ_1, σ_2, γ ($h \in C^n$ is a self-orthogonal vector of a $n\times n$ hermitian matrix A if $(hA,h) = hAh^* = 0$, i.e., if h is self-orthogonal in the indefinite metric of type (18) induced in C^n by A).

Suppose now that the pair λ, $\bar{\lambda}$ is admissible with respect to $x_1\sigma_2-x_2\sigma_1+\gamma$ and that $\hat{v}$, $\hat{u}$ are normalized so that (50) holds. Define:

$$\tilde{\gamma} = \gamma + \sigma_1(\hat{u}\hat{v})\sigma_2-\sigma_2(\hat{u}\hat{v})\sigma_1 \tag{52}$$

From the normalization conditions (50) it follows that in (48) q is purely imaginary, hence $\tilde{\gamma}$ is hermitian. It is a straightforward although tedious matter to verify the following relations:

$$(x_1\sigma_2-x_2\sigma_1+\tilde{\gamma})\left(I - \frac{(\hat{u}\hat{v})}{\lambda_k-x_k}\sigma_k\right) = \left(I-\sigma_k\frac{(\hat{u}\hat{v})}{\lambda_k-x_k}\right)(x_1\sigma_2-x_2\sigma_1+\gamma)\ , \tag{53}$$

$$k = 1,2 \quad (x_k \neq \lambda_k)$$

Recall now that for any two square matrices K_1, K_2

$$\det(I+K_1K_2) = \det(I+K_2K_1) \tag{54}$$

Thus by taking determinants it follows from (53) that

$$\det(x_1\sigma_2-x_2\sigma_1+\tilde{\gamma}) = \det(x_1\sigma_2-x_2\sigma_1+\gamma) = cf(x_1,x_2) \tag{55}$$

Hence $x_1\sigma_2 - x_2\sigma_1 + \tilde{\gamma}$ is a new self-adjoint determinantal representation of f; we say that it was obtained from the original representation $x_1\sigma_2 - x_2\sigma_1 + \gamma$ by the elementary transformation (52) based upon the admissible pair of conjugate points λ, $\bar{\lambda}$.

We note that this elementary transformation is invertible. It is easy to check that

$$\begin{cases} \hat{v}(\bar{\lambda}_1\sigma_2 - \bar{\lambda}_2\sigma_1 + \tilde{\gamma}) = 0 \\ (\lambda_1\sigma_2 - \lambda_2\sigma_1 + \tilde{\gamma})\hat{u} = 0 \end{cases} \tag{56}$$

i.e. $\ker(\bar{\lambda}_1\sigma_2 - \bar{\lambda}_2\sigma_1 + \tilde{\gamma}) = \ker(\lambda_1\sigma_2 - \lambda_2\sigma_1 + \gamma)$, $\operatorname{coker}(\lambda_1\sigma_2 - \lambda_2\sigma_1 + \tilde{\gamma}) = \operatorname{coker}(\bar{\lambda}_1\sigma_2 - \bar{\lambda}_2\sigma_1 + \gamma)$. It follows that if the points λ, $\bar{\lambda}$ are admissible with respect to $x_1\sigma_2 - x_2\sigma_1 + \gamma$, then the points $\bar{\lambda}$, λ are admissible with respect to $x_1\sigma_2 - x_2\sigma_1 + \tilde{\gamma}$, and the elementary transformation based upon the pair of points $\bar{\lambda}$, λ transforms the new representation $x_1\sigma_2 - x_2\sigma_1 + \tilde{\gamma}$ back into the original one $x_1\sigma_2 - x_2\sigma_1 + \gamma$.

To explore the connection between the original representation and the new one, we consider the corresponding vector bundles $\varepsilon(x) = \ker(x_1\sigma_2 - x_2\sigma_1 + \gamma)$, $\tilde{\varepsilon}(x) = \ker(x_1\sigma_2 - x_2\sigma_1 + \tilde{\gamma})$. Let us define

$$T_k(x) = I - \sigma_k \frac{(\hat{u}\hat{v})}{\lambda_k - x_k}, \quad k = 1, 2 \tag{57}$$

$T_k(x)$ is defined for $x_k \neq \lambda_k$, and it is easily seen that for $x_k \neq \bar{\lambda}_k$ it is invertible and

$$T_k^{-1}(x) = I + \sigma_k \frac{(\hat{u}\hat{v})}{\bar{\lambda}_k - x_k} \tag{58}$$

From (53) it follows now that for $x_k \neq \lambda_k, \bar{\lambda}_k$

$$\varepsilon(x)T_k^{-1}(x) = \tilde{\varepsilon}(x) \tag{59}$$

It can be also seen that (if $x_1 \neq \bar{\lambda}_1$, $x_2 \neq \bar{\lambda}_2$) $T_1^{-1}(x)$ and $T_2^{-1}(x)$ coincide when restricted to the vector bundle $\varepsilon(x)$.

Suppose now $v(x)$ is a rational section of $\varepsilon(x)$, then

$$\tilde{v}(x) = v(x)T_k^{-1}(x) \tag{60}$$

is a rational section of $\tilde{\varepsilon}(x)$. From (57), (58) it follows that:

$$\operatorname{div}(\tilde{v}(x)) = \operatorname{div}(v(x)) + \lambda - \bar{\lambda} \tag{61}$$

(see ref. 9 for the details). Thus we have

THEOREM 6 (ref. 9): $C\ell \operatorname{div}(\tilde{\varepsilon}(x)) =$
$C\ell \operatorname{div}(\varepsilon(x)) + C\ell\{\lambda - \bar{\lambda}\}$ □ (62)

If the curve f has genus $g > 0$ (i.e. if it is not rational), then for any two distinct points A, B on it, $A - B \not\equiv 0$ (we have seen this in the previous section for the particular case of elliptic curves). Hence $C\ell \operatorname{div}(\tilde{\varepsilon}(x)) \neq C\ell \operatorname{div}(\varepsilon(x))$. By the results of the previous section the new representation $x_1\sigma_2 - x_2\sigma_1 + \tilde{\gamma}$ is not equivalent to the original $x_1\sigma_2 - x_2\sigma_1 + \gamma$. The natural question is: given a self-adjoint determinantal representation of f, how many other (non-equivalent) representations can we obtain from it by elementary transformations based on admissible pairs of conjugate points?

For the case of our cubic $F_{\theta_1\theta_2}$ we can give a complete answer to this question. Let U_1, U_2 be two self-adjoint determinantal representations of $F_{\theta_1\theta_2}$, we shall say that they can be bridged by an elementary transformation, if there exists an elementary transformation, based upon some admissible pair of conjugate points, that transforms U_1 into a representation equivalent to U_2. We recall that the set of canonical representations (30) of $F_{\theta_1\theta_2}$ consists, in general, of two connected components.

THEOREM 7: Two canonical representations (30) of $F_{\theta_1\theta_2}$ can be bridged by an elementary transformation if and only if they both belong to the same component (in the set of canonical representations). □

PROOF: Let U, U' be two canonical representations (30), and let $f(x)$, $f'(x)$ be the rational sections (42) of the corresponding vector bundles. The elementary transformation based upon the admissible pair of conjugate (affine) points λ, $\bar{\lambda}$ transforms U into a representation equivalent to U' if and only if

$$\operatorname{div}(f'(x)) \equiv \operatorname{div}(f(x)) + \lambda-\bar{\lambda} \tag{63}$$

By (43), this may be rewritten as:

$$(1,-b',-d') - 4(0,0,1) \equiv (1,-b,-d) - 4(0,0,1) + \lambda-\bar{\lambda} \tag{64}$$

We shall use now the Corollary of Theorem 5. Let $\lambda \leftrightarrow w$, then $\bar{\lambda} \leftrightarrow \bar{w}$ (since in (14) the $\wp$-function has real invariants g_2, g_3); let $(1,-b,-d) \leftrightarrow v$, $(1,-b',-d') \leftrightarrow v'$. We conclude that:

$$v' \equiv v+w-\bar{w} = v+2i \operatorname{Im} w \pmod{2\omega, 2\omega'} \tag{65}$$

We have already considered in section 4 the value v in the period parallelogram corresponding to the affine point $(-b,-d)$ on $F_{\theta_1\theta_2}$. We have seen that the component of U in the set of canonical representations is determined by the component of v (the interval v belongs to) in Figures 5-6. i.e. by $\operatorname{Re} v$. But from (65) it follows that:

$$\operatorname{Re} v' \equiv \operatorname{Re} v \pmod{2 \operatorname{Re} \omega, 2 \operatorname{Re} \omega'} \tag{66}$$

thus v, v' and hence U, U' are in the same component. Consequently, if U and U' can be bridged by an elementary transformation, then they are in the same component.

We proceed to show the converse. Suppose U, U' are in the same component, i.e. v, v' are in the same interval in Figures 5-6. We shall denote the open interval by I_N and the closed interval (with its endpoints identified) by I_C. We want to find w in the period parallelogram so that (65) holds and $\lambda, \bar{\lambda}$ is an admissible pair of points with respect to U. Of course, since $\lambda, \bar{\lambda}$ are affine points, $w \not\equiv 0 \pmod{2\omega, 2\omega'}$. Also, it follows from (65) that $\operatorname{Re} w$ does not affect the result of the elementary transformation; we decide to fix $\operatorname{Re} w \equiv 0 \pmod{2 \operatorname{Re} \omega, 2 \operatorname{Re} \omega'}$. Thus we are looking for $w \in I_N$ satisfying:

$$v' \equiv v+2w \pmod{2\omega, 2\omega'} \tag{67}$$

Clearly, for any v in I_N or in I_C, as w changes over I_N, (67) gives (actually, gives twice) every $v' \neq v$ in

I_N or in I_C respectively. Thus the only question to be settled is the admissibility of the pair of points $\lambda = (\lambda_1,\lambda_2)$, $\bar{\lambda} = (\bar{\lambda}_1,\bar{\lambda}_2)$ corresponding to w, $\bar{w}$.

Explicit calculations with elliptic functions, that we omit, show that there are two possible cases.

a) U belongs to the compact component of the set of canonical representations, i.e. $v \in I_C$ (of course this is possible only if θ_1, $\theta_2 \in \mathbb{R}$). Then for every $w \in I_N$ the corresponding pair of points λ, $\bar{\lambda}$ is admissible with respect to U. For every canonical representation U' in the compact component, (67) enables us to choose $w \in I_N$ and the points λ, $\bar{\lambda}$ so that the elementary transformation applied to U yields a representation equivalent to U'.

b) U belongs to the non-compact component of the set of canonical representations, i.e. $v \in I_N$. Then there exist two values of w in I_N so that the corresponding pair of points λ, $\bar{\lambda}$ is unadmissible with respect to U. However it turns out that these are precisely the values of w in I_N satisfying $v+2w \equiv 0 \pmod{2\omega, 2\omega'}$. Denote them by w_0, w_1. Let U' be any canonical representation in the non-compact component and choose $w \in I_N$ so that $v+2w \equiv v' \pmod{2\omega, 2\omega'}$; then $w \neq w_0$, w_1, since $v' \not\equiv 0 \pmod{2\omega, 2\omega'}$; thus the corresponding pair of points λ, $\bar{\lambda}$ is admissible with respect to U, and the elementary transformation applied to U yields a representation equivalent to U'.

We conclude that whenever U, U' are in the same component, they can be bridged by an elementary transformation. The theorem is proved.

7. DEFINITE AND INDEFINITE REPRESENTATIONS

We want to consider now a certain notion of positivity, or definiteness, which appears quite naturally in the context of self-adjoint determinantal representations. This notion will also enable us to understand the partition of the set of canonical representations (30) into two components.

Let $U(x_0,x_1,x_2) = x_1A+x_2B+x_0C$ be a self-adjoint determinantal representation of a real curve F of degree n.

It seems natural to call U positive if one of the coefficient matrices A, B, C is positive definite. However, it may happen that while this is not the case, this can be achieved by an appropriate real projective change of coordinates (5). Hence we arrive at the following (projectively invariant) definition. A self-adjoint determinantal representation $U = x_1A+x_2B+x_0C$ of F is called *definite* if there exist $c_0, c_1, c_2 \in \mathbb{R}$ so that $c_1A+c_2B+c_0C > 0$, *indefinite* otherwise. Clearly, equivalent representations are definite or indefinite simultaneously. We note that the question whether three given hermitian matrices admit a positive definite real linear combination is interesting in itself.

It is an important fact that elementary transformations preserve definiteness. Indeed, using the notations of the previous section, it follows from (53) that for $\xi_1, \xi_2 \in \mathbb{R}$ the matrices $\xi_1\sigma_2-\xi_2\sigma_1+\gamma$ and $\xi_1\sigma_2-\xi_2\sigma_1+\tilde{\gamma}$ are congruent; hence the representations $x_1\sigma_2-x_2\sigma_1+\gamma$, $x_1\sigma_2-x_2\sigma_1+\tilde{\gamma}$ are simultaneously definite or indefinite. Moreover, we conclude that for definite representations the elementary transformations preserve the coefficients of a positive linear combination of the coefficient matrices: if $c_0, c_1, c_2 \in \mathbb{R}$ are such that $c_1\sigma_2-c_2\sigma_1+c_0\gamma > 0$, then also $c_1\sigma_2-c_2\sigma_1+c_0\tilde{\gamma} > 0$.

We can investigate now which self-adjoint determinantal representations of $F_{\theta_1\theta_2}$ are definite and which ones are indefinite. We consider the set of canonical representations (30) and we recall that it contains, in general, the non-compact component and the compact one.

THEOREM 8: In the set of canonical representations (30) of $F_{\theta_1\theta_2}$, the non-compact component consists precisely of indefinite representations, and the compact component consists precisely of definite representations. □

Hence $F_{\theta_1\theta_2}$ always possesses indefinite self-adjoint determinantal representations; if the roots $\theta_1, \theta_2 \in \mathbb{R}$, then it possesses also definite representations.

PROOF OF THEOREM 8: First let $U = x_1A+x_2B+x_0C$ be a

canonical representation in tne non-compact component. As we have seen in the proof of Theorem 7, there exist pairs of conjugate (affine) points on $F_{\theta_1\theta_2}$ which are unadmissible with respect to U. By (51) the coefficient matrices A, B, C possess a common self-orthogonal vector, hence they admit no positive definite real linear combination and U is an indefinite representation.

Consider now the compact component. Any two canonical representations there can be bridged by an elementary transformation, which preserves definiteness. Hence we pick up a certain canonical representation in the compact component and check directly that it is definite; it follows that all the representations in the compact component are definite and the theorem is proved.

Combining Theorems 7 and 8, we obtain

COROLLARY 1: Two self-adjoint determinantal representations of $F_{\theta_1\theta_2}$ can be bridged by an elementary transformation if and only if they are both definite or both indefinite. □

Since any two definite representations of $F_{\theta_1\theta_2}$ can be bridged by an elementary transformation, which preserves the coefficients of a positive linear combination of the coefficient matrices, we have

COROLLARY 2: Let $c_0, c_1, c_2 \in R$ be such that for *some* definite self-adjoint determinantal representation $\tilde{U} = x_1\tilde{A} + x_2\tilde{B} + x_0\tilde{C}$ of $F_{\theta_1\theta_2}$ $c_1\tilde{A} + c_2\tilde{B} + c_0\tilde{C} > 0$; then for *any* definite self-adjoint determinantal representation $U = x_1A + x_2B + x_0C$ of $F_{\theta_1\theta_2}$ $c_1A + c_2B + c_0C > 0$. □

Finally, we note the following. If a self-adjoint determinantal representation U is definite, then its coefficient matrices can have no common self-orthogonal vector, hence by (51) every pair of conjugate (affine) points on the curve is admissible with respect to U. The proof of Theorem 8 shows (for determinantal representations of $F_{\theta_1\theta_2}$) the converse: if a self-adjoint determinantal representation U of $F_{\theta_1\theta_2}$ is indefinite, there exists a pair of conjugate affine points on $F_{\theta_1\theta_2}$ that is unadmissible with respect to U, and the coefficient matrices have a common self-orthogonal vector in the corresponding vector bundle $\ker U$. Thus:

COROLLARY 3: A self-adjoint determinantal representation $U = x_1A + x_2B + x_0C$ of $F_{\theta_1\theta_2}$ either is definite or the coefficient matrices A, B, C have a common self-orthogonal vector in the corresponding vector bundle. □

Of course, the pairs of conjugate points on $F_{\theta_1\theta_2}$ that are unadmissible with respect to the indefinite representation U are "few enough" and "properly situated" (compare the proof of Theorem 7); they do not prevent us from obtaining all other indefinite representations from U by elementary transformations based on admissible pairs of conjugate points.

8. REAL RATIONAL CUBICS

We make some remarks on self-adjoint determinantal representations of real rational (i.e. singular) cubics (see ref. 13). By algebraic methods completely analogous to those used in section 3, we obtain for them a complete set of canonical forms. Similarly to (30), this set forms a one-dimensional real variety. The complete system of invariants of a self-adjoint determinantal representation is entirely different, however: it is determined by the behavior of the corresponding vector bundle at the singular point. The reason for the existence of a continuum of non-equivalent representations is thus a geometric one (the existence of a singular point) for real rational cubics and a topological one (the genus g=1) for real elliptic cubics.

Despite these differences it turns out that the set of canonical representations of a real rational cubic contains always a non-compact component, while in some cases it contains an additional component which is compact (this compact component is actually a point). Moreover, the non-compact component consists of indefinite representations, while the compact component consists of definite representations. Finally, for an indefinite self-adjoint determinantal representation of a real rational cubic, the coefficient matrices possess a common self-orthogonal vector in the corresponding vector bundle. We combine this fact with Corollary 3 of the previous section, and formulate it in purely "matrix terms" as:

THEOREM 9: Let A, B, C be hermitian matrices of order 3, so that $\det(x_1A+x_2B+x_0C)$ is an irreducible cubic; then either the matrices A, B, C admit a positive definite real linear combination, or they possess a common self-orthogonal vector in the corresponding vector bundle $\ker(x_1A+x_2B+x_0C)$ along the above cubic. □

It seems reasonable to conjecture that the conclusions of this theorem hold in a more general setting, perhaps even for arbitrary triples of hermitian matrices A, B, C.

I wish to express my most deep thanks to Prof. Livšic for his help and guidance.

REFERENCES

1. Ahiezer, N.I.: Elements of the Theory of Elliptic Functions, Izdat. "Nauka", Moscow, 1970 (Russian).

2. Clemens, C.H.: A Scrapbook of Complex Curve Theory, Plenum Press, New-York, 1980.

3. Cook, A.J., Thomas, A.D.: Line bundles and homogenous matrices, Quart. J. Math. Oxford (2), 30 (1979), 423-429.

4. Dixon, A.C.: Note on the reduction of a ternary quartic to a symmetrical determinant, Proc. Comb. Phil. Soc. 2 (1900-1902), 350-351.

5. Fulton, W.: Algebraic Curves, Benjamin, Reading, Massachusets, 1974.

6. Gohberg, I., Lancaster, P. and Rodman, L.: Matrices and Indefinite Scalar Products, Birkhäuser-Verlag, 1983.

7. Kravitsky, N.: Regular colligations for several commuting operators in Banach space, Integral Equations and Operator Theory, 6 (1983), 224-249.

8. Kravitsky, N.: On determinantal representation of algebraic curves, Mathematical Theory of Networks and Systems, Proceedings of the MTNS-83 International Symposium, Beer-Sheva, Israel, June 1983 (Springer-Verlag, Berlin-Heidelberg, 1984).

9. Livšic, M.S.: Cayley-Hamilton theorem, vector bundles and divisors of commuting operators, Integral Equations and Operator Theory, 6 (1983), 250-273.

10. Maltzev, A.I.: Principles of Linear Algebra, Gosudarst. Izdat. Tehn.-Teor. Lit., Moscow-Leningrad, 1948 (Russian).

11. Mumford, D.: Algebraic Geometry I, Springer-Verlag, Berlin-Heidelberg, 1976.

12. Room, T.G.: The Geometry of Determinantal Loci, Cambridge University Press, Cambridge, 1938.

13. Vinnikov, V.: Determinantal representations of real cubics and canonical forms of corresponding triples of matrices, Mathematical Theory of Networks and Systems, Proceedings of the MTNS-83 International Symposium, Beer-Sheva, Israel, June 1983 (Springer-Verlag, Berlin-Heidelberg, 1984).

14. Vinnikov, V.: A complete description of self-adjoint determinantal representations of real irreducible cubics, to appear in Proceedings of MTNS-85- 7th International Symposium on the Mathematical Theory of Networks and Systems, Stockholm, Sweden, June 1985 (to be published by North-Holland, Amsterdam).

15. Walker, R.J.: Algebraic Curves, Princeton, New Jersey, 1950.

Victor Vinnikov
Department of Mathematics
and Computer Sciences
Ben-Gurion University of the Negev
Beer Sheva, Israel

Operator Theory:
Advances and Applications Vol. 19

ℓ_2 - SYSTEMS AND THEIR SCATTERING REPRESENTATION

J.C. Willems and C. Heij

By means of scattering theory we derive special representations of a subclass of the linear, shift invariant closed subspaces of $\ell_2(\mathbb{Z},\mathbb{R}^q)$, namely the ℓ_2-behaviour of discrete time linear time invariant finite dimensional systems. Moreover, an explicit algorithm to obtain these scattering representations is described.

NOTE: An extensive version of this paper will appear in [3].

1. INTRODUCTION

Representation of linear systems is the main topic of this paper, which is organized as follows.

In section 2 we give our definition of a system as consisting of a set of trajectories. We will restrict attention to (deterministic) discrete time linear time invariant finite dimensional systems. This class of systems will be denoted by $\mathbb{B}$. More specifically, we consider the ℓ_2 - behaviour of those systems, i.e. we consider the model class $\mathbb{B}_2 := \{H; \exists \mathcal{B} \in \mathbb{B} \;\; H = \mathcal{B} \cap \ell_2\}$. We will state some results concerning characterization and representation of systems in $\mathbb{B}_2$.

Using scattering theory, in section 3 we conclude that systems in $\mathbb{B}_2$ have special representations which we will call scattering representations and which closely resemble forward and backward innovation representations of stochastic processes.

Section 4 deals with explicit construction of the scattering representations. The construction consists in taking arbitrary forward and backward state space realizations, determining the positive definite solutions of corresponding Riccati equations and transforming the parameters of the state space realizations by means of these solutions.

2. FINITE DIMENSIONAL ℓ_2 - SYSTEMS

DEFINITION. A dynamical system is a triple $\Sigma := \{T, W, \mathcal{B}\}$ with

$B \subset W^T$.

Here we consider $T = \mathbb{Z}$, $W = \mathbb{R}^q$. B is called the behaviour of the system and simply consists of a subset of the set $(\mathbb{R}^q)^{\mathbb{Z}}$ of all discrete time trajectories in $\mathbb{R}^q$. We will restrict attention to finite dimensional linear time invariant ℓ_2-systems, which nicely can be characterized as follows.

Let σ denote the left shift on $(\mathbb{R}^q)^{\mathbb{Z}}$ and T_{pwc} the topology of pointwise convergence. Define the class of systems $\mathbb{B}$ as follows.

$$\mathbb{B} := \{B \subset (\mathbb{R}^q)^{\mathbb{Z}};\ B \text{ linear},\ \sigma B = B,\ B \text{ closed in } T_{pwc}\}$$

DEFINITION. A finite dimensional (forward) state space system is a system $\Sigma_s = \{\mathbb{Z}, \mathbb{R}^m \times \mathbb{R}^n \times \mathbb{R}^q, B_s\}$ where $B_s = \{(v,x,w) \in (\mathbb{R}^m \times \mathbb{R}^n \times \mathbb{R}^q)^{\mathbb{Z}};\ \begin{pmatrix}\sigma x\\ w\end{pmatrix} = \begin{pmatrix}A & B\\ C & D\end{pmatrix}\begin{pmatrix}x\\ v\end{pmatrix}\} =: B_s(A,B,C,D)$ for given m,n,q and matrices (A,B,C,D) of appropriate dimensions.

DEFINITION. B_s is called a (forward) realization of B if $B = \{w; \exists (v,x)\ (v,x,w) \in B_s\}$. It is called minimal if both m and n are as small as possible within the class of realizations of B.

Proof of the following theorem can be found in Willems [2].

THEOREM. $B \in \mathbb{B}$ if and only if it has a (finite dimensional) realization $B_s(A,B,C,D)$.

Now consider the ℓ_2-behaviour of systems in $\mathbb{B}$. Let $\ell_2^q := \ell_2(\mathbb{Z}, \mathbb{R}^q)$.

$$\mathbb{B}_2 := \{H; \exists B \in \mathbb{B} \quad H = B \cap \ell_2^q\}$$

Every $H \in \mathbb{B}_2$ is a linear, shift invariant subspace of ℓ_2^q which is closed in the ℓ_2^q-topology. However, not any such subspace is in $\mathbb{B}_2$. Let V be a linear, shift invariant subspace of ℓ_2^q and let V' be the closure of V in $(\mathbb{R}^q)^{\mathbb{Z}}$ with respect to T_{pwc}. Then $V \in \mathbb{B}_2$ if and only if $V = V' \cap \ell_2^q$.

DEFINITION. $B_s^2(A,B,C,D) := \{(v,x,w) \in \ell_2^m \times \ell_2^n \times \ell_2^q;\ \begin{pmatrix}\sigma x\\ w\end{pmatrix} = \begin{pmatrix}A & B\\ C & D\end{pmatrix}\begin{pmatrix}x\\ v\end{pmatrix}\}$ is called a realization of $H \subset \ell_2^q$ if $H = \{w;\ \exists (v,x)\ (v,x,w) \in B_s^2\}$. It is called minimal if both m and n are as small as possible in the class of realizations of H.

THEOREM. $H \in \mathbb{B}_2$ if and only if H has a (finite dimensional) realization B_s^2.

PROPOSITION. A realization $\mathcal{B}_s^2(A,B,C,D)$ of $H \in \mathbb{B}_2$ is minimal if and only if

(i) $\mathcal{B}_s^2$ is perfectly observable (i.e. $w^+ = 0 \Rightarrow x(0) = 0$, where $w^+ := w|_{\{0,1,2,\ldots\}}$)

(ii) $\mathcal{B}_s^2$ is reachable (i.e. (A,B) is a reachable pair)

(iii) $\ker D = \{0\}$.

PROPOSITION. If $\mathcal{B}_s^2(A,B,C,D)$ is a minimal realization of $H \in \mathbb{B}_2$, then all minimal realizations are obtained by $\mathcal{B}_s^2(S(A+BF)S^{-1}, SBR, (C+DF)S^{-1}, DR)$, where S and R are nonsingular and F is arbitrary, all of appropriate dimensions.

Note S corresponds to a change of base in X, the state space, R to a change of base in V, the input space, and F corresponds to state feedback.

3. SCATTERING THEORY AND ℓ_2 - SYSTEMS

Let $H \in \mathbb{B}_2$ have minimal realization $\mathcal{B}_s^2(A,B,C,D) = \{(v,x,w);\ \begin{pmatrix}\sigma x\\ w\end{pmatrix} = \begin{pmatrix}A & B\\ C & D\end{pmatrix}\begin{pmatrix}x\\ v\end{pmatrix}\}$. There exists a map $L := L(A,B,C,D): \ell_2^m \longrightarrow \ell_2^q: v \longrightarrow w$ iff $\sigma(A) \cap \{|z| = 1\} = \emptyset$. We will call L the <u>driving operator</u> corresponding to $\mathcal{B}_s^2$.

So by means of feedback we can construct operators L such that

(i) $\operatorname{im} L = H$

(ii) $L\sigma = \sigma L$ (shift invariance)

(iii) L causal ($v^- = 0 \Rightarrow (Lv)^- = 0$), or

(iii') L anticausal ($v^+ = 0 \Rightarrow (Lv)^+ = 0$)

(iv) L has a finite dimensional realization.

So we can obtain H as the image of a linear shift invariant (anti-)causal <u>finite dimensional</u> operator. By means of scattering theory we will derive that H also can be obtained as the image of a linear shift invariant (anti-)causal <u>isometric</u> operator. In the next section we will show that this scattering representation in fact itself is finite dimensional.

The scattering theorem of Lax and Phillips [1] amounts to the following.

Let H be Hilbert, $V: H \longrightarrow H$ unitary and D_+, D_- closed subspaces of H which satisfy

(a) $VD_+ \subset D_+$, $\bigcap_{k \in \mathbb{Z}} V^k D_+ = \{0\}$, $\overline{\bigcup_{k \in \mathbb{Z}} V^k D_+} = H$

(b) $VD_- \supset D_-$, $\bigcap_{k \in \mathbb{Z}} V^k D_- = \{0\}$ $\overline{\bigcup_{k \in \mathbb{Z}} V^k D_-} = H$

Then there exist Hilbert spaces $N_{\pm}$ and operators $L_{\pm}$: $\ell_2(\mathbb{Z}, N_{\pm}) \longrightarrow H$ such that

(i) $\operatorname{im} L_{\pm} = H$

$(\tilde{ii})$ $V^* L_+ = L_+\sigma$, $V^* L_- = L_-\sigma$

$(\tilde{iii})$ $L_+\ell_2(\mathbb{Z}_+, N_+) = D_+$ $(\mathbb{Z}_+ := \{0,1,2,\ldots\})$

$(\tilde{iii}')$ $L_-\ell_2(\mathbb{Z}_-, N_-) = D_-$ $(\mathbb{Z}_- := \{\ldots,-2,-1\})$

(v) $L_{\pm}$ isometric.

Moreover, $L_{\pm}$ are unique up to isomorphisms of $N_{\pm}$.

Let $H \in \mathbb{B}_2$ and define $H_+ := \{w \in H;\ w^- = 0\}$, $H_- := \{w \in H;\ w^+ = 0\}$. One can show that $(H,D_+,D_-,V) := (H,H_+,H_-,\sigma^*)$ satisfies (a) and (b), so there exist operators $L_{\pm} : \ell_2(\mathbb{Z}, N_{\pm}) \longrightarrow \ell_2^q$ such that L_+ satisfies (i), (ii), (iii), (v) and L_- satisfies (i), (ii), (iii'), (v).

We will call L_+ the forward scattering representation (FSR) of H, L_- the backward scattering representation (BSR). In the next section we will show that L_+ and L_- also satisfy (iv). Moreover, we will give an explicit algotithm to construct L_+,L_- for $H \in \mathbb{B}_2$.

4. SCATTERING REPRESENTATIONS OF SYSTEMS IN $\mathbb{B}_2$.

We will concentrate on the construction of L_+.

THEOREM. *For* $H \in \mathbb{B}_2$, *the (causal isometric) Lax-Phillips* FSR L_+ *is finite dimensional and can be obtained from minimal (forward) realizations.*

That is, there exists a minimal realization $B_s^2(A_+,B_+,C_+,D_+)$ of H such that $L_+ = L(A_+,B_+,C_+,D_+)$, i.e. the operator which maps v_+ into w in $B_s^2(A_+,B_+,C_+,D_+) := \{(v_+,x,w);\ \begin{pmatrix}\sigma x\\ w\end{pmatrix} = \begin{pmatrix}A_+ & B_+\\ C_+ & D_+\end{pmatrix}\begin{pmatrix}x\\ v_+\end{pmatrix}\}$.

To prove the theorem, we first give necessary and sufficient conditions for L(A,B,C,D) to be an isometry. We will use the following terminology.

Let $K = K^T \in \mathbb{R}^{n\times n}$ be nonsingular, $\langle x_1,x_2\rangle_K := x_2^T K x_1$. The index of K is defined as $\operatorname{in}(K) := \max\{\dim V;\ \langle x,x\rangle_K < 0\ \forall 0 \neq x \in V\}$. Then $(\mathbb{R}^n, \langle \cdot,\cdot\rangle_K)$ is called a Pontryagin space of index in(K). A map $M: (\mathbb{R}^{d_1}, \langle\cdot,\cdot\rangle_{K_1}) \rightarrow (\mathbb{R}^{d_2}, \langle\cdot,\cdot\rangle_{K_2})$ is called a (K_1,K_2) isometry if $\forall a \in \mathbb{R}^{d_1}$ $\|Ma\|_{K_2} = \|a\|_{K_1}$.

Now let $H \in \mathbb{B}_2$ have a minimal realization $B_s^2(A,B,C,D)$ with $\sigma(A) \cap \{|z| = 1\} = \emptyset$ and let L(A,B,C,D) be the corresponding driving operator.

PROPOSITION. ([3]). L isometry $\Leftrightarrow \exists K = K^T \in \mathbb{R}^{n \times n}$ nonsingular such that $\begin{pmatrix} A & B \\ C & D \end{pmatrix}$ is a $\left(\begin{pmatrix} K & 0 \\ 0 & I_m \end{pmatrix}, \begin{pmatrix} K & 0 \\ 0 & I_q \end{pmatrix}\right)$ isometry; L causal $\Leftrightarrow K > 0$.

Now to prove the theorem, it suffices to find a minimal realization $B_s^2(A_+,B_+,C_+,D_+)$ of H such that $\exists K_+ > 0$ for which $\begin{pmatrix} A_+ & B_+ \\ C_+ & D_+ \end{pmatrix}$ is a $\left(\begin{pmatrix} K_+ & 0 \\ 0 & I_m \end{pmatrix}, \begin{pmatrix} K_+ & 0 \\ 0 & I_q \end{pmatrix}\right)$ isometry.

Let $B_s^2(A,B,C,D)$ be an arbitrary minimal realization of H. Then $B_s^2(A+BF,BR,C+DF,DR)$ also are minimal realizations. Writing out the isometry condition for L leads to the following conditions.

(ARE) $\quad K = A^TKA - (B^TKA + D^TC)^T(B^TKB + D^TD)^{-1}(B^TKA + D^TC) + C^TC$

(N) $\quad RR^T = (B^TKB + D^TD)^{-1}$

(F) $\quad F = -(B^TKB + D^TD)^{-1}(B^TKA + D^TC)$

Because (A,B) is reachable and (A,B,C,D) is perfectly observable, (ARE) has a unique solution $K_+ > 0$. Let R_+ be a corresponding solution for (N) and determine F_+ from K_+ by (F). We conclude that with $A_+ := A + BF_+$, $B_+ := BR_+$, $C_+ := C + DF_+$, $D_+ := DR_+$, $L(A_+,B_+,C_+,D_+)$ is a causal isometric driving operator. Because of unicity, $L_+ = L(A_+,B_+,C_+,D_+)$ which proves the theorem.

The following gives an algorithm to obtain L_+ for $H \in \mathbb{B}_2$.

STEP 1. For $H \in \mathbb{B}_2$ determine an arbitrary minimal realization $B_s^2(A,B,C,D)$, e.g. by a method described in Willems [2].

STEP 2. Solve (ARE) for $K_+ > 0$; determine F_+, R_+.

STEP 3. $A_+ := A + BF_+$, $B_+ := BR_+$, $C_+ := C + DF_+$, $D_+ := DR_+$, then $L_+ = L(A_+,B_+,C_+,D_+)$.

The BSR L_- can be determined as follows. If (ARE) has a solution $K_- < 0$ and if $B^TK_-B + D^TD > 0$, then determine F_- from K_- by (F) and R_- as a solution of (N). L_- is given by $L_- = L(A + BF_-, BR_-, C + DF_-, DR_-)$. If (ARE) has no solution $K_- < 0$ with $B^TK_-B + D^TD > 0$, then L_- can be determined from the time-reverse system $\tilde{H} := \{\tilde{w}; \exists w \in H\ \tilde{w} = Rw\}$ where $(Rw)(t) := w(-t)$, $t \in \mathbb{Z}$. Determine $\tilde{L}_+$ for $\tilde{H}$, then $L_- = R\tilde{L}_+R$.

These scattering representations of finite dimensional systems can be used e.g. in model reduction, see Willems and Heij [3].

REFERENCES

[1] Lax, P.D., and Phillips, R.S., Scattering theory (Academic Press, New York, 1967).

[2] Willems, J.C., From time series to linear system, Part I: Finite dimensional linear time invariant systems; Part II: Exact modeling; Part III: Approximate modeling; submitted.

[3] Willems, J.C., and Heij, C., Scattering theory and approximation of linear systems. To appear in Byrnes, C., and Lindquist, A., eds., Proceedings of MTNS-85, 7th International Symposium on the mathematical theory of networks and systems, Stockholm, June 10-14, 1985.

J.C. Willems and C. Heij
Mathematics Institute and Econometrics Institute
University of Groningen
The Netherlands

Operator Theory:
Advances and Applications, Vol. 19

BALANCED REALIZATIONS IN INFINITE DIMENSIONS

N.J. Young

Let G be an operator-valued function analytic in the complement of the closed unit disc and vanishing at ∞. It is shown that, provided the Hankel operator determined by G is bounded, there exists a "parbalanced" realization of G. This means that there exists a discrete-time linear time-invariant system, whose states are elements of a Hilbert space, having G as its transfer function and having equal controllability and observability gramians (these terms are explained below: knowledge of systems theory is not presupposed). The system constructed has contractive state operator and bounded observability and reachability maps.

In the study of a linear system it is often profitable to make use of a *realization* of the system. Suppose that only the input-output behaviour is known, and represented by a given function G analytic in the complement of the closed unit disc. We aim to construct a model of the internal behaviour of some system having the same input-output characteristics as the initial one. Among such modelling systems there is a special class called *balanced realizations* which are particularly useful in several contexts, notably model reduction [10] and robust control [4]. In the former application the aim is to replace a complex model by a simpler one without too great a loss of accuracy: this is a common requirement in control system design. Hitherto balanced realizations have mainly been used for systems having rational transfer functions G; however, there is a definite need for model reduction of systems with non-rational G. Continuous time systems containing delays are often encountered, and these give rise to transfer functions involving exponentials. A realization of such a transfer function will necessarily have an infinite-dimensional state space.

A standard method of model reduction for finite-dimensional systems is to truncate a balanced realization. The question arises as to whether this approach will also be effective in infinite dimensions. R.F. Curtain and K. Glover [1, 2, 3] have investigated this question and have obtained very promising results, both theoretical and experimental. They have convergence theorems and error estimates for the truncations, and have also calculated some examples analytically. They have, however, come across one delicate point, and that is the question of the *existence* of infinite-dimensional balanced realizations, and the boundedness properties of various operators associated therewith. For rational G it is quite straightforward to write down a balanced realization, but in the non-rational case, although it is not hard to write down matrices which appear to have the right formal properties, it takes more work to show that these matrices do define bounded Hilbert space operators. The purpose of this paper is to carry out that work. Provided G determines a compact (block) Hankel operator, G has a balanced realization with all the boundedness properties one could desire, and, in particular, with a contractive state operator: see §3, Corollary 1.

The construction is based on the "shift realization" - a neat way of producing realizations discovered independently by J.W. Helton [8] and P. Fuhrmann [5], and presented in detail in Fuhrmann's book [6]. As this paper is aimed at operator theorists not all of whom are systems specialists I shall not assume familiarity with [6], but shall briefly present the relevant definitions and explain the shift realization.

For Hilbert spaces, U, Y, $\mathcal{L}(U, Y)$ denotes the space of bounded operators from U to Y with the operator norm. A *contraction* is an operator of norm at most one. $\mathcal{L}(Y, Y)$ is abbreviated to $\mathcal{L}(Y)$. We denote by $\ell^2 \otimes Y$ the space of sequences $(y_n)_0^\infty$ such that $y_n \in Y$ and $\sum_0^\infty \|y_n\|^2 < \infty$, with the obvious inner product. All Hilbert spaces occurring will be supposed separable, so that we may identify $\ell^2 \otimes Y$ with the Hardy space $H^2 \otimes Y$ of Y-valued analytic functions in the unit disc $\mathbb{D}$ and with a subspace of the space $L^2 \otimes Y$ of square-integrable Y-valued measurable functions on the unit circle $\partial\mathbb{D}$: see [11] for full details.

Let g be a measurable $\mathcal{L}(U, Y)$-valued function on $\partial\mathbb{D}$ having the property that $g(\cdot)u \in L^2 \otimes Y$ for all $u \in U$ (this will certainly hold if g is essentially bounded). The Hankel operator

$$\Gamma_g : \mathcal{D}(\Gamma_g) \subseteq H^2 \otimes U \longrightarrow H^2 \otimes Y$$

is defined by

$$\Gamma_g f = P_+(gJf), \quad f \in \mathcal{D}(\Gamma_g),$$

where

$$P_+ : L^2 \otimes Y \longrightarrow H^2 \otimes Y$$

is the orthogonal projection and

$$J : H^2 \otimes U \longrightarrow L^2 \otimes U$$

is given by $Jf(\zeta) = f(\bar{\zeta})$, $\zeta \in \partial\mathbb{D}$. $\mathcal{D}(\Gamma_g)$ contains the space of U-valued polynomials, which is dense in $H^2 \otimes U$. Nehari's theorem [13] asserts that Γ_g extends to a bounded operator on $H^2 \otimes U$ if and only if there exists an essentially bounded measurable $\mathcal{L}(U, Y)$-valued function h on $\partial\mathbb{D}$ with Fourier series $\sum_{-\infty}^{\infty} h_n z^n$ such that $h_n = g_n$ for $n = 0, 1, 2, \cdots$, where $g \sim \sum_{-\infty}^{\infty} g_n z^n$ is the Fourier series of g (actually Nehari proved the case $U = \mathbb{C} = Y$: this is an extension due to Page [13]). Furthermore, the extension of Γ_g (which we shall again denote by Γ_g) is a compact operator if and only if the h above can be chosen to be a norm-continuous function on $\partial\mathbb{D}$ taking its values in the space $\mathcal{K}(U, Y)$ of compact linear operators from U to Y. We shall express these conditions by the following self-explanatory notation:

Γ_g is bounded if and only if $g \in P_+ L^\infty(\mathcal{L}(U, Y))$,
Γ_g is compact if and only if $g \in P_+ \mathcal{C}(\mathcal{K}(U, Y))$.

1. THE SHIFT REALIZATION

We consider linear time invariant discrete time dynamical systems in their usual formulation. In this the time variable t is considered to take non negative integer values and the state of the system at time t is represented by an element x_t of a vector space H, the *state space*. The state space is most commonly taken to be finite-dimensional, but for the study of non-rational transfer functions infinite-dimensional spaces are required, and it seems natural to represent states by elements of a Hilbert space. At time t we can exert an influence on the system by means of an input u_t, which is an element of a second Hilbert space U, and we can make an observation, represented by an element y_t of a third Hilbert space Y. The relationship between the quantities is described by the system equations:

$$x_{t+1} = Ax_t + Bu_t \qquad (1.1a)$$

$$y_t = Cx_t \qquad (1.1b)$$

where $A : H \longrightarrow H$, $B : U \longrightarrow H$, $C : H \longrightarrow Y$ are bounded linear operators. A is called the *state operator*.

We can use generating functions, or "z-transforms" in engineering parlance, to eliminate x_t from equations (1.1) and thereby obtain a description of the characteristics of the system in terms of its input-output behaviour. Suppose that $(x_t)_0^\infty$, $(y_t)_0^\infty$, $(u_t)_0^\infty$ are sequences in H, Y, U respectively which satisfy (1.1), and suppose that $u_t = 0$ for all but a finite number of values of t. Write

$$x(z) = \sum_{t=0}^{\infty} x_t z^{-t}$$

etcetera: the boundedness of A, B, C ensure that $x(z)$ and $y(z)$ converge in a neighbourhood of ∞. On multiplying (1.1) by z^{-t} and summing we find:

$$\begin{aligned} z(x(z) - x_0) &= Ax(z) + Bu(z), \\ y(z) &= Cx(z). \end{aligned}$$

On eliminating $x(z)$ we obtain

$$y(z) = C(zI - A)^{-1}Bu(z) + Cz(zI - A)^{-1}x_0,$$

valid in a neighbourhood of ∞. The analytic function

$$G(z) = C(zI - A)^{-1}B,$$

which maps the resolvent set of A into the space $\mathcal{L}(U, Y)$ of bounded linear operators from U to Y, is called the *transfer function* of the system (A, B, C).

There are various notions of stability for a system (A, B, C). For the present purpose (constructing balanced realizations in as much generality as possible) it makes sense to adopt the least restrictive notion we can, and so we say that (A, B, C) is *stable* provided that the spectrum $\sigma(A)$ of A is contained in the closed unit disc clos $\mathbb{D}$. It is then immediate that the transfer function of a stable system (A, B, C) is an analytic $\mathcal{L}(U, Y)$-valued function in $\mathbb{C} \setminus$ clos $\mathbb{D}$ which vanishes at ∞.

The *realization problem* is: given such a function G, to find a Hilbert space H and a stable system (A, B, C) with state space H having transfer function G. For rational matrix functions there is a highly developed theory of realization, (cf. [9]) originating with Kalman and making use of finite-dimensional linear algebra. The existence question for non-rational functions was solved, naturally enough, by operator theorists: P. Fuhrmann [6] and J.W. Helton [8] gave very similar treatments, at about the same time. For H^2 functions G there is a simple construction using the backward shift S^* on

$H^2 \otimes Y$ (S, as usual, denotes the operation of multiplication by the independent variable on $H^2 \otimes Y$). Here it is.

Consider first the scalar case, $U = \mathbb{C} = Y$. Let G be a scalar valued analytic function in $\mathbb{C} \setminus \operatorname{clos} \mathbb{D}$ satisfying $G(\infty) = 0$. Let

$$g(\zeta) = \frac{1}{\zeta} G\left(\frac{1}{\zeta}\right), \qquad \zeta \in \mathbb{D}.$$

Then g is analytic in $\mathbb{D}$. Suppose further that $g \in H^2$. Then, for any $\zeta \in \mathbb{D}$,

$$\begin{aligned} g(\zeta) &= \frac{1}{2\pi i} \int_{\partial \mathbb{D}} \frac{g(\lambda)}{\lambda - \zeta}\, d\lambda \\ &= \frac{1}{2\pi i} \int g(\lambda)\,(1 - \zeta\bar{\lambda})^{-1}\, d\lambda/\lambda \\ &= \frac{1}{2\pi i} \int g(\lambda)\,\overline{(1 - \bar{\zeta}S)^{-1}1}\, d\lambda/\lambda \\ &= (g, (I - \bar{\zeta}S)^{-1}1)_{H^2} \\ &= ((I - \zeta S^*)^{-1}g, 1)_{H^2}. \end{aligned}$$

Multiplying both sides by ζ and putting $z = 1/\zeta$ we obtain, for $z \in \mathbb{C} \setminus \operatorname{clos} \mathbb{D}$,

$$G(z) = ((zI - S^*)^{-1}g, 1)_{H^2}. \tag{1.2}$$

In the general case, in which U and Y are arbitrary separable Hilbert spaces, we may apply formula (1.2) to the scalar function $(G(\cdot)u, y)$, where u, y are fixed elements of U, Y respectively. On then varying y we obtain the following statement.

If $G: \mathbb{C} \setminus \operatorname{clos} \mathbb{D} \longrightarrow \mathcal{L}(U, Y)$ is such that, for every $u \in U$ the function

$$\zeta \longrightarrow g(\zeta)u$$

belongs to $H^2 \otimes Y$, where $g(\zeta) = \zeta^{-1}G(\zeta^{-1})$, then for any $z \in \mathbb{C} \setminus \operatorname{clos} \mathbb{D}$,

$$G(z)u = C(zI - S^*)^{-1}g(\cdot)u,$$

where $C: H^2 \otimes Y \longrightarrow Y$ is defined by

$$Cf = f(0).$$

Hence a realization (A, B, C) of G is obtained by taking $H^2 \otimes Y$ as state space and defining A to be the backward shift operator on $H^2 \otimes Y$.

2. THE RESTRICTED SHIFT REALIZATION

The shift realization just constructed typically has an unnecessarily large state space. Certainly, for rational matrix functions G this will be so, for a finite-dimensional state space suffices. More generally, we can often obtain a state space smaller than the whole of $H^2 \otimes Y$ by simply replacing $H^2 \otimes Y$ by the smallest subspace H for which the foregoing construction works. In view of the definition of B, H must contain the functions $\{g(\cdot)u : u \in U\}$ - the "column space" of g, and for A to act on H we also require H to be S^*-invariant. We therefore let

$$H = \text{closed linear span of } \{S^{*j}g(\cdot)u : u \in U,\ j = 0, 1, 2, \ldots\}.$$

$$A = S^*|H : H \longrightarrow H,$$

$$B : U \longrightarrow H : u \longmapsto g(\cdot)u, \tag{2.1}$$

$$C : H \longrightarrow Y : f \longmapsto f(0).$$

This gives us the *restricted shift realization* (A, B, C).

For non-rational G this realization will also be infinite-dimensional, and so the question arises as to whether it is an improvement over the shift realization. The answer is that it does have some desirable irredundance properties, known as controllability and observability. For a system (A, B, C) the *reachability operator* is the densely defined operator R from $\ell^2 \otimes U \approx H^2 \otimes U$ to H with matrix

$$[B \;\; AB \;\; A^2B \;\ldots\; A^jB \;\ldots],$$

having domain the space of finitely non-zero sequences. R maps the sequence

$$(u_k, u_{k-1}, \ldots u_1, u_0, 0, 0, \ldots.)$$

to the value of x_{k+1} obtained when x_0 is taken to be 0 and the sequence $(u_0, u_1, \ldots, u_k)$ is input. For this reason the system is said to be *controllable* if R has dense range. This means that the system can be manoeuvred from state 0 to any prescribed neighbourhood in H by a finite sequence of inputs.

Dually, the *observability operator* $\mathcal{O}$ of the system (A, B, C) maps the state x to the sequence output by the system when $x_0 = x$ and all inputs are zero. $\mathcal{O}$ has matrix $\begin{bmatrix} C \\ CA \\ CA^2 \\ \vdots \end{bmatrix}$. We should like to say that $\mathcal{O}$ maps H, or at least a dense subset of H, to $\ell^2 \otimes Y \approx H^2 \otimes Y$, but we cannot do this without stronger assumptions on A. $\mathcal{O}$ is defined on

$$\mathcal{D}(\mathcal{O}) \overset{def}{=\!=} \{x \in H : \sum_{j=0}^{\infty} \|CA^j x\|^2 < \infty\},$$

with range $\ell^2 \otimes Y \approx H^2 \otimes Y$. An element of Ker $\mathcal{O}$ produces nothing at the output and so cannot be observed. The system is said to be *observable* if Ker $\mathcal{O} = \{0\}$.

Let us work out R and $\mathcal{O}$ for the restricted shift realization (2.1). Let

$$x(\zeta) \sim \sum_0^{\infty} x_j \zeta^j \in H,$$

so that $x_j \in Y$. Then

$$\mathcal{O}x = \begin{bmatrix} C \\ CS^* \\ \vdots \end{bmatrix} x = \begin{bmatrix} x(0) \\ (S^* x)(0) \\ \vdots \end{bmatrix} = \begin{bmatrix} x_0 \\ x_1 \\ \vdots \end{bmatrix} \sim x.$$

Thus $\mathcal{O} : H \longrightarrow H^2 \otimes Y$ is the injection mapping, and so the system (A, B, C) is observable. It is immediate from the definition of H that the range of R is dense in H, and so (A, B, C) is also controllable.

It is of interest to know under what conditions on G the reachability operator R is bounded (i.e. extends to a bounded operator mapping $H^2 \otimes U$ into H), so let us study R a little further. For $u(\zeta) \sim \sum_0^{\infty} u_j \zeta^j \in H^2 \otimes U$ we have

$$\begin{aligned} R &= [B \;\; S^* B \; \ldots] \begin{bmatrix} u_0 \\ u_1 \\ \vdots \end{bmatrix} = \sum_j S^{*j} g(.)\, u_j \\ &= \sum_j P_+\left\{\bar{\zeta}^j g(\zeta) u_j\right\} \\ &= P_+(g J u) \end{aligned} \tag{2.2}$$

where $P_+ : L^2 \otimes Y \longrightarrow H^2 \otimes Y$ is the orthogonal projection and $J : H^2 \otimes U \longrightarrow L^2 \otimes U$ is defined by $Jf(\zeta) = f(\bar{\zeta})$. Thus R is (a restriction of) the Hankel operator

$$\Gamma_g : H^2 \otimes U \longrightarrow H^2 \otimes Y : f \longmapsto P_+(g J f).$$

To summarise: if $G : \mathbb{C} \setminus \operatorname{clos} \mathbb{D} \longrightarrow \mathcal{L}(U, Y)$ is such that

$G(\cdot)u \in H^2 \otimes Y$ for every $u \in U$ then the restricted shift realization of G exists and is controllable and observable; its observability operator is an isometry, while its reachability operator is a restriction of the Hankel operator Γ_g. The reachability operator is bounded if and only if $g \in P_+L^\infty(\mathcal{L}(U,Y))$, and is compact if and only if $g \in P_+\mathcal{C}(\mathcal{K}(U,Y))$.

Note that in the scalar case ($U = \mathbb{C} = Y$) there is an alternative characterization of boundedness and compactness: Γ_g is bounded if and only if $g \in BMOA$, the space of analytic functions of bounded mean oscillation, and Γ_g is compact if and only if $g \in VMOA$, the space of analytic functions of vanishing mean oscillation (see, for example [12]).

3. BALANCED REALIZATIONS

Imagine a system whose state is represented by a point on the real line, evolving in the absence of control inputs according to the rule

$$x_{t+1} = \frac{1}{2}x_t .$$

Suppose the position of the point can be controlled by its being moved a distance u and that the position of the point can be observed directly. If we choose to measure inputs and outputs in centimetres, but use light years to measure the state variable then we obtain the system equations

$$\begin{aligned} x_{t+1} &= \frac{1}{2}x_t + cu_t , \\ y_t &= c^{-1}x_t, \end{aligned}$$

where c is the speed of light in cm/year. The transfer function of the system is $\left(z - \frac{1}{2}\right)^{-1}$: it does not reflect the choice of variables in the state space. However, it should be easy to accept that the presence of inordinate magnitudes like c and c^{-1} will cause difficulties in the analysis of the system, if only in numerical computations. Formally, the c's will appear in the reachability and observability operators:

$$\begin{aligned} R(u_j) &= c \sum_j 2^{-j} u_j , \qquad && (u_j) \in \ell^2, \\ \mathcal{O}x &= c^{-1}x\left(2^{-j}\right)_0^\infty, && x \in \mathbb{R}. \end{aligned}$$

R is "large" and $\mathcal{O}$ is "small". A way of eliminating this phenomenon is to require that the "moduli" RR^* and $\mathcal{O}^*\mathcal{O}$, which are both operators on the state space, should be equal. More tangible motivation for this notion is given in [10]. It is shown in [10] that, in the finite-dimensional (stable) case one can always bring about such an equality by a change of basis in the state space. We shall tackle the analogous problem for infinite-dimensional systems.

Consider a stable system (A, B, C) having state space H, a separable Hilbert space, with A, B, C being bounded operators. Suppose further that the observability and reachability operators $R, \mathcal{O}$ extend to bounded linear operators $R : H^2 \otimes U \longrightarrow H$, $\mathcal{O} : H \longrightarrow H^2 \otimes Y$. Let

$$W = RR^* : H \longrightarrow H,$$
$$M = \mathcal{O}^*\mathcal{O} : H \longrightarrow H.$$

W, M are called the *controllability gramian* and the *observability gramian* of the system respectively. In finite dimensions, where A, B, C etc. are taken to be matrices rather than operators, the system is said to be *balanced* if $W = M = \Sigma$, where Σ is a diagonal matrix. In the present context diagonalizability of the positive operators W, M does not seem to be so relevant, so I shall say that the system (A, B, C) is *parbalanced* if $W = M$ (by analogy with "parboiled"). I shall further say that the system is *balanced* if $W = M$ and the positive operator W has a diagonal matrix with respect to some orthonormal basis of H.

The problem we shall study is: for which transfer functions G does there exist a parbalanced realization?

In order to construct parbalanced realizations we shall perform a "change of basis" in the shift realization. Suppose that G is an $\mathcal{L}(U, Y)$-valued analytic function in $\mathbb{C} \setminus \operatorname{clos} \mathbb{D}$ for which the restricted shift realization of §2 exists and has bounded reachability operator R. Proceeding purely formally, let us consider the realization

$$(\bar{A}, \bar{B}, \bar{C}) = (T^{-1}AT, T^{-1}B, CT)$$

where (A, B, C) is the restricted shift realization and T is a bounded operator with zero kernel and dense range on the state space H. We have

$$\begin{aligned} \bar{C}(zI - \bar{A})^{-1}\bar{B} &= CT(zI - T^{-1}AT)^{-1}T^{-1}B \\ &= C(zI - A)^{-1}B, \end{aligned}$$

so that $(\bar{A}, \bar{B}, \bar{C})$ is also a realization of G. Furthermore, since the observability operator $\mathcal{O}$ of (A, B, C) is an isometry, $\mathcal{O}^*\mathcal{O} = I_H$ and so from the relation

$$\bar{\mathcal{O}} = \begin{bmatrix} \bar{C} \\ \bar{C}\bar{A} \\ \vdots \end{bmatrix} = \begin{bmatrix} C \\ CA \\ \vdots \end{bmatrix} T = \mathcal{O}T$$

we get

$$\overline{M} = \mathcal{O}^*\mathcal{O} = T^*\mathcal{O}^*\mathcal{O}T = T^*T.$$

Likewise, the observability operators $\overline{R}$, R of the two realizations satisfy

$$\overline{R} = [\overline{B}\ \overline{A}\overline{B}\ \overline{A}^2\overline{B} \ldots] = T^{-1}[B\ AB\ A^2B \ldots]$$
$$= T^{-1}R,$$

so that

$$\overline{W} = \overline{R}\overline{R}^* = T^{-1}RR^*T^{-*} = T^{-1}WT^{-*},$$

where W, $\overline{W}$ are the controllability gramians of the two systems. We shall thus have

$$\overline{M} = \overline{W} \iff T^*T = T^{-1}WT^{-*}$$
$$\iff TT^*TT^* = W,$$

and this is certainly satisfied if we choose $T = W^{1/4}$. The argument suggests that the realization

$$(W^{-1/4}AW^{1/4},\ W^{-1/4}B,\ CW^{1/4}),$$

where (A, B, C) is the restricted shift realization and $W = \Gamma_g\Gamma_g^*$, will be a parbalanced realization of G. The argument is indeed valid as it stands in the event that W is surjective, for then $W^{-1/4}$ is a bounded linear operator. Typically, however, for non-rational G, W will have dense range which is a proper subset of H, so that $W^{-1/4}$ is only a densely defined unbounded operator, and it is not obvious that $W^{-1/4}AW^{1/4}$ is a bounded operator on H, nor that $W^{-1/4}B$ makes sense. Our task is to show that the construction does nevertheless work and give us the desired parbalanced realization: indeed, it turns out that $W^{-1/4}AW^{1/4}$ is actually a contraction on H.

THEOREM 1. *Let U, Y be separable Hilbert spaces and let*

$$G : \mathbb{C} \setminus \operatorname{clos} \mathbb{D} \longrightarrow \mathcal{L}(U, Y)$$

be an analytic function such that $g \in P_+L^\infty(\mathcal{L}(U, Y))$, where

$$g(\zeta) = \frac{1}{\zeta}G\left(\frac{1}{\zeta}\right), \qquad \zeta \in \mathbb{D} \tag{3.1}$$

There exists a separable Hilbert space H and a parbalanced realization $(\overline{A}, \overline{B}, \overline{C})$ of G with state space H such that $\overline{A}$ is a contraction on H and $\overline{B}: U \longrightarrow H$, $\overline{C} : H \longrightarrow Y$ are bounded operators, and such that the (equal) gramians $\overline{M}$, $\overline{W}$ of the realization are bounded operators on H.

In the above one may take H to be the closure in $H^2 \otimes Y$ of the *range of the Hankel operator*

$$\Gamma_g : H^2 \otimes U \longrightarrow H^2 \otimes Y,$$

$$\Gamma_g u = P_+(gJu), \qquad u \in H^2 \otimes U,$$

the symbols being defined as in (2.2), *and* $\overline{M}$ *and* $\overline{W}$ *to be the restriction of* $(\Gamma_g \Gamma_g^*)^{1/2}$ *to* H.

Note that the condition $g \in P_+ L^\infty(\mathcal{L}(U,Y))$ is satisfied whenever G is bounded on $\mathbb{C} \setminus \operatorname{clos} \mathbb{D}$, but is also satisfied for some unbounded G (e.g. $G(z) = \sum_1^\infty \frac{1}{nz^n}$).

The fact that the state operator is a contraction in a finite-dimensional balanced realization is due to Pernebo and Silverman [14].

PROOF. The hypotheses ensure that Γ_g is a bounded operator from $H^2 \otimes U$ to $H^2 \otimes Y$. For a fixed $u \in U$, which we identify with a constant function in $H^2 \otimes U$, we have

$$\Gamma_g u = P_+(gJu) = P_+(gu) = gu.$$

Thus $gu \in H^2 \otimes Y$ for all $u \in U$, and hence the restricted shift realization (A, B, C) of G exists. Let H, R, $\mathcal{O}$ have the same meanings as in §2 and let $W = RR^*$: note that $R = \Gamma_g$, and so R and W are bounded operators. W has zero kernel and dense range in H.

We shall show that $W^{-1/4} A W^{1/4}$ is defined on H and is a contraction. Firstly, observe that A is itself a contraction, being a restriction of the backward shift S^* on $H^2 \otimes Y$. Next notice that

$$W - AWA^* \geqslant 0.$$

Indeed, for any finitely non-zero sequence (u_j) in $\ell^2 \otimes U$ we have

$$AR(u_j) = A \sum_0^\infty A^j Bu_j = \sum_1^\infty A^j Bu_{j-1}$$

$$= RS(u_j),$$

S here denoting the forward shift on $\ell^2 \otimes U$. Thus

$$\begin{aligned} W - AWA^* &= RR^* - ARR^*A^* \\ &= RR^* - RSS^*R^* \\ &= R(I - SS^*)R^* \geqslant 0. \end{aligned}$$

We can write this relation

$$AW^{1/2}(AW^{1/2})^* \leqslant W^{1/2} W^{1/2},$$

and so, by a well known lemma (e.g. [6, p.124]) there exists a contraction Z on H such that

$$AW^{1/2} = W^{1/2}Z. \tag{3.2}$$

That is, the operator $W^{-1/2}AW^{1/2}$ is defined on the whole of H and is a contraction. We may therefore define $\overline{A}$ on the dense subspace $W^{1/4}H$ of H by

$$\begin{aligned}\overline{A} &= W^{1/4}ZW^{-1/4} \\ &= W^{1/4}(W^{-1/2}AW^{1/2})W^{-1/4} \\ &= W^{-1/4}AW^{1/4}.\end{aligned} \tag{3.3}$$

The next step is to show that the formula

$$\overline{A}^*\overline{A} = W^{-1/4}Z^*AW^{1/4}$$

makes sense and is valid on $W^{1/4}H$. From (3.3), $\overline{A}$ maps $W^{1/4}H$ into itself. From (3.2),

$$Z^*W^{1/2} = W^{1/2}A^*$$

and hence $Z^*W^{1/4}$ maps $W^{1/4}H$ into $W^{1/2}H$. Thus the operator $W^{-1/4}Z^*W^{1/4}$ is defined on $W^{1/4}H$: I claim that

$$\overline{A}^* \supseteq W^{-1/4}Z^*W^{1/4}.$$

This is defined to mean that, for all $x \in W^{1/4}H$ (= domain $W^{-1/4}Z^*W^{1/4}$) and $y \in W^{1/4}H$ (= domain $\overline{A}$),

$$(W^{-1/4}Z^*W^{1/4}x, y) = (x, \overline{A}y).$$

Writing $x = W^{1/4}\xi$, $y = W^{1/4}\eta$, we find

$$\begin{aligned}(W^{-1/4}Z^*W^{1/4}x, y) &= (Z^*W^{1/2}\xi, \eta) \\ &= (W^{1/2}A^*\xi, \eta) \\ &= (\xi, AW^{1/2}\eta) \\ &= (\xi, AW^{1/4}y) \\ &= (\xi, W^{1/4}\overline{A}y) \\ &= (W^{1/4}\xi, \overline{A}y) \\ &= (x, \overline{A}y).\end{aligned}$$

Thus $\overline{A}^*$ exists and agrees with $W^{-1/4}Z^*W^{1/4}$ on $W^{1/4}H$, and so $\overline{A}^*\overline{A}$ exists on $W^{1/4}H$ and is given by

$$\begin{aligned}\overline{A}^*\overline{A} &= W^{-1/4}Z^*W^{1/4}.\,W^{-1/4}AW^{1/4} \\ &= W^{-1/4}Z^*AW^{1/4},\end{aligned}$$

establishing the claim. It follows that the positive operator $\overline{A}^*\overline{A}$, defined on $W^{1/4}H$, has the property that $W^{1/4}(\overline{A}^*\overline{A})W^{-1/4}$ is a contraction (since both Z and A are contractions). If $W^{-1/4}$ were bounded we could immediately conclude that $\sigma(\overline{A}^*\overline{A}) \subseteq [0,1]$ and hence that $\overline{A}$ is a contraction. The conclusion is still true for unbounded $W^{-1/4}$, as we now show.

LEMMA 1. *Let M be a positive operator on a dense subspace D of a separable Hilbert space H, let $P \in \mathcal{L}(H)$ have zero kernel and let PMP^{-1} be a contraction on PD. Then M is a contraction on D.*

PROOF. Since $M \geqslant 0$, M has a self-adjoint extension $\widetilde{M}$ with domain $\widetilde{D} \supseteq D$[15, §124]. Let us show that $P\widetilde{M}P^{-1}$ is a contraction on $P\widetilde{D}$. Note that $\widetilde{M}$ is the closure of M [15, §117]. The graph of PMP^{-1} is

$$\begin{aligned}\{(x, PMP^{-1}x) : x \in PD\} &= \{(Py, PMy) : y \in D\} \\ &= (P \oplus P)\{(y, My) : y \in D\}.\end{aligned}$$

Thus, to say that PMP^{-1} is a contraction is to say that $P \oplus P$ maps the graph of M (a subspace of $H \oplus H$) to the closed subset

$$\{(\xi, \eta) : \|\eta\| \leqslant \|\xi\|\}$$

of $H \oplus H$. Since $P \oplus P$ is continuous, it will also map the closure of the graph of M, which is the graph of $\widetilde{M}$, into the same closed subset. It follows that $P\widetilde{M}P^{-1}$ is a contraction.

Let the spectral decomposition of $\widetilde{M}$ [15, §120] be

$$\widetilde{M} = \int_{\mathbb{R}^+} \lambda \, E(d\lambda).$$

Suppose that $\|M\| > 1$: then $\|\widetilde{M}\| > 1$ and so the space $E(1,\infty) \neq \{0\}$. Let the spectral decomposition of P^*P be

$$P^*P = \int_{\mathbb{R}^+} \lambda \, F(d\lambda).$$

Since $\operatorname{Ker} P = \{0\}$, we have $F(0,\infty) = H$, and hence there exists $m > 0$ such that $F(m^2,\infty)$ is not orthogonal to $E(1,\infty)$. Write $K = F(m^2,\infty)$ and note that $\|Px\| \geqslant m\|x\|$ for $x \in K$. Choose a finite interval $[a,b) \subset (1,\infty)$ such that $E[a,b)$ is not orthogonal to K. Let

$$\Pi = P_K E[a,b),$$

where P_K is the orthogonal projection $H \longrightarrow K$, and let $p = \|\Pi\|$, so that $p > 0$. For $n \in \mathbb{N}$ divide $[a,b)$ into n equal subintervals. The images of these subintervals under E constitute an orthogonal direct decomposition of $E[a,b)H$. If the restriction of Π to each of these n direct summands had norm less than $p/\sqrt{n}$ we should have $\|\Pi\| < p$, contradicting the choice of p. Thus, for at least one of the subintervals, say $[a_n, b_n)$, we have

$$\|\Pi\, E[a_n, b_n)\| \geqslant p/\sqrt{n}$$

Choose $y_n \in E[a_n, b_n)H$ such that $\|y_n\| = 1$ and

$$\|\Pi\, y_n\| > p/(2\sqrt{n}) .$$

Let $x_n = Py_n$. Write $y_n' = \Pi\, y_n$, $y_n'' = (I - \Pi)y_n$, so that $y_n' \in K$, $y_n'' \perp K$. Since K, $K^{\perp}$ are invariant under P^*P,

$$\begin{aligned}
\|Py_n\|^2 &= (P^*Py_n, y_n) \\
&= (P^*Py_n', y_n') + (P^*Py_n'', y_n'') \\
&\geqslant \|Py_n'\|^2 \\
&\geqslant m^2 \|y_n'\|^2 \\
&= m^2 \|\Pi\, y_n\|^2 \\
&> \frac{m^2 p^2}{4n} .
\end{aligned}$$

Thus

$$\|x_n\| > \frac{mp}{2\sqrt{n}} \tag{3.4}$$

Now let

$$\lambda_n = \frac{a_n + b_n}{2}, \qquad \varepsilon_n = \frac{b_n - a_n}{2} = \frac{b - a}{2n} .$$

We have

$$\begin{aligned}
\|\tilde{M} y_n - \lambda_n y_n\|^2 &= \int_{[a_n, b_n)} |\lambda - \lambda_n|^2 \big(E(d\lambda)y_n, y_n\big) \\
&\leqslant \varepsilon_n^2 \|y_n\|^2 = \varepsilon_n^2 .
\end{aligned}$$

Thus

$$\begin{aligned}
P\tilde{M}P^{-1}x_n &= P\tilde{M}y_n = P(\lambda_n y_n + \eta_n) \\
&= \lambda_n x_n + P\eta_n
\end{aligned}$$

where

$$\|\eta_n\| \leqslant \varepsilon_n .$$

Hence, using (3.4) we obtain

$$\begin{aligned}
\|P\tilde{M}P^{-1}x_n\| &\geqslant \lambda_n \|x_n\| - \|P\eta_n\| \\
&\geqslant \lambda_n \|x_n\| - \|P\| \varepsilon_n
\end{aligned}$$

$$\|P\tilde{M}P^{-1}x_n\| \geq \lambda_n\|x_n\| - \|P\|\varepsilon_n \frac{2\sqrt{n}}{mp}\|x_n\|$$
$$= \lambda_n\|x_n\| - \|P\| \frac{b-a}{n} \frac{2\sqrt{n}}{mp}\|x_n\|.$$

Since $\lambda_n > a > 1$ and $x_n \neq 0$, by choosing n sufficiently large we find that $\|P\tilde{M}P^{-1}\| > 1$, which is a contradiction. Thus $\|M\| \leqslant 1$ and the Lemma is proved.

We resume the proof of the theorem, knowing now that $\overline{A}^*\overline{A}$ is a contraction on $W^{1/4}H$, and hence that $\overline{A}$ is also. We may therefore extend $\overline{A}$ in a unique way to a contraction on H: this extension will also be denoted by $\overline{A}$.

Next we construct a bounded operator $\overline{B} : U \longrightarrow H$. The idea is that $\overline{B} = W^{-1/4}B$, but it needs to be shown that this makes sense. Pick a partial isometry $V : H^2 \otimes U \longrightarrow H$ such that $R = (RR^*)^{1/2}V = W^{1/2}V$. Let

$$\overline{B}u = W^{1/4}Vu, \qquad u \in U,$$

where u is identified with the constant function in $H^2 \otimes U$ taking the value u. Then

$$W^{1/4}\overline{B}u = W^{1/2}Vu = Ru = \Gamma_g u$$
$$= P_+(gu) = gu = Bu,$$

so that $W^{1/4}\overline{B} = B$ as desired, while $\overline{B}$ as defined is clearly a bounded operator from U to H.

To complete and summarise the definition of our candidate system $(\overline{A}, B, \overline{C})$:

H = closed linear span of $\{S^{*j}gu : u \in U,\ j = 0,1,2,\ldots\}$,

$\overline{A}$ = the unique contractive extension of $= W^{-1/4}AW^{1/4}$ to H,

$\overline{B} = W^{-1/4}B = W^{1/4}V$

$\overline{C} = CW^{1/4}$

where A, B, C are as in (2.1),

$W = RR^* = \Gamma_g\Gamma_g^*$,

$R = W^{1/2}V$, $V : H^2 \otimes U \longrightarrow H$ a partial isometry.

Note that the relation $\overline{B} = W^{-1/4}B$ shows that $\overline{B}$ is uniquely determined by the above, even though the partial isometry V may not be.

Let us verify that $(\overline{A}, \overline{B}, \overline{C})$ is indeed a realization of G. From the definition of $\overline{A}$ we have, for any $z \in \mathbb{C}$,

$$W^{1/4}(zI-\overline{A}) = (zI-A)W^{1/4}.$$

Since both A and $\overline{A}$ are contractions we can infer, for any $z \in \mathbb{C} \setminus \operatorname{clos} \mathbb{D}$, that

$$(zI-A)^{-1}W^{1/4} = W^{1/4}(zI-\overline{A})^{-1}.$$

Hence

$$\begin{aligned} W^{1/4}(zI-\overline{A})^{-1}\overline{B} &= (zI-A)^{-1}W^{1/4}\overline{B} \\ &= (zI-A)^{-1}B. \end{aligned}$$

Thus

$$\begin{aligned} \overline{C}(zI-\overline{A})^{-1}\overline{B} &= CW^{1/4}(zI-\overline{A})^{-1}\overline{B} \\ &= C(zI-A)^{-1}B \\ &= G(z), \end{aligned}$$

as expected.

Let $\overline{R}$, $\overline{\mathcal{O}}$ be the reachability and observability operators of the system $(\overline{A}, \overline{B}, \overline{C})$. For any finitely non-zero sequence $(u_j) \in \ell^2 \otimes U$ we have

$$\begin{aligned} W^{1/4}\overline{R}(u_j) &= \sum_j W^{1/4}\overline{A}^j\overline{B}u_j \\ &= \sum_j A^j W^{1/4}\overline{B}u_j \\ &= \sum_j A^j B u_j \\ &= R(u_j) \\ &= W^{1/2}V(u_j). \end{aligned}$$

Hence

$$\overline{R} = W^{1/4}V,$$

so that $\overline{R}$ does extend to a bounded operator on the whole of $H^2 \otimes U$, and the gramian

$$\overline{W} \stackrel{def}{=\!=} \overline{R}\,\overline{R}^*$$

satisfies

$$W^{1/4}\overline{R}\,\overline{R}^*W^{1/4} = RR^* = W$$

and hence

$$\overline{W} = W^{1/2}$$

$\overline{\mathcal{O}}$ is also bounded. Consider any $x \in H$: then

$$\overline{C}\overline{A}^j x = CW^{1/4}A^j x$$
$$= CA^j W^{1/4}x.$$

Thus $\overline{\mathcal{O}}x$ is defined for every $x \in H$ and

$$\overline{\mathcal{O}}x = (\overline{C}\overline{A}^j x)_j = (CA^j W^{1/4}x)_j$$
$$= \mathcal{O}W^{1/4}x.$$

Hence

$$\overline{M} \overset{def}{=} \overline{\mathcal{O}}^* \overline{\mathcal{O}}$$
$$= W^{1/4}\mathcal{O}^*\mathcal{O}W^{1/4}$$
$$= W^{1/2} = \overline{W}.$$

Thus the realization $(\overline{A}, \overline{B}, \overline{C})$ of G has bounded reachability and observability operators and has gramians

$$\overline{M} = \overline{W} = W^{1/2} = (\Gamma_g \Gamma_g^*)^{1/2}|H.$$

It is thus a parbalanced realization of G with the properties stated in the theorem. □

COROLLARY 1. *Let* $G : \mathbb{C} \setminus \text{clos}\,\mathbb{D} \longrightarrow \mathcal{L}(U, Y)$ *be analytic and compact-valued (i.e.* $G(z)$ *is a compact operator for* $z \in \mathbb{C} \setminus \text{clos}\,\mathbb{D}$*). Let* $g \in P_+ \mathcal{C}(\mathcal{K}(U, Y))$*, where* g *is given by (3.1). Then* G *has a balanced realization with contractive state operator and bounded (equal and diagonal) gramians. Thus, in particular,* G *has a balanced realization whenever* G *is analytic and matrix-valued on* $\mathbb{C} \setminus \text{clos}\,\mathbb{D}$ *and has a continuous extension to* $\mathbb{C} \setminus \mathbb{D}$.

PROOF. The hypothesis ensures that Γ_g, and hence $W^{1/2}$, be compact. Thus $\overline{M} = \overline{W} =$ a compact positive operator, and for a suitable choice of orthonormal basis in H this will have a diagonal matrix. □

For the sake of concreteness, let us write down matrices for the balanced realization obtained in this way. In the notation of §2,

$$\overline{M} = \overline{W} = W^{1/2} = (\Gamma_g \Gamma_g^*)^{1/2}.$$

Thus, to obtain a diagonal matrix representation for $\overline{M}$, $\overline{W}$ we should choose an orthonormal basis for H, the closure of Range Γ_g, consisting of eigenvectors of $\Gamma_g \Gamma_g^*$. Accordingly, let the singular values (s-numbers) of Γ_g be $s_0 \geqslant s_1 \geqslant \ldots$, repeated according to multiplicity, and let $\langle v_i, w_i \rangle$ be a

Schmidt pair for Γ_g corresponding to s_i. This means that

$$\Gamma_g v_i = s_i w_i , \quad \Gamma_g^* w_i = s_i v_i ,$$

and the (v_i), (w_i) are complete orthonormal sequences in H. We have

$$W w_i = \Gamma_g \Gamma_g^* w_i = s_i^2 w_i ,$$

and so the matrix of W with respect to the w_i is $diag\{s_i^2\}$. The matrices of $\overline{A}$, $\overline{B}$, $\overline{C}$ can be expressed as follows.

$\overline{A} \sim [a_{ij}]$ where

$$\begin{aligned} a_{ij} &= (\overline{A} w_j, w_i) = (W^{-1/4} A W^{1/4} w_j, w_i) \\ &= (S^* s_j^{1/2} w_j, s_i^{-1/2} w_i) \\ &= (s_j/s_i)^{1/2} (w_j, S w_i) \\ &= \left(\frac{s_j}{s_i}\right)^{1/2} \frac{1}{2\pi i} \int (w_j(z), z w_i(z))_Y \frac{dz}{z} , \end{aligned} \tag{3.5}$$

where the integral is anticlockwise round the unit circle.

Pick an orthonormal basis (e_i) in U, finite or infinite. With respect to the bases (e_i), (w_i) in U, H respectively

$\overline{B} \sim [b_{ij}]$ where

$$\begin{aligned} b_{ij} &= (\overline{B} e_j, w_i)_H \\ &= (W^{-1/4} B e_j, w_i) \\ &= (B e_j, s_i^{-1/2} w_i) \\ &= s_i^{-1/2} (B e_j, w_i) \\ &= s_i^{-1/2} (g e_j, w_i)_{H^2 \otimes Y} \\ &= s_i^{-1/2} \frac{1}{2\pi i} \int (g(z) e_j, w_i(z))_Y \frac{dz}{z} \end{aligned} \tag{3.6}$$

Now pick an orthonormal basis (f_i) in Y. With respect to the bases (w_i), (f_i) in H, Y respectively,

$\overline{C} \sim [c_{ij}]$ where

$$\begin{aligned} c_{ij} &= (\overline{C} w_j, f_i)_Y \\ &= (C W^{1/4} w_j, f_i) \end{aligned}$$

$$c_{ij} = s_j^{1/2}(Cw_j, f_i) = s_j^{1/2}(w_j(0), f_i)_Y \qquad (3.7)$$

To summarise:

COROLLARY 2. *Let* G *be as in Corollary 1, let* $(w_i)_0^\infty$ *be a complete orthonormal sequence of eigenvectors of* $\Gamma_g\Gamma_g^*$ *in the closure of the range of* Γ_g*. Let the eigenvalue of* $\Gamma_g\Gamma_g^*$ *corresponding to* w_i *be* s_i^2. *Let* (e_i), (f_i) *be arbitrary orthonormal bases of* U, Y *respectively. Then a balanced realization of* G *is given by the matrices* $([a_{ij}], [b_{ij}], [c_{ij}])$, *defined by (3.5) - (3.7). Each of these matrices represents a bounded linear operator, and* $[a_{ij}]$ *represents a contraction. The reachability and observability maps of the realization are also bounded.*

Note that it would be possible to write down the formulae (3.5) - (3.7) without all the foregoing theory and to check formally that the corresponding gramians are $diag\{s_0, s_1, \ldots\}$. However it would then not be clear that $[a_{ij}]$ and $[b_{ij}]$ define bounded operators nor, *a fortiori*, that $[a_{ij}]$ defines a contraction: cf. [1].

4. UNIQUENESS OF PARBALANCED REALIZATIONS

In this section we show that the parbalanced realization of a transfer function is essentially unique: this is obviously important if parbalanced realizations are to play a canonical role. In finite dimensions it is easy to show that balanced realizations are unique up to a unitary similarity (see [7] or [14]): indeed, since two balanced realizations of a rational transfer function are by definition controllable and observable, they are similar to one another, and it takes only a few lines to show that the similarity must be unitary if it is to preserve the (equal) gramians. In infinite dimensions it is no longer true that controllable and observable realizations of the same transfer function are similar to each other, as is shown by an example in [6, p.248]. Nevertheless the analogous conclusion for parbalanced realizations does hold.

Let (A_i, B_i, C_i), $i = 1, 2$ be systems having the same input and output spaces U and Y, but possibly different state spaces H_1, H_2. We shall say that the two systems are unitarily equivalent if there is a unitary operator $V\colon H_1 \longrightarrow H_2$ such that

$$A_1 = V^* A_2 V, \quad B_1 = V^* B_2, \quad C_1 = C_2 V.$$

We shall show that parbalanced systems with the same transfer function are unitarily equivalent. First we obtain the conclusion for output normal systems.

LEMMA 2. *An output normal controllable stable system, with bounded reachability operator, is unitarily equivalent to the restricted shift realization of its transfer function.*

PROOF. Let (A, B, C) be such a system and let U, Y and H be its input, output and state spaces respectively. Its transfer function is

$$G(z) = C(zI - A)^{-1}B.$$

As in §2 write $g(\zeta) = \zeta^{-1}G(\zeta^{-1})$, so that

$$g(\zeta) = \sum_{n=0}^{\infty} (CA^n B)\zeta^n.$$

Let $\mathcal{O} : H \longrightarrow H^2 \otimes Y$ be the observability operator of the system. By hypothesis, $\mathcal{O}^* \mathcal{O} = I$, so that $\mathcal{O}$ is an isometry. For $u \in U$ we have

$$g(.)u = \mathcal{O} Bu \in H^2 \otimes Y,$$

and so the restricted shift realization $(\tilde{A}, \tilde{B}, \tilde{C})$ of G exists. Its state space $\tilde{H}$ is the smallest closed S^*-invariant subspace of $H^2 \otimes Y$ containing $\{g(.)u : u \in U\}$, $\tilde{A}$ is the restriction of S^* to $\tilde{H}$, etc. (cf. (2.1)). We show that the range of $\mathcal{O}$ is precisely $\tilde{H}$. For any $j \geq 0$ and $u \in U$ we have

$$\begin{aligned} \mathcal{O} R z^j \otimes u &= \mathcal{O}(A^j Bu) = \sum_{n=0}^{\infty} (CA^{j+n}B)z^n \\ &= S^{*j} \sum_{n=0}^{\infty} (CA^n Bu)z^n \\ &= S^{*j} g(.)u \in \tilde{H}. \end{aligned} \tag{4.1}$$

Thus $\mathcal{O}$ maps the range of R into $\tilde{H}$. As (A, B, C) is controllable, the range of R is dense in H and so $\mathcal{O}$ maps H into $\tilde{H}$. Further, as j runs through the non-negative integers and u ranges over U, the right hand side of (4.1) describes a spanning set for $\tilde{H}$. It follows that the range of $\mathcal{O}$ is dense in $\tilde{H}$. As $\mathcal{O}$ is an isometry its range is complete in $\tilde{H}$, hence closed, and so equal to $\tilde{H}$.

Let $V : H \longrightarrow \tilde{H}$ be the range restriction of $\mathcal{O}$: that is, $Vx = \mathcal{O}x$ for $x \in H$. Then V is a unitary operator, and, for $x \in H$,

$$VAx = \mathcal{O}Ax = \sum_{n=0}^{\infty} (CA^n Ax)z^n$$

$$= S^* \sum_{n=0}^{\infty} (CA^n x)z^n = S^* \mathcal{O} x$$

$$= \tilde{A}Vx.$$

Thus

$$A = V^* \tilde{A} V.$$

Recall that $\tilde{C}f = f(0)$ for $f \in \tilde{H} \subseteq H^2 \otimes Y$, so that, for $x \in H$,

$$\tilde{C}Vx = \tilde{C}\mathcal{O}x = \tilde{C}\left(\sum_{n=0}^{\infty} (CA^n x)z^n\right)$$

$$= Cx.$$

Thus

$$C = \tilde{C}V.$$

And for $u \in U$ we have

$$\tilde{B}u = g(.)u = \sum_{n=0}^{\infty} (CA^n Bu)z^n$$

$$= \mathcal{O}Bu = VBu,$$

so that

$$B = V^* \tilde{B}.$$

Thus (A, B, C) is unitarily equivalent to the restricted shift system $(\tilde{A}, \tilde{B}, \tilde{C})$ as required.

THEOREM 2. *Let (A_i, B_i, C_i) be a parbalanced system with input space U, output space Y, state space H_i and bounded controllability/observability gramian Σ_i, $i = 1, 2$. If the two systems have the same transfer function then they are unitarily equivalent.*

PROOF. Exactly as in (3.2) we have

$$\Sigma_i \geqslant A_i^* \Sigma_i A_i ,$$

so that there exists a contraction $\overline{A}_i$ on H_i such that

$$\overline{A}_i \Sigma_i^{1/2} = \Sigma_i^{1/2} A_i .$$

As previously the relations

$$\overline{B}_i = \Sigma_i^{1/2} B_i , \quad \overline{C}_i = C_i \Sigma_i^{-1/2}$$

define bounded operators $U \longrightarrow H_i$ and $H_i \longrightarrow Y$. $(\overline{A}_i, \overline{B}_i, \overline{C}_i)$ is an output normal system with controllability gramian Σ_i^2, and with the same transfer function as the initial systems. By Lemma 2 the systems $(\overline{A}_i, \overline{B}_i, \overline{C}_i)$, $i = 1, 2$, are both unitarily equivalent to the restricted shift realization of a single transfer function, and hence they are unitarily equivalent to each other. That is, there exists a unitary operator $V: H_1 \longrightarrow H_2$ such that

$$\overline{A}_1 = V^*\overline{A}_2 V, \quad \overline{B}_1 = V^*\overline{B}_2, \quad \overline{C}_1 = \overline{C}_2 V. \tag{4.2}$$

The relation between the corresponding controllability gramians is then

$$\Sigma_1^2 = V^*\Sigma_2^2 V.$$

On taking one fourth powers we deduce that

$$\Sigma_1^{1/2} = V^*\Sigma_2^{1/2} V.$$

Then (4.2) gives us

$$\begin{aligned}\Sigma_1^{1/2} A_1 \Sigma_1^{-1/2} &= \overline{A}_1 = V^*\overline{A}_2 V \\ &= V^*\Sigma_2^{1/2} A_2 \Sigma_2^{-1/2} V \\ &= \Sigma_1^{1/2} V^* A_2 V \Sigma_1^{-1/2},\end{aligned}$$

from which it follows that

$$A_1 = V^* A_2 V.$$

Likewise

$$B_1 = V^* B_2, \quad C_1 = C_2 V,$$

and the two parbalanced systems are unitarily equivalent as required.

ACKNOWLEDGEMENTS.

I am grateful to Ruth Curtain and Keith Glover for helpful conversations and to Cambridge University Engineering Department for hospitality at the time this research was carried out.

REFERENCES

1. R.F. Curtain, Balanced realisations for discrete-time infinite-dimensional systems, paper presented to IFIP Congress, Budapest, 1985.

2. R.F. Curtain and K. Glover, Balanced realisations for infinite-dimensional systems, this Workshop.

3. R.F. Curtain and K. Glover, Realisation and approximation for infinite-dimensional systems (in preparation).

4. J.C. Doyle et al., Advances in Multivariable Control, ONR/Honeywell Workshop, Minneapolis 1984.

5. P.A. Fuhrmann, Realization theory in Hilbert space for a class of transfer functions, *J. Functional Analysis* **18**, 338-349, 1975.

6. P.A. Fuhrmann, *Linear systems and operators in Hilbert space*, McGraw-Hill, New York, 1981.

7. K. Glover, All optimal Hankel-norm approximations of linear multivariable systems and their L^{∞}-error bounds, *Int. J. Control*, **39**, 1115-1193, 1984.

8. J.W. Helton, Discrete time systems, operator models and scattering theory, *J. Functional Analysis* **16**, 15-38, 1974.

9. T. Kailath, *Linear systems*, Prentice Hall, Englewood Cliffs, N.J., 1980.

10. B.C. Moore, Principal component analysis in linear systems: controllability, observability and model reduction, *IEEE Trans. Automatic Control* **26**, 17-32, 1981.

11. B. Sz.-Nagy and C. Foiaş, *Harmonic analysis of operators on Hilbert space*, Akadémiai Kiadó, Budapest, 1970.

12. N.K. Nikolskii, Ha-plitz operators: a survey of some recent results, in *Operators and Function Theory*, ed. S.C. Power, Reidel Publishing Co., NATO ASI Series C, Vol. 153.

13. L.B. Page, Bounded and compact vectorial Hankel operators, *Trans. Amer. Math. Soc.* **150**, 529-539, 1970.

14. L. Pernebo and L.M. Silverman, Model reduction via balanced state space representation, *IEEE Trans. Automatic Control* **27**, 382-387, 1982.

15. F. Riesz and B.Sz.-Nagy, *Functional Analysis*, F. Ungar Publishing Co., New York 1955.

N.J. Young,
Department of Mathematics,
University of Glasgow,
University Gardens,
GLASGOW G12 8QW
SCOTLAND

WORKSHOP PROGRAM

TUESDAY, JUNE 4, 1985

9.30 Opening and welcome.

9.50 L. de Branges, *The Riemann hypothesis for Hilbert spaces of entire functions.*

10.35 M.S. Livšic, *Collective motions of systems.*

11.15 Coffee and tea.

11.30 H.S.V. de Snoo, *Characteristic functions of unitary operator colligations in Pontryagin spaces.*

12.15 L. Rodman, *Perturbations of invariant subspaces of matrices.*

12.50 Lunch.

14.00 A. Ben-Artzi, *On inversion of Toeplitz matrices.*

14.35 I. Koltracht, *Parallel algorithms for matrices with recursive structure.*

15.15 Coffee and tea.

15.30 G. Heinig, *Hankel and Toeplitz matrices generated by the Markov parameters of rational functions.*

16.15 P. Fuhrmann, *Inversion of Hankel matrices and the partial realization problem.*

17.15 Reception by invitation of the Vrije Universiteit.

WEDNESDAY, JUNE 5, 1985

9.00 B. Gramsch, *Fréchet algebras of operators and* C^{∞} *- elements in dynamical systems.*

9.45 S. Goldberg, *Time varying linear systems and integral operators along chains of projections.*

10.25 Coffee and tea.

10.40 R. Mennicken, *Boundary eigenvalue problems.*

11.25 H. Langer, *Ordinary differential operators with an indefinite weight function.*

12.10 V. Vinnikov, *Classification of self-adjoint determinantal representations of real irreducible cubics.*

12.45 Lunch.

14.00 L. Lerer, *On Toeplitz and Wiener-Hopf operators with contourwise rational matrix and operator symbols.*

14.35 C.V.K. Prabhakara Rao, *State space construction of solutions for the lossless inverse scattering problem.*

15.20 A.C.M. Ran, *Hadamard factorization theorem for a class of entire matrix-valued functions.*

16.00 Coffee and tea.

16.15 V. Protopopescu, *Recent advances in abstract kinetic theory.*

17.00 L. Roozemond, *Inversion of Wiener-Hopf integral equations of the first kind and factorization of matrix polynomials.*

THURSDAY, JUNE 6, 1985

8.15 Bus trip to the Erasmus Universiteit at Rotterdam.

9.30 Coffee and tea.

10.00 Welcome.

10.10 J.L. Rovnyak, *Some problems of best approximation with constraints.*

10.55 J.A. Ball, *Pairs of shift invariant subspaces of matrices and non-canonical LU factorization of unitary completions.*

11.40 R.L. Ellis, *The maximum distance problem.*

12.15 Lunch by invitation of the Economische Faculteit, Erasmus Universiteit.

13.30 H. Dym, *Factorization, band extension and maximum entropy.*

14.15 Coffee and tea.

15.30 Boat tour harbours Rotterdam.

17.45 Workshop dinner at Schoonhoven.

FRIDAY, JUNE 7, 1985

9.00 J.C. Willems, ℓ_2 *- systems and their scattering representation.*

9.45 P. Lancaster, *Inverse eigenvalue problems for damped vibrating systems.*

10.25 Coffee and tea.

10.40 N. Young, *An algorithm for balanced realization.*

11.25 R.F. Curtain, *Balanced realizations for infinite - dimensional systems*

12.10 J.C. Doyle, *Generalized matrix interpolation and optimal control.*

13.00 Lunch.

14.15 E. Meister, *Some classes of integral and integro - differential equations on the half-line.*

15.00 K. Clancey, *On the method of V.A. Malyšev.*

15.40 Coffee and tea.

15.55 F.-O. Speck, *Cross - factorization, least squares, and the quarter plane problem.*

LIST OF PARTICIPANTS

A. Ben Artzi (Tel-Aviv)
J.A. Ball (Blacksburg)
H. Bart (Rotterdam)
J. van der Bijl (Amsterdam)
L. de Branges (West Lafayette)
J.A. van Casteren (Antwerpen)
K. Clancey (Athens)
M. Constabel (Darmstadt)
R.F. Curtain (Groningen)
P. Dewilde (Delft)
A. Dijksma (Groningen)
J. Donig (Darmstadt)
J.C. Doyle (Minneapolis)
H. Dym (Rehovot)
R.L. Ellis (College Park)
B.A. Francis (Toronto)
P.A. Fuhrmann (Beersheva)
I. Gohberg (Tel-Aviv)
S. Goldberg (College Park)
B. Gramsch (Mainz)
G. Heinig (Karl-Marx-Stadt)
J.M. Jansen (Amsterdam)
M.A. Kaashoek (Amsterdam)
J. van de Kant (Amsterdam)
J.A. Knottnerus (Amsterdam)
I. Koltracht (Calgary)
P. Lancaster (Calgary)
H. Langer (Dresden)
L. Lerer (Haifa)
M. Livšic (Beersheva)
E. Meister (Darmstadt)
R. Mennicken (Regensburg)
G. Peeters (Leuven)
V. Protopopescu (Oak Ridge)
A.C.M. Ran (Blacksburg)
L. Rodman (Tel-Aviv)
L. Roozemond (Amsterdam)
J. Rovnyak (Charlottesville)
F. van Schagen (Amsterdam)
H.S.V. de Snoo (Groningen)
F.-O. Speck (Darmstadt)
V. Vinnikov (Beersheva)
J.C. Willems (Groningen)
H.J. Woerdeman (Amsterdam)
N. Young (Glasgow)
H. Zwart (Groningen)